Biology: Exploring Life

SECOND EDITION

Biology: Exploring Life

Gil Brum
California State Polytechnic University, Pomona

Larry McKane
California State Polytechnic University, Pomona

Gerry Karp
Formerly of the University of Florida, Gainesville

JOHN WILEY & SONS, INC.

New York / Chichester / Brisbane / Toronto / Singapore

iv

Acquisitions Editor *Sally Cheney*
Developmental Editor *Rachel Nelson*
Marketing Manager *Catherine Faduska*
Associate Marketing Manager *Deb Benson*
Senior Production Editor *Katharine Rubin*
Text Designer/"Steps" Illustration Art Direction *Karin Gerdes Kincheloe*
Manufacturing Manager *Andrea Price*
Photo Department Director *Stella Kupferberg*
Senior Illustration Coordinator *Edward Starr*
"Steps to Discovery" Art Illustrator *Carlyn Iverson*
Cover Design *Meryl Levavi*
Text Illustrations *Network Graphics/Blaize Zito Associates, Inc.*
Photo Editor/Cover Photography Direction *Charles Hamilton*
Photo Researchers *Hilary Newman, Pat Cadley, Lana Berkovitz*
Cover Photo *James H. Carmichael, Jr./The Image Bank*

This book was set in New Caledonia by Progressive Typographers and printed and bound by Von Hoffmann Press.
The cover was printed by Lehigh. Color separations by Progressive Typographers and Color Associates, Inc.

Recognizing the importance of preserving what has been written, it is a policy of
John Wiley & Sons, Inc. to have books of enduring value published in the
United States printed on acid-free paper, and we exert our best efforts to that end.

The paper in this book was manufactured by a mill whose forest management
programs include sustained yield harvesting of its timberlands. Sustained yield
harvesting principles ensure that the number of trees cut each year does not
exceed the amount of new growth.

Library of Congress Cataloging in Publication Data:
Brum, Gilbert D.
    Biology : exploring life/Gil Brum, Larry McKane, Gerry Karp.—2nd ed.
        p.   cm.
    Includes bibliographical references and index.
    ISBN 0-471-54408-6 (cloth)
    1. Biology.   I. McKane, Larry.   II. Karp, Gerald.   III. Title.
  QH308.2.B78   1993
  574—dc20   93-23383
                                                                  CIP

Unit I ISBN 0-471-01827-9 (pbk)
Unit II ISBN 0-471-01831-7 (pbk)
Unit III ISBN 0-471-01830-9 (pbk)
Unit IV ISBN 0-471-01829-5 (pbk)
Unit V ISBN 0-471-01828-7 (pbk)
Unit VI ISBN 0-471-01832-5 (pbk)

Printed in the United States of America

10 9 8 7 6 5 4 3

For the Student, we hope this book helps you discover the thrill of exploring life and helps you recognize the important role biology plays in your everyday life.

To Margaret, Jan, and Patsy, who kept loving us even when we were at our most unlovable.

To our children, Jennifer, Julia, Christopher, and Matthew, whose fascination with exploring life inspires us all. And especially to Jenny—we all wish you were here to share the excitement of this special time in life.

Preface to the Instructor

Biology: Exploring Life, Second Edition is devoted to the process of investigation and discovery. The challenge and thrill of understanding how nature works ignites biologists' quests for knowledge and instills a desire to share their insights and discoveries. The satisfactions of knowing that the principles of nature can be understood and sharing this knowledge are why we teach. These are also the reasons why we created this book.

Capturing and holding student interest challenges even the best of teachers. To help meet this challenge, we have endeavored to create a book that makes biology relevant and appealing, that reveals biology as a dynamic process of exploration and discovery, and that emphasizes the widening influence of biologists in shaping and protecting our world and in helping secure our futures. We direct the reader's attention toward principles and concepts to dispel the misconception of many undergraduates that biology is nothing more than a very long list of facts and jargon. Facts and principles form the core of the course, but we have attempted to show the *significance* of each fact and principle and to reveal the important role biology plays in modern society.

From our own experiences in the introductory biology classroom, we have discovered that

- emphasizing principles, applications, and scientific exploration invigorates the teaching and learning process of biology and helps students make the significant connections needed for full understanding and appreciation of the importance of biology; and

- students learn more if a book is devoted to telling the story of biology rather than a recitation of facts and details.

Guided by these insights, we have tried to create a process-oriented book that still retains the facts, structures, and terminology needed for a fundamental understanding of biology. With these goals in mind, we have interwoven into the text

1. an emphasis on the ways that science works,
2. the underlying adventure of exploration,
3. five fundamental biological themes, and
4. balanced attention to the human perspective.

This book should challenge your students to think critically, to formulate their own hypotheses as possible explanations to unanswered questions, and to apply the approaches learned in the study of biology to understanding (and perhaps helping to solve) the serious problems that affect every person, indeed every organism, on this planet.

THE DEVELOPMENT STORY

The second edition of *Biology: Exploring Life* builds effectively on the strengths of the First Edition by Gil Brum and Larry McKane. For this edition, we added a third author, Gerry Karp, a cell and molecular biologist. Our complementary areas of expertise (genetics, zoology, botany, ecology, microbiology, and cell and molecular biology) as well as awards for teaching and writing have helped us form a balanced team. Together, we exhaustively revised and refined each chapter until all three of us, each with our different likes and dislikes, sincerely believed in the result. What evolved from this process was a satisfying synergism and a close friendship.

THE APPROACH

The elements of this new approach are described in the upcoming section "To the Student: A User's Guide." These pedagogical features are embedded in a book that is written in an informal, accessible style that invites the reader to explore the process of biology. In addition, we have tried to keep the narrative focused on *processes*, rather than on static facts, while creating an underlying foundation that helps students make the connections needed to tie together the information into a greater understanding than that which comes from memorizing facts alone. One way to help students make these connections is to relate the fundamentals of biology to humans, revealing the human perspective in each biological principle, from biochemicals to ecosystems. With each such insight, students take a substantial step toward becoming the informed citizens that make up responsible voting public.

We hope that, through this textbook, we can become partners with the instructor and the student. The biology teacher's greatest asset is the basic desire of students to understand themselves and the world around them. Unfortunately, many students have grown detached from this natural curiosity. Our overriding objective in creating this book was to arouse the students' fascination with exploring life, building knowledge and insight that will enable them to make real-life judgments as modern biology takes on greater significance in everyday life.

THE ART PROGRAM

The diligence and refinement that went into creating the text of *Biology: Exploring Life,* Second Edition characterizes the art program as well. Each photo was picked specifically for its relevance to the topic at hand and for its aesthetic and instructive value in illustrating the narrative concepts. The illustrations were carefully crafted under the guidance of the authors for accuracy and utility as well as aesthetics. The value of illustrations cannot be overlooked in a discipline as filled with images and processes as biology. Through the use of cell icons, labeled illustrations of pathways and processes, and detailed legends, the student is taken through the world of biology, from its microscopic chemical components to the macroscopic organisms and the environments that they inhabit.

SUPPLEMENTARY MATERIALS

In our continuing effort to meet all of your individual needs, Wiley is pleased to offer the various topics covered in this text in customized paperback "splits." For more details, please contact your local Wiley sales representative. We have also developed an integrated supplements package that helps the instructor bring the study of biology to life in the classroom and that will maximize the students' use and understanding of the text.

The *Instructor's Manual,* developed by Michael Leboffe and Gary Wisehart of San Diego City College, contains lecture outlines, transparency references, suggested lecture activities, sample concept maps, section concept map masters (to be used as overhead transparencies), and answers to study guide questions.

Gary Wisehart and Mark Mandell developed the test bank, which consists of four types of questions: fill-in questions, matching questions, multiple-choice questions, and critical thinking questions. A computerized test bank is also available.

A comprehensive visual ancillary package includes four-color transparencies (200 figures from the text), *Process of Science* transparency overlays that break down various biological processes into progressive steps, a video library consisting of tapes from Coronet MTI, and the *Bio Sci* videodisk series from Videodiscovery, covering topics in biochemistry, botany, vertebrate biology, reproduction, ecology, animal behavior, and genetics. Suggestions for integrating the videodisk material in your classroom discussions are available in the instructor's manual.

A comprehensive study guide and lab manual are also available and are described in more detail in the User's Guide section of the preface.

Acknowledgments

*I*t was a delight to work with so many creative individuals whose inspiration, artistry, and vital steam guided this complex project to completion. We wish we were able to acknowledge each of them here, for not only did they meet nearly impossible deadlines, but each willingly poured their heart and soul into this text. The book you now hold in your hands is in large part a tribute to their talent and dedication.

There is one individual whose unique talent, quick intellect, charm, and knowledge not only helped to make this book a reality, but who herself made an enormous contribution to the content and pedagogical strength of this book. We are proud to call Sally Cheney, our biology editor, a colleague. Her powerful belief in this textbook's new approaches to teaching biology helped instill enthusiasm and confidence in everyone who worked on it. Indeed, Sally is truly a force of positive change in college textbook publishing – she has an uncommon ability to think both like a biologist and an editor; she knows what biologists want and need in their classes and is dedicated to delivering it; she recognizes that the future of biology education is more than just publishing another look-alike text; and she is knowledgeable and persuasive enough to convince publishers to stick their necks out a little further for the good of educational advancement. Without Sally, this text would have fallen short of our goal. With Sally, it became even more than we envisioned.

Another individual also helped make this a truly special book, as well as made the many long hours of work so delightful. Stella Kupferberg, we treasure your friendship, applaud your exceptional talent, and salute your high standards. Stella also provided us with two other important assets, Charles Hamilton and Hilary Newman. Stella and Charles tirelessly applied their skill, and artistry to get us images of incomparable effectiveness and beauty, and Hilary's diligent handling helped to insure there were no oversights.

Our thanks to Rachel Nelson for her meticulous editing, for maintaining consistency between sometimes dissimilar writing styles of three authors, and for keeping track of an incalculable number of publishing and biological details; to Katharine Rubin for expertly and gently guiding this project through the myriad levels of production, and for putting up with three such demanding authors; to Karin Kincheloe for a stunningly beautiful design; to Ishaya Monokoff and Ed Starr for orchestrating a brilliant art program; to Network Graphics, especially John Smith and John Hargraves, who executed our illustrations with beauty and style without diluting their conceptual strength or pedagogy, and to Carlyn Iverson, whose artistic talent helped us visually distill our "Steps to Discovery" episodes into images that bring the process of science to life.

We would also like to thank Cathy Faduska and Alida Setford, their creative flair helped us to tell the story behind this book, as well as helped us convey what we tried to accomplish. And to Herb Brown, thank you for your initial confidence and continued support. A very special thank you to Deb Benson, our marketing manager. What a joy to work with you, Deb, your energy, enthusiasm, confidence, and pleasant personality bolstered even our spirits.

We wish to acknowledge Diana Lipscomb of George Washington University for her invaluable contributions to the evolution chapters, Judy Goodenough of the University of Massachusetts, Amherst, for contributing an outstanding chapter on Animal Behavior, and Dorothy Rosenthal for contributing the end–of–chapter "Critical Thinking Questions."

To the reviewers and instructors who used the First Edition, your insightful feedback helped us forge the foundation for this new edition. To the reviewers, and workshop and conference participants for the Second Edition, thank you for your careful guidance and for caring so much about your students.

Dennis Anderson, *Oklahoma City Community College*
Sarah Barlow, *Middle Tennessee State University*
Robert Beckman, *North Carolina State University*
Timothy Bell, *Chicago State University*
David F. Blaydes, *West Virginia University*
Richard Bliss, *Yuba College*
Richard Boohar, *University of Nebraska, Lincoln*
Clyde Bottrell, *Tarrant County Junior College*
J. D. Brammer, *North Dakota State University*
Peggy Branstrator, *Indiana University, East*
Allyn Bregman, *SUNY, New Paltz*
Daniel Brooks, *University of Toronto*

Gary Brusca, *Humboldt State University*
Jack Bruk, *California State University, Fullerton*
Marvin Cantor, *California State University, Northridge*
Richard Cheney, *Christopher Newport College*
Larry Cohen, *California State University, San Marcos*
David Cotter, *Georgia College*
Robert Creek, *Eastern Kentucky University*
Ken Curry, *University of Southern Mississippi*
Judy Davis, *Eastern Michigan University*
Loren Denny, *southwest Missouri State University*
Captain Donald Diesel, *U. S. Air Force Academy*
Tom Dickinson, *University College of the Cariboo*

Mike Donovan, *Southern Utah State College*
Robert Ebert, *Palomar College*
Thomas Emmel, *University of Florida*
Joseph Faryniarz, *Mattatuck Community College*
Alan Feduccia, *University of North Carolina, Chapel Hill*
Eugene Ferri, *Bucks County Community College*
Victor Fet, *Loyola University, New Orleans*
David Fox, *Loyola University, New Orleans*
Mary Forrest, *Okanagan University College*
Michael Gains, *University of Kansas*
S. K. Gangwere, *Wayne State University*
Dennis George, *Johnson County Community College*
Bill Glider, *University of Nebraska*
Paul Goldstein, *University of North Carolina, Charlotte*
Judy Goodenough, *University of Massachusetts, Amherst*
Nels Granholm, *South Dakota State University*
Nathaniel Grant, *Southern Carolina State College*
Mel Green, *University of California, San Diego*
Dana Griffin, *Florida State University*
Barbara L. Haas, *Loyola University of Chicago*
Richard Haas, *California State University, Fresno*
Fredrick Hagerman, *Ohio State University*
Tom Haresign, *Long Island University, Southampton*
W. R. Hawkins, *Mt. San Antonio College*
Vernon Hendricks, *Brevard Community College*
Paul Hertz, *Barnard College*
Howard Hetzle, *Illinois State University*
Ronald K. Hodgson, *Central Michigan University*
W. G. Hopkins, *University of Western Ontario*
Thomas Hutto, *West Virginia State College*
Duane Jeffrey, *Brigham Young University*
John Jenkins, *Swarthmore College*
Claudia Jones, *University of Pittsburgh*
R. David Jones, *Adelphi University*
J. Michael Jones, *Culver Stockton College*
Gene Kalland, *California State University, Dominiquez Hills*
Arnold Karpoff, *University of Louisville*
Judith Kelly, *Henry Ford Community College*
Richard Kelly, *SUNY, Albany*
Richard Kelly, *University of Western Florida*
Dale Kennedy, *Kansas State University*
Miriam Kittrell, *Kingsborough Community College*
John Kmeltz, *Kean College New Jersey*
Robert Krasner, *Providence College*
Susan Landesman, *Evergreen State College*
Anton Lawson, *Arizona State University*
Lawrence Levine, *Wayne State University*
Jerri Lindsey, *Tarrant County Junior College*
Diana Lipscomb, *George Washington University*
James Luken, *Northern Kentucky University*

Ted Maguder, *University of Hartford*
Jon Maki, *Eastern Kentucky University*
Charles Mallery, *University of Miami*
William McEowen, *Mesa Community College*
Roger Milkman, *University of Iowa*
Helen Miller, *Oklahoma State University*
Elizabeth Moore, *Glassboro State College*
Janice Moore, *Colorado State University*
Eston Morrison, *Tarleton State University*
John Mutchmor, *Iowa State University*
Jane Noble-Harvey, *University of Delaware*
Douglas W. Ogle, *Virginia Highlands Community College*
Joel Ostroff, *Brevard Community College*
James Lewis Payne, *Virginia Commonwealth University*
Gary Peterson, *South Dakota State University*
MaryAnn Phillippi, *Southern Illinois University, Carbondale*
R. Douglas Powers, *Boston College*
Robert Raikow, *University of Pittsburgh*
Charles Ralph, *Colorado State University*
Aryan Roest, *California State Polytechnic Univ., San Luis Obispo*
Robert Romans, *Bowling Green State University*
Raymond Rose, *Beaver College*
Richard G. Rose, *West Valley College*
Donald G. Ruch, *Transylvania University*
A. G. Scarbrough, *Towson State University*
Gail Schiffer, *Kennesaw State University*
John Schmidt, *Ohio State University*
John R. Schrock, *Emporia State University*
Marilyn Shopper, *Johnson County Community College*
John Smarrelli, *Loyola University of Chicago*
Deborah Smith, *Meredith College*
Guy Steucek, *Millersville University*
Ralph Sulerud, *Augsburg College*
Tom Terry, *University of Connecticut*
James Thorp, *Cornell University*
W. M. Thwaites, *San Diego State University*
Michael Torelli, *University of California, Davis*
Michael Treshow, *University of Utah*
Terry Trobec, *Oakton Community College*
Len Troncale, *California State Polytechnic University, Pomona*
Richard Van Norman, *University of Utah*
David Vanicek, *California State University, Sacramento*
Terry F. Werner, *Harris-Stowe State College*
David Whitenberg, *Southwest Texas State University*
P. Kelly Williams, *University of Dayton*
Robert Winget, *Brigham Young University*
Steven Wolf, *University of Missouri, Kansas City*
Harry Womack, *Salisbury State University*
William Yurkiewicz, *Millersville University*

Gil Brum
Larry McKane
Gerry Karp

Brief Table of Contents

PART 1 / Biology: The Study of Life   1

Chapter 1   Biology: Exploring Life   3
Chapter 2   The Process of Science   29

PART 2 / Chemical and Cellular Foundations of Life   45

Chapter 3   The Atomic Basis of Life   47
Chapter 4   Biochemicals: The Molecules of Life   65
Chapter 5   Cell Structure and Function   87
Chapter 6   Energy, Enzymes, and Metabolic Pathways   115
Chapter 7   Movement of Materials Across Membranes   135
Chapter 8   Processing Energy: Photosynthesis and Chemosynthesis   151
Chapter 9   Processing Energy: Fermentation and Respiration   171
Chapter 10   Cell Division: Mitosis   193
Chapter 11   Cell Division: Meiosis   211

PART 3 / The Genetic Basis of Life   227

Chapter 12   On the Trail of Heredity   229
Chapter 13   Genes and Chromosomes   245
Chapter 14   The Molecular Basis of Genetics   265
Chapter 15   Orchestrating Gene Expression   291
Chapter 16   DNA Technology: Developments and Applications   311
Chapter 17   Human Genetics: Past, Present, and Future   333

PART 4 / Form and Function of Plant Life   355

Chapter 18   Plant Tissues and Organs   357
Chapter 19   The Living Plant: Circulation and Transport   383

Chapter 20   Sexual Reproduction of Flowering Plants   399
Chapter 21   How Plants Grow and Develop   423

PART 5 / Form and Function of Animal Life   445

Chapter 22   An Introduction to Animal Form and Function   447
Chapter 23   Coordinating the Organism: The Role of the Nervous System   465
Chapter 24   Sensory Perception: Gathering Information about the Environment   497
Chapter 25   Coordinating the Organism: The Role of the Endocrine System   517
Chapter 26   Protection, Support, and Movement: The Integumentary, Skeletal, and Muscular Systems   539
Chapter 27   Processing Food and Providing Nutrition: The Digestive System   565
Chapter 28   Maintaining the Constancy of the Internal Environment: The Circulatory and Excretory Systems   585
Chapter 29   Gas Exchange: The Respiratory System   619
Chapter 30   Internal Defense: The Immune System   641
Chapter 31   Generating Offspring: The Reproductive System   661
Chapter 32   Animal Growth and Development: Acquiring Form and Function   685

PART 6 / Evolution   711

Chapter 33   Mechanisms of Evolution   713
Chapter 34   Evidence for Evolution   741
Chapter 35   The Origin and History of Life   759

P A R T 7 / *The Kingdoms of Life:*
Diversity and Classification   781

Chapter 36   The Monera Kingdom and Viruses   783
Chapter 37   The Protist and Fungus Kingdoms   807
Chapter 38   The Plant Kingdom   829
Chapter 39   The Animal Kingdom   857

P A R T 8 / *Ecology and Animal Behavior*   899

Chapter 40   The Biosphere   901
Chapter 41   Ecosystems and Communities   931

Chapter 42   Community Ecology: Interactions
between Organisms   959
Chapter 43   Population Ecology   983
Chapter 44   Animal Behavior   1005

Appendix A   Metric and Temperature Conversion
Charts   A-1
Appendix B   Microscopes: Exploring the
Details of Life   B-1
Appendix C   The Hardy-Weinberg Principle   C-1
Appendix D   Careers in Biology   D-1

Contents

PART 1
Biology: The Study of Life   1

Chapter 1 / Biology: Exploring Life   3

Steps to Discovery:
Exploring Life—The First Step   4

Distinguishing the Living from the Inanimate   6
Levels of Biological Organization   11
What's in a Name   11
Underlying Themes of Biology   15
Biology and Modern Ethics   25

Reexamining the Themes   25
Synopsis   26

Chapter 2 / The Process of Science   29

Steps to Discovery:
What Is Science?   30

The Scientific Approach   32
Applications of the Scientific Process   35
Caveats Regarding "The" Scientific Method   40

Synopsis   41

BIOETHICS

Science, Truth, and Certainty:
Is a Theory "Just a Theory"?   40

PART 2
Chemical and Cellular Foundations
of Life   45

Chapter 3 / The Atomic Basis of Life   47

Steps to Discovery:
The Atom Reveals Some of Its Secrets   48

The Nature of Matter   50
The Structure of the Atom   51
Types of Chemical Bonds   55
The Life-Supporting Properties of Water   59
Acids, Bases, and Buffers   61

Reexamining the Themes   62
Synopsis   63

BIOLINE

Putting Radioisotopes to Work   54

THE HUMAN PERSPECTIVE

Aging and Free Radicals   56

Chapter 4 / Biochemicals: The
Molecules of Life   65

Steps to Discovery:
Determining the Structure of Proteins   66

The Importance of Carbon in Biological Molecules   68
Giant Molecules Built From Smaller Subunits   69
Four Biochemical Families   69
Macromolecular Evolution and Adaptation   83

Reexamining the Themes   84
Synopsis   85

THE HUMAN PERSPECTIVE

Obesity and the Hungry Fat Cell   76

Chapter 5 / Cell Structure and Function 87

Steps to Discovery:
The Nature of the Plasma Membrane 88

Discovering the Cell 90
Basic Characteristics of Cells 91
Two Fundamentally Different Types of Cells 91
The Plasma Membrane: Specialized for
Interaction with the Environment 93
The Eukaryotic Cell: Organelle
Structure and Function 95
Just Outside the Cell 108
Evolution of Eukaryotic Cells: Gradual
Change or an Evolutionary Leap? 110

Reexamining the Themes 111
Synopsis 112

THE HUMAN PERSPECTIVE

Lysosome-Related Diseases 101

Chapter 6 / Energy, Enzymes, and Metabolic Pathways 115

Steps to Discovery:
The Chemical Nature of Enzymes 116

Acquiring and Using Energy 118
Enzymes 123
Metabolic Pathways 127
Macromolecular Evolution and Adaptation
of Enzymes 131

Reexamining the Themes 132
Synopsis 132

BIOLINE

Saving Lives by Inhibiting Enzymes 126

THE HUMAN PERSPECTIVE

The Manipulation of Human Enzymes 131

Chapter 7 / Movement of Materials Across Membranes 135

Steps to Discovery:
Getting Large Molecules Across Membrane Barriers 136

Membrane Permeability 138
Diffusion: Depending on the Random
Movement of Molecules 138
Active Transport 142
Endocytosis 144

Reexamining the Themes 148
Synopsis 148

THE HUMAN PERSPECTIVE

LDL Cholesterol, Endocytosis, and Heart Disease 146

Chapter 8 / Processing Energy: Photosynthesis and Chemosynthesis 151

Steps to Discovery:
Turning Inorganic Molecules into Complex Sugars 152

Autotrophs and Heterotrophs:
Producers and Consumers 154
An Overview of Photosynthesis 155
The Light-Dependent Reactions: Converting Light
Energy into Chemical Energy 156
The Light-Independent Reactions 163
Chemosynthesis: An Alternate Form of Autotrophy 167

Reexamining the Themes 168
Synopsis 169

BIOLINE

Living on the Fringe of the Biosphere 167

THE HUMAN PERSPECTIVE

Producing Crop Plants Better Suited
to Environmental Conditions 165

Chapter 9 / Processing Energy: Fermentation and Respiration 171

Steps to Discovery:
The Machinery Responsible for ATP Synthesis 172

Fermentation and Aerobic Respiration:
A Preview of the Strategies 174
Glycolysis 175
Fermentation 178
Aerobic Respiration 180

Balancing the Metabolic Books: Generating and
Yielding Energy   185
Coupling Glucose Oxidation to Other Pathways   188

> *Reexamining the Themes*   190
> *Synopsis*   190

BIOLINE

The Fruits of Fermentation   179

THE HUMAN PERSPECTIVE

The Role of Anaerobic and Aerobic Metabolism
in Exercise   186

Chapter 10 / Cell Division: Mitosis   193

Steps to Discovery:
Controlling Cell Division   194

Types of Cell Division   196
The Cell Cycle   199
The Phases of Mitosis   202
Cytokinesis: Dividing the Cell's Cytoplasm
and Organelles   206

> *Reexamining the Themes*   207
> *Synopsis*   208

THE HUMAN PERSPECTIVE

Cancer: The Cell Cycle Out of Control   202

Chapter 11 / Cell Division: Meiosis   211

Steps to Discovery:
Counting Human Chromosomes   212

Meiosis and Sexual Reproduction: An Overview   214
The Stages of Meiosis   216
Mitosis Versus Meiosis Revisited   222

> *Reexamining the Themes*   223
> *Synopsis*   223

THE HUMAN PERSPECTIVE

Dangers That Lurk in Meiosis   221

PART 3
The Genetic Basis of Life   227

Chapter 12 / On the Trail of Heredity   229

Steps to Discovery:
The Genetic Basis of Schizophrenia   230

Gregor Mendel: The Father of Modern Genetics   232
Mendel Amended   239
Mutation: A Change in the Genetic Status Quo   241

> *Reexamining the Themes*   242
> *Synopsis*   242

Chapter 13 / Genes and Chromosomes   245

Steps to Discovery:
The Relationship Between Genes and Chromosomes   246

The Concept of Linkage Groups   248
Lessons from Fruit Flies   249
Sex and Inheritance   253
Aberrant Chromosomes   257
Polyploidy   260

> *Reexamining the Themes*   261
> *Synopsis*   262

BIOLINE

The Fish That Changes Sex   254

THE HUMAN PERSPECTIVE

Chromosome Aberrations and Cancer   260

Chapter 14 / The Molecular Basis
of Genetics   265

Steps to Discovery:
The Chemical Nature of the Gene   266

Confirmation of DNA as the Genetic Material   268
The Structure of DNA   268
DNA: Life's Molecular Supervisor   272
The Molecular Basis of Gene Mutations   286

> *Reexamining the Themes*   287
> *Synopsis*   288

THE HUMAN PERSPECTIVE

The Dark Side of the Sun   280

Chapter 15 / Orchestrating Gene Expression 291

Steps to Discovery:
Jumping Genes: Leaping into the Spotlight 292

Why Regulate Gene Expression? 294
Gene Regulation in Prokaryotes 298
Gene Regulation in Eukaryotes 302
Levels of Control of Eukaryotic Gene Expression 303

Reexamining the Themes 308
Synopsis 309

BIOLINE
RNA as an Evolutionary Relic 306

THE HUMAN PERSPECTIVE
Clones: Is There Cause for Fear? 296

Chapter 16 / DNA Technology: Developments and Applications 311

Steps to Discovery:
DNA Technology and Turtle Migration 312

Genetic Engineering 314
DNA Technology I: The Formation and Use of Recombinant DNA Molecules 321
DNA Technology II: Techniques That Do Not Require Recombinant DNA Molecules 325
Use of DNA Technology in Determining Evolutionary Relationships 329

Reexamining the Themes 329
Synopsis 330

BIOLINE
DNA Fingerprints and Criminal Law 328

THE HUMAN PERSPECTIVE
Animals That Develop Human Diseases 320

BIOETHICS
Patenting a Genetic Sequence 315

Chapter 17 / Human Genetics: Past, Present, and Future 333

Steps to Discovery:
Developing a Treatment for an Inherited Disorder 334

The Consequences of an Abnormal Number of Chromosomes 336
Disorders that Result from a Defect in a Single Gene 338
Patterns of Transmission of Genetic Disorders 343
Screening Humans for Genetic Defects 346

Reexamining the Themes 350
Synopsis 351

BIOLINE
Mapping the Human Genome 340

THE HUMAN PERSPECTIVE
Correcting Genetic Disorders by Gene Therapy 348

PART 4
Form and Function of Plant Life 355

Chapter 18 / Plant Tissues and Organs 357

Steps to Discovery:
Fruit of the Vine and the French Economy 358

The Basic Plant Design 360
Plant Tissues 362
Plant Tissue Systems 365
Plant Organs 372

Reexamining the Themes 379
Synopsis 379

BIOLINE
Nature's Oldest Living Organisms 366

THE HUMAN PERSPECTIVE
Agriculture, Genetic Engineering, and Plant Fracture Properties 375

Chapter 19 / The Living Plant: Circulation and Transport 383

Steps to Discovery:
Exploring the Plant's Circulatory System 384

Xylem: Water and Mineral Transport 386
Phloem: Food Transport 393

Reexamining the Themes 396
Synopsis 397

BIOLINE
Mycorrhizae and Our Fragile Deserts 388

THE HUMAN PERSPECTIVE
Leaf Nodules and World Hunger: An Unforeseen Connection 396

Chapter 20 / Sexual Reproduction of Flowering Plants   399

Steps to Discovery:
What Triggers Flowering?   400

Flower Structure and Pollination   402
Formation of Gametes   407
Fertilization and Development   408
Fruit and Seed Dispersal   412
Germination and Seedling Development   416
Asexual Reproduction   417
Agricultural Applications   418

Reexamining the Themes   419
Synopsis   420

BIOLINE
The Odd Couples: Bizarre Flowers and
Their Pollinators   404

THE HUMAN PERSPECTIVE
The Fruits of Civilization   414

Chapter 21 / How Plants Grow and Develop   423

Steps to Discovery:
Plants Have Hormones   424

Levels of Control of Growth and Development   426
Plant Hormones: Coordinating Growth and
Development   427
Timing Growth and Development   434
Plant Tropisms   438

Reexamining the Themes   440
Synopsis   440

THE HUMAN PERSPECTIVE
Saving Tropical Rain Forests   428

PART 5
Form and Function of Animal Life   445

Chapter 22 / An Introduction to Animal Form and Function   447

Steps to Discovery:
The Concept of Homeostasis   448

Homeostatic Mechanisms   450
Unity and Diversity Among Animal Tissues   451
Unity and Diversity Among Organ Systems   455
The Evolution of Organ Systems   457
Body Size, Surface Area, and Volume   460

Reexamining the Themes   462
Synopsis   463

Chapter 23 / Coordinating the Organism: The Role of the Nervous System   465

Steps to Discovery:
A Factor Promoting the Growth of Nerves   466

Neurons and Their Targets   468
Generating and Conducting Neural Impulses   470
Neurotransmission: Jumping the Synaptic Cleft   474
The Nervous System   479
Architecture of the Human Central Nervous System   481
Architecture of the Peripheral Nervous System   489
The Evolution of Complex Nervous Systems from
Single-Celled Roots   491

Reexamining the Themes   493
Synopsis   494

BIOLINE
Deadly Meddling at the Synapse   478

THE HUMAN PERSPECTIVE
Alzheimer's Disease: A Status Report   484

BIOETHICS
Blurring the Line Between Life and Death   489

Chapter 24 / Sensory Perception: Gathering Information About the Environment   497

Steps to Discovery:
Echolocation: Seeing in the Dark   498

The Response of a Sensory Receptor to a Stimulus   500
Somatic Sensation   500
Vision   502
Hearing and Balance   506

Chemoreception: Taste and Smell   509
The Brain: Interpreting Impulses into Sensations   510
Evolution and Adaptation of Sense Organs   511

Reexamining the Themes   514

Synopsis   514

BIOLINE

Sensory Adaptations to an Aquatic Environment   512

THE HUMAN PERSPECTIVE

Two Brains in One   504

Chapter 25 / Coordinating the Organism: The Role of the Endocrine System   517

Steps to Discovery:
The Discovery of Insulin   518

The Role of the Endocrine System   520
The Nature of the Endocrine System   521
A Closer Look at the Endocrine System   523
Action at the Target   533
Evolution of the Endocrine System   534

Reexamining the Themes   536

Synopsis   537

BIOLINE

Chemically Enhanced Athletes   528

THE HUMAN PERSPECTIVE

The Mysterious Pineal Gland   533

Chapter 26 / Protection, Support, and Movement: The Integumentary, Skeletal, and Muscular Systems   539

Steps to Discovery:
Vitamin C's Role in Holding the Body Together   540

The Integument: Covering and Protecting the Outer Body Surface   542
The Skeletal System: Providing Support and Movement   544

The Muscles: Powering the Motion of Animals   551
Body Mechanics and Locomotion   558
Evolution of the Skeletomuscular System   560

Reexamining the Themes   561

Synopsis   562

THE HUMAN PERSPECTIVE

Building Better Bones   550

Chapter 27 / Processing Food and Providing Nutrition: The Digestive System   565

Steps to Discovery:
The Battle Against Beri-Beri   566

The Digestive System: Converting Food into a Form Available to the Body   568
The Human Digestive System: A Model Food-Processing Plant   568
Evolution of Digestive Systems   576

Reexamining the Themes   582

Synopsis   583

BIOLINE

Teeth: A Matter of Life and Death   570

THE HUMAN PERSPECTIVE

Human Nutrition   578

Chapter 28 / Maintaining the Constancy of the Internal Environment: the Circulatory and Excretory Systems   585

Steps to Discovery:
Tracing the Flow of Blood   586

The Human Circulatory System: Form and Function   588
Evolution of Circulatory Systems   603
The Lymphatic System: Supplementing the Functions of the Circulatory System   606
Excretion and Osmoregulation: Removing Wastes and Maintaining the Composition of the Body's Fluids   606
Thermoregulation: Maintaining a Constant Body Temperature   614

Reexamining the Themes   615

Synopsis   615

BIOLINE

The Artificial Kidney   611

THE HUMAN PERSPECTIVE

The Causes of High Blood Pressure   591

Chapter 29 / Gas Exchange: The Respiratory System   619

Steps to Discovery:
Physiological Adaptations in Diving Mammals   620

Properties of Gas Exchange Surfaces   622
The Human Respiratory System: Form and Function   623
The Exchange of Respiratory Gases and the
Role of Hemoglobin   627
Regulating the Rate and Depth of Breathing
to Meet the Body's Needs   629
Adaptations for Extracting Oxygen from
Water Versus Air   632
Evolution of the Vertebrate Respiratory System   636

Reexamining the Themes   637
Synopsis   637

THE HUMAN PERSPECTIVE
Dying for a Cigarette   630

Chapter 30 / Internal Defense: The Immune System   641

Steps to Discovery:
On the Trail of a Killer: Tracking the AIDS Virus   642

Nonspecific Mechanisms: A First Line of Defense   644
The Immune System: Mediator of Specific
Mechanisms of Defense   646
Antibody Molecules: Structure and Formation   650
Immunization   654
Evolution of the Immune System   655

Reexamining the Themes   658
Synopsis   658

BIOLINE
Treatment of Cancer with Immunotherapy   648

THE HUMAN PERSPECTIVE
Disorders of the Human Immune System   656

Chapter 31 / Generating Offspring: The Reproductive System   661

Steps to Discovery:
The Basis of Human Sexuality   662

Reproduction: Asexual Versus Sexual   664
Human Reproductive Systems   667
Controlling Pregnancy: Contraceptive Methods   677
Sexually Transmitted Diseases   681

Reexamining the Themes   682
Synopsis   682

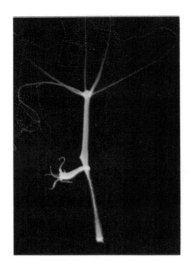

BIOLINE
Sexual Rituals   668

THE HUMAN PERSPECTIVE
Overcoming Infertility   679

BIOETHICS
Frozen Embryos and Compulsive Parenthood   680

Chapter 32 / Animal Growth and Development: Acquiring Form and Function   685

Steps to Discovery:
Genes That Control Development   686

The Course of Embryonic Development   688
Beyond Embryonic Development: Continuing Changes in
Body Form   699
Human Development   700
Embryonic Development and Evolution   705

Reexamining the Themes   706
Synopsis   707

THE HUMAN PERSPECTIVE
The Dangerous World of a Fetus   704

PART 6
Evolution   711

Chapter 33 / Mechanisms of Evolution   713

Steps to Discovery:
Silent Spring Revisited   714

The Basis of Evolutionary Change   717
Speciation: The Origin of Species   731
Patterns of Evolution   734
Extinction: The Loss of Species   736
The Pace of Evolution   737

Reexamining the Themes 738

Synopsis 738

BIOLINE

A Gallery of Remarkable Adaptations 728

Chapter 34 / Evidence for Evolution 741

Steps to Discovery:
An Early Portrait of the Human Family 742

Determining Evolutionary Relationships 744
Evidence for Evolution 746
The Evidence of Human Evolution:
The Story Continues 752

Reexamining the Themes 756

Synopsis 757

Chapter 35 / The Origin and History of Life 759

Steps to Discovery:
Evolution of the Cell 760

Formation of the Earth: The First Step 762
The Origin of Life and Its Formative Stages 763
The Geologic Time Scale: Spanning Earth's History 764

Reexamining the Themes 775

Synopsis 778

BIOLINE

The Rise and Fall of the Dinosaurs 772

PART 7
The Kingdoms of Life: Diversity and Classification 781

Chapter 36 / The Monera Kingdom and Viruses 783

Steps to Discovery:
The Burden of Proof: Germs and Infectious Diseases 784

Kingdom Monera: The Prokaryotes 787
Viruses 800

Reexamining the Themes 804

Synopsis 805

BIOLINE

Living with Bacteria 799

THE HUMAN PERSPECTIVE

Sexually Transmitted Diseases 794

Chapter 37 / The Protist and Fungus Kingdoms 807

Steps to Discovery:
Creating Order from Chaos 808

The Protist Kingdom 810
The Fungus Kingdom 820

Reexamining the Themes 826

Synopsis 827

Chapter 38 / The Plant Kingdom 829

Steps to Discovery:
Distinguishing Plant Species: Where Should the Dividing Lines Be Drawn? 830

Major Evolutionary Trends 833
Overview of the Plant Kingdom 834
Algae: The First Plants 835
Bryophytes: Plants Without Vessels 842
Tracheophytes: Plants with Fluid-Conducting Vessels 843

Reexamining the Themes 854

Synopsis 854

BIOLINE

The Exceptions: Plants That Don't Photosynthesize 836

THE HUMAN PERSPECTIVE

Brown Algae: Natural Underwater Factories 840

Chapter 39 / The Animal Kingdom 857

Steps to Discovery:
The World's Most Famous Fish 858

Evolutionary Origins and Body Plans 861
Invertebrates 863
Vertebrates 886

Reexamining the Themes 895

Synopsis 895

BIOLINE

Adaptations to Parasitism among Flatworms 872

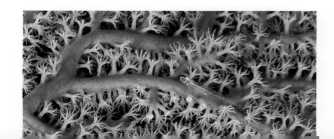

PART 8
Ecology and Animal Behavior   899

Chapter 40 / The Biosphere   901

Steps to Discovery:
The Antarctic Ozone Hole   902

Boundaries of the Biosphere   904
Ecology Defined   905
The Earth's Climates   905
The Arenas of Life   908

Reexamining the Themes   927

Synopsis   928

THE HUMAN PERSPECTIVE

Acid Rain and Acid Snow: Global Consequences
of Industrial Pollution   918

BIOETHICS

Values in Ecology and Environmental Science: Neutrality
or Advocacy?   909

Chapter 41 / Ecosystems and Communities   931

Steps to Discovery:
The Nature of Communities   932

From Biomes to Microcosms   934
The Structure of Ecosystems   934
Ecological Niches and Guilds   939
Energy Flow Through Ecosystems   942
Biogeochemical Cycles: Recycling Nutrients
in Ecosystems   947
Succession: Ecosystem Change and Stability   952

Reexamining the Themes   955

Synopsis   956

BIOLINE

Reverberations Felt Throughout an Ecosystem   936

THE HUMAN PERSPECTIVE

The Greenhouse Effect: Global Warming   950

Chapter 42 / Community Ecology: Interactions Between Organisms   959

Steps to Discovery:
Species Coexistence: The Unpeaceable Kingdom   960

Symbiosis   962
Competition: Interactions That Harm
Both Organisms   963

Interactions that Harm One Organism and
Benefit the Other   965
Commensalism: Interactions that Benefit One Organism
and Have No Affect on the Other   976
Interactions That Benefit Both Organisms   977

Reexamining the Themes   979

Synopsis   979

Chapter 43 / Population Ecology   983

Steps to Discovery:
Threatening a Giant   984

Population Structure   986
Factors Affecting Population Growth   988
Factors Controlling Population Growth   996
Human Population Growth   997

Reexamining the Themes   1002

Synopsis   1002

BIOLINE

Accelerating Species Extinction   997

THE HUMAN PERSPECTIVE

Impacts of Poisoned Air, Land, and Water   993

Chapter 44 / Animal Behavior   1005

Steps to Discovery:
Mechanisms and Functions of Territorial Behavior   1006

Mechanisms of Behavior   1008
Learning   1011
Development of Behavior   1014
Evolution and Function of Behavior   1015
Social Behavior   1018
Altruism   1022

Reexamining the Themes   1026

Synopsis   1027

BIOLINE

Animal Cognition   1024

**Appendix A   Metric and Temperature
Conversion Charts   A-1**
**Appendix B   Microscopes: Exploring the Details of
Life   B-1**
Appendix C   The Hardy-Weinberg Principle   C-1
Appendix D   Careers in Biology   D-1
Glossary   G-1
Photo Credits   P-1
Index   I-1

To The Student:
A User's Guide

*B*iology is a journey of exploration and discovery, of struggle and breakthrough. It is enlivened by the thrill of understanding not only what living things do but also how they work. We have tried to create such an experience for you.

Excellence in writing, visual images, and broad biological coverage form the core of a modern biology textbook. But as important as these three factors are in making difficult concepts and facts clear and meaningful, none of them reveals the excitement of biology—the adventure that unearths what we know about life. To help relate the true nature of this adventure, we have developed several distinctive features for this book, features that strengthen its biological core, that will engage and hold your attention, that reveal the human side of biology, that enable every reader to understand how science works, that stimulate critical thinking, and that will create the informed citizenship we all hope will make a positive difference in the future of our planet.

Steps to Discovery

The process of science enriches all parts of this book. We believe that students, like biologists, themselves, are intrigued by scientific puzzles. Every chapter is introduced by a "Steps to Discovery" narrative, the story of an investigation that led to a scientific breakthrough in an area of biology which relates to that chapter's topic. The "Steps to Discovery" narratives portray biologists as they really are: human beings, with motivations, misfortunes, and mishaps, much like everyone experiences. We hope these narratives help you better appreciate biological investigation, realizing that it is understandable and within your grasp.

Throughout the narrative of these pieces, the writing is enlivened with scientific work that has provided knowledge and understanding of life. This approach is meant not just to pay tribute to scientific giants and Nobel prize winners, but once again to help you realize that science does not grow by itself. Facts do not magically materialize. They are the products of rational ideas, insight, determination, and, sometimes, a little luck. Each of the "Steps to Discovery" narratives includes a painting that is meant primarily as an aesthetic accompaniment to the adventure described in the essay and to help you form a mental picture of the subject.

S TEPS TO DISCOVERY
A Factor Promoting the Growth of Nerves

*R*ita Levi-Montalcini received her medical degree from the University of Turin in Italy in 1936, the same year that Benito Mussolini began his anti-Semitic campaign. By 1939, as a Jew, Levi-Montalcini had been barred from carrying out research and practicing medicine, yet she continued to do both secretly. As a student, Levi-Montalcini had been fascinated with the structure and function of the nervous system. Unable to return to the university, she set up a simple laboratory in her small bedroom in her family's home. As World War II raged throughout Europe, and the Allies systematically bombed Italy, Levi-Montalcini studied chick embryos in her bedroom, discovering new information about the growth of nerve cells from the spinal cord into the nearby limbs. In her autobiography *In Praise of*

Imperfection, she writes: "Every time the alarm sounded, I would carry down to the precarious safety of the cellars the Zeiss binocular microscope and my most precious silver-stained embryonic sections." In September 1943, German troops arrived in Turin to support the Italian Fascists. Levi-Montalcini and her family fled southward to Florence, where they remained in hiding for the remainder of the w[...]

After the war ended, Levi-Montalcini continued [...] research at the University of Turin. In 1946, she acce[...] an invitation from Viktor Hamburger, a leading expe[...] the development of the chick nervous system, to co[...] Washington University in St. Louis to work with him f[...] semester; she remained at Washington University [...] years.

A chick embryo and one of its nerve cells helped scientists discover nerve growth factor (NGF).

One of Levi-Montalcini's first projects was the reexamination of a previous experiment of Elmer Bueker, a former student of Hamburger's. Bueker had removed a limb from a chick embryo, replaced it with a fragment of a mouse connective tissue tumor, and found that nerve fibers grew into this mass of implanted tumor cells. When Levi-Montalcini repeated the experiment she made an unexpected discovery: One part of the nervous system of these experimental chick embryos—the sympathetic nervous system—had grown five to six times larger than had its counterpart in a normal chick embryo. (The sympathetic nervous system helps control the activity of internal organs, such as the heart and digestive tract.) Close examination revealed that the small piece of tumor tissue that had been grafted onto the embryo had caused sympathetic nerve fibers to grow "wildly" into all of the chick's internal organs, even causing some of the blood vessels to become obstructed by the invasive fibers. Levi-Montalcini hypothesized that the tumor was releasing some soluble substance that induced the remarkable growth of this part of the nervous system. Her hypothesis was soon confirmed by further experiments. She called the active substance **nerve growth factor (NGF)**.

The next step was to determine the chemical nature of NGF, a task that was more readily performed by growing the tumor cells in a culture dish rather than an embryo. But Hamburger's laboratory at Washington University did not have the facilities for such work. To continue the project, Levi-Montalcini boarded a plane, with a pair of tumor-bearing mice in the pocket of her overcoat, and flew to Brazil, where she had a friend who operated a tissue culture laboratory. When she placed sympathetic nervous tissue in the proximity of the tumor cells in a culture dish, the nervous tissue sprouted a halo of nerve fibers that grew toward the tumor cells. When the tissue was cultured in the absence of NGF, no such growth occurred.

For the next 2 years, Levi-Montalcini's lab was devoted to characterizing the substance in the tumor cells that possessed the ability to cause nerve outgrowth. The work was carried out primarily by a young biochemist, Stanley Cohen, who had joined the lab. One of the favored approaches to studying the nature of a biological molecule is to determine its sensitivity to enzymes. In order to determine if nerve growth factor was a protein or a nucleic acid, Cohen treated the active material with a small amount of snake venom, which contains a highly active enzyme that degrades nucleic acid. It was then that chance stepped in.

Cohen expected that treatment with the venom wo[...] ther destroy the activity of the tumor cell fraction (i[...] was a nucleic acid) or leave it unaffected (i[...] protein). To Cohen's surprise, treatment wi[...] *increased* the nerve-growth promoting activity of the [...] rial. In fact, treatment of sympathetic nerve tissue wit[...] venom alone (in the absence of the tumor extract) ind[...] the growth of a halo of nerve fibers! Cohen soon discov[...] why: The snake venom possessed the same nerve gr[...] factor as did the tumor cells, but at much higher concen[...] tion. Cohen soon demonstrated that NGF was a prote[...]

Levi-Montalcini and Cohen reasoned that since sn[...] venom was derived from a *modified* salivary gland, th[...] other salivary glands might prove to be even better sour[...] of the protein. This hypothesis proved to be correct. Wh[...] Levi-Montalcini and Cohen tested the salivary glands fro[...] male mice, they discovered the richest source of NGF yet, [...] source 10,000 times more active than the tumor cells an[...] ten times more active than snake venom.

A crucial question remained: Did NGF play a role i[...] the normal development of the embryo, or was its ability to stimulate nerve growth just an accidental property of the molecule? To answer this question, Levi-Montalcini and Cohen injected embryos with an antibody against NGF, which they hoped would inactivate NGF molecules wherever they were present in the embryonic tissues. The embryos developed normally, with one major exception: They virtually lacked a sympathetic nervous system. The researchers concluded that NGF must be important during normal development of the nervous system; otherwise, inactivation of NGF could not have had such a dramatic effect.

By the early 1970s, the amino acid sequence of NGF had been determined, and the protein is now being synthesized by recombinant DNA technology. During the past decade, Fred Gage, of the University of California, has found that NGF is able to revitalize aged or damaged nerve cells in rats. Based on these studies, NGF is currently being tested as a possible treatment of Alzheimer's disease. For their pioneering work, Rita Levi-Montalcini and Stanley Cohen shared the 1987 Nobel Prize in Physiology and Medicine.

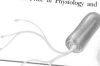

*M*any students are overwhelmed by the diversity of living organisms and the multitude of seemingly unrelated facts that they are forced to learn in an introductory biology course. Most aspects of biology, however, can be thought of as examples of a small number of recurrent themes. Using the thematic approach, the details and principles of biology can be assembled into a body of knowledge that makes sense, and is not just a collection of disconnected facts. Facts become ideas, and details become parts of concepts as you make connections between seemingly unrelated areas of biology, forging a deeper understanding.

All areas of biology are bound together by evolution, the central theme in the study of life. Every organism is the product of evolution, which has generated the diversity of biological features that distinguish organisms from one another and the similarities that all organisms share. From this basic evolutionary theme emerge several other themes that recur throughout the book:

- **Relationship between Form and Function**
- **Biological Order, Regulation, and Homeostatis**
- **Acquiring and Using Energy**
- **Unity Within Diversity**
- **Evolution and Adaptation**

We have highlighted the prevalent recurrence of each theme throughout the text with an icon, shown above. The icons can be used to activate higher thought processes by inviting you to explore how the fact or concept being discussed fits the indicated theme.

FORM AND FUNCTION
The length and shape of this tiger lily and that of the hummingbird's beak match, enabling the hummingbird to gather nutritious nectar from the flower base, while the flower deposits or collects pollen for reproduction.

ACQUIRING AND USING ENERGY
All organisms acquire and use energy whether by directly trapping the energy in sunlight or by harvesting energy stored in the bodies of plants or animals.

ORDER, REGULATION AND HOMEOSTASIS
Whether active or dormant, enveloped by snow or blistering heat, all organisms must maintain order and regulate internal conditions to remain alive.

UNITY WITHIN DIVERSITY
Despite the remarkable variation in different kinds of organisms on earth, all organisms are composed of one or more cells.

EVOLUTION AND ADAPTATION
Evolution has produced the astounding variety of life on earth. Like the lizard hidden on the bark of this tree, each kind of organism possesses adaptations that enable it to survive and reproduce in its particular habitat.

Reexamining the Themes

*E*ach chapter concludes with a "Reexamining the Themes" section, which revisits the themes and how they emerge within the context of the chapter's concepts and principles. This section will help you realize that the same themes are evident at all levels of biological organization, whether you are studying the molecular and cellular aspects of biology or the global characteristics of biology.

When two organisms have the same protein, the difference in amino acid sequence of that protein can be correlated with the evolutionary relatedness of the organisms. The amino acid sequence of hemoglobin, for example, is much more similar between humans and monkeys—organisms that are closely related—than between humans and turtles, who are only distantly related. In fact, the evolutionary tree that emerges when comparing the structure of specific proteins from various animals very closely matches that previously constructed from fossil evidence.

The fact that the amino acid sequences of proteins change as organisms diverge from one another reflects an underlying change in their genetic information. Even though a DNA molecule from a mushroom, a redwood tree, and a cow may appear superficially identical, the sequences of nucleotides that make up the various DNA molecules are very different. These differences reflect evolutionary changes resulting from natural selection (Chapter 34).

Virtually all differences among living organisms can be traced to evolutionary changes in the structure of their various macromolecules, originating from changes in the nucleotide sequences of their DNA. (See CTQ #7.)

REEXAMINING THE THEMES

Relationship between Form and Function

The structure of a macromolecule correlates with a particular function. The unbranched, extended nature of the cellulose molecule endows it with resistance to pulling forces, an important property of plant cell walls. The hydrophobic character of lipids underlies many of their biological roles, explaining, for example, how waxes are able to provide plants with a waterproof covering. Protein function is correlated with protein shape. Just as a key is shaped to open a specific lock, a protein is shaped for a particular molecular interaction. For example, the shape of each polypeptide chain of hemoglobin enables a molecule of oxygen to fit perfectly into its binding site. A single alteration in the amino acid sequence of a hemoglobin chain can drastically reduce the molecule's oxygen-carrying capacity.

Biological Order, Regulation, and Homeostasis

Both blood sugar levels and body weight in humans are controlled by complex homeostatic mechanisms. The level of glucose in your blood is regulated by factors acting on the liver, which stimulate either glycogen breakdown (which increases blood sugar) or glycogen formation (which decreases blood sugar). Your body weight is, at least partly, determined by factors emanating from fat cells which either increase metabolic rate (which tends to decrease body weight) or slow down metabolic rate (which tends to increase body weight).

Acquiring and Utilizing Energy

The chemical energy that fuels biological activities is stored primarily in two types of macromolecules: polysaccharides and fats. Polysaccharides, including starch in plants and glycogen in animals, function primarily in the short-term storage of chemical energy. These polysaccharides can be rapidly broken down to sugars, such as glucose, which are readily metabolized to release energy. Gram-for-gram, fats contain even more energy than polysaccharides and function primarily as a long-term storage of chemical energy.

Unity within Diversity

All organisms, from bacteria to humans, are composed of the same four families of macromolecules, illustrating the unity of life—even at the biochemical level. The precise nature of these macromolecules and the ways they are organized into higher structures differ from organism to organism, thereby building diversity. Plants, for example, polymerize glucose into starch and cellulose, while animals polymerize glucose into glycogen. Similarly, many proteins (such as hemoglobin) are present in a variety of organisms, but the precise amino acid sequence of the protein varies from one species to the next.

Evolution and Adaptation

Evolution becomes very apparent at the molecular level when we compare the structure of macromolecules among diverse organisms. Analysis of the amino acid sequences of proteins and the nucleotide sequences of nucleic acids reveals a gradual change over time in the structure of macromolecules. Organisms that are closely related have proteins and nucleic acids whose sequences are similar than are those of distantly related organisms. To a large degree, the differences observed among diverse organisms derives from the evolutionary differences in nucleic acid and protein sequences.

The segregation of alleles and their independent assortment during meiosis increase genotype diversity by promoting new combinations of genes. But the shuffling of existing genes alone does not explain the presence of such a vast diversity of life. If all organisms descended from a common ancestor, with its relatively small complement of genes, where did all the genes present in today's millions of species come from? The answer is mutation.

Most mutant alleles are detrimental; that is, they are more likely to disrupt a well-ordered, smoothly functioning organism than to increase the organism's fitness. For example, a mutation might change a gene so that it produces an inactive enzyme needed for a critical life function. Occasionally, however, one of these stable genetic changes creates an advantageous characteristic that fitness of the offspring. In this way, mutation raw material for evolution and the diversification earth.

One of the requirements for genes is stability: remain basically the same from generation to generation; the fitness of organisms would rapidly deteriorate same time, there must be some capacity for change; otherwise, there would be no potential tion. Alterations in genes do occur, albeit ra changes (mutations) represent the raw m tion. (See CTQ #7.)

REEXAMINING THE THEMES

Biological Order, Regulation, and Homeostasis

Mendel discovered that the transmission of genetic factors followed a predictable pattern, indicating that the processes responsible for the formation of gametes, including the segregation of alleles, must occur in a highly ordered manner. This orderly pattern can be traced to the process of meiosis and the precision with which homologous chromosomes are separated during the first meiotic division. Mendel's discovery of independent assortment can also be connected with the first meiotic division, when each pair of homologous chromosomes becomes aligned at the metaphase plate in a manner that is independent of other pairs of homologues.

Unity within Diversity

All eukaryotic, sexually reproducing organisms follow the same "rules" for transmitting inherited traits. Although Mendel chose to work with peas, he could have come to the same conclusions had he studied fruit flies or mice or had he scrutinized a family's medical records on the transmission of certain genetic diseases, such as cystic fibrosis. Alt the mechanism by which genes are transmitted is univ the genes themselves are highly diverse from one org to the next. It is this genetic difference among species forms the very basis of biological diversity

Evolution and Adaptation

Mendel's findings provided a critical link in our knowedge of the mechanism of evolution. A key tenet in theory of evolution is that favorable genetic variations crease the likelihood that an individual will survive to productive age and that its offspring will exhibit these sam favorable characteristics. Mendel's demonstration that units of inheritance pass from parents to offspring without being blended revealed the means by which advantageous traits could be preserved in a species over many genera tions. The subsequent discovery of genetic change by mu tation revealed how new genes appeared in a population, thus providing the raw material for evolution.

SYNOPSIS

Gregor Mendel discovered the pattern by which inherited traits are transmitted from parents to offspring. Mendel discovered that inherited traits were controlled by pairs of factors (genes). The two factors for a given trait in an individual could be identical (homozygous) or different (heterozygous). In heterozygotes, one of the gene variants (alleles) may be dominant over the other, recessive allele. Because of dominance, the appearance (phenotype) of the heterozygote (genotype of *Aa*) is identical to that of the homozygote with two dominant alleles

Students will naturally find many ways in which the material presented in any biology course relates to them. But it is not always obvious how you can use biological information for better living or how it might influence your life. Your ability to see yourself in the course boosts interest and heightens the usefulness of the information. This translates into greater retention and understanding.

To accomplish this desirable outcome, the entire book has been constructed with you—the student—in mind. Perhaps the most notable feature of this approach is a series of boxed essays called "The Human Perspective" that directly reveals the human relevance of the biological topic being discussed at that point in the text. You will soon realize that human life, including your own, is an integral part of biology.

PART 2 / Chemical and Cellular Foundations of Life

◁ THE HUMAN PERSPECTIVE ▷
Obesity and the Hungry Fat Cell

FIGURE 1
Actor Robert DeNiro in (left) a scene from the movie Raging Bull and (right) a recent photograph.

It has become increasingly clear in recent years that people who are exceedingly overweight—that is, obese—are at increased risk of serious health problems, including heart disease and cancer. By most definitions, a person is obese if he or she is about 20 percent above "normal" or desirable body weight. Approximately 35 percent of adults in the United States are considered obese by this definition, twice as many as at the turn of the century. Among young adults, high blood pressure is five times more prevalent and diabetes three times more prevalent in a group of obese people than in a group of people who are at normal weight. Given these statistics, together with the social stigma facing the obese, there would seem to be strong motivation for maintaining a "normal" body weight. Why, then, are so many of us so overweight? And, why is it so hard to lose them back? The answers go beyond our fondness for high-calorie foods.

Excess body fat is stored in fat cells (adipocytes) located largely beneath the skin. These cells can change their volume more than a hundredfold, depending on the amount of fat they contain. As a person gains body fat, his or her fat cells become larger and larger, accounting for the bulging, sagging body shape. If the person becomes sufficiently overweight, and their fat cells approach their maximum fat-carrying capacity, chemical messages are sent through the blood, causing formation of new fat cells that are "hungry" to begin accumulating their own fat. Once a fat cell is formed, it may expand or contract in volume, but it appears to remain in the body for the rest of the person's life.

Although the subject remains controversial, current research findings suggest that body weight is one of the properties subject to physiologic regulation in humans. Apparently, each person has a particular weight that his or her body's regulatory machinery acts to maintain. This particular value—whether 40 kilograms (80 pounds) or 200 kilograms (400 pounds)—is referred to as the person's **set-point**.

People maintain their body weight at a relatively constant value by balancing energy intake (in the form of food calories) with energy expenditure (in the form of calories burned by metabolic activities or excreted). Obese individuals are thought to have a higher set-point than do persons of normal weight. In many cases, the set-point value appears to have a strong genetic component. For instance, studies reveal there is no correlation between the body mass of adoptees and their adoptive parents, but there is a clear relationship between adoptees and their biological parents, with whom they have not lived.

The existence of a body-weight set-point is most evident when the body weight of a person is "forced" to deviate from the regulated value. Individuals of normal body weight who are fed large amounts of high-calorie foods under experimental conditions tend to gain increasing amounts of weight. If these people cease their energy-rich diets, however, they return quite rapidly to their previous levels, at which point further weight loss stops. This is illustrated by actor Robert DeNiro, who reportedly gained about 50 pounds for the filming of the movie "Raging Bull" (Figure 1), and then lost the weight prior to his next acting role. Conversely, a person who is put on a strict, low-calorie diet will begin to lose weight. The drop in body weight soon triggers a decrease in the person's resting metabolic rate; that is, the amount of calories burned when the person is not engaged in physical activity. The drop in metabolic rate is the body's compensatory measure for the decreased food intake. In other words, it is the body's attempt to halt further weight loss. This effect is particularly pronounced among obese people who diet and lose large amounts of weight: Their pulse rate and blood pressure drop markedly, their fat cells shrink to "ghosts" of their former selves, and they tend to be continually hungry. If these obese individuals go back to eating a normal diet, they tend to regain the lost weight rapidly. The drive of these formerly obese persons to increase their food intake is probably a response to chemical signals emanating from the fat cells as they shrink below their previous size.

---

630 • PART 5 / Form and Function of Animal Life

◁ THE HUMAN PERSPECTIVE ▷
Dying for a Cigarette?

...average, smoking cigarettes will cut ...imately 6 to 8 years off your life, ...han 5 minutes for every cigarette ... Cigarette smoking is the greatest ... preventable death in the United ... according to a 1991 report by the ... for Disease Control (CDC), ...,000 Americans die each year ...ng-related causes. Smoking ac-...87 percent of all lung-cancer ...smokers are more susceptible ...the esophagus, larynx, mouth, ...l bladder than are nonsmok-...reased incidence of lung ...among smokers compared to ...hown in Figure 1a, and ...d by quitting is shown in ...effects of smoking on lung ...Figure 2. Atherosclero-...and peptic ulcers also ...g greater frequency than ...rs. For example, long-...5 times more likely to ...terial disease than are ...sema (a condition ...culty in breathing) ...ction of lung tissue, ...ammation of the air-...e prevalent among

...ger other people. ...ponsible for the ...nocent bystand-...re the same air ...passive (invol-...own; second-...seriously ill ...ers have dou-...ry infections ...posed to to-...ng married ...us; 20 per-...mong non-...ibutable to inhaling other

people's tobacco smoke. Another "innocent bystander" is a fetus developing in the uterus of a woman who smokes. Smoking increases the incidence of miscarriage and stillbirth and decreases the birthweight of the infant. Once born, these babies suffer twice as many respiratory infections as do babies of nonsmoking mothers.

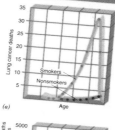

(a)

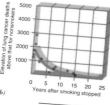

(b)

Why is smoking so bad for your health? The smoke emitted from a burning cigarette contains more than 2,000 identifiable substances, many of which are either irritants or carcinogens. These compounds include carbon monoxide, sulfur dioxide, formaldehyde, nitrosamines, toluene, ammonia, and radioactive isotopes. Autopsies of respiratory tissues from smokers (and from nonsmokers who have lived for long periods with smokers) show widespread cellular changes, including the presence of precancerous cells (cells that may become malignant, given time) and a marked reduction in the number of cilia that play a vital role in the removal of bacteria and debris from the airways.

Of all the compounds found in tobacco (including smokeless varieties), the most important is nicotine, not because it is carcinogenic, but because it is so addictive. Nicotine is addictive because it acts like a neurotransmitter by binding to certain acetylcholine receptors (page 477), stimulating postsynaptic neurons. The physiological effects of this stimulation include the release of epinephrine, an increase in blood sugar, an elevated heart rate, and the constriction of blood vessels, causing elevated blood pressure. A smoker's nervous system becomes "accustomed" to the presence of nicotine and decreases the output of the natural neurotransmitter. As a result, when a person tries to stop smoking, with the sudden absence of nicotine, together with the decreased level of the natural transmitter, decreases stimulation of postsynaptic neurons, which creates a craving for a cigarette—a "nicotine fit." Ex-smokers may be so conditioned to the act of smoking that the craving for cigarettes can continue long after the physiological addiction disappears.

*T*he "Biolines" are boxed essays that highlight fascinating facts, applications, and real-life lessons, enlivening the mainstream of biological information. Many are remarkable stories that reveal nature to be as surprising and interesting as any novelist could imagine.

◁ B I O L I N E ▷
DNA Fingerprints and Criminal Law

On February 5, 1987, a woman and her 2-year-old daughter were found stabbed to death in their apartment in the New York City borough of the Bronx. Following a tip, the police questioned a resident of a neighboring building. A small bloodstain was found on the suspect's watch, which was sent to a laboratory for DNA fingerprint analysis. The DNA from the white blood cells in the stain was amplified using the PCR technique and was digested with a restriction enzyme. The restriction fragments were then separated by electrophoresis, and a pattern of labeled fragments was identified with a radioactive probe. The banding pattern produced by the DNA from the suspect's watch was found to be a perfect match to the pattern produced by DNA taken from one of the victims. The results were provided to the opposing attorneys, and a pretrial hearing was called in 1989 to discuss the validity of the DNA evidence.

During the hearing, a number of expert witnesses for the prosecution explained the basis of the DNA analysis. According to these experts, no two individuals, with the exception of identical twins, have the same nucleotide sequence in their DNA. Moreover, differences in DNA sequence can be detected by comparing the lengths of the fragments produced by restriction-enzyme digestion of different DNA samples. The patterns produce a DNA fingerprint" (Figure 1) that is as unique to an individual as is a set of conventional fingerprints lifted from a glass. In 1989, DNA fingerprints had already been used in more than 200 criminal cases in the United States and had been hailed as the most important development in forensic science (the application of medical facts

FIGURE 1
Alec Jeffreys of the University of Leicester, England, examining a DNA fingerprint. Jeffreys was primarily responsible for developing the DNA fingerprint technique and was the scientist who confirmed the death of Josef Mengele.

to legal problems) in decades. The widespread use of DNA fingerprinting evidence in court had been based on its general acceptability in the scientific community. According to a report from the company performing the DNA analysis, the likelihood that the same banding patterns could be obtained by chance from two *different* individuals in the community was only one in 100 million.

What made this case (known as the Castro case, after the defendant) memorable and distinct from its predecessors was that the defense also called an expert witnesses to scrutinize the data and to present their opinions. While these experts confirmed the capability of DNA fingerprinting to identify an individual out of a huge population, they found serious technical flaws in the analysis of the DNA samples used by the prosecution. In an unprecedented occurrence, the experts who had earlier testified *for the prosecution* agreed that the DNA analysis in this case was unreliable and should not be used as evidence! The problem was not with the technique itself but in the way it had been carried out in this particular case. Consequently, the judge threw out the evidence.

In the wake of the Castro case, the use of DNA fingerprinting to decide guilt or innocence has been seriously questioned. Several panels and agencies are working to formulate guidelines for the licensing of forensic DNA laboratories and the certification of their employees. In 1992, a panel of the National Academy of Sciences released a report endorsing the general reliability of the technique but called for the institution of strict standards *to be set by scientists*.

Meanwhile, another issue regarding DNA fingerprinting has been raised and hotly debated. Two geneticists, Richard Lewontin of Harvard University and Daniel Hartl of Washington University, coauthored a paper published in December 1991, suggesting that scientists do not have enough data on genetic variation within different racial or ethnic groups to calculate the odds that two individuals—a suspect and a perpetrator of the crime—are one and the same on the basis of an identical DNA fingerprint. The matter remains an issue of great concern in both the scientific and legal communities and has yet to be resolved.

◁ B I O L I N E ▷
The Fish That Changes Sex

In vertebrates, gender is generally a biologically inflexible commitment: An individual develops into either a male or a female as dictated by the sex chromosomes acquired from one's parents. Yet, even among vertebrates, there are organisms that can reverse their sexual commitment. The Australian cleaner fish (Figure 1), a small animal that sets up "cleaning stations" to which larger fishes come for parasite removal, can change its gender in response to environmental demands. Most male cleaner fish travel alone rather than with a school. Except for a single male, schools of cleaner fish are comprised entirely of females. Although it might seem logical to conclude that maleness engenders solo travel, it is actually the other way around: Being alone fosters maleness. A cleaner fish that develops away from a school *becomes* a male, whereas the same fish developing in a school would have become a female.

FIGURE 1
The small Australian wrasse (cleaner fish) is seen on a much larger grouper.

But what of the one male in the school—the one with the harem? He may have developed as a solo fish and then found a school in need of his spermatogenic services. But there is another way a school may acquire a male. If the male in a school dies (or is removed experimentally), one of the females, the one at the top of a behavioral hierarchy that exists in each school, becomes uncharacteristically aggressive and takes over the behavioral role of the missing male. She begins to develop male gonads, and within a few weeks, the female becomes a reproductively competent male, indistinguishable from other males. Furthermore, the sex change is reversible. If a fully developed male enters the school during the sexual transition, the almost-male fish developmentally backpedals, once again assuming the biological and behavioral role of a female.

⚠ Not all organisms follow the mammalian pattern of sex determination. In some animals, most notably birds, the opposite pattern is found: The female's cells have an X and a Y chromosome, while the male's cells have two Xs. An exception to this rule of a strict relation between sex and chromosomes is discussed in the Bioline: The Fish That Changes Sex. Although some plants possess sex chromosomes and gender distinctions between individuals, most have only autosomes; consequently, each individual produces both male and female parts.

SEX LINKAGE

For fruit flies and humans alike, there are hundreds of genes on the X chromosome that have no counterpart on the smaller Y chromosome. Most of these genes have nothing to do with determining gender, but their effect on phenotype usually *depends on* gender. For example, in females, a recessive allele on one X chromosome will be masked (and not expressed) if a dominant counterpart resides on the other X chromosome. In males, it only takes one recessive allele on the single X chromosome to determine the individual's phenotype since there is no corresponding allele on the Y chromosome. Inherited characteristics determined by genes that reside on the X chromosome are called **X-linked characteristics.**

So far, some 200 human X-linked characteristics have been described, many of which produce disorders that are found almost exclusively in men. These include a type of heart-valve defect (*mitral stenosis*), a particular form of mental retardation, several optical and hearing impairments, muscular dystrophy, and red-green colorblindness (Figure 13-8).

One X-linked recessive disorder has altered the course of history. The disease is **hemophilia,** or "bleeder's disease," a genetic disorder characterized by the inability to produce a clotting factor needed to halt blood flow quickly following an injury. Nearly all hemophiliacs are males. Although females can inherit two recessive alleles for hemophilia, this occurrence is extremely rare. In general, women who have acquired the rare defective allele are heterozygous **carriers** for the disease. The phenotype of a carrier

Several ethical issues are discussed in the Bioethics essays which add provocative pauses throughout the text. Biological Science does not operate in a vacuum but has profound consequences on the general community. Because biologists study life, the science is peppered with ethical considerations. The moral issues discussed in these essays are neither simple nor easy to resolve, and we do not claim to have any certain answers. Our goal is to encourage you to consider the bioethical issues that you will face now and in the future.

Coordinating the Organism: The Role of the Nervous System / CHAPTER 23 • 489

◁ BIOETHICS ▷
Blurring the Line between Life and Death
By ARTHUR CAPLAN
Division of the Center for Biomedical Ethics at the University of Minnesota

Theresa Ann Campo Pearson didn't have a very long life. When she died in 1992, she was only 10 days old. Despite her short life, she became the center of a very strange, sad, and wrenching ethical controversy. Theresa died because her brain had failed to form. She had anencephaly, a condition in which only the brainstem, located at the top of the spinal cord, is present. Her parents wanted to donate Theresa's organs; the courts said no. Some people found it strange that Theresa's parents, Laura Campo and Justin Pearson, did not get their way. Why not allow donation, when every day in North America a baby dies because there is no heart, lung, or liver available for transplantation?

Anencephaly is best described as completely "unabling," not disabling. Children born with anencephaly cannot think, feel, sense, or be aware of the world. Many are stillborn; the majority of the rest die within days of birth. A mere handful live for a few weeks. Theresa's parents

knew all this. But rather than abort the pregnancy, they chose to have their baby. In fact, the baby was born by Caesarean section, at least partly in the hope that it would be born alive, thereby making organ donation possible. When Theresa died at Broward General Medical Center in Fort Lauderdale, Florida, however, no organs were taken. Two Florida courts ruled that the baby could not be used as a source of organs unless she was brain-dead, and Theresa Ann Campo was never pronounced brain-dead.

Brain death refers to a situation in which the brain has irreversibly lost all function and activity. Babies born with anencephaly have some brain function in their brainstem so, while they cannot think or feel, they are alive. According to Florida law—and the law in more than 40 other states—only those individuals declared brain-dead can donate organs. The courts of Florida had no other option but to deny the request for organ donation.

One obvious solution is to change the law so that states could decide that organs can be removed upon parental consent from either those who are born brain-dead or from babies who are born with anencephaly. Another solution is to rewrite the definition of death to say that death occurs either when the brain has totally ceased to function or if a baby is born anencephalic. Do you feel that either of these changes should be made? Some may argue that medicine will fudge the line between life and death in order to get organs for transplant. Do you agree with this concern? How do you think redefining death will affect a person's decision to check off the donation box on the back of a driver's license? Do you think people may worry that if they are known to be potential donors they won't be aggressively treated at the hospital? In your opinion, would changing the definition of death to include anencephaly be beneficial or deleterious?

Like the brain, the spinal cord is composed of white matter (myelinated axons) and gray matter (dendrites and cell bodies). However, the arrangement of these types of matter is reversed in the spinal cord, compared to their arrangement in the brain: The spinal cord's white matter surrounds the gray matter (Figure 23-16).

The human central nervous system is the most complex and highly evolved assembly of matter. Among its functions are the processing of sensory information collected from both the external and internal environment; the regulation of internal physiological activities; the coordination of complex motor activities; and the endowment of such intangible "mental" qualities as emotions, creativity, language, and the ability to think, learn, and remember. (See CTQ #6.)

ARCHITECTURE OF THE PERIPHERAL NERVOUS SYSTEM

The peripheral nervous system provides the neurological bridge between the central nervous system and the various parts of the body. The peripheral nervous system is made up of paired nerves that extend into the periphery from the CNS at various levels along the body. Each nerve is composed of a large bundle of myelinated axons surrounded by a connective tissue sheath. Twelve pairs of **cranial nerves** emerge from the central stalk of the human brain, and 31 pairs of **spinal nerves** extend from the spinal cord out between the vertebrae of humans (Figure 23-16). For the most part, the cranial nerves *innervate* (supply nerves to) tissues and organs of the head and neck, whereas the spinal nerves innervate the chest, abdomen, and limbs.

Additional Pedagogical Features

We have worked to assure that each chapter in this book is an effective teaching and learning instrument. In addition to the pedagogical features discussed above, we have included some additional tried-and-proven-effective tools.

KEY POINTS

Key points follow each major section and offer a condensation of the relevant facts and details as well as the concepts discussed. You can use these key points to reaffirm your understanding of the previous reading or to alert you to misunderstood material before moving on to the next topic. Each key point is tied to a Critical Thinking Question found at the end of the chapter; together, they encourage you to analyze the information, taking it beyond mere memorization.

Sample page (Chapter 18)

Plant Tissues and Organs / CHAPTER 18 • 361

➠ Many plants replenish old and dying cells with vigorous new cells. But since each plant cell has a surrounding cell wall (Chapter 7) old plant cells do not just wither and disappear when they die. Instead, dead plant cells leave cellular "skeletons" where they once lived. As a result, the longer a plant lives, the more complex its anatomy becomes. **Annuals** are plants that live for 1 year or less, such as corn and marigolds. Because they live for such a brief period, these plants do not completely replace old cells. As a result, annuals are anatomically less complex than are **biennials**—plants that live for 2 years—and **perennials**—herbs, shrubs, and trees that live longer than 2 years. Biennials (carrots, Queen Anne's lace) and perennials (rosebushes, apple trees) are able to live longer than annuals because they produce new cells to replace those that cease functioning or die, providing a continual supply of young, vigorous cells.

In this chapter, we will focus on the body construction of flowering plants, the most familiar, most evolutionarily advanced, and structurally complex of any group in the plant kingdom. All flowering plants are **vascular plants;** that is, they contain specialized cells that circulate water, minerals, and food (organic molecules) throughout the plant. Botanists divide flowering plants into two main groups: **dicotyledons**, or dicots (*di* = two, *cotyledon* = embryonic seed leaf), and **monocotyledons**, or monocots (*mono* = one). Table 18-1 illustrates the many differences that distinguish dicots from monocots and will be used as a reference throughout the chapter.

SHOOTS AND ROOTS

The flowering plant body is a study in contradictions. A typical plant grows through the soil and the air simultaneously, two very different habitats with very different conditions. As a result, the two main parts of the plant differ dramatically in form (anatomy) and function (physiology): The underground **root system** anchors the plant in the soil and absorbs water and nutrients, while the aerial **shoot system** absorbs sunlight and gathers carbon dioxide for photosynthesis (Figure 18-2). The shoot system also produces stems, leaves, flowers, and fruits. Interconnected vascular tissues transport materials between the aerial shoot system and the underground root system. These connections allow water and minerals absorbed by the root to be conducted to shoot tissues, and for food produced by the shoot to be transported to root tissues. We will discuss the various components of these two systems in more detail later in the chapter.

Over 90 percent of all plant species are flowering plants. Flowering plants are the most recently evolved plant group, having undergone rapid evolution during the past 1 million to 2 million years as environmental conditions on land became more variable. (See CTQ #2.)

TABLE 18-1

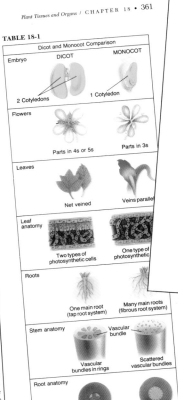

Dicot and Monocot Comparison	DICOT	MONOCOT
Embryo	2 Cotyledons	1 Cotyledon
Flowers	Parts in 4s or 5s	Parts in 3s
Leaves	Net veined	Veins parallel
Leaf anatomy	Two types of photosynthetic cells	One type of photosynthetic
Roots	One main root (tap root system)	Many main roots (fibrous root system)
Stem anatomy	Vascular bundles in rings	Scattered vascular bundles
Root anatomy	Xylem in center	Pith in center
Secondary growth	Yes	No

Sample page (page 288)

288 • PART 3 / *The Genetic Basis of Life*

the corresponding polypeptide. The cumulative effect of gradual changes in polypeptides over evolutionary time has been the generation of life's diversity.

Evolution and Adaptation

➠ Evolutionary change from generation to generation depends on genetic variability. Much of this variability arises from reshuffling maternal and paternal genes during meiosis, but somewhere along the way new genetic information must be introduced into the population. New genetic information arises from mutations in existing g[...] Some of these mutations arise during replication; o[...] occur as the result of unrepaired damage as the DNA i[...] "sitting" in a cell. Mutations that occur in an individ[...] germ cells can be considered the raw material on w[...] natural selection operates; whereas harmful mutatio[...] duce offspring with a reduced fitness, beneficial mutat[...] produce offspring with an increased fitness.

SYNOPSIS

Experiments in the 1940s and 1950s established conclusively that DNA is the genetic material. These experiments included the demonstration that DNA was capable of transforming bacteria from one genetic strain to another; that bacteriophages injected their DNA into a host cell during infection; and that the injected DNA was transmitted to the bacteriophage progeny.

DNA is a double helix. DNA is a helical molecule consisting of two chains of nucleotides running in opposite directions, with their backbones on the outside, and the nitrogenous bases facing inward like rungs on a ladder. Adenine-containing nucleotides on one strand always pair with thymine-containing nucleotides on the other strand, likewise for guanine- and cytosine-containing nucleotides. As a result, the two strands of a DNA molecule are complementary to one another. Genetic information is encoded in the specific linear sequence of nucleotides that make up the strands.

DNA replication is semiconservative. During replication, the double helix separates, and each strand serves as a template for the formation of a new, complementary strand. Nucleotide assembly is carried out by the enzyme DNA polymerase, which moves along the two strands in opposite directions. As a result, one of the strands is synthesized continuously, while the other is synthesized in segments that are covalently joined. Accuracy is maintained by a proofreading mechanism present within the polymerase.

Information flows in a cell from DNA to RNA to protein. Each gene consists of a linear sequence of nucleotides that determines the linear sequence of amino acids in a polypeptide. This is accomplished in two majo[...] steps: transcription and translation.

During transcription, the information spelled out b[...] the gene's nucleotide sequence is encoded in a mole[...] cule of messenger RNA (mRNA). The mRNA contain[...] a series of codons. Each codon consists of three nucleotides Of the 64 possible codons, 61 specify an amino acid, and the other 3 stop the process of protein synthesis.

During translation, the sequence of codons in the mRNA is used as the basis for the assembly of a chai[...] of specific amino acids. Translating mRNA mes[...] occurs on ribosomes and requires tRNAs, which ser[...] decoders. Each tRNA is folded into a cloverleaf structu[...] with an anticodon at one end—which binds to a comple[...] mentary codon in the mRNA—and a specific amino acid a[...] the other end—which becomes incorporated into the growing polypeptide chain. Amino acids are added to their appropriate tRNAs by a set of enzymes. The sequential interaction of charged tRNAs with the mRNA results in th[...] assembly of a chain of amino acids in the precise order dictated by the DNA.

Mutation is a change in the genetic message. Gene mutations may occur as a single nucleotide substitution, which leads to the insertion of an amino acid different from that originally encoded. In contrast, the addition of one or two nucleotides throws off the reading frame of the ribosome as it moves along the mRNA, leading to the incorporation of incorrect amino acids "downstream" from the point of mutation. Exposure to mutagens increases the rate of mutation.

SYNOPSIS

The synopsis section offers a convenient summary of the chapter material in a readable narrative form. The material is summarized in concise paragraphs that detail the main points of the material, offering a useful review tool to help reinforce recall and understanding of the chapter's information.

REVIEW QUESTIONS

Along with the synopsis, the Review Questions provide a convenient study tool for testing your knowledge of the facts and processes presented in the chapter.

224 • PART 2 / Chemical and Cellular Foundations of Life

Key Terms

zygote (p. 214)
meiosis (p. 214)
life cycle (p. 214)
germ cell (p. 214)
somatic cell (p. 214)
meiosis I (p. 216)

reduction division (p. 216)
synapsis (p. 216)
tetrad (p. 216)
crossing over (p. 216)
genetic recombination (p. 216)
synaptonemal complex (p. 218)

maternal chromosome (p. 219)
paternal chromosome (p. 219)
independent assortment (p. 219)
meiosis II (p. 219)

Review Questions

1. Match the activity with the phase of meiosis in which it occurs.

   a. synapsis
   b. crossing over
   c. kinetochores split
   d. independent assortment
   e. homologous chromosomes separate
   f. cytokinesis

   1. prophase I
   2. metaphase I
   3. anaphase I
   4. telophase I
   5. prophase II
   6. anaphase II
   7. telophase II

2. How do crossing over and independent assortment increase the genetic variability of a species?

3. Why is meiosis I (and not meiosis II) referred to as the reduction division?

4. Suppose that one human sperm contains x amount of DNA. How much DNA would a cell just entering meiosis contain? A cell entering meiosis II? A cell just completing meiosis II? Which of these three cells would have a haploid number of chromosomes? A diploid number of chromosomes?

Critical Thinking Questions

1. Why are disorders, such as Down syndrome, that arise from abnormal chromosome numbers, characterized by a number of seemingly unrelated abnormalities?

2. A gardener's favorite plant had white flowers and long seed pods. To add some variety to her garden, she transplants some plants of the same type, but with pink flowers and short seed pods from her neighbor's garden. To her surprise, in a few generations, she grows plants with white flowers and short seed pods and plants with pink flowers and long seed pods, as well as the original combinations. What are two ways in which these new combinations could have arisen?

3. Set up the meiosis template in the diagram below on a large sheet of paper. Then use pieces of colored yarn or pipe cleaners to simulate chromosomes and make a model of the phases of meiosis. (See template on opposite page)

4. Would you expect two genes on the same chromosome, such as yellow flowers and short stems, always to be exchanged during crossing over? How might they remain together in spite of crossing over?

5. Suppose paternal chromosomes always lined up on the same side of the metaphase plate of cells in meiosis I. How would this affect genetic variability of offspring? Would they all be identical? Why or why not?

Additional Readings

Chandley, A. C. 1988. Meiosis in man. Trends in Gen. 4:79–83. (Intermediate)

Hsu, T. C. 1979. Human and mammalian cytogenetics. New York: Springer-Verlag. (Intermediate)

John, B. 1990. Meiosis. New York: Cambridge University Press. (Advanced)

Moens, P. B. 1987. Meiosis. Orlando: Academic. (Advanced)

Patterson, D. 1987. The causes of Down syndrome. Sci. Amer. Feb:52–60. (Intermediate-Advanced)

White, M. J. D. 1973. The chromosomes. Halsted. (Advanced)

STIMULATING CRITICAL THINKING

Each chapter contains as part of its end material a diverse mix of Critical Thinking Questions. These questions ask you to apply your knowledge and understanding of the facts and concepts to hypothetical situations in order to solve problems, form hypotheses, and hammer out alternative points of view. Such exercises provide you with more effective thinking skills for competing and living in today's complex world.

ADDITIONAL READINGS

Supplementary readings relevant to the Chapter's topics are provided at the end of every chapter. These readings are ranked by level of difficulty (introductory, intermediate, or advanced) so that you can tailor your supplemental readings to your level of interest and experience.

Careers in Biology

*T*he appendices of this edition include "Careers in Biology," a frequently overlooked aspect of our discipline. Although many of you may be taking biology as a requirement for another major (or may have yet to declare a major), some of you are already biology majors and may become interested enough to investigate the career opportunities in life sciences. This appendix helps students discover how an interest in biology can grow into a livelihood. It also helps the instructor advise students who are considering biology as a life endeavor.

APPENDIX
◄ D ►

Careers in Biology

Although many of you are enrolled in biology as a requirement for another major, some of you will become interested enough to investigate the career opportunities in life sciences. This interest in biology can grow into a satisfying livelihood. Here are some facts to consider:

- Biology is a field that offers a very wide range of possible science careers
- Biology offers high job security since many aspects of it deal with the most vital human needs: health and food
- Each year in the United States, nearly 40,000 people obtain bachelor's degrees in biology. But the number of newly created and vacated positions for biologists is increasing at a rate that exceeds the number of new graduates. Many of these jobs will be in the newer areas of biotechnology and bioservices.

Biologists not only enjoy job satisfaction, their work often changes the future for the better. Careers in medical biology help combat diseases and promote health. Biologists have been instrumental in preserving the earth's life-supporting capacity. Biotechnologists are engineering organisms that promise dramatic breakthroughs in medicine, food production, pest management, and environmental protection. Even the economic vitality of modern society will be increasingly linked to biology.

Biology also combines well with other fields of expertise. There is an increasing demand for people with backgrounds or majors in biology complexed with such areas as business, art, law, or engineering. Such a distinct blend of expertise gives a person a special advantage.

The average starting salary for all biologists with a Bachelor's degree is $22,000. A recent survey of California State University graduates in biology revealed that most were earning salaries between $20,000 and $50,000. But as important as salary is, most biologists stress job satisfaction, job security, work with sophisticated tools and scientific equipment, travel opportunities (either to the field or to scientific conferences), and opportunities to be creative in their job as the reasons they are happy in their career.

Here is a list of just a few of the careers for people with degrees in biology. For more resources, such as lists of current openings, career guides, and job banks, write to Biology Career Information, John Wiley and Sons, 605 Third Avenue, New York, NY 10158.

A SAMPLER OF JOBS THAT GRADUATES HAVE SECURED IN THE FIELD OF BIOLOGY*

Agricultural Biologist	Bioanalytical Chemist	Brain Function	Environmental Center
Agricultural Economist	Biochemical/Endocrine	Researcher	Director
Agricultural Extension	Toxicologist	Cancer Biologist	Environmental Engineer
Officer	Biochemical Engineer	Cardiovascular Biologist	Environmental Geographer
Agronomist	Pharmacology Distributor	Cardiovascular/Computer	Environmental Law Specialist
Amino-acid Analyst	Pharmacology Technician	Specialist	Farmer
Analytical Biochemist	Biochemist	Chemical Ecologist	Fetal Physiologist
Anatomist	Biogeochemist	Chromatographer	Flavorist
Animal Behavior	Biogeographer	Clinical Pharmacologist	Food Processing Technologist
Specialist	Biological Engineer	Coagulation Biochemist	Food Production Manager
Anticancer Drug Research	Biologist	Cognitive Neuroscientist	Food Quality Control
Technician	Biomedical	Computer Scientist	Inspector
Antiviral Therapist	Communication Biologist	Dental Assistant	Flower Grower
Arid Soils Technician	Biometerologist	Ecological Biochemist	Forest Ecologist
Audio-neurobiologist	Biophysicist	Electrophysiology/	Forest Economist
Author, Magazines & Books	Biotechnologist	Cardiovascular Technician	Forest Engineer
Behavioral Biologist	Blood Analyst	Energy Regulation Officer	Forest Geneticist
Bioanalyst	Botanist	Environmental Biochemist	Forest Manager

Written by Gary Wisehart and Michael Leboffe of San Diego City College, the *Study Guide* has been designed with innovative pedagogical features to maximize your understanding and retention of the facts and concepts presented in the text. Each chapter in the *Study Guide* contains the following elements.

Concepts Maps

In Chapter 1 of the *Study Guide*, the beginning of a concept map stating the five themes is introduced. In each subsequent chapter, the concept map is expanded to incorporate topics covered in each chapter as well as the interconnections between chapters and the five themes. "Connector" phrases are used to link the concepts and themes, and the text icons representing the themes are incorporated into the concept maps.

Go Figure!

In each chapter, questions are posed regarding the figures in the text. Students can explore their understanding of the figures and are asked to think critically about the figures based on their understanding of the surrounding text and their own experiences.

Self-Tests

Each chapter includes a set of matching and multiple-choice questions. Answers to the Study Guide questions are provided.

Concept Map Construction

The student is asked to create concept maps for a group of terms, using appropriate connector phrases and adding terms as necessary.

*B*iology: Exploring Life, Second Edition is supplemented by a comprehensive *Laboratory Manual* containing approximately 60 lab exercises chosen by the text authors from the National Association of Biology Teachers. These labs have been thoroughly class-tested and have been assembled from various scientific publications. They include such topics as

- Chaparral and Fire Ecology: Role of Fire in Seed Germination (*The American Biology Teacher*)
- A Model for Teaching Mitosis and Meiosis (*American Biology Teacher*)

- Laboratory Study of Climbing Behavior in the Salt Marsh Snail (*Oceanography for Landlocked Classrooms*)
- Down and Dirty DNA Extraction (*A Sourcebook of Biotechnology Activities*)
- Bioethics: The Ice-Minus Case (*A Sourcebook of Biotechnology Activities*)
- Using Dandelion Flower Stalks for Gravitropic Studies (*The American Biology Teacher*)
- pH and Rate of Enzymatic Reactions (*The American Biology Teacher*)

Biology:
Exploring Life

Biology: The Study of Life

To understand life you must explore the obvious and the subtle, as well as all levels in between. This translucent jellyfish represents the organismal level of biological organization. But to understand how a jellyfish or any other organism survives, grows, and reproduces, biologists must study all levels of organization, including the organs, tissues, cells, and even molecules that make up an organism, as well as the population, community and ecosystem in which the organism lives.

PART
· 1 ·

◄ 1 ►

Biology: Exploring Life

STEPS TO DISCOVERY
Exploring Life—the First Step

DISTINGUISHING THE LIVING FROM THE INANIMATE

Organisms are Highly Complex and Organized

Organisms Are Composed Of Cells

Organisms Acquire And Utilize Energy

Organisms Produce Offspring Similar To Themselves

Organisms Are Built According To Genetic Instructions

Organisms Grow In Size And Change by Development

Organisms Respond To Stimuli

Organisms Carry Out A Variety Of Chemical Reactions

Organisms Maintain A Relatively Constant Internal Environment

LEVELS OF BIOLOGICAL ORGANIZATION

WHAT'S IN A NAME

Assigning Scientific Names

Classification Of Organisms

UNDERLYING THEMES OF BIOLOGY

The Relationship Between Form and Function

Biological Order, Regulation, and Homeostasis

Acquiring and Using Energy

Unity within Diversity

Evolution and Adaptation

BIOLOGY AND MODERN ETHICS

Exploring Life — the First Step

*B*iology is not magic. It does create some very impressive illusions, however, such as its amazing "disappearing acts." A leaf-shaped butterfly lands on an oakbranch and "disappears," becoming indistinguishable from the real leaves of the tree. A succulent plant growing close to the ground looks more like a rock than a living organism. Unseen by hungry animals, these organisms escape being eaten by hiding in plain sight. In England before the 1800s, where most trees were covered with light-colored lichens, the common white peppered moth was very adept at such disappearing acts. All it had to do was land on one of these mottled white tree surfaces, and the moth became virtually invisible. But then, disaster struck for these light-colored moths. The industrial revolution gained its full stride, and the white peppered moth performed a different kind of disappearing act, one that almost lasted forever.

Blackened by industrial smoke, the bright landing places of the white peppered moths changed to dark, sooty surfaces. The white peppered moths were now easily spotted by birds as they "hid" on their former sanctuaries. In these new conditions, they no longer had a competitive edge, and their numbers plunged toward the vanishing point. Yet, as the lighter-colored moths were eliminated, something unusual began to happen. Rarely seen before the industrial revolution, dark-colored peppered moths began to grow in number. Soon the population of peppered moths

Before the Industrial Revolution, dark pepper moths were eaten more frequently by birds than white moths as they rested on

was back to its former prevalence, but the new moths were dark—perfectly camouflaged on the newly blackened trees of industrialized England. Somehow the peppered moth species had "switched colors," and the species continued to survive.

To understand how the moths changed in response to their environment, we first must discuss what they *did not* do. These moths were not like chameleons: An individual moth could no more change its color to match its background than you can. It was the *species,* not the individual moths, that changed color so that, by the mid-1950s, virtually all peppered moths in industrialized Britain were dark. If they could not recognize their dark background and change colors, where did the black moths come from? The answer is *genetic variability.*

Most of the traits among the individuals in any species are similar, but there are also many genetic differences. People, for example, have two eyes, a nose, fingernails, and hundreds of other features that illustrate our similarities. Yet, we all look different from one another—even from our own parents—because of genetic variability, that is, differences in the genes possessed by different organisms. Genes are coded bits of information in cells that determine an organism's traits. Copies of these genes are passed on from parents to their offspring, who thereby inherit traits characteristic of their parents. Occasionally a spontaneous change, a *mutation,* will occur in a gene, which causes the offspring to inherit a new trait. In a population of billions of light-colored moths, for example, a few dark offspring will inevitably be produced by mutation of a pigment gene. Before the industrial revolution, however, few of these dark varieties were ever found because they were easily seen against the light surfaces and snatched up by birds. But around the mid-1800s, environmental conditions changed. The light moths had become the easy targets, and the dark moths became the "invisible" and predominant variety. Had there not been genetic variation, there would have been no dark peppered moths, and the species would now be extinct in these industrialized areas.

There is another surprising chapter in this story. If you go to industrialized England today, you will once again see the white peppered moth. Modern pollution controls have cleaned up the air, and the surfaces of the trees are once again brightly colored. The lighter moths once again have the competitive edge, and the dark moth is rarely found. The process of change continues in the peppered moth, and in all other types of organisms, according to the dictates of environmental conditions.

The case of the peppered moth illustrates how a species can change over time. The process that has changed the population of peppered moths is the same one responsible for generating the millions of species produced by three and a half billion years of biological evolution. Thus, the study of the peppered moth provides a vivid portrayal of how changes in the environment can change the genetic composition of a species. The mechanism by which new species evolve is discussed later in this chapter, and some of the overwhelming evidence for its occurrence is described in Chapter 34.

Our modern knowledge of evolution helps us understand life. It enables us to answer such childlike questions as "Why do we have houseflies?" as well as more global questions about the possibility of human extinction. Evolution enriches the study of life by making rational sense of it. It helps us understand where organisms and their properties come from and why they exist. As you progress through this book, you will find that the theme of evolution illuminates all areas of biology. It will help you piece together this information into a satisfying body of understanding that will enable you to make better sense of the world and what happens among living things.

the light bark of trees. As tree trunks were darkened by industrial smoke, the situation reversed.

You became a biologist long before you opened this book. It happens to everyone sometime during childhood, a time that marks an awakening of the fascination we have for all organisms. Children explore ant colonies and inspect sprouting seeds. They watch hypnotically as a spider spins its web. For some, the fascination grows and becomes more focused. Casual curiosity gives way to the irresistible lure of exploration, and an interesting pastime ripens into an exciting search for knowledge and discovery. These are biology's "lifers"—scientists who are engaged in a lifelong adventure, a quest to make rational sense of living phenomena.

Biology is the study of life. It is a multidimensional, dynamic, creative activity that replaces mystery with understanding. No list of terms or facts can produce understanding, any more than a pile of unassembled gears and springs can explain how a clock works. Individual biological facts reveal little about how life works; they are mere threads that must be woven together by concepts or principles to arrive at an understanding of living phenomena. Biologists are detectives who use bits of information as clues to solve the complex mysteries of life. Their work also yields practical bonuses, such as controlling diseases, increasing crop yields, and proposing measures to preserve our environment.

Biology is an intellectual adventure that will take you from the landscape of molecules and cells to the global community of organisms. In distilling this information for use as a general introduction, we have struggled to retain the challenge and excitement of biology. We hope the discoveries you make in this pursuit will rekindle your fascination with biological phenomena and reaffirm your connection with all living things.

▼ ▼ ▼

DISTINGUISHING THE LIVING FROM THE INANIMATE

Most of us have little trouble identifying something as being alive or inanimate. The diverse organisms shown in Figure 1-1 are easily distinguished from nonliving environmental components. But try to define what distinguishes the two, and you may find yourself at a loss. You'll discover that many of the properties that flag an object as living cannot be found in all organisms or may be exhibited by some nonliving things. For example, you may have selected the ability to move as a basic "characteristic of life." But is a redwood tree

an inanimate object simply because you cannot observe any movement? Or, conversely, is a river alive because its movement is evident? Because of this difficulty, rather than trying to define life, biologists describe it—usually as a list of properties that characterize all living things:

- *Organization* Organisms maintain a high degree of complexity and order.
- *Cells* Organisms are composed of one or more cells.
- *Energy* Organisms acquire and use energy.
- *Reproduction* Organisms produce offspring similar to themselves.

FIGURE 1-1

Defining life. As simple as it is to identify each object in the photograph as either alive or inanimate, *justifying* your choices can be much more difficult. Most characteristics popularly associated with living things (growth, movement, reproduction, consumption of food) may also be properties of nonliving entities. Clouds and mineral crystals grow; rivers, air, and clouds move; fire consumes "food" (fuel) as it "grows" and "reproduces" giving rise to "progeny" fires. Yet none of these is alive. Defining "life" requires a combination of many properties, not one of which can, by itself, be considered *the* criterion of life.

- *Heredity* Organisms contain a genetic blueprint that dictates their characteristics.
- *Growth and development* Organisms grow in size and change in appearance and abilities.
- *Responsiveness* Organisms respond to changes in their environment.
- *Metabolism* Organisms carry out a variety of controlled chemical reactions.
- *Homeostasis* Organisms maintain a relatively constant internal environment, despite fluctuations in their external environment.

As you examine these properties in more detail in the next few pages, bear in mind that they not only "define" life, they are also inseparably linked to its success — its nonstop presence on this planet for about 3.5 billion years. Mountains crumble, continents collide, climates change, yet life on earth persists in spite of the changes (Figure 1-2).

ORGANISMS ARE HIGHLY COMPLEX AND ORGANIZED

Complexity is a measure of the number of parts that make up an object and the precision by which the parts are *organized*. An automobile is more complex than a bicycle and less complex than a space shuttle. These differences in complexity reflect the relative capabilities of these objects. Living organisms (Figure 1-3*a*) are vastly more complex than any space ship and are capable of a vastly greater variety of activities.

ORGANISMS ARE COMPOSED OF CELLS

As we will see in Chapter 5, cells are the functional units of life—all living organisms are composed of cells. For some **unicellular** organisms—those that consist of one cell— the cell *is* the organism. Most organisms, however, are **multicellular** (Figure 1-3*b*), consisting of hundreds to trillions of cells, depending on the organism's size and complexity.

(a)

(b)

FIGURE 1-2

The enduring nature of life. *(a)* This moss-covered skull testifies that life goes on in spite of the death of an individual. *(b)* This bristlecone pine has endured cold, lack of soil, and slashing winds in this same spot for thousands of years.

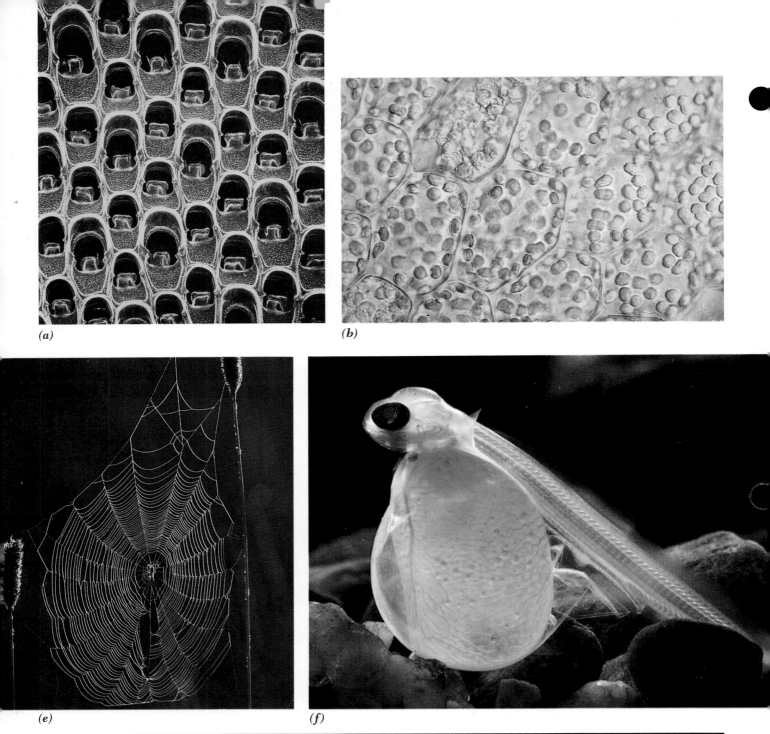

(a)   *(b)*

(e)   *(f)*

FIGURE 1-3

A number of life's properties illustrated.   *(a) Organization.* The skeletal remains of these marine bryozoans reveal a high degree of order and organization.   *(b) Cells.* A multicellular organism is a cooperative constellation of cells, each of which can perform all the activities associated with life. *(c) Acquisition of energy.* This spider is about to harvest energy from a fly.   *(d) Reproduction.* The resemblance within this family of swallow-tailed bee eaters is the dual product of reproduction and heredity.   *(e) Heredity.* The construction of a spiderweb is a task that requires no learning—each spider constructs the web on the very first attempt. Like the spider's color, shape, and other physical characteristics, web-spinning behavior is a genetically programmed trait inherited from the spider's

(c)

(d)

(g)

(h)

(i)

parents. *(f) Growth and development.* All organisms develop from a single cell. This week-old salmon embryo is acquiring the form of an adult fish but still obtains all its nutrients from the yolk sac protruding from its midsection. *(g) Responsiveness.* The final form of these sculptured tree roots reflects their ability to respond to the presence of a giant boulder. *(h) Metabolism.* The glow produced by this marine animal (ctenophore) is due to the metabolic release of energy from its food molecules, one of the thousands of chemical activities that occur in an organism. *(i) Homeostasis.* Even when exposed to subzero air temperatures, thermal homeostasis is maintained in these snow monkeys, whose body temperatures remain steady in spite of the cold.

ORGANISMS ACQUIRE AND UTILIZE ENERGY

Manufacturing and maintaining complexity requires the constant input of *energy*. Your car, for example, required energy in the manufacture and assembly of its parts; it requires additional energy to keep it in working order and to fuel your travels. Living organisms also require an input of energy to build and maintain their tissues and fuel their activities. Some organisms, such as plants, acquire the necessary energy by trapping sunlight and converting it to a form they can use. Other organisms acquire energy by feeding on the tissues of other life forms. For example, the spider in Figure 1-3c is about to harvest the energy stored in the body of a fly. The original source of energy for both of these animals was sunlight absorbed by plants.

ORGANISMS PRODUCE OFFSPRING SIMILAR TO THEMSELVES

New life is generated by the process of *reproduction*. Reproduction not only produces more organisms, it produces organisms that are very similar to their parent(s). Humans always produce humans, and octopuses always produce octopuses. Similar individuals are produced generation after generation, and thus the characteristics and functions of a particular kind of organism persist long after the parents have died (Figure 1-3d).

ORGANISMS ARE BUILT ACCORDING TO GENETIC INSTRUCTIONS

Offspring resemble their parents because they *inherit* a set of *genetic instructions* as part of the process of reproduction. The genetic instructions that we inherited from our mother and father consist of a vast collection of individual *genes* that determine the shapes of our faces, the intricate interconnections among our billions of nerve cells, and our propensity to form social units. The extent to which the genetic blueprint dictates biological activity can be appreciated by watching a spider as it plays out the intricate, genetically programmed behavior responsible for constructing a web (Figure 1-3e).

ORGANISMS GROW IN SIZE AND CHANGE BY DEVELOPMENT

Organisms cannot survive and perpetuate their kind without growing at some time during their life. *Growth*, an increase in size, is usually accompanied by *development*, a change in an organism's form and function. Although some simpler organisms, notably bacteria, show few developmental changes as they increase in size, virtually all other organisms change their form and capabilities dramatically as they grow (Figure 1-3f). An acorn develops into an oak

tree; a caterpillar changes into a butterfly; a fertilized human egg develops into a person capable of contemplating his or her own nature and origin.

ORGANISMS RESPOND TO STIMULI

The ability to respond to *stimuli,* changes within an organism or in its external environment, is literally a matter of life or death. Responses may help an organism escape predators, capture prey, optimize its exposure to sunlight, move away from detrimental environmental conditions, move toward a source of water or other resources, locate mates, change growth patterns according to season, and perform many other activities necessary for survival. Some responses must be very rapid to succeed. The angler fish, for instance, responds to the presence of a smaller fish so rapidly that its strike cannot be seen; the prey seemingly disappears. Plants generally respond to stimuli more slowly than animals, but their response is just as crucial to their survival (Figure 1-3g).

ORGANISMS CARRY OUT A VARIETY OF CHEMICAL REACTIONS

Metabolism is the sum total of all the chemical reactions occurring within an organism. Even the simplest bacterial cell is capable of hundreds of different chemical transformations, none of which occurs to any significant degree in the inanimate world. Virtually all chemical changes in organisms require **enzymes**—molecules that increase the rate at which a chemical reaction occurs. Some enzymes participate in the breakdown of food molecules; others mediate the controlled release of usable energy; still others contribute to the assembly of substances required to build more tissue (Figure 1-3h).

ORGANISMS MAINTAIN A RELATIVELY CONSTANT INTERNAL ENVIRONMENT

An organism can remain alive only as long as the properties of its internal environment remain within a certain range. If, for example, an organism's cellular fluids become too salty, too warm, or too acidic, or if they retain too high a level of toxic waste products, the cells will die, and consequently the entire organism will die. Organisms possess self-regulatory mechanisms that allow them to maintain a relatively constant internal environment despite being bombarded by changing external conditions. This characteristic of life is known as **homeostasis.** The temperature of your body, for example, is normally maintained at a steady 37°C (98.6°F), even in very cold or very hot surroundings. Homeostatic control centers in your brain detect deviations from the normal temperature and "automatically" instruct the body to shiver, which generates heat, or to sweat, which cools the body. Bacteria, plants, bread mold, animals—indeed all organisms—possess homeostatic control mechanisms (Figure 1-3*i*).

Life defies a simple definition. Life exists only when a particular combination of properties exists in a single entity. The same combination occurs in all living organisms. (See CTQ #1.)

LEVELS OF BIOLOGICAL ORGANIZATION

An object, animate or nonliving, is composed of relatively simple building blocks that are assembled into increasingly complex subunits, which ultimately combine to form the final complex structure. This book, for example, consists of ink and paper, which combine to form letters printed on pages (a higher level of organization). Individual letters combine to form a more complex structure (a word), which groups with other words to form the next level of complexity, a sentence. These levels of organization increase in steps until the final book is formed. Living organisms can also be placed within such a *hierarchy of organization,* in which simpler structures combine to form the more complex structures of the next level of organization, which in turn interact to form even more complex units (Figure 1-4).

All matter is built of protons, electrons, and neutrons. These three kinds of subatomic (smaller than atoms) particles combine to form many different types of atoms. Atoms, in turn, can combine with one another in a nearly infinite variety of ways to form molecules of limitless variety. In living organisms, molecules join together to form subcellular components (including structures called *organelles*) that are assembled into *cells.* Cells occupy a very special level of organization; a cell is the simplest unit in the biological hierarchy that is, in itself, alive.

The attainment of the cellular organization reveals something profound and fascinating about hierarchies of organization: *Increasing complexity not only generates a higher order of structural organization, but also creates new properties that exceed the sum of the parts used to form the structure.* To return to our book analogy, when individual letters are combined to form a word, the structure at the next level of organization, an additional property emerges —the word's meaning. No new materials were added; the new property is strictly a function of higher organization. Words are then combined to form the next level, a sentence, from which another property emerges—a statement. The levels of organization increase until the final structure—a manuscript—is created at a level complex enough for a story to be created.

Life itself is such an emergent property, a phenomenon that first appears at the level of the cell and that exceeds the sum of the parts. If not organized to form a cell, the components of a cell are incapable of generating or sustaining life. Subcellular organelles and cytoplasm removed from one another simply deteriorate—they cannot maintain their ordered state, reproduce, nor respond to stimuli. They are not alive. Yet together, as an intact cell, all the properties of life emerge.

As highly ordered as organisms are, they too are subunits of even more complex levels of organization. Individuals of the same species inhabiting the same area constitute a **population.** Different populations in a particular area interact with each other to form **communities,** which form part of a particular **ecosystem.** Ecosystems consist of the community in an area plus the nonliving environment, for example, the water, rocks, and mud, together with the bacteria, plants, and animals at the bottom of a lake. All the world's ecosystems combine to form the **biosphere.**

This hierarchy of life provides one of the foundations on which this book is organized. We begin at the atomic and molecular levels, work our way through a variety of subjects at the cellular level, turn to a discussion of organs and organ systems, then move on to an appreciation of the diverse forms of organisms on earth, culminating in a look at the ways organisms interact with each other and their environment.

From atoms to the biosphere, life can be investigated at different levels of complexity and organization. With each increasing step in complexity, new properties emerge that could not exist at lower states of organization. It is these emergent properties that epitomize the significance of the hierarchy of biological organization. Life itself is such a property. (See CTQ #2.)

WHAT'S IN A NAME

The sheer complexity of organisms seems so bewildering that some people believe life to be unfathomable by the rational mind. Biologists constantly chip away at this notion by exploring and discovering real (and fully understandable) explanations for phenomena. Our ability to solve these natural mysteries would be severely handicapped, however, without a system that creates meaningful order out of the overwhelming variety of organisms to be studied. This orderly system enables us to distinguish clearly between different types of organisms, and to classify and assign names to the almost 2 million known types of living organisms, as well as their extinct ancestors. This formal system of naming, cataloguing, and describing organisms is the science of **taxonomy.**

Taxonomy perpetuates order by examining each type of organism and describing its properties, defining the set of criteria that distinguishes that particular type of organism

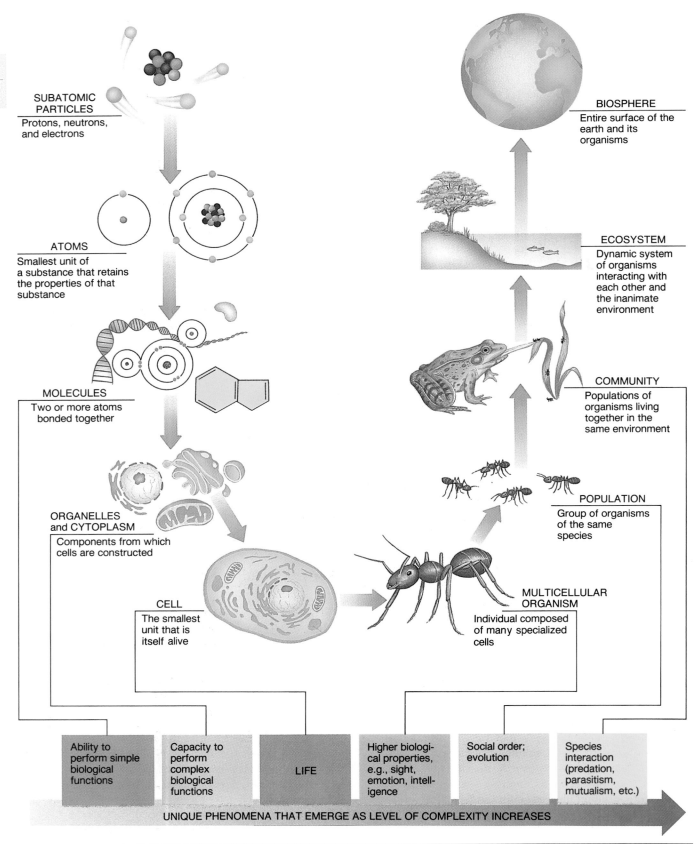

SUBATOMIC
PARTICLES
Protons, neutrons,
and electrons

BIOSPHERE
Entire surface of the
earth and its
organisms

ATOMS
Smallest unit of
a substance that retains
the properties of that
substance

ECOSYSTEM
Dynamic system
of organisms
interacting with
each other and
the inanimate
environment

MOLECULES
Two or more atoms
bonded together

COMMUNITY
Populations of
organisms living
together in the
same environment

ORGANELLES
and CYTOPLASM
Components from which
cells are constructed

POPULATION
Group of organisms
of the same
species

CELL
The smallest
unit that is
itself alive

MULTICELLULAR
ORGANISM
Individual composed
of many specialized
cells

| Ability to perform simple biological functions | Capacity to perform complex biological functions | LIFE | Higher biological properties, e.g., sight, emotion, intelligence | Social order; evolution | Species interaction (predation, parasitism, mutualism, etc.) |

UNIQUE PHENOMENA THAT EMERGE AS LEVEL OF COMPLEXITY INCREASES

FIGURE 1-4

Levels of biological organization. Each single-step jump increases structural complexity and may
also generate a unique property distinct from the structure.

from all other types. (Taxonomists have described more than 1.8 million distinct species, which represents only a fraction of the estimated total of 5 million to 50 million species alive on earth today.) From this body of descriptive information, taxonomists create the two orderly systems that are so necessary to biology:

- A *system of nomenclature* that assigns a specific name to each type of organism, a label that doesn't vary from biologist to biologist.

- A *system of classification* that groups organisms into categories according to similarities in major properties. This formal system has a dual purpose. First, it provides a standard set of criteria that can be used to ascertain the identity of an organism from its observable properties. Second, the classification scheme reveals the degree of kinship, and thus ancestral relationships, among different groups of organisms (the importance of this will become more apparent when we discuss evolution later in this chapter).

ASSIGNING SCIENTIFIC NAMES

All species are assigned a name consisting of two latinized words according to our **binomial system of nomenclature** (binomial loosely means "two names"). Latinizing these names may strike you as overly formal and difficult, but it would be infinitely more confusing if each scientific name reflected the native language of the biologist who provided the label. Some scientific names would be in English, others would be written with Chinese characters, and a few others would contain only Arabic letters. The use of Latin standardizes the language of nomenclature so that all species receive names expressed in the same language, using the same alphabet. *Homo sapiens,* for example, is recognized worldwide as the scientific name for the human species.

The first word in an organism's pair of names always identifies its *genus,* a group that contains closely related species. The second word, called the *specific epithet,* singles out from within that genus one kind of organism, the species. No two species have the same binomial name.

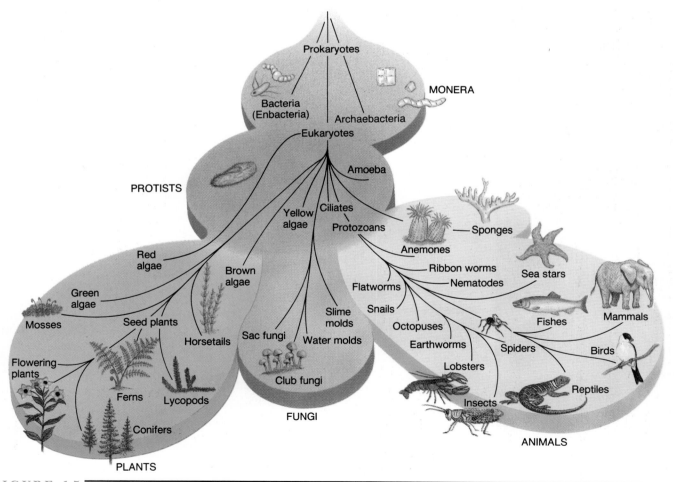

FIGURE 1-5
The five-kingdom scheme of biological classification. This diagram shows the relationship among the five kingdoms and the relative numbers of species that each contained. The animal kingdom, with its 1.3 million species, is the largest kingdom; in other words, it contains the greatest diversity.

CLASSIFICATION OF ORGANISMS

Taxonomists classify organisms by sorting them into groups according to traits that reveal *phylogenetic* relationships. This means that the organisms in the same group have a more similar ancestral history and are therefore more closely related than are organisms in different groups. ("Phylogenetic" refers to *evolutionary* relationships, revealing how recently two types of organisms shared a common ancestor.)

Phylogenetic relationships are found in all taxonomic categories. The lowest level of category is the species. Each of the millions of species contain individual organisms that are most similar to one another. The next level consists of larger groups (each one called a **genus**) containing one or more closely related species. The third level is created by grouping similar genera (plural for genus) into *families*. For example, all cats—from house cats to tigers—form a single family. The sequence continues in this manner, each level

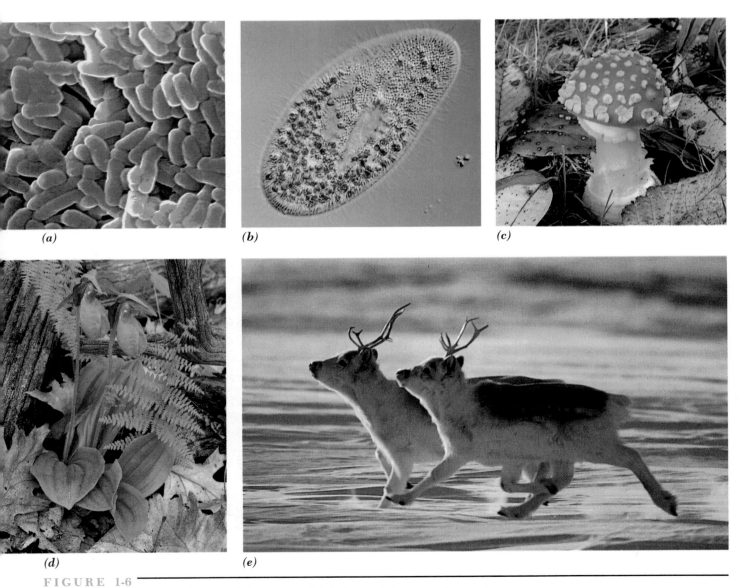

(a)   *(b)*   *(c)*

(d)   *(e)*

FIGURE 1-6

Representative organisms from each kingdom. *(a)* Rod-shaped bacteria—Monera kingdom; *(b) Paramecium,* a unicellular animal-like organism—Protista (protist) kingdom; *(c)* poisonous *Amanita* mushroom—Fungi (fungus) kingdom; *(d)* Pink moccasin flower—Plantae (plant) kingdom; *(e)* a pair of young caribou—Animalia (animal) kingdom.

composed of groups formed by clustering the categories of the previous level. Related families form an *order;* related orders form a *class;* related classes form a *division* (or *phylum,* in animals); and related divisions (or phyla) form the highest level of biological classification, a *kingdom.*[1]

The currently accepted scheme of classification maintains that all organisms on earth belong to one of five kingdoms (Figures 1-5 and 1-6). The simplest organisms, bacteria and their relatives, belong to the Monera kingdom; the most complex belong to the Animal kingdom (the characteristics of each kingdom and the evolutionary relationships within each are discussed in Chapters 35 through 38).

The genus and specific epithet to which a particular type of organism belongs provide the binomial name that identifies that species. Organisms are grouped into a classification scheme that provides order and consistency to the way we view the realm of life, allowing us to communicate about biology, and giving us a better understanding of how life evolved to its current state. (See CTQ #4.)

UNDERLYING THEMES OF BIOLOGY

As you progress through your introductory biology course, you will learn about a diverse array of subjects, ranging from the atomic and molecular basis of life to the behavior of plants and animals. Yet despite such an immense spectrum, a number of general themes pervade all areas of the biological sciences. These common themes form a foundation for which much of the information gathered by biologists can be understood. As you progress from one major topic to the next in this book, you will notice that a portion of each discussion focuses on these common threads (these passages are accompanied by an icon that alerts you to the theme being emphasized). At the close of each chapter, you will have the opportunity to review how some of the material presented in the chapter exemplifies one or more of these basic themes. These themes (Figure 1-7) are

- the relationship between form and function;
- biological order, regulation, and homeostasis;
- unity within diversity;
- acquiring and using energy;
- evolution and adaptation.

THE RELATIONSHIP BETWEEN FORM AND FUNCTION

It is evident that the tools we produce in our factories serve useful functions because of their structure. A wrench is obviously better suited for loosening a nut than is a screwdriver. Organisms also manufacture "tools" and "machinery" whose functions are closely correlated with their structure. Human hands, for example, are well suited for grasping; the pointed canine teeth of a wolf are suitable for seizing prey and tearing the flesh from bone; and the canopy of leaves that forms at the top of a tree perfectly matches the plant's need to maximize its exposure to sunlight, its sole source of energy.

The relationship between form and function is evident at all levels of biological organization. Once you have read about the internal structure of the kidney in Chapter 28, for example, you will see how its form—the spatial arrangement of microscopic tubules and blood vessels—makes it possible for the organ to remove waste products, excess salts, and water from blood and to concentrate them in a liquid that is easily discharged from the body. At the molecular level, you will see how the shape of a protein (its form) allows it to carry out a particular function, whether to promote a chemical reaction, support part of a cell, or hunt down and destroy a particular virus.

BIOLOGICAL ORDER, REGULATION, AND HOMEOSTASIS

The enormous complexity needed by an organism to sustain its own life goes hand in hand with a high degree of *order,* a term that suggests precision in the arrangement of components and the management of processes. There is nothing special, for example, about the atoms that make up your body. It is the unique *arrangement* of these atoms that makes you different from every other organism on the planet. If the atoms in your body were arranged differently, they might construct a tree or perhaps a porpoise.

Maintaining biological order requires *regulation.* Without such regulation, maintaining homeostasis would be impossible. Even eating a bag of salty potato chips could be fatal—the salt concentration in your body could increase to intolerable levels, and your cells would die. Fortunately, your kidney's regulatory mechanisms remove excess salt from the bloodstream and discard it in urine.

Virtually every biological activity—from the expression of a cell's genes, to the blood pressure in the major arteries, to sexual behavior—is subject to some form of regulation. Humans, as well as most other familiar animals, have two organ systems devoted primarily to the regulation of bodily activities (these are the endocrine and nervous systems). Without biological regulation we would not be able to breathe, focus our eyes, or maintain an internal environment conducive to life. Instead, there would be chaos and disorder rather than the ordered complexity needed for life. If even a single biochemical reaction in a cell were suddenly to escape the cell's regulatory mechanisms, the product of that reaction could accumulate to higher and higher concentrations until it jeopardized the entire biochemical operations of that cell, and possibly the entire organism.

---

[1] A newly proposed taxonomic group, called a *domain,* contains kingdoms that possess similar cell types. Three domains have taxonomically solid foundations but, at the time of this printing, have not yet been officially adopted.

FORM AND FUNCTION
The length and shape of this tiger lily and that of the hummingbird's beak match, enabling the hummingbird to gather nutritious nectar from the flower base, while the flower deposits or collects pollen for reproduction.

ORDER, REGULATION AND HOMEOSTASIS
Whether active or dormant, enveloped by snow or blistering heat, all organisms must maintain order and regulate internal conditions to remain alive.

UNITY WITHIN DIVERSITY
Despite the remarkable variation in different kinds of organisms on earth, all organisms are composed of one or more cells.

ACQUIRING AND USING ENERGY
All organisms acquire and use energy whether by directly trapping the energy in sunlight or by harvesting energy stored in the bodies of plants or animals.

EVOLUTION AND ADAPTATION
Evolution has produced the astounding variety of life on earth. Like the lizard hidden on the bark of this tree, each kind of organism possesses adaptations that enable it to survive and reproduce in its particular habitat.

ACQUIRING AND USING ENERGY

Energy obtained from the environment provides the fuel needed to run the processes that make life possible. Regardless of whether an organism harvests its energy from the sun or obtains it by consuming nutrients from another organism, to run out of energy is to run out of life. Energy is needed for the biochemical reactions occurring inside cells, reactions that construct the organism and repair or replace damaged components. Energy helps maintain our constant body temperatures, whether we are walking in the desert or swimming in frigid water. Energy limitations determine the number of eggs that can be laid by an individual frog. The dynamics of energy acquisition also explain why there are more than ten times as many rabbits in the world as there are predators that feed on them. These are just a few of the countless roles of energy that make it an inescapable theme in our explorations of life.

UNITY WITHIN DIVERSITY

We share this planet with millions of other species. With such a bewildering variety of organisms, one might identify "diversity" as the central theme in biology. Despite this diversity, however, there is a remarkable amount of "sameness" among all organisms, even among such radically different organisms as humans, spiders, ferns, and bacteria.

The unity of life is particularly apparent when we examine the fundamental levels of biological organization. For example, a person and a mushroom seem to share few similarities at the level of the whole organism, but both are quite similar on the cellular and molecular levels. Both are composed of cells. Both are built from the same classes of biological molecules, as are all other organisms (Chapter 4). Both use enzymes to promote the chemical reactions of their metabolism. The universal use of enzymes (Chapter 6) to promote chemical reactions is another clear example of life's basic unity. Many of these chemical reactions are identical in all organisms.

Humans and mushrooms, like all other organisms, use the same molecule (DNA) to store genetic information. In both organisms, these instructions are contained within discrete units of genetic information, the genes. Furthermore, all organisms use the same genetic "language" to encode these instructions. The universal nature of the genetic code is a particularly striking example of biological unity, a testimony to the common ancestry of all species.

EVOLUTION AND ADAPTATION

The last of the five themes—evolution and adaptation—constitutes the most important unifying principle in biology. It explains why the biosphere is populated by millions of species rather than just one type of organism. At the same time, this principle accounts for the unmistakable unity that exists among even the most diverse organisms. The principle of **evolution** states simply that species change over time. As a result of evolutionary changes, new species of organisms emerge while older forms of life may disappear. In other words, life evolves.

Evolution explains how all forms of life that have ever existed are members of one extended, genetically related "family" of organisms. You and that mushroom are similar in so many fundamental ways because you shared the same ancestor far back in time, and both of you retain many of the same molecular, genetic, and cellular endowments provided by that ancestor. All of the diverse species on earth are descendants of primitive cells that were the earliest life forms. It is this common ancestry that accounts for the unity of life.

Biological success for a particular type of organism depends on how well suited the organism is to its environment. An organism can survive only in environments that supply all its essential needs, and then only if the organism possesses the "equipment" needed for acquiring those environmental resources. Even then, it can succeed only if it can survive the adverse conditions to which it is exposed. In other words, an organism must be *adapted* to its particular set of environmental conditions (Figure 1-8). **Adaptations** are traits that improve the suitability of an organism to its environment.

Adaptations can be identified as part of every organism you see. The body shape and coloration of some insects (such as the peppered moths discussed earlier) better enable them to avoid detection by birds or other predators. The trunk of an elephant enables the animal to reach a supply of edible leaves on high branches that are unavailable to most of the competition (other leaf-eating, ground-based animals). The thick fur coat of an arctic fox provides the insulation needed to allow the animal to sleep on a bed of ice without expending metabolic energy keeping itself warm.

Pick out a trait that is part of your own body. The trait you have chosen is probably some type of *anatomic* (structural) adaptation, such as a grasping hand, an eyelid that protects the surface of the eye, or a foot that can balance the weight of the entire body. A few of you may have selected a property that reveals a more subtle *physiologic* trait (one concerning biological processes and functions), such as eyesight or ability to taste. Some of you may have even identified a *behavioral* adaptation, such as automatic breathing or blinking. In fact, many structural or physiologic adaptations of organisms would be useless without an accompanying behavioral adaptation. Your respiratory structures, for example, would be of little value if you didn't inherit breathing behavior as well. Spiders not only possess silk-producing glands, for example, but also inherit the "automatic" (instinctive) behavioral adaptations that direct them to weave a characteristic web and instruct them in its use for capturing prey.

Every topic we discuss in this book has an evolutionary component. As we describe certain anatomic, physiologic, or behavioral characteristics of various organisms, keep in

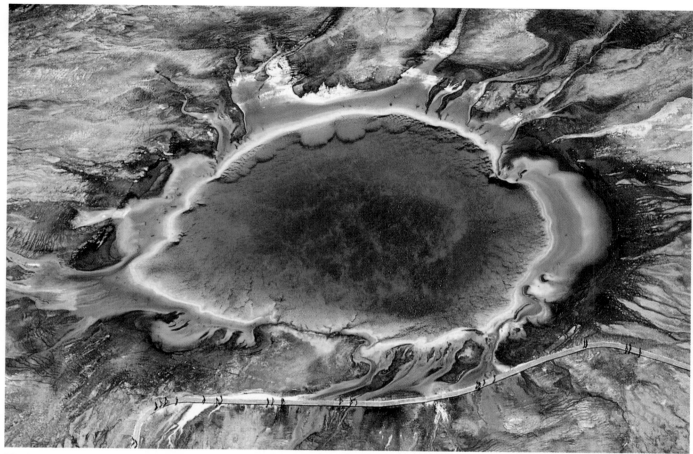

FIGURE 1-8

Adaptations to temperature differences are revealed by the concentric bands of color in this aerial photograph of a hot sulfur pool at Yellowstone National Park. The pond, which is hottest in the center and coolest at its rim, is colored by enormous populations of microorganisms concentrated in each ring-shaped thermal zone. Each distinctly pigmented type of microorganism is specifically adapted to (and limited to) one narrow temperature range.

mind how these features better suit the organism to its environment, increasing the chance for survival of both the individual and the species. We will see how some of these adaptations have become modified over evolutionary time as environmental conditions have changed and new species have evolved. For example, the tiny bones in your middle ear, which transmit sound through a recess in your skull, evolved from bones that were originally parts of the jaws of ancestral fishes that lived long before the first four-legged animal ever set forth on dry land.

Evolution is the cornerstone of our understanding of biological phenomena. The events that led to the formulation of an explanation of the mechanism by which evolution occurs constitute one of the most important stories in the natural sciences.

A Voyage That Altered Our Concept of the Origin of Species

What would you do if someone offered you a free trip around the world and all the person asked in return is that you pursue your favorite hobby during the trip? Such an opportunity presented itself to a 22-year-old university graduate named Charles Darwin when one of his professors recommended that Darwin be appointed "naturalist" for a scientific expedition around the world.

It was 1831, and Charles Darwin (Figure 1-9a) had just graduated from the University of Cambridge with a rather undistinguished academic record. As a young boy, he had been an avid insect and plant collector, an interest that became even more focused during his college years. Darwin's zeal is illustrated by a passage he later wrote: "One

(a)

FIGURE 1-9

Charles Darwin and the voyage of the *HMS Beagle*.
(a) Portrait of the young Charles Darwin. (b) Map of the route taken by the *Beagle* during its approximately 5-year trip around the world.

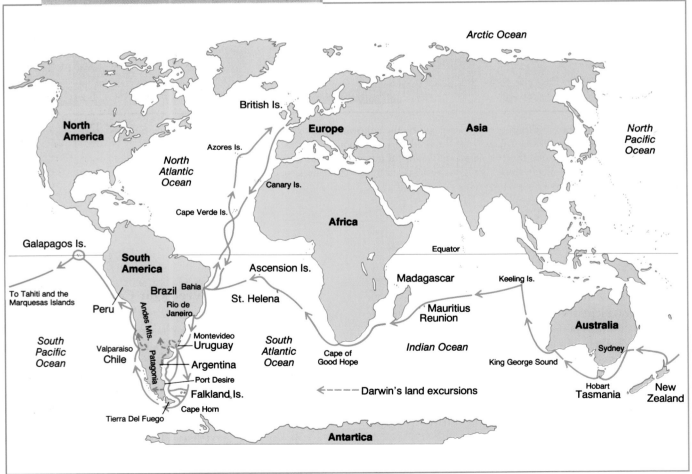

(b)

day, on tearing off some old bark, I saw two rare beetles and seized one in each hand; then I saw a third and new kind, which I could not bear to lose, so that I popped the one which I held in my right hand into my mouth. Alas, it ejected some intensely acrid fluid, which burnt my tongue so that I was forced to spit the beetle out, which was lost, as well as the third one."

Even as a college graduate, Darwin was still unsure of his lifelong profession. He had abandoned the idea of following in the footsteps of his physician father and had decided to become a clergyman in the Church of England when this "once in a lifetime" offer was proposed to him. An enthusiastic Darwin accepted the commission on the *HMS Beagle*, a small (25-meter, or 75-foot-long) three-masted surveying ship that was to map the coastlines and harbors of South America (Figure 1-9*b*). As the ship's naturalist, it was Darwin's job to collect and organize thousands of specimens from around the world.

Darwin left England with the firm belief that all of the world's plants and animals had been created directly by the hand of God. As he traveled and explored new lands, Darwin began to question this view of a static world whose life and landscape had been fixed since the time of creation. The questions began even before the *Beagle* had reached its first port of call. During the first few days, as Darwin lay in his hammock suffering from seasickness, he read the newly published *Principles of Geology*, a textbook written by Charles Lyell. Lyell's book meticulously presented evidence that the earth was much older than previously believed and that it had gradually changed over long periods by such natural forces as mountain building, erosion, volcanoes, flooding, and earthquakes.

Within the first month of the voyage, Darwin himself was able to observe evidence in support of Lyell's contentions. As the ship sailed into the harbor at the Cape Verde Islands, Darwin saw a cliff rising about 15 meters above the sea. He discovered that the cliff was composed of limestone that contained embedded seashells of the type seen along the shore. It was evident that the sea bed had been lifted upward as part of the cliffs. Later in the voyage, Darwin found marine deposits located high in the Andes Mountains, hundreds of miles from the closest shore. He saw firsthand evidence of the ability of natural forces to change the landscape after experiencing a severe earthquake off the South American coast. Darwin found that the land at one site had risen nearly a meter as a result of the quake. At a nearby site, the captain of the ship had discovered rotting mussels clinging to the rocks 3 meters above the high-tide level.

Evidence for a gradual change in the organisms living in the area was less dramatic than were the geologic observations, but they were just as convincing. A few miles inland from the east coast of Argentina, Darwin found a fossil bed containing a giant sloth, a hippopotamus-like animal, and a giant armadillo. It was clear to Darwin that these were the remnants of extinct creatures, yet they were clearly related

in form to those living today. One of these creatures, Darwin wrote, was "perhaps one of the strangest animals ever discovered." It was the size of an elephant with teeth that suggested it fed by gnawing, much like modern-day rats.

Darwin saw the clear anatomic relationships between these extinct animals and those living today and wondered how it was that these ancient animals had disappeared. The captain of the *Beagle* suggested that they had been left off Noah's ark during the great flood and had been drowned as the waters covered the land. Darwin, however, considered another explanation. Perhaps they had been driven to extinction by animals that had invaded the South American continent from the north following the formation of a land bridge connecting the North and South American land masses. His proposal proved to be remarkably farsighted.

Darwin's most important observations were made as he explored the Galapagos Islands, a small archipelago (group of islands) located approximately 900 kilometers off the coast of Ecuador and named after the giant tortoises that are found only on these islands. The volcanic islands had been discovered several hundred years earlier and were a site where ships replenished their supply of fresh water and captured the strange tortoises for fresh meat on a long sea journey.

Having visited other volcanic islands along the journey, Darwin noted that the plants and animals of such habitats tended to be similar to those living on the nearby mainland. Islands off the coast of Africa, for example, contained animals similar to those found on the African mainland, whereas the Galapagos contained animals similar to those inhabiting the South American mainland. This was particularly curious for the Galapagos, since the climate and geography of the islands were much different from those of the mainland. Darwin wondered why animals of similar appearance would be living in such different habitats.

One of the most important scientific observations in history concerned a group of dull-looking birds—the now-famous Darwin's finches. Darwin observed and collected 13 species of finches that were found only on the Galapagos Islands (and one on nearby Cocos Island). Although these birds were similar to one another in overall body form, they differed in their "lifestyle" and the shape of their beaks (Figure 1-10). Some of the species lived on the ground, others in the trees. Some had strong, thick beaks adapted for crushing seeds, while others had beaks that were especially suited for feeding on flowers or insects. Among the insect eaters was the so-called woodpecker finch, which, unlike a true woodpecker that catches insects with its long tongue, digs insects out of the tree bark using a cactus spine held in its beak. Darwin wondered why animals with such different feeding habits looked so much alike.

The observations made on the Galapagos Islands were an important component in Darwin's view that species were not created in their final, unchangeable form but instead had evolved from other species. Darwin eventually concluded that all of the island finches were descendants of one

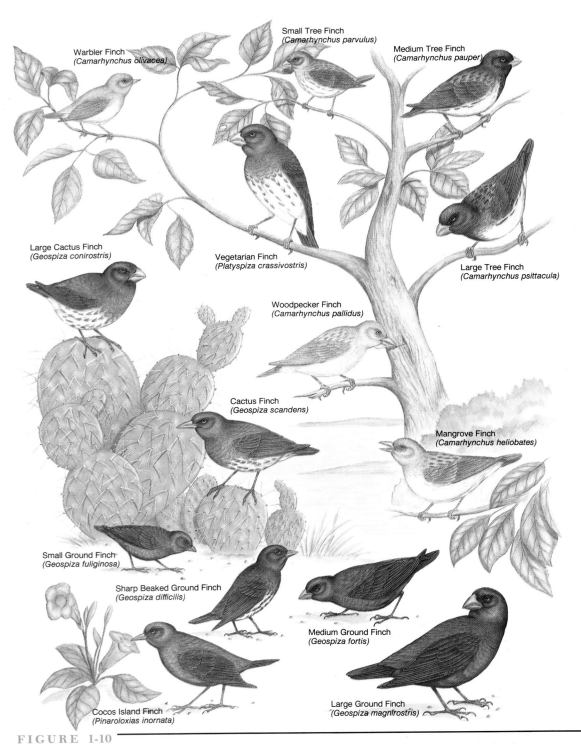

FIGURE 1-10

Darwin's finches. Darwin concluded that all 14 species of finches inhabiting the Galapagos Islands (and nearby Cocos Island) had evolved from a single ancestral finch that came to the islands from the mainland. The finches differ primarily in beak size and shape. These distinctions adapt them to different habitats and food sources. They include seed-eating ground finches; nectar-feeding finches; insect-eating finches; and one remarkable tool-using tree finch—the woodpecker finch—that bores into wood with its beak and then uses a cactus spine or twig to excavate its prey.

of the mainland species of ground finch that had drifted several hundred kilometers to one of the islands. Unlike the mainland, the islands were essentially devoid of competitors, so the immigrants were able to establish a thriving population. But how were individuals of one species able to give rise to the various species now present on the islands? In other words, how does evolution work?

Darwin's Theory of Evolution by Natural Selection

Soon after Darwin returned to England he happened to read an essay that had been written about 40 years earlier by the Reverend Thomas Malthus. Malthus pointed out that a single pair of humans have the reproductive potential to produce billions of people in a relatively small number of generations. It was evident that the size of the population was held in check as a consequence of mortality brought about by famine, war, and disease.

Darwin realized that this same potential for overpopulation was present in all populations. Any species of animal —from slow-breeding elephants to minute insects —could, given an absence of mortality, generate enough individuals in a relatively short period of time to cover the earth many times over. Yet, the numbers of individuals of most species remain relatively constant from one year to the next. Darwin concluded that there must be a "struggle for existence" among animals such that only a small percentage of those conceived actually live to maturity. Darwin was not suggesting that there was a physical struggle between individuals but rather a competition between individuals of a community and a struggle to survive the potentially adverse factors in their environment, such as lack of food or water, predation, parasitism, disease, cold, heat, flooding, and salinity.

Assuming there is such a struggle, what determines which members of the population survive and which are eliminated? If all the individuals were identical, then all would have the same chance of survival. But Darwin was aware that members of a plant or animal population are no more identical than are members of a human population (Figure 1-11). In other words, there is **variability** within a population. Variation among animals may be reflected in the color of the coat, body size, an ability to withstand high temperature, the choreography of a mating dance, or any other type of anatomic, physiologic, or behavioral characteristic. As Darwin pondered the matter, it became clear that some members of the population possessed characteristics that gave them an increased chance of survival relative to other members of the population. The survivors might have a more efficient style of gathering food, or a particularly high resistance to a common parasite, or a little faster gait. These animals would be better adapted to their environment and thus more likely to survive.

It was evident to Darwin that survival itself was not the most important consideration; rather, surviving *to reproduce successfully* was the critical factor. Those organisms that are best suited to survive tend to produce a greater number of offspring. According to Darwin, the environment "selects" those organisms that will survive and reproduce, while it eliminates those organisms that do not. Darwin called this process **natural selection.** Viewed in this way, natural selection is equivalent to *differential reproduction*, the production of more offspring by individuals that are better adapted to their environment.

Although the biologists of Darwin's time were totally ignorant of the mechanism of heredity, they knew from observation that offspring tend to inherit the characteristics

FIGURE 1-11
Genetic variation in the human population.

of their parents. Consequently, the offspring of survivors will tend to have traits that make them suited to survive and reproduce in the same environment. As a result of differential reproduction, after several generations a population will tend to collect genetic traits that make its members better adapted to its environment. Individuals with traits less suited would be less likely to successfully compete and would therefore leave fewer (if any) offspring. This change in genetic composition of a population from generation to generation is the very essence of the evolutionary process. Giraffes, for example, have long necks because individuals that happened to have longer necks were more successful in obtaining food than were their short-necked competitors. Longer-necked members of the population were more likely to survive and have offspring which, like their parents, would have longer necks.

Natural selection produces organisms that are adapted to their environment. Environments do not remain constant over long periods of time, however; climates change, fires destroy vegetation, new parasites or predators may appear, food supplies change, and so forth. Consequently, an animal that is successful within a particular habitat at one time might be poorly adapted at some other time. Darwin realized that the natural variation that exists within a population provides the basis for evolutionary change. Consider, for example, a hypothetical population of fleas that fed on the blood of the bison that roamed the American plains during the early nineteenth century. As long as the bison were plentiful, natural selection would favor those individuals who were most efficient at finding and piercing the skin of these large animals as they grazed on the prairie grasses. As the numbers of bison sharply diminished, those fleas that could locate and survive on the blood of horses or dogs might be favored by natural selection, and the population would gradually shift toward this new variety of insect. Given sufficient time, the gradual accumulation by differential reproduction of small genetic differences among individual fleas could radically transform a species' characteristics. If the populations of a species become isolated from one another so they cannot interbreed, then a strong likelihood exists that the populations will evolve into separate species. After several billion years, this divergence of species has produced the diversity of life that we see today, including the human species.

A Surprise in the Mail

Darwin returned from his voyage around the world in 1836. He spent the next few years examining the specimens he had collected, reading the writings of others, and preparing manuscripts for the journal of his voyage, including a treatise on barnacles and a book on coral atolls (coral islands surrounding a lagoon); he also pondered the implications of his thoughts on evolution. Although Darwin discussed his ideas about evolution and natural selection with his friends in the scientific community, it wasn't until 1842 that he set

them down on paper in a brief essay that was not sent out for publication. A few years later he wrote a longer essay and, at the urging of his friends, began preparing a manuscript for a book on the subject.

In June 1858, more than 20 years after the voyage of the *Beagle* and after 11 chapters of his book had been written, Darwin received a letter in the mail from Alfred Russel Wallace, a fellow naturalist who had been working in Malaysia. Contained in this letter—and in the short manuscript that soon followed—was the outline of a theory of evolution by natural selection that was virtually identical to that being formulated by Darwin. Ironically, Wallace was writing to ask Darwin if he would forward his manuscript to Charles Lyell for presentation to the Linnaean Society. Darwin was shocked by the communication and consulted with Lyell and another colleague, Joseph Hooker, as to what course of action he should take. Lyell and Hooker decided that they would present the work of both men jointly at a meeting of the Society, which they did in 1858. Rather than becoming competitors, or even enemies, Darwin and Wallace became lifelong friends. Darwin rapidly finished the remainder of the manuscript, and *The Origin of the Species* was published on November 24, 1859. The response was thunderous. All 1,250 copies of the book were sold on the first day it was available. Discussions, protests, and personal attacks followed—continuing to this very day.

A Summary of Darwin's Theory of Evolution by Natural Selection

The concept that organisms are related because of common descent had been discussed by a number of philosophers and naturalists before the nineteenth century. Charles Darwin, however, was the first to provide a feasible mechanism—natural selection—to explain how biological evolution might have occurred. Darwin arrived at this mechanism by the following logic:

1. All species have the reproductive potential to overpopulate the earth, yet populations of organisms remain relatively stable over time.

2. Consequently, a large percentage of the members of a population must die at an early age.

3. There is variation among the members of a population such that different individuals have different traits.

4. It follows that the members of a population whose traits make them better adapted to their environment at a particular time will be more likely to survive to reproductive age and to produce more offspring than will members that are less well adapted. This is the basis of natural selection.

5. Since offspring inherit the traits of their parents, those that have successful parents tend to acquire successful traits. Consequently, a population will change (evolve) over time, its members acquiring new traits that make them better adapted to a changing environment.

6. Given sufficient time, this process of evolution by natural selection can account for the formation of new species and thus for the diversity of life on earth, both past and present.

All organisms are the products of evolution, the central theme underlying biology. Evolution has generated patterns that are so recurrent among organisms, they join the list of life's fundamental themes. These themes help tie together the facts of biology to provide a deeper understanding of life. (See CTQ #5.)

BIOLOGY AND MODERN ETHICS

Our modern understanding of biological concepts and the application of these concepts is already changing the planet and all its inhabitants, and the influence continues to grow dramatically as we approach the next century. Although these changes hold tremendous promise and hope for a healthier environment for all organisms, new biological technologies come with their share of ethical controversy. Some applications seem clearly unethical, such as using technology to develop biological weapons (strains of viruses or bacteria that can incapacitate or kill people). Yet many of these issues are not so easily classified as either ethical or unacceptable. Consider the following questions:

- Should scientists be allowed to patent life?
- Do the benefits of genetic engineering outweigh the potential risks that this new technology poses?

- Should we use our biological know-how to make abortions safer and more accessible, for example, by using a "morning-after" pill that prevents pregnancy after an egg is fertilized?
- Do economic freedoms and needs justify the destruction of habitats and subsequent extinction of species?
- Although treating disease by replacing faulty genes with good ones is not so controversial, is it ethical to use the same technology to introduce genes into healthy people, genes that enhance a particular desirable characteristic, such as intelligence or athletic prowess?
- Who should decide whether or not organs can be removed from brain-dead patients so that they can be donated to save others?
- Should people with AIDS be allowed to work in health-care professions?

These questions represent just the tip of the bioethical iceberg. An informed public is the best protection against clearly unethical applications or abuses of biological knowledge. Information and open discussion is also essential before the more debatable issues can be resolved. Regardless of how they are resolved, these bioethical decisions will permanently influence the world and all its inhabitants. To introduce you to some of these ethical dilemmas that we must all confront, we have included a series of essays that appear in several of this book's chapters. The ethical pendulums in these essays may leave you feeling indecisive, wanting some definitive piece of information that will tip the scales one way or the other. Alternatively, you may find yourself in impassioned discussions with classmates and family. As long as you can support your feelings with facts, even if those feelings are ambiguous, then you are participating in shaping the future.

REEXAMINING THE THEMES

The Relationship between Form and Function

Every component of an organism has a form that suits its function. Each enzyme has a shape that allows it to combine with a particular chemical and change it to another chemical. Similarly, an elephant's trunk suits its task of stripping leaves from tall trees and reaching water far below the elephant's head. It is much easier to understand biological concepts if the relationship between a structure and its function is firmly established.

Biological Order, Regulation, and Homeostasis

Maintaining the high degree of organization required for life depends on hundreds of homeostatic control mechanisms. Virtually every metabolic reaction in an organism needs to be regulated to prevent excesses or deficiencies of its activity. The saltiness of the internal environment must be stable or the organisms will die. All organisms have mechanisms for detecting imbalances and correcting them to maintain a stable internal environment conducive to life.

Acquiring and Using Energy

All organisms require a constant supply of energy to maintain biological order, to grow, to repair damage, to move substances within the organism, and to run other life-sustaining activities. Plants and similar organisms acquire their energy directly from the sun. Other species harvest energy from the tissues of organisms on which they feed. Regardless of the energy-acquiring strategy, the original source of energy for virtually all organisms is the sun.

Unity within Diversity

The diversity of life on earth is extraordinary; the biosphere supports at least 1.8 million species and perhaps 50 million species. Yet all organisms have fundamental similarities, such as all being composed of one or more cells, all inherit a set of genetic instructions inscribed in the same genetic "language," all possess homeostatic control mechanisms, and all use enzymes to direct metabolic activity, to name just a few. Although universals emerge at every level in the hierarchy of life, they are particularly abundant at the fundamental levels.

Evolution and Adaptation

All organisms and their traits are the products of evolution. Organisms that are well adapted to their environment survive and reproduce, while poorly adapted individuals leave far fewer offspring. As conditions change, however, different individuals are reproductively favored, and species with different characteristics emerge. The "invisible" peppered moth, the long-necked giraffe, the intelligent human, indeed all species evolved in this way.

SYNOPSIS

Organisms possess a combination of properties that is not found anywhere in the inanimate world. Organisms are highly organized. They are composed of one or more cells; they acquire and use energy; they produce offspring similar to themselves; they contain a genetic blueprint that dictates their characteristics; they grow in size and change in appearance and abilities; they respond to changes in their environment; they carry out a variety of controlled chemical reactions; and they maintain a relatively constant internal environment.

Organisms can be described at various levels of organization. At the lowest, simplest levels, life consists of subatomic particles organized into specific atoms, and atoms organized into specific molecules. At the other end of the spectrum, communities made up of various species are organized into ecosystems which, in turn, make up the biosphere. With each step-increase in complexity, new properties emerge. Life, for example, emerges at the cellular level.

While exploring life, a number of themes can be identified which reoccur at various levels of organization. These themes form a foundation for understanding some of the recurring patterns of life. Five such themes are identified throughout the text. (1) The parts of an organism, whether a molecule or an organ, have a structure that is specifically suited for their function. (2) Living objects, from cells to multicellular organisms, possess regulatory mechanisms that ensure the maintenance of a relatively constant, ordered internal state. (3) Organisms require energy to carry out the activities that make life possible. (4) Despite the enormous diversity of organisms on earth, all organisms share many common features that reveal their descent from a common ancestor. (5) Species change over time, and new species are generated. This process of biological evolution produces organisms that are well suited (adapted) to the temporary conditions of their environment.

Evolution explains both the unity and diversity of life. Darwin arrived at his theory of evolution by natural selection by considering a series of observations and conclusions about individual organisms and populations (summarized in the last paragraph of the chapter).

Key Terms

biology (p. 6)
unicellular (p. 7)
multicellular (p. 7)
metabolism (p. 10)
enzyme (p. 10)
homeostasis (p. 10)

population (p. 11)
community (p. 11)
ecosystem (p. 11)
biosphere (p. 11)
taxonomy (p. 11)
binomial system of nomenclature (p. 13)

genus (p. 14)
evolution (p. 18)
adaptation (p. 18)
variability (p. 23)
natural selection (p. 23)

Review Questions

1. Describe the various properties that characterize life, and identify some inanimate objects that exhibit one or more of these properties. Can you identify any inanimate objects that possess *all* of these properties combined?

2. Rearrange the following list from the simpler to the more complex: ecosystems, subatomic particles, cells, organs, the biosphere, populations, molecules.

3. How did Lyell's book on geology and Malthus's book on populations influence Darwin's thoughts on evolution?

4. Describe Darwin's explanation for the presence of similar-looking finches with different types of beaks and feeding habits on different Galapagos islands.

Critical Thinking Questions

1. The Nobel Prize winner Albert Szent-Gyorgyi once wrote, " . . . there is only one life and one living matter, however different its structures, colorful its functions, and varied its appearance. We are all but recent leaves on the same old tree of life and even if this life has adapted itself to new functions and conditions, it uses the same old basic principles over and over again." Discuss the meaning of this statement in light of the ideas presented in the first section of this chapter.

2. During the early history of biology, a debate ensued between the "mechanists," who believed that life would eventually be understood entirely in terms of its physical and chemical properties, and the "vitalists," who believed that life is characterized by a special property not explainable solely in terms of its physical and chemical makeup. Is the emergence of new properties at higher levels of complexity consistent with a mechanistic view, a vitalistic view, or neither? Explain.

3. Create an analogy that reveals how novel properties emerge as levels of order increase. Begin by drawing an organizational hierarchy for one of the following: composing a song, building a house, writing a novel. Identify the intangible "extra" properties that emerge as the level of complexity increases. Why would the structures be of little value without these intangible "extras"?

4. Create a classification scheme for vehicles. Begin by observing and identifying as many "species" as you can. Group similar species into "genera," and so on, until you reach the top of the taxonomic hierarchy. Ask a friend to use your scheme to identify his/her vehicle. Do all vehicles fall clearly into one group or another? (Compare this to the classification of living organisms.)

5. Identify a characteristic in yourself that represents each of the five themes identified in this book as being fundamental to exploring life.

6. Considering the process of natural selection, how does greater variation within a species strengthen the chance that the species will avoid extinction following the appearance of some new deadly environmental condition (such as the appearance of a new virus or a marked drop in temperature)?

7. The necks of giraffes were used as an example of adaptation. Explain how the combination of genetic variation, competition for food, and natural selection accounts for the development of this adaptation, and why there are no shortnecked giraffes.

Additional Readings

Asimov, I. 1980. *A short history of biology.* New York: American Museum of Science. (Introductory)

Barlow, C. 1991. *From Gaia to selfish genes: Selected writings in the life sciences.* Cambridge: MIT Press. (Advanced)

Bronowski, J. 1973. *The ascent of man.* Boston: Little, Brown. (Intermediate)

Darwin, C. 1859. *The origin of species by means of natural selection.* (recent edition) New York: Random House. (Intermediate)

Desmond, A., and J. Moore. 1992. *Darwin.* New York: Warner. (Introductory)

Gould, S. J. 1987. Darwinism defined: the difference between fact and theory. *Discover* Jan: 64–70. (Intermediate)

Mayr, E. 1991. *One long argument: Charles Darwin and the genesis of modern evolutionary thought.* Cambridge: Harvard. (Advanced)

Moorehead, A. 1969. *Darwin and the Beagle.* New York: Harper & Row. (Introductory)

Perutz, M. F. 1989. *Is science necessary: Essays on science and scientists.* New York: E. P. Dutton. (Intermediate)

The Process of Science

STEPS
TO
DISCOVERY
What is Science?

THE SCIENTIFIC PROCESS

Conducting an Experiment to Test a
Hypothesis

Interpreting Experimental Results

Formulating New Hypotheses

Additional Tests, Conflicting Results, and
Final Confirmation

From Hypothesis to Theory

APPLICATIONS OF THE SCIENTIFIC PROCESS

Accidents and Scientific Discovery

Testing Additives and Drugs on Animals
and Humans

CAVEATS REGARDING "THE" SCIENTIFIC METHOD

BIOETHICS

Science, Truth, and Certainty:
Is a Theory "Just A Theory?"

S TEPS TO DISCOVERY

What Is Science?

You have probably heard the time-honored saying that if you get wet on a cold day, before long you will be suffering from a cold. For many of us, our experience seems to verify this statement: We get caught in a cold rain or step in a puddle, and within a couple of days we have a cold. If this has happened to you, can you conclude that getting chilled causes colds? In other words, if two events occur close together, is that proof that the first event *caused* the second?

The answer is no. In fact, there is no relationship between getting cold and catching a cold. The reason these two events often seem to occur together is because both tend to happen with the onset of wintry weather. Actually, colds are more common in the winter because people spend more time in buildings when the weather is cold or rainy, and being indoors increases close exposure to other people, many of whom have colds and are shedding the virus. In addition, cold wintry air dries the mucus layer that lines the nasal passages, exposing the underlying cells to viral attack. Getting chilled does *not* make you more vulnerable to the common cold; however, that is just a popular myth.

Let's examine a few other time-honored beliefs plus some recent newsworthy claims:

- Shaving makes hair grow back thicker.

- Eating bean sprouts and similar "health foods" promotes good health and increases longevity. Casual observation often reveals that sprout eaters feel healthier and experience fewer health problems than do people who eat no sprouts or "health foods."

- Smoking does not cause diseases that cut short the life span. Defenders of tobacco companies claim that the long lives of some smokers proves that smoking does not cause diseases that shorten life. George Burns, for example, has

Scientists work both in the laboratory and in the field to unravel the secrets of biological activities that accompany seasonal

lived into his nineties, in spite of his many years of cigar smoking.

Each of these examples, plus countless other conclusions that find their way into our lives, suffers a serious shortcoming in reasoning: They are all products of *anecdotal evidence*—information that is based on personal experiences and testimonials. Such stories lead many people to jump to conclusions that a cause-and-effect relationship exists between events that actually may be related simply by coincidence or clustered by some underlying and less obvious set of events. Let's reexamine each of the above conclusions and see how anecdotes often lead to questionable conclusions and misinformation.

• When people shave an area of their body, the hair often grows back thicker, and the more they shave an area, the thicker the hair gets. But shaving usually begins with the onset of puberty, and the effects of puberty increases body hair growth over a period of years, so hair gradually grows in thicker, even if shaving is avoided. It is puberty, not shaving, that increases hair growth.

• People who regularly eat bean or alfalfa sprouts are generally health-conscious individuals who also tend to eat little red meat and to exercise frequently. Any or all of these factors may be responsible for the alleged better health of sprout eaters; the sprouts themselves are not responsible. In fact, Bruce Ames, a University of California researcher who invented a valuable test for detecting cancer-causing activity in various substances, has found that sprouts, mushrooms, and so-called health foods are actually loaded with chemicals that could cause cancer.

• People who believe that the continued survival of a 90-year-old smoker proves that tobacco smoke doesn't shorten life are more likely to have a false sense of security about smoking. Scientific studies (as opposed to anecdotal evidence) have proven that smoking is the leading cause of preventable death in the United States and other industrialized countries. We can explain George Burns's longevity in the same way we explain how some people survive falls out of airborne planes without parachutes: They are very lucky. We certainly would not conclude that falling out of airplanes without a parachute is harmless. Again, anecdotal evidence fails to reveal the critical information; that is, how long George Burns would live had he *not* smoked cigars?

Anecdotal evidence is the heart of misinformation. It often leads to absurdly irrational beliefs, such as black cats causing bad luck or ostriches sticking their heads in the ground to escape danger (such behavior would quickly lead to a severe ostrich shortage). Anecdotal evidence has also misled people in their attempts to understand themselves and the nature of life. One particularly prevalent anecdote asserts that the fact that we don't understand all aspects of life proves that life is a mystery beyond human understanding.

Scientists, however, believe that all phenomena in the universe have rational, verifiable explanations. We will never know everything there is to know about life and the universe, but we continue to expand our understanding by making careful observations, asking questions, and seeking answers. Yet, none of these activities alone has provided us with our current understanding of how organisms work. Observations alone, like anecdotal evidence, provide descriptive information, but they rarely reveal the mechanisms responsible for biological activities or the ways in which the intricate processes of biology affect one another. For example, the changing of seasons from warm, summer-like conditions to cold, wintry days and nights is accompanied by a profound change in the biological activity of countless organisms: Tree leaves turn brightly colored and fall; many animals hibernate, while others migrate to a warmer climate; still other animals develop a thick winter coat of fur. But is it the onset of cold weather that triggers these biological changes or the shortening of the days as winter approaches? In other words, what is the *cause* of this observed effect? Furthermore, *how* does that factor bring about the observed change?

The answers to such questions are not obtained by simple observation. Rather, they require the combined input of several different methods for acquiring and examining information, assembled into what amounts to a scientific approach. Using scientific methods provides us with a way of not only understanding how life and the universe are put together and how they interact but of verifying the accuracy of the information we discover and the explanations we propose. While this approach can help scientists evaluate and explain nature, it can also be used to solve problems in our personal lives and help us understand for ourselves.

In this chapter, we provide a number of examples that illustrate how the results of one study become the starting point of subsequent studies so that scientific knowledge builds from one confirmed explanation to another and from one generation of scientists to the next. This pattern of sequential scientific growth is revisited in the "Steps to Discovery" feature that launches each chapter. You will also become acquainted with the self-correcting nature of science, a built-in self-check that assures the eventual detection and elimination of incorrect or biased results.

The job of understanding how living things work will never be completed. Scientific findings always suggest new questions and predictions, so the path of discovery and exploration is endless. That is part of the excitement of biology—scientists will never work themselves out of a job. There will always be questions to answer.

---

changes, such as bird migration, leaf color changes, leaf fall, and butterfly development.

"*T*he skies were overcast, and the city was expecting heavy rain and frogs." This might have been a typical forecast in 1668 since at that time it was commonly believed that frogs developed from falling drops of rain. Although it was evident that mammals, such as pigs and dogs, were born, and that birds were hatched from eggs, the origin of smaller creatures, such as frogs and flies, was quite obscure. After all, no one had yet seen the development of an organism from a microscopic egg. The notion that most living organisms arose directly from inanimate materials was known as **spontaneous generation** and was popular among scientists and nonscientists alike for hundreds of years. It seemed particularly evident that flies arose directly from decayed meat, since everyone had observed rotting meat covered with maggots—the larval stage of flies.

One of the first skeptics of spontaneous generation was Francesco Redi, an Italian physician and naturalist. Redi was familiar with the recent accounts of the famous English physician William Harvey, who, during a dissection of one of the King's deer, had found a tiny fetus that appeared much like a miniature adult deer. Harvey had concluded that animals develop from seeds or eggs that were too small to be seen. Redi had a natural curiosity about life and was willing to question the validity of spontaneous generation.

▼   ▼   ▼

THE SCIENTIFIC APPROACH

By recognizing the possibility that spontaneous generation might be an erroneous concept, Redi had taken the first step on the path to scientific discovery. This is the way scientists typically begin an investigation; they learn about the information currently available on a subject, and they make observations that are relevant to the matter at hand. Redi had observed that maggots tended to "appear" in places where adult flies could also be found. Based both on Harvey's findings and his own observations, Redi proposed an alternative explanation for the origin of maggots: "The flesh of dead animals cannot engender worms unless the eggs of the living be deposited therein." A tentative proposal of this type is called a **hypothesis.** Two hallmarks of a good hypothesis are that

- it is consistent with observations collected up to that point; and
- its validity can be tested by experimentation (or further observation).

Redi came up with an experimental plan by which he could test his hypothesis that maggots arose only from eggs deposited by flies and not from meat (Figure 2-1).

CONDUCTING AN EXPERIMENT TO TEST A HYPOTHESIS

In his experiment, Redi put a dead snake, some fish, some eels, and a slice of veal separately into four large, wide-mouthed vessels and sealed the openings with wax. He then placed the same materials into another set of four vessels which he left open to the air.

RECORDING RESULTS OF THE EXPERIMENT

Within a matter of days, Redi observed that the decaying meat within each of the open vessels was teeming with maggots, and flies were observed coming and going at will. In contrast, the meat in the sealed vessels was in the same state of decay, but no maggots were evident.

INTERPRETING EXPERIMENTAL RESULTS

Redi concluded that the closed vessels failed to produce maggots because flies were unable to reach the meat. But Redi recognized a serious flaw in this interpretation: What if maggots failed to develop in the sealed vessels because of a lack of fresh air rather than the absence of flies? It is essential in any scientific experiment to be able to determine *cause and effect.* In Redi's initial experiment, the two sets of vessels differed by more than one **variable**—a condition that is subject to change. In this case, the two variables were the presence or absence of flies and the presence or absence of air. To determine which of these two variables was the *cause* of the results, Redi performed another experiment. He once again set up two sets of vessels containing meat, but this time both vessels were left open to the air; one set was left totally uncovered, while the other set was covered with a fine layer of gauze whose holes were too small to allow flies to pass through.

In the design of a proper experiment it is essential to allow only one condition to vary—this is the one variable whose role the investigator is attempting to evaluate (in this case, access by flies). Otherwise, it is impossible to determine which variable is causing the result. In modern scientific terminology, the gauze-covered set of vessels is called

(a)   **(b)**

FIGURE 2-1

Redi tests the validity of spontaneous generation. *(a)* In his initial experiment, Redi placed meats in two groups of vessels. The vessels in one group were left open to the air, while the vessels in the other group were sealed with wax. Maggots appeared only in the vessels exposed to air. *(b)* To eliminate the possibility that maggots failed to appear in the sealed vessels because of the absence of fresh air, Redi conducted a second series of experiments. Again he prepared two groups of vessels containing meat, but instead of sealing one group of vessels, he covered the openings with a layer of fine gauze through which flies could not pass. Once again, maggots appeared in the vessels in which flies could reach the decaying meat but not in the meat isolated from flies. (Conclusion: Flies do not spontaneously generate; it takes flies to make flies.)

the **control group** because the variable being tested—access by flies—is absent. The uncovered vessels constitute what is called the **experimental group** because they are exposed to the variable. If Redi's conclusions about his first experiment were correct, maggots would not develop in the gauze-covered vessels, even though the meat had been fully exposed to air, because that one crucial variable—access by flies—was being restricted. After a few days, Redi indeed observed that the meat in the uncovered vessel produced maggots, while the meat in the gauze-covered vessels remained free of these larvae. The results of the experiment supported his hypothesis that maggots appeared only from the deposited eggs of flies.

FORMULATING NEW HYPOTHESES

Scientists typically go beyond the specific results obtained from an experiment or a series of observations and use the data to explain a more general phenomenon. In other words, they formulate a more comprehensive hypothesis, which can be subjected to additional testing. Based on his experiments with flies, for example, Redi extended his observations, hypothesizing that *all living beings* "come from seeds of the plants or animals themselves."

Hypotheses form the foundation on which science grows, the fuel that drives new investigations. For Redi's expanded hypothesis to be correct, it would have to apply to all types of organisms, not just flies living in Florence. If someone were to demonstrate even one exception—one clear case of spontaneous generation—then Redi's hypothesis would have to be rejected or significantly modified.

ADDITIONAL TESTS, CONFLICTING RESULTS, AND FINAL CONFIRMATION

Another important characteristic of scientific findings is their repeatability. When papers are published, scientists present their methods so that other investigators will know exactly how the procedures were carried out. Redi had convinced the world that macroscopic organisms, such as frogs and flies, arose from the eggs of parents. The discovery of microorganisms about 300 years ago, however, rekindled the notion of spontaneous generation. These simpler microscopic life forms were believed to arise spontaneously from the remains of dead organisms or from other nonliving materials, substances in pond water, rain puddles, or a bowl of chicken soup. Did Redi's hypothesis hold true for microorganisms as well?

During the mid-1700s, a pair of similar experiments were performed that yielded totally opposite conclusions. In England, John Needham prepared a beef broth for growing bacteria. He briefly boiled the broth to kill any microorganisms, poured it into a test tube, and sealed the tube with a cork. Within a few days the tube was swarming with bacteria. He concluded that the bacteria must have formed from the remains of heat-killed microorganisms and proclaimed the theory of spontaneous generation to be proven. Lazzaro Spallanzani, an Italian naturalist, heard of Needham's work and performed a similar experiment, but he found no evidence of bacterial growth.

Needham's experimental results failed to pass the test of repeatability. Spallanzani believed that Needham had not been careful enough to kill all the microorganisms or to avoid contamination of his glassware. (Needham was also biased toward the concept of spontaneous generation, whereas Spallanzani was not.) Spallanzani overcame the shortcomings of Needham's experiment by boiling the broth for longer periods of time, then immediately sealing the flask. Spallanzani's critics argued that as a result of the extensive boiling procedure, he had destroyed a "vital principle" in the broth that was needed to support the spontaneous generation of organisms. More importantly, Spallanzani sealed the boiled flask so that air could not enter. Unlike Redi, he devised no way to eliminate this extra variable.

In 1860, Louis Pasteur devised an experiment in which he sterilized a preparation of nutrient broth by boiling, yet he still allowed it to remain in contact with fresh air. The

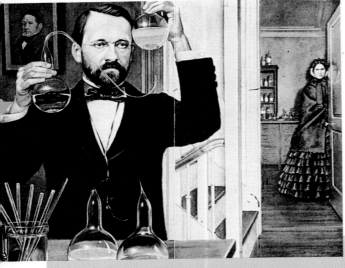

FIGURE 2-2

Pasteur's experiment disproves spontaneous generation. Pasteur poured nutrient broth into the flask, boiled it extensively, and allowed it to cool. The flask remained open to the outside air but remained free of microorganisms. After many days, Pasteur tilted the flask so that the sterilized broth would run into the neck and contact airborne particles that entered the nutrient medium. The flask soon became densely populated with microorganisms.

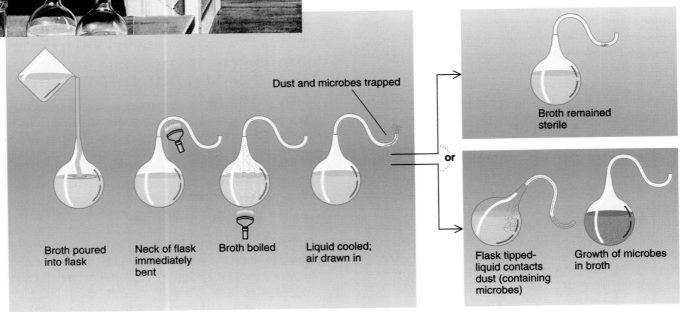

Broth poured into flask

Neck of flask immediately bent

Broth boiled

Liquid cooled; air drawn in

Dust and microbes trapped

or

Broth remained sterile

Flask tipped-liquid contacts dust (containing microbes)

Growth of microbes in broth

experiment was similar in principle to that performed by Redi nearly 200 years earlier. Pasteur added broth to a flask then melted the glass neck of the flask and shaped it into a long S-shape (Figure 2-2). The contents of the flask and the glass neck were sterilized by heat, then cooled and allowed to remain open to the outside environment. Even though fresh air could pass through the neck into the flask, all particles of matter suspended in the air settled in the "trap" of the S-curved tube, so no airborne organisms could reach the broth. Under these conditions, bacteria failed to grow in the flask. The absence of bacterial growth was not due to the destruction of some "vital principle" by heat, or by the absence of air; rather, the lack of growth was due to the absence of contamination by airborne bacteria. This fact was clearly established by Pasteur when he carried out the same experiment but, after several days of no growth, tilted the flask so that some of the sterilized broth would run into the S-shaped neck and contact the bacteria that had collected in the trap. Within 18 hours, the broth was swarming with bacteria. Pasteur had shown that living organisms do not spontaneously appear; it takes an organism to make an organism.

FROM HYPOTHESIS TO SUPPORTED THEORY

When a hypothesis has been repeatedly verified by observation and experimentation and combines with other confirmed hypotheses to help explain an important aspect of a field of investigation, the collection of related hypotheses are considered a supported **theory.** Most scientists define the term "theory" much differently from the general public, which often speaks of theories as "speculations" or "guesses." Yet, even among scientists, there is debate over the use of the term "theories." For example, should the spontaneous generation notion be called a theory? Many scientists say yes—it is a disproven theory. In this book, we will distinguish between supported theories and disproven or unsubstantiated theories. Because supported theories are backed by repeated tests, they are held with great confidence by the scientific community.

It is important to understand scientists' attitudes toward theories. It is said that a scientific theory can never be proved, only disproved. For theories to continue being accepted as correct, they must continue to be consistent with the results of new experiments or related phenomena. For instance, one of biology's most important supported theories is the theory of evolution—the theory that species change over time. The theory of evolution is no less reliable than is the atomic "theory" or the "germ theory of infectious disease." As with atoms, which cannot be directly observed, we cannot directly observe the production of new species by evolution. Biologists have gathered a tremendous amount of evidence in support of biological evolution, however, and not one major piece of evidence has ever been obtained that suggests it has not occurred. This is a very important point.

Conversely, the so-called theory of creationism proposes no testable hypotheses and thus cannot be proven right or wrong. Proponents simply accept it on faith.

The theory of evolution however, suggests many predictions that can be tested. In the words of novelist and biologist H. G. Wells, "Every mammal...is held to be descended from a reptilian ancestor. Suppose in the early Coal Measures [a period hundreds of millions of years ago], before ever a reptile existed, we found the skull of a horse or a lion. Then the whole vision of Evolution would vanish. A single human tooth...in a coal seam would demolish the entire fabric of modern biology. But never do we find any such anachronisms. The order of descent is always observed."

The scientific approach yields explanations and understanding. It is a process that combines questioning, careful observation, rigorous experimentation, and repeated verification to improve our knowledge of ourselves and the universe. (See CTQ #2.)

APPLICATIONS OF THE SCIENTIFIC PROCESS

There is another side of scientific research: It often leads to practical applications, some of which have improved our lives. Pasteur's work, which demonstrated that relatively simple measures of sterilization could prevent bacterial growth, had almost immediate practical ramifications. The British surgeon Sir Joseph Lister grasped the importance of Pasteur's work and instituted the revolutionary procedure of cleansing surgeons' hands and instruments prior to their performing an operation. The procedures produced a striking drop in the incidence of fatal postsurgical infections. Midwives and others involved in delivering babies adopted similar procedures, and the entire practice of medicine took a giant stride toward becoming safer and more effective.

ACCIDENTS AND SCIENTIFIC DISCOVERY

Most great scientific achievements are the result of creativity, diligence, and many long hours of dedicated research. However, many revolutionary discoveries were the products of unplanned events in which the observer was skilled enough to recognize the importance of the accidental occurrence. One such "accident" occurred in 1928 to Sir Alexander Fleming (Figure 2-3).

Discovering Penicillin

Fleming was a Scotsman who had studied medicine in London and was influenced by one of his professors to enter the field of bacteriology. When World War I broke out, Fleming enlisted in the army and was assigned to a bacteriologic

(a)

(b)

(c)

FIGURE 2-3

The history of penicillin. *(a)* Sir Alexander Fleming working in his lab. *(b)* Colony of *Penicillium*
growing on the surface of culture media. The amber droplets collecting on the mold's surface contain
the antibiotic penicillin. *(c)* Penicillin became available for fighting infection during World War II.
Countless lives were saved by its use. The drug was so scarce, however, that it had to be "recycled."
Penicillin was extracted from urine collected from patients receiving treatment and then reused.

research lab. During the war, Fleming became aware of
how woefully inadequate antiseptics (such as hydrogen
peroxide and carbolic acid) were in preventing wounds
from becoming infected. After watching untold numbers of
patients die from infected wounds, Fleming set out on a
lifelong search for an agent that would effectively kill microbes while, at the same time, remain relatively nontoxic to
human patients. As of 1928, only one such substance had
been discovered—an arsenic-containing compound that
was effective against certain bacteria, particularly the one
that causes syphilis. The substance had only limited value,
however.

In 1928, Fleming was working with a new antiseptic,
mercuric chloride, which was much too toxic to be taken
internally. Fleming's small lab in a London hospital was
always cluttered with petri dishes from previous experiments. One day, while talking to a colleague, Fleming
picked up one of these bacteria-inoculated dishes and noted
that it was contaminated with a mold—not an uncommon
occurrence in a bacteriology lab. While many researchers
might have simply discarded the dish, Fleming alertly observed that in this particular case the mold was surrounded
by a ring in which there were no bacterial colonies. Fleming
immediately removed a bit of the mold with a scalpel and

transferred it to a fresh tube of culture medium. He wanted to be sure he had a sample of this mysterious mold that could destroy bacterial cells growing in its vicinity.

Fleming cultured samples of the mold, which he identified as a member of the genus *Penicillium,* and collected samples of the medium in which the mold had been growing. These samples of mold-juice, as they were called, were found to be highly effective in killing bacteria. More importantly, the mold-juice could be injected into mice or rabbits without any evidence of toxicity. In the words of A. Maurois, Fleming's biographer, "Fleming had for a long time been hunting for a substance which should be able to kill the pathogenic microbes without damage to the patient's cell. Pure chance deposited this substance on his bench. But, had he not been waiting for 15 years, he would not have recognized the unknown visitor for what it was." Fleming then turned all of his attention away from studying bacteria toward working with mold-juice.

Determining the Value of Penicillin

Fleming enlisted a number of colleagues to purify the active substance, which he named penicillin. But penicillin resisted all early attempts to purify it. Meanwhile, Fleming began the first tests to see how effectively the mold-juice could suppress human infections. Although penicillin showed some promise when applied to a skin infection, the substance was not available in sufficient concentration to prove its real value. For the next few years, neither the scientific nor the medical community showed any interest in Fleming's substance. Remarkably, it wasn't until 1939, about the time England declared war with Germany, that Howard Florey, an Australian pathologist, and Ernst Chain, a Jewish chemist who had escaped Nazi Germany, began a concerted effort to purify penicillin at Oxford University.

As often happens in scientific research, Florey and Chain were aided by a new technique that allowed them to perform experiments that could not have been carried out earlier. The technique they used was called *lyophilization,* the technical name for "freeze-drying," the same process used to prepare instant coffee. Freeze-drying greatly facilitated the purification of unstable compounds, since cold conditions generally retard chemical deterioration. Using this technique, the mold-juice was frozen solid, and the water evaporated from the solid state under a vacuum, leaving the penicillin in a powdered state.

After tests on mice infected with various bacteria proved to be successful, the first test on a human was conducted in England in 1941. As was not uncommon at the time, the patient was dying from "blood poisoning," which had been caused by bacteria initially infecting a small scratch on his face. After several injections of purified penicillin, the patient showed dramatic improvement. Unfortunately, the supply of purified penicillin was quickly exhausted; the infection regained its ferocity and quickly claimed the patient's life.

From the Laboratory to the Factory

It was evident to the Oxford group that England did not have the facilities at the start of the war to produce sufficient quantities of this important new drug. Florey traveled to the United States carrying with him samples of the mold in hopes that he could persuade a U.S. pharmaceutical company to begin mass-producing penicillin. He handed over the samples, together with his data for purification, without any attempt to patent the process or to ensure financial gain for himself or his colleagues.

Within a matter of months, the U.S. company had greatly improved its ability to produce and harvest penicillin so that reasonable yields were being obtained. Fleming's mold was not very productive, however, and the search for a more generous species of *Penicillium* began. One company employee, nicknamed Moldy Mary, was responsible for daily trips to the markets of Peoria, Illinois, to look for rotting foods on which mold was growing. One day, Mary brought back a moldy cantaloupe that turned out to be growing a particularly productive species of *Penicillium.* Together with the new strain and improved culture conditions, the company was soon producing enough penicillin to save the lives of thousands of wounded soldiers of the Allied armies. The drug has been saving lives ever since.

In 1945, Fleming, Florey, and Chain were awarded the Nobel Prize for discovering and purifying the first **antibiotic** (a microbe-killing substance produced by a fungus or a bacterium). Subsequent research revealed that penicillin kills bacteria by interfering with the formation of their surrounding cell walls. Since human cells have no such walls, penicillin is not dangerous to the human body. Some people are allergic to penicillin, however. For these people, the drug may be fatal (sometimes within 10 minutes) rather than lifesaving. This is why people are asked whether they are allergic to penicillin before the antibiotic is prescribed.

TESTING ADDITIVES AND DRUGS ON ANIMALS AND HUMANS

Penicillin became commercially available at a time when there were very few rules and regulations governing the use of new drugs and products. That climate has changed in recent years, and most substances are extensively screened for their toxicity and *carcinogenic* (cancer-causing) properties. To conduct controlled experiments that measure these risks, biologists divide a *test population* of organisms into an experimental group and a control group. For example, consider tests to determine whether the food additive red dye #2 poses a cancer risk (Figure 2-4). A test population of rats was divided into an experimental group, which received food containing red dye #2, and a control group, which received exactly the same food but without red dye #2. All other variables (such as light, temperature, quantity of food, and amount of water) were identical for the two groups. A number of such studies demonstrated that red dye #2 did

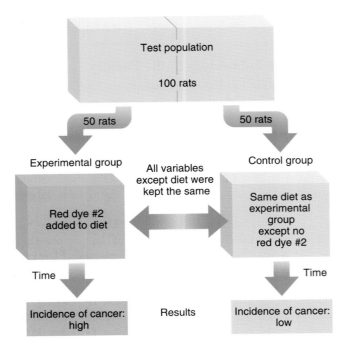

FIGURE 2-4

Does red dye #2 increase the incidence of cancer? Around the turn of the century, it was not unusual to test the health effects of food preservatives (and other chemicals) on human subjects, even children. But today not only is the use of "human guinea pigs" considered inhumane, it is also impractical since it can take 20 to 40 years for the dangerous effect of a chemical to become apparent. Because they are anatomically and physiologically similar to humans, shorter-lived rodents have now replaced human subjects in most scientific tests. Using such guidelines, scientists studied whether red dye #2 increases cancer in laboratory rats. All factors were identical in both the test group and the control group except for the addition of red dye #2 to the test group's food. Rats ingesting the dye developed an abnormally high level of cancer. Because there is a strong correlation between some causes of cancers in humans and in rats, the FDA banned red dye #2 as a food additive.

(a)   *(b)*

FIGURE 2-5

Modern biologists exploring life. Although biologists rarely become politicians or economic leaders, they nonetheless help shape the future of the world. Some scientists actively seek solutions to specific problems, such as this biologist investigating the virus that causes AIDS *(a).* Others seek answers simply for the sake of knowledge and increased understanding—these marine biologists, for example, studying the ecology of the ocean floor *(b).* Discoveries by such "basic researchers," however, often have hundreds of practical applications. As you will see throughout this book, scientific findings continue to dramatically influence our lives, from providing products of huge economic value to informing us how we must alter our global lifestyles in order to preserve the planet's life supporting capacity.

indeed increase the incidence of cancer in rats. Because rats and humans are biologically similar, the Food and Drug Administration (FDA) banned red dye #2 as a food additive.

Controlled experiments on people often include procedures to reduce the chance that personal bias will influence the results. Despite attempts to include as many controls as possible, research with humans is invariably open to debate. A well-publicized series of studies concerning the possible preventive and curative powers of vitamin C illustrates some of the complications involved in using human subjects. During the 1970s, one of the great chemists of the century, Linus Pauling, began proclaiming the benefits of vitamin C, first as a preventive of the common cold and later as a treatment for cancer. Pauling based his anticancer claims on work carried out in Scotland, in which 100 people with advanced colorectal cancers received large daily doses of vitamin C, while a larger control group made up of people of similar sex, age, and state of disease received the same treatment without vitamin C. The results of these studies suggested that vitamin C might prolong the life of patients with advanced cancer and might even cause some cancers to regress entirely. The pronouncements concerning the Scottish work were greeted with great controversy, pitting Pauling on one side against the disbelieving medical establishment on the other. Because of all the publicity, a group of researchers led by Charles Moertel at the Mayo Clinic in Minnesota decided to pursue a similar study. In contrast to the work in Scotland, the Mayo researchers carried out their study using a *double-blind* procedure. In a simple "blind" test, one group of patients receives a pill containing the drug to be tested, while a control group receives a *placebo*, a similar-looking pill that, unknown to them, contains no drug. When the researchers themselves are unaware of which subject is receiving the actual drug versus the placebo, the test is called a *double-blind test*. Double-blind tests eliminate the bias that may influence the results when researchers (or the subjects) strongly expect or hope for a particular result.

Moertel and his colleagues found no evidence that vitamin C prolonged the life of their patients with advanced cancer. But Pauling argued that the lack of a curative effect of the vitamin in the Mayo study was due to the patients' prior treatment by chemotherapy, which had damaged their immune systems and prevented them from responding positively to the vitamin C supplements. In contrast, the patients in the Scottish study had never received prior chemotherapy.

The public interest in megadoses of vitamin C continued, and a few years later Moertel and the Mayo group announced the results of another double-blind study, this time involving advanced colorectal cancer patients who had received no prior chemotherapy. (The clinicians felt that since this type of cancer did not respond well to chemotherapy anyway, it would not be unethical to withhold chemotherapy from the patients in this study.) Once again, the Mayo group found no benefit to patients receiving large

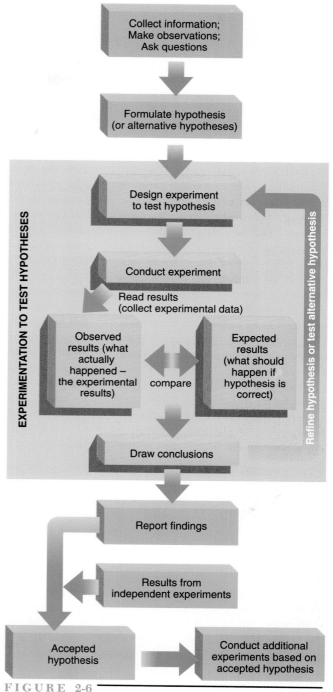

FIGURE 2-6

The scientific method. Many scientific investigations start with an observation of a phenomenon, a discussion with colleagues, or ideas triggered by listening to the presentation of a scientific paper or by reading a research article in a scientific journal. Such activities stimulate the scientist to formulate one or more hypotheses to explain a particular set of observations. These hypotheses lead to a plan for gathering experimental data that either supports a hypothesis or causes it to be rejected. Failure of one hypothesis often leads to another hypothesis, which is then tested through additional experimentation.

◁ BIOETHICS ▷

Science, Truth, and Certainty: Is a Theory "Just A Theory?"

By ANN S. CAUSEY
Prescott College

In recent times, science has come under fire for its apparent failure to provide what the general public increasingly insists it provide: the path to truth and certainty. We want to know why scientists have not yet established the truth of the theory of evolution by natural selection, though it's been over a century since Darwin proposed it. We want answers to such questions as was the record heat of the summer of 1988 a result of human caused global warming? Does cigarette smoking cause lung cancer? And, why can't scientists agree on a simple yes or no answer to such empirical questions?

In the absence of definite answers lies license for some to distort science and to manipulate public perception of it. Since scientists cannot state with certainty that global warming has begun, many critics have dismissed warnings of impending climate change as "just theory." Tobacco companies continue to rely on the lack of causal certainty in studies linking smoking and cancer, as they defend themselves against product-related lawsuits. And, creationists insist that until and unless Darwinian evolution is proved beyond any and all doubt, creationism should be taught alongside evolution in public schools.

The important question is why scientists have not been able to put such criticisms of scientific theory to rest once and for all. Why can't science provide the certain knowledge—the truth—that the general public seeks? The answer lies partly in the nature of science itself and partly in the problematic nature of the concept of truth. To construct the answer, let's examine these issues in more detail.

One of the primary tools of the scientific method is *inductive reasoning;* that is, reasoning from particular observations to general conclusions. The nature of induction, however, actually limits the extent of justification that any scientific "belief" can earn. The problem is that the conclusions we get through inductive reasoning are always less certain than are the premises from which they are drawn. For instance, from observations of the color of thousands of different crows, we may induce (hypothesize) that all crows are black. Of course, we can determine the color of any particular crow with certainty; yet we can never observe all crows that have ever existed or might in the future exist. Thus, our conclusion (all crows are black) based on our many observations, while highly probable, can never be absolutely certain. Of course,

doses of vitamin C. And, once again, Pauling responded with stinging criticism of the study. He argued that unlike the Scottish study, where patients had been kept on vitamin C for as long as they lived (as long as 12 years), the group at the Mayo Clinic had been treated with the vitamin for less than 3 months. The controversy rages on today and will continue until a definitive study on human subjects clearly proves or disproves the value of vitamin C as a therapeutic agent against cancer. Thousands of such questions, both practical and theoretical, propel scientific inquiry as modern biologists continue to investigate the world of life (Figure 2-5).

Science does not proceed along predetermined paths, and the end product cannot always be predicted. The results of scientific research often have practical applications that were never conceived of by the original researchers. (See CTQ #4.)

CAVEATS REGARDING "THE" SCIENTIFIC METHOD

The vitamin C research shows quite clearly that the results of scientific studies are often open to differences of interpretation. Readers should always be somewhat skeptical, particularly when the results are being reported second-hand in a newspaper or popular magazine.

It should also be borne in mind that the scientific method, as summarized in Figure 2-6, is only a general description of one way scientists approach their studies. Even though it is systematic, the scientific approach also allows for creativity and flexibility; there is no absolute path for discovering the answer to a question. Some scientific studies might start with an observation, others might start with a question based on previous work. Sometimes a scientific study might begin with a hypothesis to be tested, while other times an accidental discovery may trigger a finding that leads to a hypothesis. As you will see throughout

the more black crows we find, the more probable or confirmed our hypothesis is. However, no amount of observation will allow us to say with absolute certainty that *all* crows are black. And, just one observation of a non-black crow is sufficient to invalidate our hypothesis completely. Such is the nature of science: Even the best and most widely accepted hypotheses and theories generated by inductive reasoning have a built-in element of uncertainty.

Not only can we never claim absolute justification by way of certainty for any theory, we cannot even agree on what *constitutes* truth in a belief or theory. Philosophers split into three main camps on this issue. One camp, proponents of the coherence theory of truth, claims that truth is a property of a related group of consistent (or coherent) statements. Thus, the truth of any one belief depends on how well it coheres, or is consistent, with the other members of a larger generally accepted belief system. Mathematics is a good example of the coherence theory of truth in operation; from a few basic statements, an entire system of mathematical "truth" is constructed. Similarly, in science, theories gain respectability when they are coherent with a larger body of accepted judgements.

Another camp contends that truth consists of correspondence between a belief and a fact or actual state of affairs. Thus, a belief is true if it corresponds to reality, reality consisting of some realm of facts existing objectively and independent of us. This is undoubtedly the most popular theory of truth, and one that is heavily relied on in the sciences. Most scientists do, in practice, assume that a good theory explains how things really, objectively are in nature.

Finally, the third camp insists that what makes a theory true is neither its coherence with other accepted theories nor its correspondence with some alleged objective reality, but rather its usefulness. Pragmatism, as it is called, finds true theories simply by finding those that work best in predicting, explaining, and facilitating further study. When a theory stops working, it is no longer true. We thus make our own truths as we develop and use theories. Truth, for pragmatists, is dynamic and changing; most scientists agree that these traits are essential to good theories as well.

Thus, it is true that a theory is never certain, and even truth is subject to individual interpretation. While we can never remove all elements of uncertainty from scientific theories, that uncertainty is insignificant in the face of the overwhelming amount of evidence a theory must accumulate in its favor in order to win the confidence of the scientific community. Furthermore, although scientists do not have a special claim to absolute truth, they do incorporate the important elements of coherence, correspondence, and pragmatic truth in their work. A good theory will likely be "true" in all three senses.

Thus, repeatedly verified scientific theories, such as the link between smoking and cancer, the mechanism of evolution by natural selection, and the effects of pollution on global climate, warrant a very high degree of confidence in their accuracy. Lack of scientific certainty does not justify a lack of public confidence in science; the uncertainty is part and parcel of the methodology of science itself and is essential if theories are to remain dynamic and subject to revision as new evidence accumulates.

this book, each scientific "fact" has a unique and often fascinating story of exploration and discovery.

Regardless of which path is taken, scientific knowledge grows over time, with new discoveries blossoming from the seeds planted by previous researchers. Isaac Newton summarized this dynamic when he wrote, "If I have seen farther than other men, it is by standing on the shoulders of giants."

SYNOPSIS

Scientists believe that the world is understandable and that rational explanations exist for all phenomena. The basic objective of science is to understand life and the universe in a way that can be validated.

Scientific investigations commonly follow an investigative strategy we call the scientific method. The process includes making careful observations, learning what is known about the subject, and asking questions; formulating hypotheses that might explain a phenomenon or answer a question; designing and conducting experiments to test the validity of hypotheses; analyzing results; drawing conclusions as to whether to accept, reject, or modify the hypotheses being tested.

When a hypothesis has been repeatedly reinforced by experimentation, it may become a supported theory, a term that indicates its widespread acceptance. Unlike a single hypothesis, a theory combines many ideas into a unifying principle. For a theory to continue to be accepted, it must remain compatible with new evidence.

Experiments must be performed with proper controls so that the effect of a single experimental variable can be determined. Controlled experiments help distinguish cause-and-effect relationships from coincidence and re-

duce the influence of human bias. The use of a control allows the investigator to determine the effect of one factor at a time.

Scientific research often leads to practical applications. This was illustrated in this chapter by the discovery and production of penicillin.

Key Terms

spontaneous generation (p. 32)
hypothesis (p. 32)
variable (p. 32)

control group (p. 33)
experimental group (p. 33)

theory (p. 36)
antibiotic (p. 37)

Review Questions

1. Why is it important to have a control group when carrying out experiments? Why was Redi's initial experiment using sealed vessels an inappropriate control?

2. How did Pasteur's use of long s-shaped-necked flasks serve as an adequate control in testing the existence of spontaneous generation of microorganisms?

3. How is it that penicillin can be used safely in humans while preventing the growth of bacteria, which are also living organisms?

4. What is meant by the term "double-blind" experiment, and why are such experiments used in science?

Critical Thinking Questions

1. Classify the following statements as products of simple observation, speculation, or the scientific approach. How might each statement that falls into the first two categories be scientifically tested?
   a. The sun rises in the east.
   b. Since the sun always rises in the east, it will rise in the east tomorrow.
   c. I took vitamins every day this year and did not catch a cold. Therefore, vitamins prevent colds.
   d. People with protein deficiencies are more vulnerable to infectious diseases.

2. Select an article from a tabloid that makes some claims of a "scientific" nature. Evaluate how scientific the claims are, using the following questions as a guide: (1) Are the claims based on observations? (2) Have the observations been made by one person or many? Are they repeatable? (3) Is the person(s) a reliable observer? (4) Are the claims testable? (5) If tests have been performed, did they involve controls?

3. What might be a practical application of each of the following scientific findings?
   a. The body responds to the presence of foreign materials by launching an immune response that attacks that foreign invader (but none other).
   b. Every living cell in a plant has the genetic capacity to become any other type of cell in that plant. In fact, one actively growing cell taken from a mature plant can grow into an exact duplicate of the plant from which it was taken.

4. Among the many discoveries of the twentieth century that have changed our lives, seven are related to biology. Select one of those listed below and research the basic scientific advances that made the discovery possible. What other practical applications have resulted from the basic scientific advance(s)? Which, if any, of the practical applications were predicted? (1) discovery of blood types, (2) development of hybrid corn, (3) use of DDT, (4) synthesis of sex hormones, (5) use of chlor-

promazine and lithium for treating mental illness, (6) discovery of structure of DNA, (7) discovery of early prehuman fossils.

5. Design a *controlled* experiment to test the hypothesis that high doses of vitamin C reduce the incidence of colds. Can you use a test population of human subjects? What are the normal limitations, and how will they affect the results?

6. How would you apply the scientific method to test the accuracy of the following statements?
   a. An ostrich sticks its head in the ground to avoid impending danger.
   b. Disasters occur in groups of three.
   c. Large doses of vitamin E improve one's energy.
   d. Walking under a ladder will bring bad luck.
   e. Camels carry water in their humps.

Additional Readings

Asimov, I. 1984. *Asimov's new guide to science.* New York: Basic Books. (Introductory)

Gibbs, A., and A. E. Lawson. 1992. The nature of scientific thinking as reflected by the work of biologists and by biology textbooks. *Am. Biol. Teacher* 54 (3):137–152. (Intermediate)

Goldberg, A. M., and J. M. Frazier. 1989. Alternatives to animals in toxicity testing. *Sci. Amer.* Aug: 24–30. (Intermediate)

Kohn, A. 1984. *Fortune or failure: missed opportunities and chance discoveries in science.* London: Blackwell. (Introductory)

MacFarlane, G. 1984. *Alexander Fleming. The man and the myth.* London: Hogarth Press. (Introductory)

Maurois, A. 1959. *The life of Sir Alexander Fleming.* London: Jonathan Cape. (Introductory)

Moberg, C. L. 1991. Penicillin's forgotten man: Norman Heatley. *Science* 253:734–735. (Intermediate)

Richards, E. 1991. *Vitamin C and cancer: Medicine or politics.* New York: St. Martin's. (Introductory)

Serafini, A. 1989. *Linus Pauling: A man and his science.* New York: Paragon House.(Introductory)

Zuckerman, H., J. R. Cole, and J. T. Bruer, eds. 1991. *The outer cycle: Women in the scientific community.* New York: Norton. (Introductory)

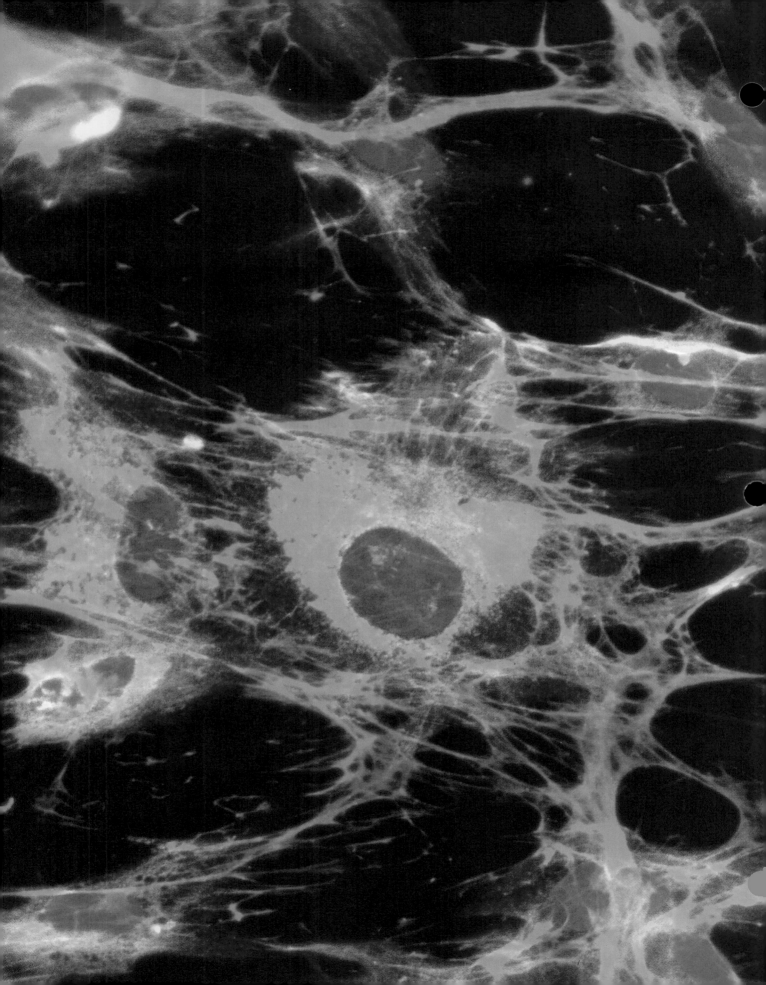

Chemical and Cellular Foundations of Life

The multitude of functions necessary for life require complexity, organization, and specialization—all of which are embodied in cells. Part of the cellular organization required for life is revealed in this photograph of a human skin cell. The green threads are microtubules, which help give the cell its shape and organization and participate in cell division. The blue threads are the cell's DNA. The remaining threads are proteins on the cell surface that enable it to interact with its environment.

PART
· 2 ·

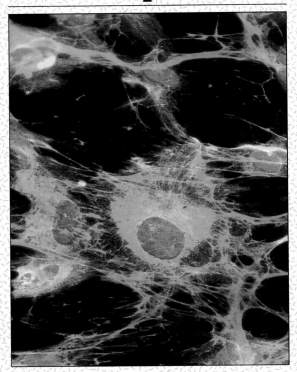

The Atomic Basis of Life

STEPS
TO
DISCOVERY
The Atom Reveals Some of Its Secrets

THE NATURE OF MATTER

THE STRUCTURE OF THE ATOM

How Atoms Differ from One Another

TYPES OF CHEMICAL BONDS

Covalent Bonds

Noncovalent Bonds

THE LIFE-SUPPORTING PROPERTIES OF WATER

The Effect of Water On Nonpolar Molecules

Water as a Solvent

Capillary Action and Surface Tension of Water

The Thermal Properties of Water

ACIDS, BASES, AND BUFFERS

The Nature and Importance of pH

BIOLINE

Putting Radioisotopes to Work

THE HUMAN PERSPECTIVE

Aging and Free Radicals

The Atom Reveals Some of
Its Secrets

*I*n the early 1890s, James Clerk Maxwell, one of the most prominent physicists of the time, authoritatively proclaimed that all of the basic principles of physics had been discovered; all that was left was to fill in the holes. Within a few years, however, the entire field of physics would undergo a revolution that altered our entire concept of the structure and properties of matter. This revolution began in 1896 as the French scientist Henri Becquerel was studying *phosphorescence*—the property of certain materials, including uranium salts, to glow in the dark after having been exposed to sunlight. A year earlier, the discovery of X-rays had been reported, and Becquerel had hypothesized that phosphorescent materials might emit X-rays after their ex-

posure to light. If this were true, he should have been able to detect the X-rays by their ability to pass through an opaque shield and to expose a photographic plate on the other side of the barrier.

On a clear day in Paris, Becquerel tried to do just that. He placed crystals of uranium salts on top of a photographic plate that had been wrapped with black paper to prevent its exposure to light. He put the package out in the sun for a few hours, then developed the plate in the darkroom, revealing a clear smudge where the "rays" from the uranium salts had penetrated the black paper and exposed the photographic film. Becquerel thought he had confirmed his hypothesis, but he wanted to repeat the experiment and

As discovered by the Curies, radioactivity results from particles hurled from the nucleus of a disintegrating atom.

prepared a similar package to be sure. The weather changed, however, and the skies became heavily overcast. Becquerel filed away the covered plate and uranium salts in a drawer, waiting for the sun to reappear. After several days of inclement weather, he impatiently developed the plate. To his astonishment, he found a very bright image of the uranium salts revealed on the film. Becquerel realized the importance of this accidental discovery. He concluded that since the "rays" emitted from the uranium salts were not a result of energy absorbed from the sun after all, they must be a spontaneous emission of the uranium itself. Becquerel had discovered a new property of matter.

Becquerel's report caught the interest of a young French physicist Pierre Curie, who had recently married Marie Sklodowska, a Polish immigrant studying physics at the Sorbonne in Paris. Marie Curie decided to conduct her doctoral thesis on the nature of Becquerel's "rays." She soon discovered that uranium was not the only source of these "rays," which she named *radioactivity*. In fact, the crude uranium ore she was working with also contained several other substances that were actually much more radioactive than uranium. She named one of these newly discovered substances *polonium,* after her native homeland. The other substance, which she called *radium,* after the Latin word meaning "ray," proved to be over a million times more radioactive than uranium. Because of the scarcity of radium in the uranium ore, it took Marie and Pierre several years, working in a small, poorly equipped shed, to extract 400 milligrams of pure radium from more than a ton of ore.

The Curies discovered both the destructive and curative properties of radium. Pierre placed a bandage containing a few crystals of radium on his arm for a few hours; the skin soon reddened, and an open sore developed, which took months to heal. This observation provided the first evidence that rays emitted from radium atoms were capable of killing living cells, and led to the use of radium to treat skin cancers. Radium crystals were placed in proximity to the cancer, causing the death of the cancerous cells and their replacement by normal tissue. Radioactive elements are still part of the arsenal used in the treatment of cancer.

The radioactive materials were affecting the health of the Curies, however, who took virtually no precautions. For example, spills of extremely radioactive materials were simply mopped up by hand with rags, which were thrown in the garbage. Consequently, Marie's hands became hardened and numb from working with the materials. For their work on discovering radioactivity, the Curies and Becquerel shared the 1903 Nobel Prize in physics, and Marie Curie won a second Nobel prize in chemistry 8 years later for her discovery of radium and polonium.

The work of the Curies also proved to be the first step in demonstrating that atoms could have an unstable structure and were not the indivisible, indestructible, minuscule spheres they were proclaimed to be. This was soon confirmed by the New Zealander Ernest Rutherford, who discovered that the rays emitted by a sample of the Curies' radium included at least two types of particles, one relatively large and positively charged, the other very small, carrying a negative charge and having very different properties. Rutherford showed that these particles were hurled out as a radioactive radium atom disintegrated with explosive force. With each explosion, the loss of particles transformed one type of atom into a different type. A radioactive uranium atom was transformed into a radioactive radium atom, which lost more particles (and discharged more energy) to become a radioactive atom of radon. The sequential deterioration continued until what had started out as an atom of uranuim was eventually transformed into an atom of nonradioactive (and thus stable) lead. In a sense, radioactivity was a fulfillment of the alchemist's dream of converting one element into another. Whereas the alchemists had tried to convert lead into gold, Rutherford had discovered that nature works in the opposite direction by spontaneously producing lead from elements vastly more valuable than gold. For his work on atomic structure and radioactive decay, Rutherford received the Nobel Prize prize in physics in 1908.

THE NATURE OF MATTER

*T*o explore life adequately, one must venture into an invisible realm—the realm of chemicals and chemical reactions, where minuscule particles combine with one another to forge everything in the universe, including that which is alive. Chemistry provides an explanation for biological properties at their most fundamental and essential level, that which ultimately accounts for what we are and what we do as living entities. Probing the chemical basis of life is a first step in understanding biology. It reveals how different particles can be arranged and rearranged to form an unlimited variety of organisms.

▼   ▼   ▼

When we look around at objects in our environment we find a seemingly endless diversity of materials. All matter, however, is composed of a limited number of basic substances, or **elements.** The diverse nature of materials stems from the variety of ways these elements can combine with one another to form more complex substances with new and different properties. The first reasonably accurate list of 28 elements was compiled in the 1780s by Antoine-Laurent Lavoisier. (His list might have grown even longer had he not been beheaded in the French revolution of 1794.) Today, scientists have identified 109 elements; 92 of these occur in nature, the remaining 17 have been synthesized by scientists. Four of these elements (carbon, hydrogen, oxygen, and nitrogen) make up over 90 percent of your body's weight.

FIGURE 3-1

Different but similar. Each of these strikingly different organisms is composed almost entirely of the same group of chemical elements.

Elements are designated by either one or two letters derived from their English or Latin names. These include carbon (symbolized by the capital letter C), hydrogen (H), oxygen (O), nitrogen (N), phosphorus (P), sulfur (S), sodium (Na, from the Latin, *natrium*), potassium (K, from the Latin, *kalium*), chlorine (Cl), magnesium (Mg), iron (Fe, from the Latin *ferrum*), and calcium (Ca). The same 12 elements (Table 3-1) mentioned above make up over 99 percent of the living matter of all the diverse organisms on earth (Figure 3-1).

There are only about 100 basic substances (elements) that can be combined in various ways to make up everything in the universe. (See CTQ #2.)

THE STRUCTURE OF THE ATOM

In 1810, John Dalton, an English schoolteacher, formulated the atomic theory of the structure of **matter,** matter being any material substance that occupies space and has weight. Dalton conceived of matter as being composed of tiny solid spheres, or **atoms** (Figure 3-2). Atoms are the smallest units of matter unique to a particular element. A century later, Ernest Rutherford began to explore the nature of the atom by allowing radioactive particles to be fired at a thin piece of gold foil. To his surprise, virtually all of the particles (over 99.9 percent) sailed right through the foil. Rutherford hypothesized that if atoms were solid like billiard balls, as previously thought, virtually all of the radioactive particles should have been deflected. He concluded that atoms consisted largely of empty space; virtually all of the weight of an atom was compacted into an extremely small volume, which he called the **nucleus.**

Based on other observations, Rutherford concluded that the compact nucleus of an atom contains one or more positively charged *subatomic* particles he called **protons.** The atomic nucleus therefore carries a positive charge—one unit of charge for each proton. Unknown to Rutherford, the nucleus of an atom also contains electrically neutral (uncharged) subatomic particles called **neutrons,** which were not discovered for another 21 years. Protons and neutrons have approximately the same mass, about $0.0000000000000000000000166$ (1.6×10^{-24}) grams. The third type of subatomic particle, called an **electron,** is about 1/1,800 the mass of a proton and is negatively charged. In Rutherford's hypothesis, the electrons were distributed in the empty space that surrounded each nucleus. He concluded that an atom is held together as a unit by the electrical attraction exerted between the oppositely charged protons and electrons.

TABLE 3-1
MOST COMMON ELEMENTS IN THE HUMAN BODY

	Percentage by weight	Atomic Number	Atomic Mass[a]
Oxygen (O)	65	8	16
Carbon (C)	18	6	12
Hydrogen (H)	10	1	1
Nitrogen (N)	3	7	14
Calcium (Ca)	2	20	40
Phosphorus (P)	1.1	15	31
Potassium (K)	0.35	19	39
Sulfur (S)	0.25	16	32
Sodium (Na)	0.15	11	23
Chlorine (Cl)	0.15	17	35
Magnesium (Mg)	0.05	12	24
Iron (Fe)	0.004	26	56

[a] Atomic number and atomic mass are discussed in the following section of the text.

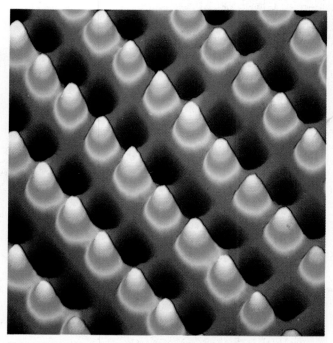

FIGURE 3-2

Photograph of rows of individual atoms of the element antimony as seen through a *scanning tunneling electron microscope.*

HOW ATOMS DIFFER FROM ONE ANOTHER

All atoms consist of the same three types of subatomic particles. An atom of one element is distinguished from those of all other elements by its **atomic number,** that is, the number of protons in its nucleus. For example, the atomic number of hydrogen is 1: All hydrogen atoms have a single proton and, conversely, all atoms with one proton are hydrogen atoms. All carbon atoms have six protons, all oxygen atoms have eight protons, and so forth (see Figure 3-4).

Atoms of the Same Element with Different-Sized Nuclei

The combined numbers of neutrons and protons in the nucleus make up the weight of an atom, or its **atomic mass.** While the number of protons is identical among all atoms of a given element, the number of neutrons can vary. Atoms having the same number of protons—the same atomic number—but different numbers of neutrons—different atomic mass—are said to be **isotopes.** Hydrogen, for example, exists as three different isotopes; all hydrogen atoms contain one proton but may have zero, one, or two neutrons. The atomic mass of these isotopes is denoted as ^1H, ^2H, and ^3H, respectively. In nature, a sample of an element will contain a mixture of various isotopes. Despite the fact that only one in every 6,000 hydrogen atoms normally contains a neutron, relatively pure *deuterium* (a hydrogen atom containing one neutron (^2H)) can be prepared by industrial procedures for use in forming "heavy water"—so called because the extra neutron makes the water more dense than the "ordinary" variety. Heavy water is incapable of supporting life. For example, seeds watered solely with heavy water will not sprout.

Hydrogen atoms with two neutrons (termed *tritium*) are **radioactive;** that is, they are physically unstable and tend to disintegrate spontaneously. Radioactive atoms such as tritium play an indispensable role in biological and medical research (see Bioline: Putting Radioisotopes to Work).

The Arrangement of Electrons in an Atom

While Ernest Rutherford was concentrating on the nucleus, a colleague of his, Niels Bohr, was studying the structural arrangement of an atom's electrons. In 1913, Bohr proposed that electrons whirl around the nucleus in orbits, not unlike the way the planets of our solar system move around the sun (Figure 3-3). Bohr proposed that each orbit—or **shell**—is situated at a specific distance from the nucleus and has a maximum number of electrons it can hold. When a shell becomes filled, any additional electrons must go into the next shell farther away from the nucleus. The innermost shell can contain only two electrons. The next two shells can contain eight each. Thus, the six electrons of the carbon atom (Figure 3-4) are distributed in two shells in a 2:4 arrangement. That is, the innermost shell holds a pair of electrons, and the second shell holds the other four electrons. Sulfur's 16 electrons form a 2:8:6 arrangement; this formation differs from that of chlorine (17 electrons) by having one fewer electron in its third shell. The electronic distribution of a number of common atoms is illustrated in

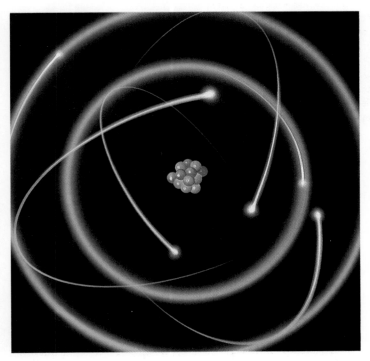

FIGURE 3-3

Simplified version of a typical atom. The nucleus of this carbon atom contains six protons and six neutrons. Six electrons orbit the nucleus in two shells. Although useful in understanding atomic structure and the chemical reactions of an atom, this simplified model (called a *Bohr atom* after the Danish physicist who proposed it) fails to show the enormous relative distance between nucleus and electrons and inaccurately represents the shape of the shells. According to modern physics, the shells consist of orbitals of spherical or dumbbell shape, in which an electron has a high probability of being located at any given instant. These orbitals are more like "clouds" encompassing the area occupied by the electrons traveling around the nucleus at about the speed of light.

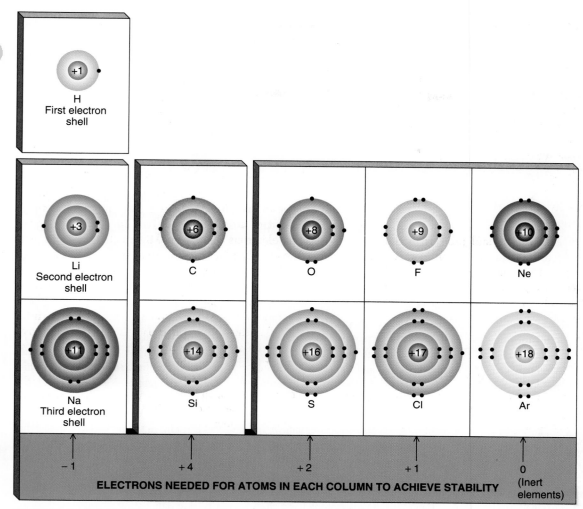

FIGURE 3-4

A representation of the arrangement of electrons in a number of common atoms. Each shell has a limited number of electrons it can hold. Electrons in each shell are grouped in pairs to illustrate that each orbital of a shell can hold two electrons. The number of outer-shell electrons is a primary determinant of the properties of elements. Atoms with similar number of outer-shell electrons have similar properties. Lithium (Li) and sodium (Na), for example, have one outer-shell electron, and both are highly reactive metals. Carbon (C) and silicon (Si) atoms can each bond with four different atoms. Because of its size, however, a carbon atom can bond to other carbon atoms, forming long-chained organic molecules, whereas silicon is unable to form comparable molecules. Neon (Ne), and argon (Ar) have filled outer shells, making these atoms highly nonreactive; they are referred to as inert gases.

Figure 3-4. As we will see shortly, the structure of these atoms determines their chemical reactivities.

By the 1920s, it was apparent that Bohr's model of the atom was overly simplistic and that electrons do not follow fixed circular orbits as depicted in Figure 3-3. Rather, they sweep around the nucleus in loosely defined "clouds," called **orbitals.** Orbitals are roughly defined by their boundaries, which may have a spherical or dumbbell shape.

Each orbital is capable of containing a maximum of two electrons. The innermost shell contains a single orbital, the second shell contains four orbitals (thus eight electrons), the third shell also contains four orbitals (thus eight electrons); and so forth.

Electrons contain energy; specifically, those of inner shells contain less energy than do those of outer shells. Electrons do not always stay in a particular shell but can

Like microscopes and telescopes, the use of radioactive isotopes, or **radioisotopes,** has opened the door to a world that humans cannot explore with their naked eyes, hands, or ears. Radioisotopes have the chemical properties of ordinary isotopes, but they reveal their presence by the radiation they emit. Thus, radioisotopes can be easily tracked and measured through the entire course of an experiment using a radiation-sensitive measuring device. Chemicals, such as sugars or amino acids, can be made with radioactive atoms and then incorporated into living cells for experimental purposes. Using radioisotopes, scientists have traced the complex chemical odysseys of molecules as they are processed by living organisms. For example, growing plants in an atmosphere enriched with radioactive carbon dioxide ($^{14}CO_2$) allowed scientists to determine how photosynthesis leads to the production of the complex molecules that make up the plant's leaves, stems, and roots (see Chapter 8).

Radioisotopes are also used in the diagnosis and treatment of disease. Radioactive iodine (^{131}I), for example, is used to detect abnormal thyroid function. The thyroid is a large endocrine gland in the neck which produces iodine-containing hormones that affect metabolism. In a diagnostic test, the person is asked to drink a solution containing small amounts of radioactive iodine, and the accumulation of the radioactivity in the thyroid is monitored. If the thyroid gland is producing too much hormone, a higher-than-normal level of radioactivity will accumulate in the gland. Much higher concentrations of ^{131}I are sometimes administered in order to *destroy* thyroid tissue nonsurgically in individuals with overactive glands.

Each type of radioisotope has a characteristic **half-life**—the time required for half of the radioisotope to decay into its stable, nonradioactive form. For example, ^{14}C has a half-life of 5,730 years, after which only half of the original amount of ^{14}C remains. During the next 5,730 years, 50 percent of the remaining amount "decays," and so on. Knowledge about half-lives has been used to provide accurate geologic "clocks" that reveal the approximate age of very old objects, such as fossils. The age of an object is determined by measuring the ratio of one isotope to another. In all *living* organisms, the ratio of ^{14}C (radioactive carbon) to ^{12}C (the most abundant nonradioactive isotope of carbon) is the same as that of atmospheric carbon dioxide. When an organism dies, however, it stops accumulating carbon. As the dead organism's ^{14}C decays, the relative proportion of ^{14}C to ^{12}C declines steadily according to the isotope's half-life. Because the rate of decay is constant, the isotopic ratio identifies the object's age. With a half-life of several thousand years, ^{14}C is useful in dating rocks or fossils that are less than 50,000 years old. Older objects can be dated using other radioisotopes, such as ^{40}K, which decays much more slowly (half-life of 1.3 billion years). Utilizing radiodating techniques, the skull of the saber-toothed tiger seen in the accompanying photo is shown to be about 1 million years old.

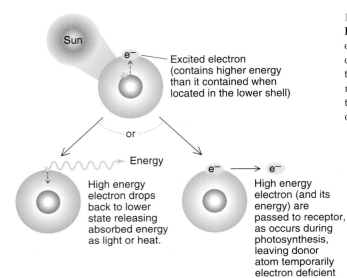

FIGURE 3-5

Electron shells and energy content. When an atom absorbs energy, an electron may jump to a shell located farther from the nucleus. This "excited" electron has two possible fates: It can drop back to a lower shell, releasing the absorbed energy back into the environment, or it can be passed on to another atom which acts as an "electron acceptor." This latter option occurs when light energy is captured during photosynthesis.

move temporarily from an inner to an outer shell by absorbing additional energy (Figure 3-5). This is what occurs, for example, when a molecule of the plant pigment chlorophyll absorbs light energy during photosynthesis (Chapter 8). Virtually every organism on earth depends on the energy captured by these electrons during photosynthesis.

Matter is made up of atoms, which are composed of only three, even smaller particles—positively charged protons, uncharged neutrons, and negatively charged electrons. The number and variety of these three particles determine the physical and chemical properties of the atom. (See CTQ #3.)

TYPES OF CHEMICAL BONDS

The atoms of most elements can interact with other atoms to form larger, more complex structures. The force of attraction that holds one atom to another is a **chemical bond.** We can distinguish between two types of chemical bonds: covalent and noncovalent.

COVALENT BONDS

Most atoms tend to establish stable partnerships with other atoms, forming larger complexes called **molecules.** A molecule containing more than one type of element (as most do) is called a **compound.** The two or more atoms that make up a molecule are joined together by **covalent bonds,** a type of bond in which pairs of electrons are shared between atoms. Examples of covalent bonding are presented in Figure 3-6. The formation of a covalent bond between two atoms is governed by a fundamental principle —namely, that an atom is most stable when its outermost electron shell is filled. One of the consequences of an unfilled outer shell is discussed in The Human Perspective: Aging and Free Radicals.

The number of bonds an atom can form depends on the number of electrons needed to fill its outer shell. For example, a hydrogen atom, with its single electron, can fill its outer shell by combining with another hydrogen atom, forming a hydrogen molecule (H_2). The two hydrogen atoms are linked together by a *single* covalent bond. In a single covalent bond, both electrons of the shared pair orbit around both nuclei of the bonded atoms, thus satisfying each atom's requirement for a filled outer shell. An oxygen atom has six outer shell electrons, leaving room for two additional electrons in the shell. An oxygen atom can satisfy this deficiency by combining with two hydrogen atoms, forming water (H_2O) and releasing energy. For confirmation that energy is released when hydrogen and oxygen atoms form a chemical bond, one need only watch the films of the explosion of the dirigible *Hindenburg* as it was about to land in New Jersey in 1937. The hydrogen gas (H_2) that

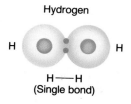

Hydrogen

H—H
(Single bond)

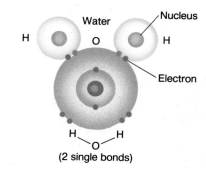

Water

Nucleus

Electron

H   H
  \ /
   O
(2 single bonds)

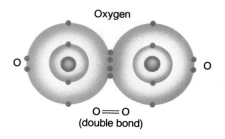

Oxygen

O＝O
(double bond)

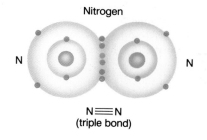

Nitrogen

N≡N
(triple bond)

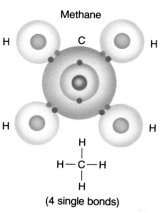

Methane

H
|
H—C—H
|
H
(4 single bonds)

FIGURE 3-6
Examples of covalent bonding. The shared electron pairs (in color) orbit around the nuclei of both atoms forming the bond.

◁ THE HUMAN PERSPECTIVE ▷
Aging and Free Radicals

Why do humans have a maximum lifespan of approximately 100 years while our close relatives, the chimpanzees, live only about half this length of time? Many biologists believe that aging results from the gradual accumulation of damage to our body tissues, which disrupts the order that is required to maintain life. The most destructive damage probably occurs to DNA. Alterations in DNA lead to the production of faulty genetic messages that promote gradual cellular deterioration. How does cellular damage occur, and why should it occur more rapidly in a chimpanzee than a human?

Atoms are stabilized when their shells are filled with electrons. Recall that electron shells consist of orbitals, each of which can hold a maximum of two electrons. Atoms or molecules that have orbitals containing a single unpaired electron are called **free radicals.** Free radicals are often formed when a covalent bond is broken in a way that each portion keeps one half of the shared electron pair. Because free radicals are so reactive, they are ex-

tremely destructive to biological tissues. For example, water—our life-sustaining medium—can be converted into free radicals when exposed to radiation from the sun

$$H_2O \rightarrow HO\cdot + H\cdot$$

("·" indicates a free radical)

Even more common is the superoxide radical ($O_2^{\cdot-}$) which is formed when a molecule of oxygen picks up an extra electron. Superoxide radicals are inevitably formed from some of the oxygen molecules we breathe, and they may be a major contributor to the aging process.

It has been hypothesized that, compared to animals with shorter lifespans, animals with longer lifespans, such as humans, have more effective mechanisms for protecting themselves against free radicals and for repairing cellular damage that has already occurred. Several kinds of substances can provide protection from free radicals. Cells contain enzymes that can convert free radicals into noninjurious mo-

lecules. Substances called *antioxidants* can also chemically alter free radicals; these include vitamin E, C, and beta-carotene (the orange pigment in carrots and other vegetables, and the parent compound for vitamin A). It has been suggested that dietary supplements containing antioxidants may help in the fight against free-radical-induced damage. In fact, some nutritionists argue that the drop in the number of cases of stomach cancer in the United States over the past few decades is due to increased consumption of antioxidants, such BHA and BHT, which are used as preservatives in many foods. In contrast, certain dietary components, such as polyunsaturated fats (page 75), might be harmful because they may be converted to free radicals as they are metabolized. This could be the basis for a possible link between polyunsaturated fats and an increase in certain types of cancer. Free radicals have also been linked to an increased risk of cardiovascular disease because they promote the buildup of plaques within arteries.

---

filled the *Hindenburg* reacted with oxygen gas (O_2) present in the atmosphere, forming a cloud of H_2O and releasing enough energy to destroy the aircraft in a blazing inferno.

Oxygen can satisfy its two-electron deficit by bonding with an identical atom, forming an oxygen molecule (O_2). The two atoms of molecular oxygen are bound together by a *double bond,* in which two pairs of electrons are shared. Some elements can even form *triple bonds* composed of three pairs of electrons, as illustrated by molecular nitrogen (N_2). Quadruple bonds do not exist.

Carbon atoms—with four electrons in their outer shell—can form four single covalent bonds. Carbon, for example, can share its electrons with four atoms of hydrogen to form methane (CH_4), a gas produced by a variety of swamp-dwelling microbes as well as by the bacteria inhabiting your large intestine. Methane is also the major component of the natural gas supplied by your gas company. The

pairs of electrons that form the single bonds in a methane molecule (Figure 3-6) are shared rather equally between the carbon and hydrogen atoms. This is because neither the hydrogen nor the carbon nucleus is strongly electron attracting. In some molecules, however, the nucleus of one atom is more electron attracting than is its partner and exerts a greater pull on the shared electrons. Among the atoms most commonly present in biological molecules, nitrogen, oxygen, phosphorus, and sulfur are strongly electron attracting.

Polar versus Nonpolar Molecules

Let's examine a molecule of water. Water's single oxygen atom attracts electrons much more forcefully than do either of its hydrogen atoms. As a result, the electrons tend to be more closely associated with the oxygen atom, which becomes somewhat negatively charged. The two hydrogen atoms, which are electron deficient, become somewhat positively charged.

$$\underset{+H}{\overset{O^-}{}} \quad H_+$$

Molecules that contain an unequal charge distribution, such as water, are referred to as **polar molecules;** they possess distinct positive and negative regions, or *poles*. In contrast, **nonpolar molecules** lack regions of electric charge. Nonpolar molecules of biological importance consist almost exclusively of carbon and hydrogen atoms and include methane, fats, and waxes.

 The difference in structure between polar and nonpolar molecules is a key determinant of the properties of biologically important molecules. As we will see shortly, for example, the polarity within the water molecule is the reason water is able to dissolve so many different substances, a property that is essential for life.

NONCOVALENT BONDS

While covalent bonds are responsible for holding the atoms of a molecule together, noncovalent bonds occur *between* molecules (Figure 3-7) or between different parts of a large biological molecule. Noncovalent bonds are not dependent on shared electrons but rather on attractive forces between positive and negative charges. Noncovalent bonds play a key role in maintaining the intricate three-dimensional shape of large molecules, such as proteins and deoxyribonucleic acid (DNA), the genetic material found in organisms. Noncovalent bonds are also crucial for holding molecules in larger complexes together (Figure 3-7). Noncovalent bonds include *ionic bonds* and *hydrogen bonds*.

Ionic Bonds: Attractions between Charged Atoms

Some atoms are so electron-attracting that they can capture electrons from other atoms. For example, when the elements sodium (a silver-colored metal) and chlorine (a toxic gas) are mixed, the single electron in the outer shell of each sodium atom migrates to an electron-attracting chlorine atom. As a result, these two materials are transformed into sodium chloride—neither a gas nor a metal, but ordinary table salt.

$$2Na^{\cdot} + :\!\overset{\cdot\cdot}{Cl}\!:\!\overset{\cdot\cdot}{Cl}\!: \rightarrow \ 2Na:\!\overset{\cdot\cdot}{Cl}\!: \ \rightarrow 2Na^+ + 2:\!\overset{\cdot\cdot}{Cl}\!:^-$$

| Sodium metal | Chlorine gas | Transient bond while electron transfer occurs | Sodium ions | Chloride ions |

Since the chloride atom has an extra electron (relative to the number of protons in its nucleus), it has a single negative charge (Cl^-). The sodium atom, which has lost an electron (leaving it with an extra proton relative to the number of electrons) has a single positive charge (Na^+). Such electri-

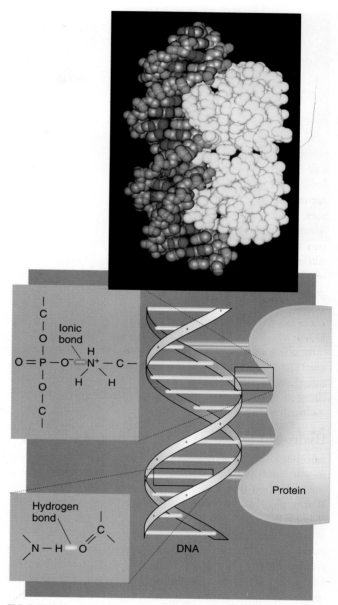

FIGURE 3-7

Noncovalent bonds play an important role in holding two or more molecules together into a complex. In the computer-simulated model depicted here, a molecule of protein (yellow atoms) is bound to a molecule of DNA by noncovalent ionic bonds. The ionic bond forms between a positively charged nitrogen atom and a negatively charged oxygen atom. The DNA molecule itself consists of two separate strands held together by noncovalent hydrogen bonds. Although a single noncovalent bond is relatively weak and easily broken, large numbers of these bonds between two molecules, as between two strands of DNA, make the overall complex quite stable.

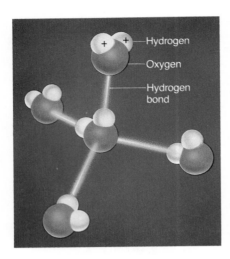

Hydrogen

Oxygen

Hydrogen
bond

FIGURE 3-8

Hydrogen bond formation between adjacent water molecules. Positively charged hydrogens of one polar water molecule align next to negative portions of another water molecule. These weak bonds are really attractions between oppositely charged portions of adjacent molecules. In each of the hydrogen bonds depicted, a hydrogen atom is being shared between two oxygen atoms. The length and strength of a bond are related properties. The hydrogen bond between an H and O atom is longer than the covalent bond between an H and O atom because it is a weaker bond.

cally charged atoms are called **ions.** Oppositely charged ions attract one another, forming a noncovalent linkage called an **ionic bond.** Ionic bonds often form between charged groups that are part of large, complex biological molecules (Figure 3-7).

Hydrogen Bonds: Sharing Hydrogen Atoms

Another kind of chemical bond holds polar molecules close to one another. This linkage—the **hydrogen bond**—is formed when two molecules share an atom of hydrogen. Water clearly illustrates how this attractive force works (Figure 3-8). Adjacent water molecules are drawn together at their oppositely charged regions so that, in effect, the oxygen atoms of two neighboring water molecules share a common hydrogen atom. The sharing is not equal, however, for the covalent bond that links the hydrogen to one of the oxygens is about 20 times stronger than is the noncovalent hydrogen bond that links it to the other oxygen.

Hydrogen bonding is a common characteristic of polar molecules. Hydrogen bonds help to stabilize giant protein molecules, and they also hold the two strands of a DNA molecule together (Figure 3-7). Even though individual hydrogen bonds are weak, when they are present in large numbers, as in a DNA molecule, their forces become additive and, taken as a whole, they provide a structure with considerable stability.

Atoms interact to form complexes that are held together by chemical bonds. These interactions occur because the interacting atoms achieve a greater degree of stability than they would possess independently. This increased stability is achieved when electrons are either shared between atoms or transferred completely from one atom to another. (See CTQ #4.)

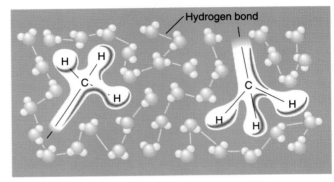

Hydrogen bond

Hydrophobic interactions

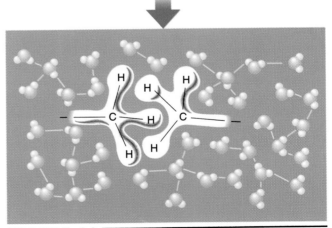

FIGURE 3-9

In a hydrophobic interaction, the nonpolar (hydrophobic) molecules are pushed together as a result of the bonds formed between the polar water molecules.

THE LIFE-SUPPORTING PROPERTIES OF WATER

Water is abundant on earth; we drink it, swim in it, and spray it on our lawns. We tend to take water for granted, even though it constitutes about 70 percent of our body weight. Life on earth is totally dependent on water, and water is probably essential to the existence of life anywhere in the universe. The life-supporting properties of water can be traced to the structure of the H_2O molecule, providing an example of the relationship between structure and function at the atomic level. We have seen how the polarity of the O—H bonds in a water molecule produces positively and negatively charged regions which, in turn, allow water molecules to form hydrogen bonds. The life-supporting properties of water stem largely from its extensive capacity to form hydrogen bonds.

THE EFFECT OF WATER ON NONPOLAR MOLECULES

If you use oil-and-vinegar salad dressings you know that you have to shake and pour the dressing rapidly or you'll end up with a lot more oil on your salad than vinegar. The reason that the oil separates from diluted vinegar, or that fat droplets form on the top of a bowl of chicken soup, has to do with the interactions between the molecules. In a bowl of chicken soup, molecules that have a polar structure, such as the amino acids derived from the meat's protein, can form hydrogen bonds with water molecules. Because of this property, polar molecules are said to be **hydrophilic**, or "water-loving" (*hydro* = water, *philic* = loving). Nonpolar molecules, such as the lipids (Chapter 4) in salad oil or chicken fat, are essentially insoluble in water because they lack charged regions that would be attracted to the poles of water molecules. Consequently, when nonpolar compounds are mixed with water, the mutual attraction between the water molecules forces the nonpolar, **hydrophobic**, or "water-fearing" (*phobia* = fearing) substances into aggregates, such as fat droplets, which minimizes their exposure to their polar surroundings (Figure 3-9). **Hydrophobic interactions** of this type play a key role in holding cell membranes together (see Figure 4-10).

WATER AS A SOLVENT

One of water's most important life-sustaining properties is its effectiveness as a solvent. A **solvent** is a substance in which another material—the **solute**—dissolves by dispersing as individual molecules or ions. The resulting product is a **solution**. Most biologically important substances, particularly polar molecules (such as sugars and amino acids) and ions, such as table salt, are highly soluble in water (Figure 3-10)—more so, in fact, than in any other solvent.

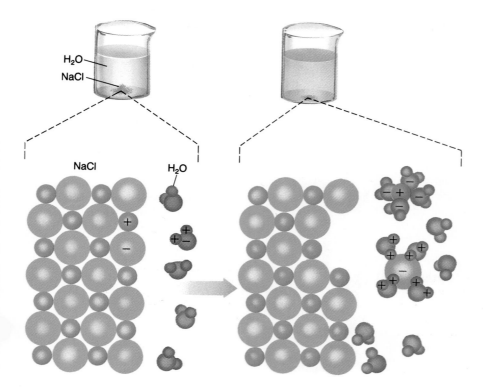

H_2O
NaCl

NaCl H_2O

FIGURE 3-10

The solvent properties of water. Ions in the salt (NaCl) crystal attract the part of the water molecule that bears the opposite charge. The strength of this attraction exceeds the ions' attraction to their oppositely charged neighbors in the crystal, and water wins the molecular "tug of war." The ions go into solution, and the salt crystal (the solute) dissolves in the water (the solvent).

Water is such an efficient solvent that by the time a raindrop completes its fall, it has become a solution containing a number of dissolved gases, mainly nitrogen, oxygen, and carbon dioxide. The molecular oxygen (O_2) that dissolves in lakes, streams, and oceans is sufficient to supply huge communities of fish and other oxygen-dependent underwater dwellers with oxygen (Figure 3-11).

CAPILLARY ACTION AND SURFACE TENSION OF WATER

Water molecules will adhere to any hydrophilic substance. Glass, for example, has a hydrophilic surface that attracts water molecules with enough force to overcome the pull of gravity. You can see the results of this attraction if you dip a narrow glass tube into water. The water seems to "crawl" up the tube, pulled along by its attraction to the hydrophilic surface. As these outer water molecules rise, they pull along adjacent molecules to which they are hydrogen-bonded, so that the water in the center of the tube also rises. This movement of water is called **capillary action.** Capillary action plays an important role in the movement of water through the soil.

Just as water molecules adhere to hydrophilic surfaces, they adhere to one another via hydrogen bonds, a property called *cohesion.* When present at a surface, this cohesion between water molecules creates a film that resists being separated. This property of water, called **surface tension,** allows a water strider seemingly to defy gravity and to walk across the surface of a pond (Figure 3-12). Surface tension can also create problems. Our lungs, for example, consist of microscopic chambers covered by a thin film of water. As we inhale, the surface tension of this film tends to prevent the expansion of these chambers. To overcome this effect, our lungs produce a substance called *surfactant,* which lowers the surface tension of the film, aiding the breathing process.

THE THERMAL PROPERTIES OF WATER

We all know it takes a long time when we're waiting for a pot of water to boil. The reason it takes so long is that water is a very good heat absorber. Heat, or thermal energy, is measured in calories. One **calorie** is defined as the amount of heat needed to raise the temperature of 1 gram of water by 1 degree Celsius. It takes only about one-half of a calorie to raise the temperature of ethyl alcohol by a degree, and even less for most other substances. This energy-absorbing property of water is also a result of its extensive hydrogen bonding. Much of the energy that is absorbed by water as it is heated is utilized in breaking hydrogen bonds and therefore does not contribute to molecular motion (which we measure as temperature). Because of this property, organisms living in large bodies of water are protected from rapid, potentially lethal changes in body temperature despite sudden changes in the temperature of the air.

Water is also an efficient cooling agent. In order to evaporate, molecules of water must absorb a large amount of thermal energy, cooling the liquid that is left behind. This property is exploited by your body when you sweat. As the sweat evaporates, it removes thermal energy from the body surface, causing a drop in body temperature. Sweating is especially effective in dry environments that encourage rapid evaporation. In humid conditions, sweat accumulates rather than evaporating, which accounts for the common complaint, "It's not the heat, it's the humidity."

Ice Formation

As the temperature of water drops, the movement of the molecules slows down, and the molecules crowd closer to one another. In other words, as water gets colder, it becomes denser. As the temperature drops below 4°C (38°F), however, a most remarkable phenomenon occurs: The movement toward increased density reverses itself, and the

FIGURE 3-11

External gills atop this striking nudibranch (sea slug) extract enough dissolved oxygen from water to allow the animal to live permanently in the ocean without having to surface for air.

FIGURE 3-12

Surface tension of water provides enough support to allow a water strider to walk on the liquid's surface as though it were a flexible membrane.

water *expands* in volume. Expansion continues as water is cooled from 4°C to 0°C, at which point it freezes to form ice. The lowered density of ice (when compared to liquid water) explains why ice floats and why a blanket of ice covers the surface of frozen lakes and oceans. This frozen crust provides insulation against additional heat loss, allowing the water beneath the ice to stay warm enough to remain a liquid. This phenomenon plays a key role in maintaining aquatic life forms through cold winter months, preventing them from freezing to death.

Life depends on water. Because of its structure, a water molecule has the ability to form several hydrogen bonds. This enables water to force hydrophobic molecules together, to dissolve many different molecules, and to absorb large amounts of heat—three properties on which life depends. (See CTQ #5.)

ACIDS, BASES, AND BUFFERS

Protons are not only found within atomic nuclei, they are also released into the medium whenever a hydrogen atom loses an electron. Let's consider acetic acid—the major ingredient of vinegar—which can undergo the following reaction:

Acetic acid → Acetate ion + Proton (hydrogen ion)

In this reaction, the hydrogen atom of the acetic-acid molecule has transferred its electron to the adjacent oxygen atom (giving the oxygen a negative charge), while the bare proton has dissociated into the medium as a hydrogen ion (H^+). Any molecule capable of releasing a hydrogen ion is defined as an **acid.** The proton released by the acetic acid molecule in the previous reaction does not remain in the free state; instead it combines with another molecule. Possible reactions involving a proton include:

- Combination with a water molecule to form a hydronium ion (H_3O^+).

$$H^+ + H_2O \rightarrow H_3O^+$$

- Combination with a hydroxyl ion (OH^-) to form a molecule of water.

$$H^+ + OH^- \rightarrow H_2O$$

- Combination with an amino group ($—NH_2$) in a protein to form a charged amine.

$$H^+ + —NH_2 \rightarrow —NH_3^+$$

Any molecule that is capable of accepting a hydrogen ion is defined as a **base.** In the first reaction indicated above, for example, water is acting as a base by accepting a proton. Water is also capable of acting as an acid by donating a proton in the following reaction

$$H_2O \rightarrow H^+ + OH^-$$

Thus, some compounds, including water, can act as either an acid or a base, depending on conditions.

If we compare various acids, we will find that they vary considerably in how readily they give up their protons. The more easily the proton is lost, the stronger the acid. For example, hydrogen chloride, the acid produced in your stomach, is a very strong acid. Most acids produced in organisms, such as acetic acid or citric acid (the acid found in lemons and oranges), are relatively weak.

THE NATURE AND IMPORTANCE OF pH

The acidity of a solution is measured by the concentration of hydrogen ions and is expressed in terms of **pH** (Figure 3-13). The pH scale ranges from 0 to 14. Pure water has a pH of 7.0, which is the pH value of a *neutral* solution. If an acid is added to pure water, the hydrogen ion concentration increases, causing the solution to become acidic, which is measured as a *lower* pH. Conversely, if a base, such as sodium hydroxide, is added to pure water, the hydrogen ion concentration decreases, causing the solution to become *basic* (alkaline), which is measured as a *higher* pH.

Most biological processes are acutely sensitive to pH, as evidenced by the devastating effects that acid rain is having on the forests in many parts of the world. The reason is that changes in hydrogen ion concentration affect the ionic state of biological molecules. For example, as the

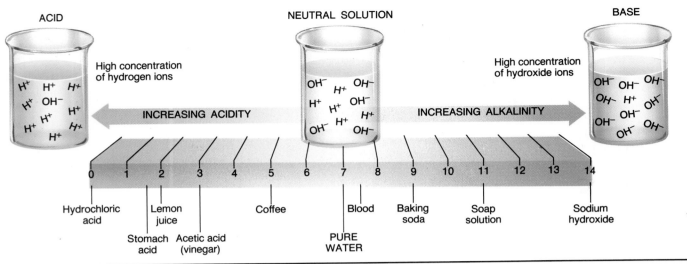

FIGURE 3-13

The pH scale. The acidity or alkalinity of a solution is indicated by a value between 0 and 14. The number is a measure of the concentration of hydrogen ions in a solution; the lower the number, the higher the H⁺ concentration. The concentration of hydrogen ions in solution is always inversely related to the concentration of hydroxyl ions; as one goes up, the other goes down. The concentrations of these two ions are equal at one point on the scale, namely pH 7, which is termed a *neutral* solution. Acidic solutions have pH values below 7, alkaline (basic) solutions are greater than 7. Each unit on the scale represents a tenfold change in the hydrogen ion concentration. For example, lemon juice (pH = 2) is ten times more acidic than is vinegar (pH = 3).

hydrogen ion concentration increases, more of the —NH_2 groups in proteins become —NH_3^+, which can disrupt the shape and activity of the entire protein. Even slight changes in pH can impede the chemical reactions on which life depends. For instance, if not corrected, a drop in the pH of the blood can lead to coma and death. Organisms protect themselves from pH fluctuations with **buffers**—chemicals that couple with free hydrogen and hydroxide ions, thereby resisting changes in pH. The blood, for example, contains a certain concentration of bicarbonate ions. If the hydrogen ion concentration should suddenly rise (as occurs during exercise), the bicarbonate ions combine with the excess protons, removing them from solution.

$$HCO_3^- \ + \ H^+ \ \rightarrow H_2CO_3$$

Bicarbonate ion   Hydrogen ion   Carbonic acid

Without buffers to protect our internal fluids from becoming too acidic or alkaline, we would not survive.

The structure and function of many biologically important molecules is influenced by the concentration of hydrogen ions in the surrounding solution, which determines the solution's acidity. An excess or deficiency of hydrogen ions impairs life processes. (See CTQ #6.)

REEXAMINING THE THEMES

Relationship between Form and Function

The chemical properties of atoms are determined by their structure, most importantly, by the number of electrons present in their outermost shell. The electronic structure of an atom determines whether or not it will bond to

other atoms to form molecules, how many bonds it can form with other atoms, and whether it will gain or lose electrons to form an ion. These chemical properties of atoms form the basis for understanding the behavior of the molecules that make life possible. The life-sustaining properties of water, for example, can be traced to the molecule's tendency to form hydrogen bonds, which is the result of an unequal sharing of electrons in the covalent bonds that hold the hydrogen and oxygen atoms together.

Acquiring and Using Energy

Electrons play a central role in the capture and transfer of energy which is essential for life. For example, energy present in sunlight is captured by electrons during photosynthesis when they are boosted from an inner, lower energy shell to an outer, higher energy shell. Solar energy captured by this mechanism is converted to the energy that fuels the activities of virtually every organism on earth.

Unity within Diversity

Remarkably, all of the diverse materials that exist in the universe are composed of the same three types of subatomic particles—protons, neutrons, and electrons. This enormous diversity results from the variety of ways in which these three subatomic particles are combined to form different kinds of atoms, which in turn combine to form different kinds of molecules. Although there are only three kinds of subatomic particles and about 100 kinds of atoms (elements), there is a virtually infinite diversity of molecules.

SYNOPSIS

Organisms are composed of a small number of elements whose atoms are built from protons, neutrons, and electrons. An atom contains a positively charged central nucleus of protons and neutrons encircled by negatively charged electrons. Distinct forms of the same element that differ from each other only in their number of neutrons are called *isotopes*. Electrons are ordered into electron shells, each of which can hold a limited number of electrons. The number of protons in an atom equals the number of electrons. If an atom gains or loses an electron, it becomes a charged ion.

Covalent bonds hold atoms together to form molecules, whereas noncovalent bonds hold molecules together in larger complexes. Covalent bonds are stable partnerships formed when atoms share their outer-shell electrons, each participant gaining a filled shell. If electrons in a bond are shared unequally by the component atoms, the molecule has a polar character.

Noncovalent bonds are formed by attractions between positive and negative regions of nearby molecules. Noncovalent bonds include ionic bonds (formed between oppositely charged ionic groups) and hydrogen bonds.

Water has unique properties on which life depends. The covalent bonds that make up a water molecule are highly polarized. As a result, water is an excellent solvent, capable of forming hydrogen bonds with virtually all polar molecules. The hydrogen bonding of water molecules to one another endows water with important thermal properties. For example, water absorbs very large amounts of heat per degree of temperature rise, the evaporation of water requires very large amounts of heat, and water becomes less dense as it approaches the freezing point.

When dissolved in water, acids increase the relative proportion of hydrogen ions (which lowers pH), and bases decrease the relative proportion of hydrogen ions (which elevates pH). Neutral solutions have a pH of 7.0, the same as pure water. Some substances resist changes in pH; these buffers help protect cytoplasm and tissue fluids from pH fluctuations.

Key Terms

element (p. 50)
matter (p. 51)
atom (p. 51)
nucleus (p. 51)
proton (p. 51)
neutron (p. 51)
electron (p. 51)
atomic number (p. 52)
atomic mass (p. 52)
isotope (p. 52)

radioactive (p. 52)
shell (p. 52)
orbital (p. 53)
radioisotope (p. 54)
half-life (p. 54)
chemical bond (p. 55)
molecule (p. 55)
compound (p. 55)
covalent bond (p. 55)

free radical (p. 56)
polar molecule (p. 57)
nonpolar molecule (p. 57)
ion (p. 58)
ionic bond (p. 58)
hydrogen bond (p. 58)
hydrophilic (p. 59)
hydrophobic (p. 59)
hydrophobic interactions (p. 59)

solvent (p. 59)
solution (p. 59)
capillary action (p. 60)
surface tension (p. 60)
calorie (p. 60)
acid (p. 61)
base (p. 61)
pH (p. 61)
buffer (p. 62)

Review Questions

1. Explain what is *wrong* (if anything) with the description of each of the following key terms:

   element   all atoms with the same combined number of protons and neutrons

   radioisotope   the most stable form of an element

   shell   always filled to its electron capacity

   ionic bond   chemical partnership in which atoms share electrons

   covalent bonds   may be single, double, triple, or quadruple

   hydrophobic compounds   polar molecules that form hydrogen bonds with each other and dissolve in water

   pH   measurement of acidity or alkalinity; the neutral pH is 0.0

2. There are three beakers on the table. One holds pure water, one holds water to which table salt (NaCl) has been added, and one holds water to which salad oil has been added. In which beaker(s) would you expect hydrogen bonds? Ionic bonds? Hydrophobic interactions?

3. Why do polar molecules, such as table sugar, dissolve so readily in water? Why does sweating help cool your body? Why does ice float on water?

4. If you were to add hydrochloric acid to water, what effect would this have on the hydrogen ion (or H_3O^+) concentration? On the pH? On the ionic charge of any proteins in solution?

Critical Thinking Questions

1. Geiger counters are instruments that detect the presence of radioactive materials. What property of radioactive elements do you suppose a Geiger counter utilizes?

2. Many different substances can be formed by combinations of atoms of the 100 naturally occurring elements. How many different types of molecules could be formed that contain only three atoms if one type of atom can be used more than once? Would all of these possibilities occur in nature? Explain why or why not.

3. Oxygen atoms have eight protons in their nucleus. How many electrons do they have? How many orbitals are in the inner electron shell? How many electrons are in the outer shell? How many more electrons can the outer shell hold before it is filled? Do all oxygen atoms have the same number of neutrons? How many neutrons does a radioactive ^{18}O atom have?

4. What types of bonds (covalent single, covalent double, ionic, or hydrogen) would you expect to find in each of the following: KCl, CH_2O, NH_4OH?

5. In 1913, the biochemist Lawrence Henderson stated that "water is the one fit substance" to support life. What evidence can you present to support Henderson's statement?

6. An enzyme in saliva, which digests starch to sugar, functions at a neutral pH. What would you expect to happen to the function of this enzyme when it enters the stomach, which has a pH of around 2.0? Why?

7. Imagine that you went outside on a hot, dry day and dipped one hand into a bowl of ethyl alcohol and the other hand into a bowl of water and then removed both hands into the air. Which hand would lose the most thermal energy? Why?

Additional Readings

Gardner, R. 1982. *Water: The life sustaining resource.* Julian Messner. New York: (Intermediate)

Pflaum, R. 1989. *Grand obsession: Madame Curie and her world.* New York: Doubleday. (Introductory)

Pais, A. 1991. *Niels Bohr's times in physics, philosophy, and polity.* Oxford: Clarendon. (Introductory)

Snyder, C. 1992. *The extraordinary chemistry of ordinary things.* New York: Wiley. (Introductory)

Yablonsky, H. A. 1975. *Chemistry.* New York: Crowell. (Introductory)

Biochemicals:
The Molecules of Life

**STEPS
TO
DISCOVERY**
Determining the Structure of Proteins

**THE IMPORTANCE OF CARBON IN
BIOLOGICAL MOLECULES**

**GIANT MOLECULES BUILT FROM
SMALLER SUBUNITS**

FOUR BIOCHEMICAL FAMILIES

Carbohydrates

Lipids

Proteins

Nucleic Acids (DNA and RNA)

**MACROMOLECULAR EVOLUTION
AND ADAPTATION**

THE HUMAN PERSPECTIVE
Obesity and the Hungry Fat Cell

S TEPS TO DISCOVER Y
Determining the Structure of Proteins

Of the various types of molecules that make up the fabric of living organisms, proteins have the most complex structure and carry out the most demanding functions. At the end of World War II, in 1945, we still knew very little about the structure of these key molecules; 15 short years later, the scientific world had an accurate, detailed picture of the way these molecular giants were constructed. During this remarkable decade and a half, a handful of investigators, using very different experimental techniques, not only changed our view of proteins but dramatically improved our understanding of life at the molecular level.

Scientists in the mid-1940s knew that proteins were made up of small molecules, called *amino acids*, linked together to form chains called *polypeptides*. What they didn't know was whether or not the sequence of amino acids in a polypeptide chain was of critical importance, and if so, *how* was it important? Frederick Sanger, a young protein chemist at Cambridge University in England, began a study to determine the order of the amino acids that make up the protein insulin. Insulin was a natural choice. Not only was it one of the few proteins available commercially (it was being extracted from the pancreas of domestic pigs and used in the treatment of diabetes), it was also a much easier protein to sequence since it contained fewer amino acids than did most proteins (Sanger eventually found it was composed of only 51 amino acids).

A polypeptide, which consists of a linear chain of specific amino acids, folds to form a protein that has a complex, but precise, shape.

When Sanger began his studies, determining the sequence of amino acids linked together in a chain required a tedious, chemical process. Most importantly, only short chains of about five amino acids could be sequenced at a time. Consider an analogy in which a chain of letters is linked together forming a sequence such as

TCEJBUSETIROVAFYMSYGOLOIB.

How would you determine this sequence if you were able to identify pieces containing only five letters in a row? If you were able to cut the chain of letters at random positions, forming overlapping fragments (such as tir*ov*, *ov*afy), you should be able to figure out the entire sequence.

TCE**JB**
   JBUS*E*
      *SE***TIR**
         TIR*OV*
            OVAFY
_____

TCEJBUSETIROVAFY ← the deduced sequence
(up to this point)

This is precisely the approach taken by Sanger. He chemically cut each purified polypeptide into random fragments, then *sequenced* the fragments and pieced together the linear order of amino acids in the entire chain. From start to finish, the work took Sanger and his colleagues nearly 10 years.

Sanger's sequences showed that the insulin molecule had a precise order of amino acids. Examination of other proteins showed them to have their own stable amino acid sequences, each of which differed from that of insulin. There were no predetermined rules of chemistry that dictated the sequence of amino acids in polypeptides; each sequence was unique and could only be revealed experimentally. Sanger had provided the foundation for the future understanding of protein function based on the chemical properties of their amino acids.

While Sanger was working on amino acid sequences in England, Linus Pauling of the California Institute of Technology was working on the three-dimensional organization of the amino acids in a polypeptide. Pauling utilized molecular models to determine how the amino acids could stably fit together, much like assembling pieces of a jigsaw puzzle. One day in 1948, while he was visting in England, Pauling was suffering from a bad cold and decided to stay in his room. Using a pencil, paper, and straightedge, he drew the atoms and chemical bonds of a polypeptide chain. As he was folding the paper in various positions, he discovered that when the polypeptide chain was arranged in a structure resembling a "spiral staircase," the hydrogen bonds from the amino acids of the chain fit perfectly into place, providing the chain with maximum stability. Pauling called the spiral structure an *alpha-helix* (see Figure 4-15*b*).

Meanwhile, John Kendrew and Max Perutz of Cambridge University were attempting to determine the shape of an entire protein molecule. They relied on a technique called X-ray crystallography, in which crystals of a purified substance are bombarded with X-rays. The X-rays are deflected by the molecules in the crystal and strike a photographic plate, producing a pattern of dots which, when interpreted mathematically, reveals the structure of the molecule that produced the pattern. Pauling had used X-ray crystallography in his early analysis of amino acids, but this was the first time the technique was applied to a giant, complex protein.

The first protein chosen by Kendrew and Perutz for study by X-ray crystallography was *myoglobin*, a protein that stores oxygen in muscle tissue. The picture revealed a compact molecule whose polypeptide chain was folded back on itself in a complex, irregular arrangement. It was discovered that not only do proteins have a unique amino acid sequence (as Sanger had demonstrated in the case of insulin), they also have a unique three-dimensional shape that fits their specific function—in myoglobin's case, oxygen binding. In addition, eight segments of the polypeptide chain were composed of Pauling's alpha-helixes, which together accounted for approximately 70 percent of the amino acids in the protein (see Figure 4-15*c*). Pauling was delighted. He was also delighted to receive the Nobel Prize for Chemistry in 1954. Four years later, Sanger received a Nobel Prize of his own, followed by Perutz and Kendrew in 1962.

*E*ach cell of a living organism is a miniature chemical plant that constantly churns out an array of molecules much more varied and vastly more complex than any chemical plant built by human hands. These molecules range in size from a few atoms to gigantic structures that, if extended, would stretch the length of your finger. Among these molecules are the most complex compounds on earth, perhaps in the universe. All of these biological molecules, from the smallest to the very largest, share a common characteristic—a chemical "skeleton" composed of carbon.

▼ ▼ ▼

THE IMPORTANCE OF CARBON IN BIOLOGICAL MOLECULES

At one time, scientists referred to molecules that could be manufactured only by living organisms as *organic chemicals*, in contrast to *inorganic chemicals*, which are found in the inanimate world. Chemists eventually discovered that all of these so-called organic compounds contained the element carbon and that many of them could be synthesized in the laboratory (Figure 4-1). These developments led to the redefining of **organic** molecules as simply "carbon containing." Those organic molecules that are produced by living cells are called **biochemicals** to distinguish them from the vast array of molecules now synthesized by organic chemists.

⊙ Carbon's chemical properties make it the ideal central element on which life is based. Unlike any other type of atom, carbons have the ability to bond with one another to form long chains. These chains of carbon atoms provide the chemical backbone of biological molecules, a backbone that may be linear, cyclic, or branched:

$$C-C-C-C-C-C$$
Linear

Cyclic

Branched

Recall from the previous chapter that a carbon atom has four unpaired electrons in its outermost shell and thus is capable of forming four single covalent bonds. Consequently, even after forming bonds with other carbons to form a chain, each carbon in the skeletons depicted above is still capable of bonding to other atoms. Consider a molecule of glucose

FIGURE 4-1
The sails of these boats are made of a synthetic, carbon-containing polymer. Thousands of different organic molecules not present in living organisms are now being synthesized by chemists.

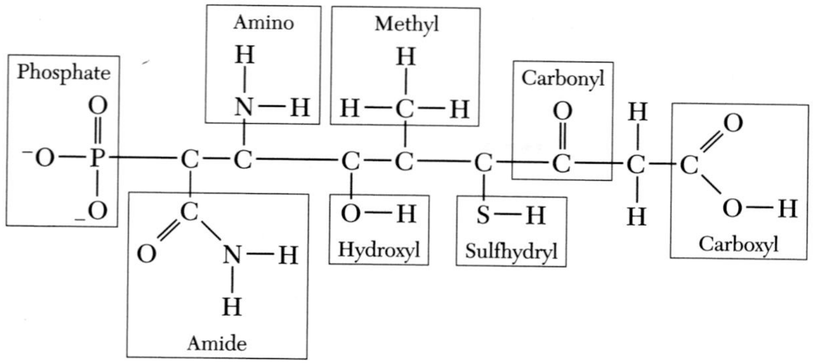

FIGURE 4-2

A gallery of functional groups attached to the carbon skeleton of a hypothetical molecule. Functional groups give organic molecules much of their chemical properties and determine the types of reactions in which the molecule can participate.

In addition to being bonded to carbon and hydrogen atoms, the carbons of a glucose molecule are also bonded to an atom that is part of a functional group. A **functional group** is a particular grouping of atoms which gives an organic molecule its chemical properties and reactivity. Some of the most common functional groups are illustrated in the hypothetical molecule depicted in Figure 4-2. Most functional groups are charged or highly polar, making organic molecules much more soluble in aqueous solutions and more chemically reactive than are those composed of only carbon and hydrogen. Functional groups account for the great diversity of organic molecules. This is illustrated by comparing similar molecules with different functional groups. Ethane (CH_3CH_3), for example, is a toxic, flammable gas. Substitute a hydroxyl group (—OH) for one of the hydrogen atoms, and the molecule (CH_3CH_2OH) becomes the intoxicating liquid, ethyl alcohol. Substitute a carboxyl group (—COOH), and the molecule (CH_3COOH) becomes acetic acid, better known as vinegar. Substitute a sulfhydryl group (—SH) and you have formed a strong, foul-smelling agent, ethyl mercaptan (CH_3CH_2SH), used by biochemists in studying enzyme reactions.

Of the 92 naturally occurring elements, only carbon has the size and bonding properties suited for forming the structural skeleton of biological molecules. The chemical properties of biological molecules are largely determined by the particular groups of atoms (functional groups) bound to the carbon skeleton. (See CTQ #2.)

GIANT MOLECULES BUILT FROM SMALLER SUBUNITS

The chemicals that form the structure and perform the functions of life are highly organized molecules. Compared to small molecules like those just mentioned, many of these organic compounds are enormous, containing hundreds to millions of carbon atoms. These enormous chemicals are called **macromolecules** (*macro* = large), and they possess special properties absent in their smaller relatives. Because of their size and the intricate shapes they can assume, these molecular giants can perform complex tasks with great precision and efficiency. Without them, complex biological activities would not be possible.

Macromolecules are constructed by assembling together small molecular subunits (Figure 4-3a), often in a linear process that resembles coupling railroad cars onto a train. This coupling process is referred to as **condensation.** Each subunit is called a **monomer** (*mono* = one; *mer* = part), and the macromolecule is referred to as a **polymer** (*poly* = many). The reverse process, in which the polymer is disassembled into its individual monomers, is called **hydrolysis** (*hydro* = water, *lysis* = split) because the bond that joins two monomers in a chain is split by the insertion of a water molecule between the two units (Figure 4-3b).

Cells produce huge macromolecules constructed from large numbers of individual subunits. The size and shape of these molecular giants contributes to their ability to perform specific tasks with rigorous precision. (See CTQ #3.)

FOUR BIOCHEMICAL FAMILIES

Macromolecules fall into four fundamental families of organic compounds: *carbohydrates, lipids, proteins,* and *nucleic acids.* The basic structure and function of each family of macromolecule is very similar in all organisms, from bacteria to humans, providing another example of the unity of life. It is not until you look very closely at the specific sequences of monomers that make up these various macromolecules that the diversity among organisms becomes apparent. The basic functions of the various macromolecules are summarized in Table 4-1.

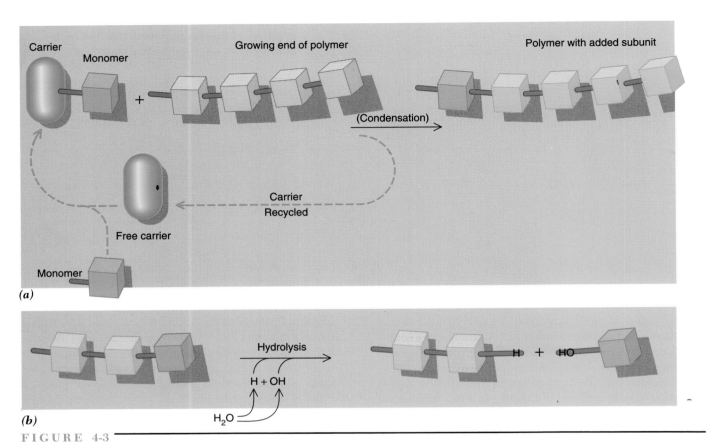

FIGURE 4-3

Monomers and polymers. *(a)* Biological macromolecules consist of monomers (subunits) linked to-gether by covalent bonds. The assembly of macromolecules does not occur simply as a result of reac-tions between free monomers. Rather, each monomer is first activated by attachment to a "carrier" molecule that subsequently transfers the monomer to the end of the growing macromolecule. *(b)* Disassembly of a macromolecule occurs by hydrolysis of the bonds that join the monomers together. Hydrolysis is the splitting of a bond by water.

CARBOHYDRATES

Carbohydrates are a group of substances which includes *simple* sugars and all larger molecules constructed of sugar subunits. Carbohydrates function primarily as storehouses of chemical energy and as durable building materials for biological construction. As illustrated in Figure 4-4, glucose and fructose are six-carbon sugars, while ribose is a five-carbon sugar. Glucose and fructose are the major ingre-dients found in honey, which has been used as a sweetener for thousands of years, long before the cultivation of sugar cane. Of the two sugars, fructose tastes several times sweeter than glucose. You might have noticed that some beverages and other foods contain "high-fructose corn syrup." This may suggest that corn is a rich source of fruc-tose, but this is not the case. In actuality, corn contains glucose, which is commercially converted to fructose by an industrial process that utilizes a common enzyme.

Individual sugars, called *monosaccharides,* can be co-valently linked together to form larger molecules. A mole-cule composed of only two sugar units is a *disaccharide* (*di* = two, *saccharide* = sugar). Lactose, a disaccharide found in milk, provides a nursing baby with most of its energy. Lactose is composed of the sugars glucose and ga-lactose. The best known (and sweetest) disaccharide is su-crose, common table sugar, which is composed of glucose and fructose. Sucrose is a major component of the sap of plants, and it carries energy from one part of the plant to another.

Polysaccharides: Macromolecular Carbohydrates

By the middle of the nineteenth century, it was known that the blood of people suffering from diabetes had a sweet taste due to an elevated level of glucose. Claude Bernard,

TABLE 4-1

MAJOR MACROMOLECULES FOUND IN LIVING SYSTEMS

Macromolecule	Constituents	Some Major Functions	Examples
Polysaccharide (large carbohydrates)	Sugars	Energy storage: physical structure	Starch; glycogen; cellulose
Lipid			
Triglycerides	Fatty acids and glycerol	Energy storage; thermal insulation; shock absorption	Fat; oil
Phospholipids	Fatty acids, glycerol, phosphate, and an R group[a]	Foundation for membranes	Plasma membrane
Waxes	Fatty acids and long-chain alcohols	Waterproofing; protection against infection	Cutin; suberin; ear wax; beeswax
Nucleic acid	Ribonucleotides; deoxyribonucleotides	Inheritance; ultimate director of metabolism	DNA; RNA
Protein	Amino acids	Catalysts for metabolic reactions; hormones; oxygen transport; physical structure	Hormones (oxytocin, insulin); hemoglobin; keratin; collagen; and a class of proteins called enzymes

[a] R group = a variable portion of a molecule.

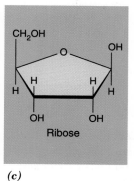

(a)   (b)   (c)

FIGURE 4-4

The structure of simple sugars. The *simple sugars* have the general formula $(CH_2O)n$, where the value of n typically ranges from three to seven. **(a)** Four structural conventions for representing glucose ($n = 6$). You might note that the carbon backbone of glucose was depicted as a straight chain on page 68, but is shown as a ring here. Sugars with five or more carbon atoms undergo a type of "self-reaction" where one end of the sugar molecule bonds with the other end, forming a closed, or ring-containing, molecule. In (1), all atoms of the molecule are shown; in (2), the carbons are omitted since their position is understood; in (3), only the skeleton is indicated, with no atomic detail shown; and (4) depicts the convention used throughout this textbook. Each ball represents a carbon. **(b)** Fructose, like glucose, is a six-carbon sugar, but it self-reacts to form a different type of ring. Note that even though glucose and fructose have the same formula ($C_6H_{12}O_6$), they have a different arrangement of atoms; they are said to be *isomers*. **(c)** Ribose is a five-carbon sugar ($n = 5$). Ribose is a major component of nucleotides and nucleic acids, which will be discussed later in the chapter.

one of the great physiologists of the period, was looking for the cause of diabetes by investigating the source of blood sugar. It was assumed at the time that any sugar present in a human or animal had to have been previously consumed in the diet. Working with dogs, Bernard found that even if the animals were placed on a diet totally lacking carbohydrates, their blood still contained a normal amount of glucose. Clearly, glucose could be formed in the body from other types of compounds.

Upon further investigation, Bernard found that glucose enters the blood from the liver. He also found that liver tissue contains an insoluble polymer of glucose he named *glycogen*. Bernard concluded that various food materials (such as proteins) were carried to the liver where they were chemically converted to glucose and stored as glycogen. Then, as the body needed sugar for fuel, the glycogen in the liver was transformed to glucose, which was released into the bloodstream to satisfy glucose-depleted tissues. This is an example of homeostasis, a concept that was first developed by Bernard and which holds that organisms maintain fairly constant internal conditions (see Chapter 22). In Bernard's hypothesis, the balance between glycogen formation

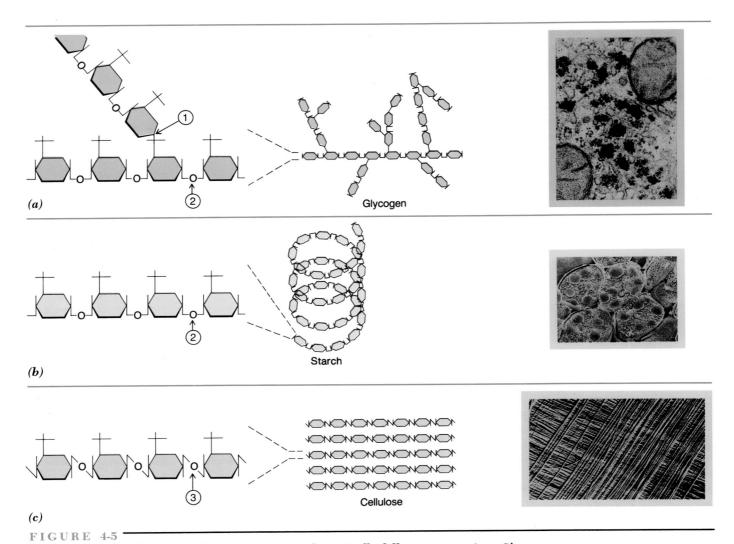

FIGURE 4-5

Three polysaccharides, identical sugar monomers, dramatically different properties. Glycogen *(a)*, starch *(b)*, and cellulose *(c)*, are each composed entirely of glucose subunits, yet their chemical and physical properties (and thus their functions) are very different due to the distinct ways that the monomers are linked together (three different types of linkages are indicated by the circled numbers). Glycogen molecules are the most highly branched; starch molecules assume a helical (spiral) arrangement; and cellulose molecules are unbranched and highly extended and are bundled together to form very tough fibers suited for their structural role. Electron micrographs show glycogen granules in a liver cell, starch grains (amyloplasts) in a plant seed, and cellulose fibers in a plant cell wall.

Strenuous exercise requires energy which is stored as the polysaccharide glycogen.

and glycogen breakdown in the liver was the prime determinant in maintaining the relatively constant (homeostatic) level of glucose in the blood.

Polysaccharides As Energy Stores: Glycogen and Starch Bernard's analysis proved to be correct. The molecule he named *glycogen* is a type of **polysaccharide**—a long chain of sugar units joined together as a polymer. Subsequent studies revealed that many different animals bank their surplus chemical energy in glycogen, a highly branched polysaccharide composed entirely of glucose monomers (Figure 4-5a). In humans, glycogen is stored and used as fuel in a wide variety of tissues, including muscles, but only the liver serves as a glucose supplier for the rest of the body. Muscles typically contain enough glycogen to fuel about 30 minutes of strenuous activity (Figure 4-6).

Another polymer of glucose is *starch*, the polysaccharide most commonly used by plants for energy storage. Potatoes and cereals, for example, consist primarily of starch. Like glycogen, starch is a polymer composed entirely of glucose monomers, but it is much less branched (Figure 4-5b). Even though animals don't produce starch, they can readily digest it by hydrolysis. In fact, starch is the primary source of energy in the human diet in most parts of the world.

Structural Polysaccharides: Cellulose and Chitin
Two of the most important structural polysaccharides are *cellulose* and *chitin*. Cellulose is the earth's most abundant polysaccharide (Figure 4-7), forming the tough fibers present in wood and plant cell walls (Chapter 5). Cotton textiles owe their durability to the long, unbranched cellu-

lose molecules (Figure 4-5c), which are ideally constructed to resist pulling forces. Animals lack an enzyme capable of digesting cellulose, so this vast reserve remains unavailable to them as a direct source of energy. Ironically, cellulose is composed entirely of glucose, the same monomer found in starch, one of the most easily digested polymers. The two polysaccharides differ in the way the glucose monomers are linked together, however (Figure 4-5b,c). Because of this minor structural difference, the bonds of starch are readily hydrolyzed by an enzyme in our digestive tracts, while the cellulose molecules pass through intact, providing fiber that aids in the formation and elimination of feces.

Unlike animals, a variety of microorganisms possess the necessary enzyme for digesting cellulose. If not for these microscopic cellulose decomposers, the world would be permanently littered with dead bodies of plants. A number of animals are able to take advantage of the cellulose in their diet by harboring cellulose-digesting microorganisms within their digestive tract. Cows, for example, rely on bacteria and protozoa living within a special digestive chamber (the rumen) to digest the grass and hay that make up much of their diet. Even termites depend on cellulose-digesting microbes living in their stomachs to digest their meal of microscopic wood particles.

Another important structural polysaccharide—chitin—is a polymer of a nitrogen-containing sugar and is a major component of the outer covering (*exoskeleton*) of spiders, crustaceans, and insects (Figure 4-8). Chitin gives the exoskeleton a tough, resilient quality, not unlike that of certain plastics. Insects and crustaceans owe much of their biologic success to this highly adaptive polysaccharide covering.

FIGURE 4-7

A showcase of structural polysaccharides. The organisms in this picture reveal a few of the natural roles of polysaccharides. The fibrous polysaccharide that supports and shapes the pitcher plant is cellulose. The internal organs and skin of these frogs are secured in place by structural polysaccharides that strengthen connective tissue. Although not shown, insects that fall prey to the carnivorous pitcher plant possess external skeletons consisting of the rigid polysaccharide chitin.

FIGURE 4-8

The body of this grasshopper is covered with a glistening outer skeleton consisting largely of the polysaccharide chitin.

LIPIDS

Lipids are a diverse group of organic molecules whose only common property is their inability to dissolve in water—a property that explains many of their varied biological functions. The hydrophobic character of lipids is revealed by their molecular structure: The carbon atoms of lipids are bonded as much to other carbon atoms as to hydrogen atoms. Lipids include fats, oils, phospholipids, steroids, and waxes.

Fats

Fats consist of three fatty acids coupled to a single molecule of glycerol. As illustrated in Figure 4-9, **fatty acids** are long, water-insoluble chains composed primarily of —CH_2— units. Each fatty acid in a fat molecule is linked to one of the three carbons of the glycerol "backbone." Fats are very rich in chemical energy; a gram of fat contains over twice the energy content of a gram of carbohydrate (for reasons discussed in Chapter 6). Unlike carbohydrate, which functions primarily as a short-term, rapidly available energy source, fat reserves are utilized to store energy on a long-term basis. It is estimated that an average person contains about 0.5 kilograms of glycogen. During the course of a strenuous day's activity, a person can virtually deplete his or her body's entire store of glycogen. In contrast, the average person contains approximately 16 kilograms of fat and, as we all know, it can take a very long time to deplete our store of this material. The subject of fat production is discussed in The Human Perspective: Obesity and the Hungry Fat Cell.

Fats occur in either a solid or a liquid state (liquid fats are termed *oils*). In general, the greater the number of *unsaturated* (double) bonds that exist between the carbons of the fatty acid chains, the less well these long chains can be packed together. This lowers the temperature at which the lipid melts. The profusion of double bonds in vegetable oils accounts for their liquid state—both in the plant cell and on the grocery shelf—and for their label, "polyunsaturated" (many double bonds) fats. In contrast, almost all the linkages between the carbons of animal fats are single bonds; that is, they are *saturated* with hydrogens, causing the material to remain a solid at room temperature. Polyunsaturated fats in the human diet may be less likely to promote cardiovascular disease than saturated animal fats, such as those found in lard and butter, but the matter remains the subject of research.

Phospholipids

Most **phospholipids** are similar in structure to fats except that the glycerol is attached to only two fatty acid chains instead of three. The third glycerol carbon is joined to a phosphate group, which, in turn, is linked to one of a variety of small polar groups (Figure 4-10). The end of the phospholipid containing the phosphate and polar group is soluble in water, while the opposite end containing the fatty acid tails is hydrophobic and "shuns" the aqueous medium. This bipolar nature of phospholipids allows these molecules

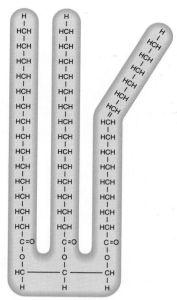

FIGURE 4-9

A fat molecule consists of three long-chain fatty acids joined to a glycerol (lower portion of the fat molecule).

to form bimolecular sheets (*bilayers*) and plays a key role in their function as a major component of cell membranes (see Chapter 5).

Steroids

Steroids are molecules that are built around a characteristic four-ringed skeleton (Figure 4-11). One of the most common steroids is *cholesterol*, a component of animal cell membranes and a precursor for the synthesis of a number of *hormones*—chemical messengers sent from one part of the body to other parts, orchestrating many of the body's processes. For example, sexual maturation is coordinated by steroid hormones: *testosterone* in males and *estrogen* in females. Cholesterol is absent from plant cells, which is why vegetable oils are considered cholesterol-free. This designation is particularly important for people with elevated levels of blood cholesterol because they are at a substantially higher risk of developing *atherosclerosis*, a dangerous "hardening of the arteries," which is a leading cause of heart disease. The presence of high cholesterol levels in the blood promotes the formation of cholesterol-containing buildup (*plaques*) on the inner lining of the arteries. Over periods of many years, these plaques can narrow the diameter of the arteries, greatly reducing blood flow and triggering the formation of blood clots, a major cause of heart attacks (see The Human Perspective, Chapter 7).

Waxes Similar in structure to fats, **waxes** contain many more fatty acids linked to a longer chain backbone. Waxes provide a waterproof covering over the leaves and stems of many plants, preventing the organism from losing precious water. Waxes also serve in construction of the honeycombs of a beehive, as a protective material in the human ear canal, and as a waterproof component spread over the feathers of birds.

◁ THE HUMAN PERSPECTIVE ▷
Obesity and the Hungry Fat Cell

FIGURE 1
Actor Robert DeNiro in (*left*) a scene from the movie *Raging Bull* and (*right*) a recent photograph.

It has become increasingly clear in recent years that people who are exceedingly overweight—that is, obese—are at increased risk of serious health problems, including heart disease and cancer. By most definitions, a person is obese if he or she is about 20 percent above "normal" or desirable body weight. Approximately 35 percent of adults in the United States are considered obese by this definition, twice as many as at the turn of the century. Among young adults, high blood pressure is five times more prevalent and diabetes three times more prevalent in a group of obese people than in a group of people who are at normal weight. Given these statistics, together with the social stigma facing the obese, there would seem to be strong motivation for maintaining a "normal" body weight. Why, then, are so many of us so overweight? And, why is it so hard to lose unwanted pounds and yet so easy to gain them back? The answers go beyond our fondness for high-calorie foods.

Excess body fat is stored in fat cells (*adipocytes*) located largely beneath the skin. These cells can change their volume more than a hundredfold, depending on the amount of fat they contain. As a person gains body fat, his or her fat cells become larger and larger, accounting for the bulging, sagging body shape. If the person becomes sufficiently overweight, and their fat cells approach their maximum fat-carrying capacity, chemical messages are sent through the blood, causing formation of new fat cells that are "hungry" to begin accumulating their own fat. Once a fat cell is formed, it may expand or contract in volume, but it appears to remain in the body for the rest of the person's life.

Although the subject remains controversial, current research findings suggest that body weight is one of the properties subject to physiologic regulation in humans. Apparently, each person has a particular weight that his or her body's regulatory machinery acts to maintain. This particular value—whether 40 kilograms (80 pounds) or 200 kilograms (400 pounds)—is referred to as the person's **set-point.**

People maintain their body weight at a relatively constant value by balancing energy intake (in the form of food calories) with energy expenditure (in the form of calories burned by metabolic activities or excreted). Obese individuals are thought to have a higher set-point than do persons of normal weight. In many cases, the set-point value appears to have a strong genetic component. For instance, studies reveal there is no correlation between the body mass of adoptees and their adoptive parents, but there is a clear relationship between adoptees and their biological parents, with whom they have not lived.

The existence of a body-weight set-point is most evident when the body weight of a person is "forced" to deviate from the regulated value. Individuals of normal body weight who are fed large amounts of high-calorie foods under experimental conditions tend to gain increasing amounts of weight. If these people cease their energy-rich diets, however, they return quite rapidly to their previous levels, at which point further weight loss stops. This is illustrated by actor Robert DeNiro, who reportedly gained about 50 pounds for the filming of the movie "Raging Bull" (Figure 1), and then lost the weight prior to his next acting role. Conversely, a person who is put on a strict, low-calorie diet will begin to lose weight. The drop in body weight soon triggers a decrease in the person's resting metabolic rate; that is, the amount of calories burned when the person is not engaged in physical activity. The drop in metabolic rate is the body's compensatory measure for the decreased food intake. In other words, it is the body's attempt to halt further weight loss. This effect is particularly pronounced among obese people who diet and lose large amounts of weight: Their pulse rate and blood pressure drop markedly, their fat cells shrink to "ghosts" of their former selves, and they tend to be continually hungry. If these obese individuals go back to eating a *normal* diet, they tend to regain the lost weight rapidly. The drive of these formerly obese persons to increase their food intake is probably a response to chemical signals emanating from the fat cells as they shrink below their previous size.

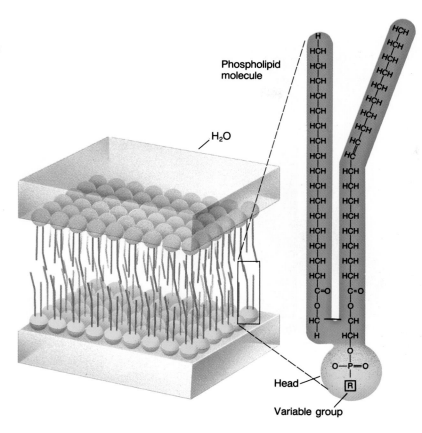

Phospholipid
molecule

H₂O

Head

Variable group

FIGURE 4-10

Phospholipids—molecules that are both soluble and insoluble in water. The polar phosphate "head" is attracted to water, while the nonpolar fatty acid "tails" extend in the opposite direction, away from water. This "molecular schizophrenia" results in molecules that, in cells, align so that they develop double-layered sheets. All life depends on cell membranes formed from these phospholipid sheets.

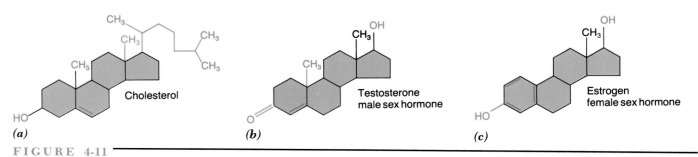

(a) Cholesterol

(b) Testosterone male sex hormone

(c) Estrogen female sex hormone

FIGURE 4-11

The structure of steroids. *(a)* All steroids share the basic four-ring skeleton *(gray brown)*. (Although not shown, a carbon atom occupies the point of each angle.) The orange parts of the molecule are unique to this steroid, cholesterol. In humans, the seemingly minor differences in chemical structure between testosterone *(b)* and estrogen *(c)* generate profound biological differences. Testosterone induces male characteristics, such as a deep voice and facial hair; estrogen stimulates the development of female characteristics, such as breast enlargement.

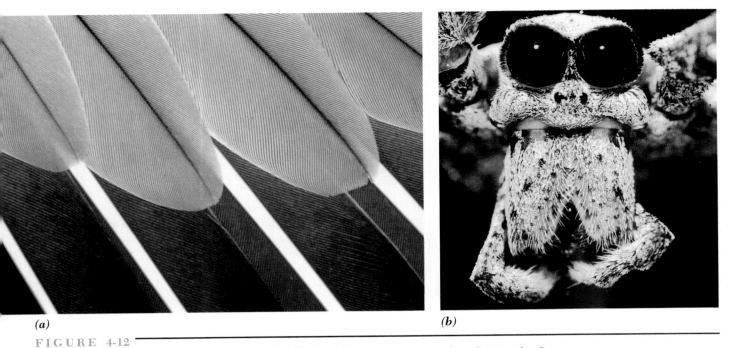

(a)   *(b)*

FIGURE 4-12
This protein gallery shows two of the thousands of biological structures composed predominantly of protein. These include: *(a)* the fabric of feathers used for thermal insulation, flight, and sex recognition among birds and *(b)* the lenses of eyes, as in this net-casting spider.

FIGURE 4-13
Proteins are assembled from amino acids that are joined by peptide bonds. Each peptide bond forms by the linkage of an amino group from one amino acid and a carboxyl group from the neighboring amino acid. A string of amino acids joined by peptide bonds is called a polypeptide chain. The formation of a polypeptide is one of the most complex molecular processes in biology and involves the participation of many different components.

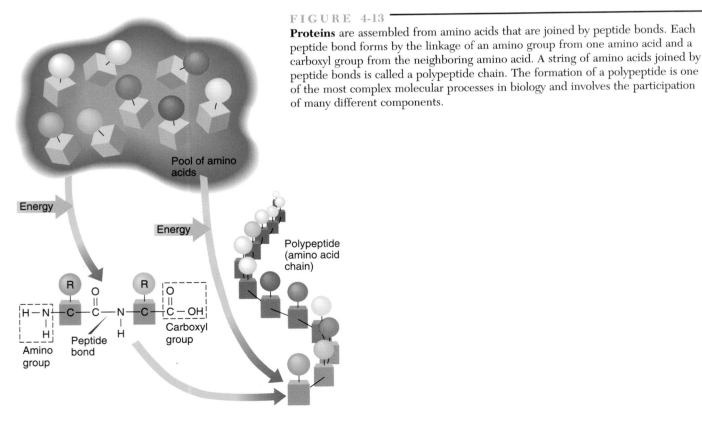

Proteins

If you were to drive off all the water from an organism's body, more than half the dried remains would consist of protein. It is estimated that the typical mammalian cell has at least 10,000 different proteins. **Proteins** are the macromolecules of the cell that "make things happen." Among their many functions, proteins determine much of what gets into and out of each cell, they regulate the expression of genes, and they form the machinery for biological movements, ranging from muscle contraction to the swimming of a sperm up the female reproductive tract. Protein constitutes the major structural component of hair, feathers, fingernails, skin, ligaments, tendons, and thousands of other biological structures (Figure 4-12). Another essential group of proteins is *enzymes,* the mediators of metabolism. Thousands of different enzymes collaborate to direct the development and maintenance of every organism.

The Building Blocks of Proteins

As we described in the beginning of the chapter, proteins are polymers made of amino acid monomers. Each protein has a unique sequence of amino acids which gives that molecule its unique properties. As a group, proteins have so many different functions because different proteins have strikingly different amino acid sequences. Many of the capabilities of a protein can be understood by examining the properties of its constituent amino acids. The same 20 amino acids are found in virtually all proteins, whether in a virus or a human. There are two aspects of amino acid structure to consider: that which is common to all of them, and that which is unique to each.

All free (unlinked) amino acids have a carboxyl group and an amino group (Figure 4-13), separated by a single carbon atom; these are the common parts of the amino acids. The remainder of each amino acid—the **R group** (Figure 4-13)—is variable among the 20 building blocks.

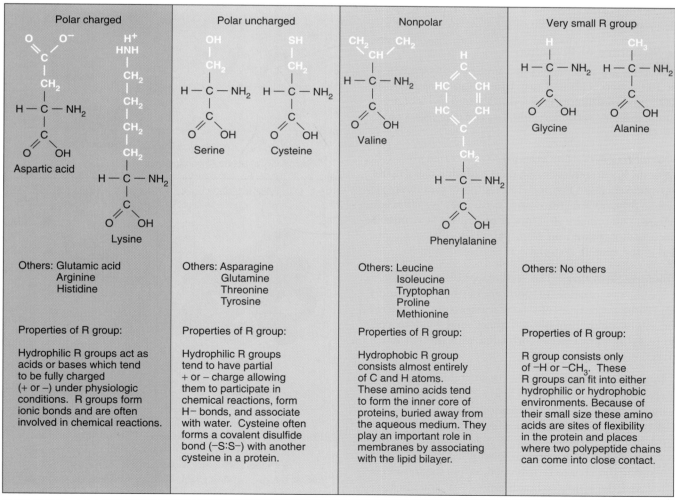

FIGURE 4-14

The structure and properties of amino acids.

Amino acids are conveniently classified on the basis of the polar (hydrophilic) versus nonpolar (hydrophobic) nature (page 59) of their R groups. They fall roughly into four categories: polar charged, polar uncharged, nonpolar, and those with a very small R group (Figure 4-14). The nonpolar amino acids tend to be clustered in the center of a protein, away from the surrounding water, giving that region an almost oil-like character. If all the amino acids are considered together, there is a large variety of reactions in which they can participate and a great many types of bonds they can form. The activities of particular amino acids, situated at particular sites within a protein, account for the function of that protein.

Proteins are synthesized by joining each amino acid to two other amino acids, forming a long, continuous, unbranched polymer called a **polypeptide** chain. The amino acids are joined by **peptide bonds**—linkages formed by joining the carboxyl group of one amino acid to the amino group of its neighbor, with the elimination of a molecule of water. The backbone of a polypeptide chain (illustrated by the gray cubes of Figure 4-13) is composed of the common parts of the string of amino acids, while the R groups (illustrated by the colored spheres of Figure 4-13) project out as "side groups."

The Structure of Proteins

Proteins are large molecules; most proteins contain at least 100 amino acids and may contain up to as many as 20,000. Protein structure is described at four levels of organization: primary, secondary, tertiary, and quaternary (Figure 4-15).

Primary Structure   The *primary structure* of a protein is the specific linear sequence of amino acids that make up its polypeptide chain(s) (Figure 4-15). Different proteins have different primary structures and, consequently, different functions. The primary structure of proteins is determined by information passed on from parents to offspring, encoded in the genetic material.

Secondary Structure   The secondary structure defines the spatial organization of portions of the polypeptide chain. Three major types of secondary structure are recognized: alpha helix, beta-pleated sheet, and random coil (Figure 4-15); each can be correlated with protein function.

1. In an *alpha (α) helix*, the backbone of the polypeptide assumes the form of a spiral held together by hydrogen bonds. *Alpha-keratin,* the major protein of wool, consists largely of alpha-helix, which gives the material extensibility. When a fiber of wool is stretched, the hydrogen bonds are broken, disrupting the helix and allowing the fiber to be extended. When the tension is relieved, the hydrogen bonds can reform, and the fiber snaps back to its original length.

2. In a *beta (β)-pleated sheet,* two or more sections of the polypeptide chain lie side by side, forming an accor-

dionlike sheet that is held together by hydrogen bonds. *Fibroin,* the major protein of silk, is composed largely of pleated sheets stacked on top of one another. Unlike the fibers of wool, those of silk cannot be stretched because the polypeptide chains are already extended.

3. Any portion of a polypeptide chain not organized into a helix or a sheet is said to be in a *random coil.* For example, the sites in the myoglobin (Figure 4-15)

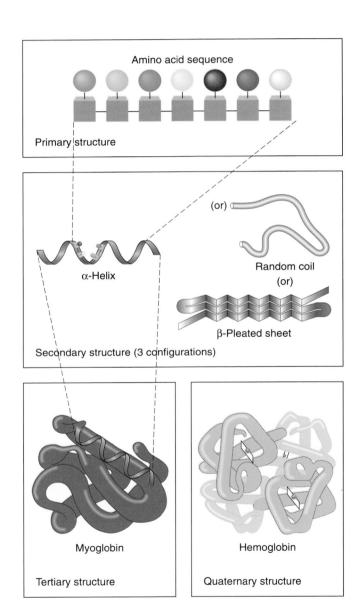

FIGURE 4-15

Four levels of protein structure.   Primary structure describes the specific sequence of amino acids of a polypeptide chain. Secondary structure describes the conformation of a portion of the polypeptide chain. Tertiary structure describes the manner in which an entire polypeptide chain is folded. Proteins consisting of more than one polypeptide chain have quaternary structure, which describes the way the chains are arranged within the protein.

where the polypeptide makes a sharp turn are regions of random coil. The random coils tend to be the most flexible portions of a polypeptide and often represent sites of greatest activity.

Tertiary Structure   The *tertiary structure* describes the shape of an entire polypeptide chain. Like myoglobin, the polypeptide chains of most proteins are folded and twisted to form a globular-shaped molecule with a complex internal organization (Figure 4-15c). Each protein has a precise shape that enables it to carry out a precise function. The polypeptide chain is held in this complex shape largely by noncovalent bonds (ionic bonds, hydrogen bonds) and hydrophobic interactions. Proteins are not rigidly fixed structures but flexible molecules capable of relatively large-scale, internal movements, called **conformational changes.** These changes in shape contribute to their various activities. The movement of your body, for example, results from the additive effect of millions of conformational changes taking place within the proteins of your muscles.

Quaternary Structure   Many proteins are composed of more than one polypeptide chain. The spatial arrangement of the combined chains describes these proteins' *quaternary structure*. Hemoglobin, for example, is an iron-containing blood protein consisting of four polypeptide chains (Figure 4-15d), each of which can bind and transport a single molecule of oxygen.

Determination of Protein Shape

How does the complex, folded form of a protein take shape within a cell? The first important insights into this question were obtained in the late 1950s by Christian Anfinsen and his co-workers at the National Institutes of Health. Anfinsen and his colleagues were working on an enzyme called *ribonuclease.* In these experiments (Figure 4-16), purified preparations of ribonuclease were treated with agents that caused the protein molecules to unfold, losing their secondary and tertiary structure. This type of disruption is known as **denaturation.** It was found that once the denaturing agents were removed from the preparation, the disorganized ribonuclease molecules *spontaneously* reformed themselves into active enzyme molecules. The results of these experiments indicated that the three-dimensional shape of a protein may "spring" directly from the protein's primary structure. In other words, once a given linear sequence of amino acids is strung together, the protein generally spontaneously folds into the proper shape.

The importance of amino acid sequence in determining the form (and function) of a protein is dramatically illustrated by the consequences of changing just one amino acid in hemoglobin. Substituting a hydrophobic, uncharged amino acid (valine) for a hydrophilic, negatively charged amino acid (glutamic acid) distorts the protein and seriously impairs its ability to carry oxygen. Under certain conditions, the distorted hemoglobin elongates red blood cells into a sickle shape (Figure 4-17) that clogs vessels. This potentially fatal disease is called *sickle cell anemia,* and it is a chronic inherited anemia that occurs primarily in people of African descent.

NUCLEIC ACIDS (DNA AND RNA)

One of the properties distinguishing the living from the inanimate is the ability of the living to reproduce offspring that have characteristics similar or identical to those of their parent(s). The instructions required to "build" a new indi-

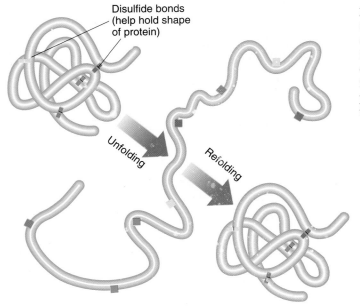

Disulfide bonds (help hold shape of protein)

Unfolding

Refolding

FIGURE 4-16

Self-folding. When the polypeptide chain of a ribonuclease molecule is experimentally unfolded, it is capable of spontaneously reforming its natural three-dimensional shape when the agents that caused denaturation are removed. This experiment provided the first evidence that it is primarily the sequence of amino acids in a polypeptide that dictates the way the polypeptide becomes folded to form the active molecule.

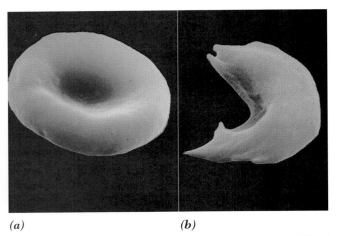

F I G U R E 4-17
The effects of a single amino acid change in a polypeptide chain of hemoglobin. *(a)* A normal human red blood cell. *(b)* A red blood cell from a person with sickle cell anemia. The cell becomes sickle shaped when the level of oxygen in the blood drops. These abnormally shaped cells can clog small blood vessels, causing pain and life-threatening crises.

vidual are transmitted to the offspring encoded in the polymer **deoxyribonucleic acid (DNA)**—the material of which genes are made. DNA is primarily a storehouse of genetic information (Figure 4-18). The hereditary messages stored in the DNA blueprint govern cellular activities through the formation of molecules of **ribonucleic acid (RNA).** Both DNA and RNA are **nucleic acids**—macromolecules constructed as a long chain (strand) of **nucleotide** monomers. For the time being, we will describe the basic structure of nucleic acids, using RNA as the representative molecule. We will describe the more complex three-dimensional structure of DNA, along with the story of its discovery, in Chapter 14.

Each nucleotide in a strand of RNA consists of three parts: (1) a five-carbon sugar called *ribose*, (2) a *phosphate* group, and (3) a *nitrogenous (nitrogen-containing)* base (Figure 4-19*a*). While the sugar and phosphate groups are identical in all nucleotides, there are four different types of nitrogenous bases in RNA: uracil, cytosine, guanine, and adenine. Nucleotides are covalently joined to one another through sugar–phosphate linkages forming a single linear strand of RNA (Figure 4-19*b*).

✪ Nucleic acid structure and function illustrate the importance of biological order at the molecular level. The

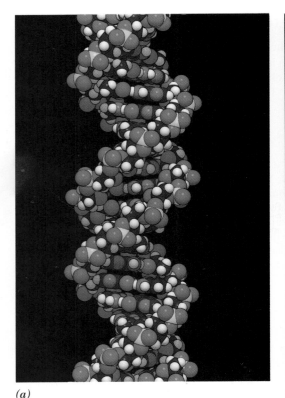

F I G U R E 4-18
DNA: The material of the genes. *(a)* A structural model of a DNA molecule. *(b)* Even complex behaviors, such as the defense of territory and mating rituals of these Japanese cranes, are encoded within the DNA that makes up the animals' genes.

information in a nucleic acid is encoded in the ordered linear sequence of its nucleotides. There is a flow of information in a cell from DNA to RNA to protein. The sequence of nucleotides in the DNA of the genes determines the sequence of nucleotides in a molecule of RNA, which in turn determines the sequence of amino acids in a protein. Consequently, a change in the sequence of nucleotides in the DNA can lead to a change in the sequence of amino acids in the corresponding protein. This is the basis for the single change in amino acid sequence in the sickle cell version of hemoglobin.

Nucleotides are not only important as monomers in nucleic acids, they have important functions in their own right. Most of the energy being put to use at this very instant within your body is derived from the nucleotide **adenosine triphosphate (ATP)**. The structure of ATP and its key role in cellular metabolism is discussed in Chapter 6.

Just as different organisms are adapted to different habitats, and different organs in your body are adapted for different physiologic activities, different types of macromolecules are adapted to perform different chemical functions. Carbohydrates are sugar-containing molecules that provide readily available energy and durable building materials. Lipids are a varied group of energy-rich, water-insoluble molecules; they function in long-term energy storage and as insoluble walls within cellular membranes. Proteins derive their properties from a precise sequence of their amino acid subunits; they carry out most of the diverse, precision-requiring biological activities. Nucleic acids are composed of strands of nucleotide subunits whose precise sequence serves as a storehouse of genetic information. (See CTQ #5.)

MACROMOLECULAR EVOLUTION AND ADAPTATION

■■▶ Recall from Chapter 1 that adaptations are traits that improve the suitability of an organism to its environment. Macromolecules are biochemical adaptations that are subject to natural selection and evolutionary change in the same way as are other types of characteristics. Some macromolecules evolve very slowly. Cellulose, for example, is a structural component of cell walls in virtually all plants from the simplest algae to the most complex trees. Similarly, glycogen is almost universally present as the storage form of glucose in animals. In contrast, proteins show marked changes from organism to organism, both in form and function; most of the proteins in your body are not found in simpler animals, such as sponges and worms. For example, while hemoglobin is the oxygen-carrying protein of all vertebrates (animals with backbones), from fish to mammals, many other animals utilize a totally different protein for this same function. Octopus blood, for example, contains a bright blue, copper-containing protein called *hemocyanin* which is totally unrelated to hemoglobin.

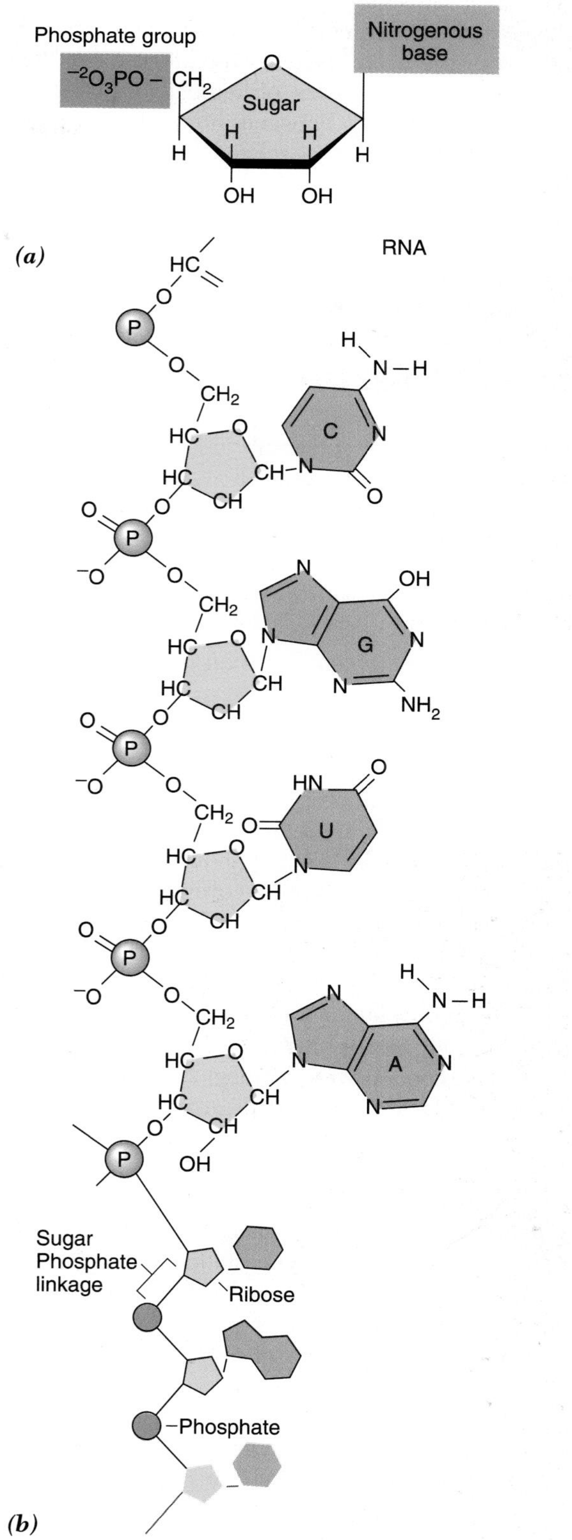

FIGURE 4-19

Nucleotides and nucleic acids. *(a)* A nucleotide consists of a sugar (ribose in RNA), a phosphate group, and a nitrogenous base (either uracil, cytosine, guanine, or adenine in RNA). *(b)* Nucleotides are joined to one another by sugar–phosphate linkages to form long strands of nucleic acids.

When two organisms have the same protein, the difference in amino acid sequence of that protein can be correlated with the evolutionary relatedness of the organisms. The amino acid sequence of hemoglobin, for example, is much more similar between humans and monkeys—organisms that are closely related—than between humans and turtles, who are only distantly related. In fact, the evolutionary tree that emerges when comparing the structure of specific proteins from various animals very closely matches that previously constructed from fossil evidence.

The fact that the amino acid sequences of proteins change as organisms diverge from one another reflects an underlying change in their genetic information. Even though a DNA molecule from a mushroom, a redwood tree, and a cow may appear superficially identical, the sequences of nucleotides that make up the various DNA molecules are very different. These differences reflect evolutionary changes resulting from natural selection (Chapter 34).

> **Virtually all differences among living organisms can be traced to evolutionary changes in the structure of their various macromolecules, originating from changes in the nucleotide sequences of their DNA. (See CTQ #7.)**

REEXAMINING THE THEMES

Relationship between Form and Function

The structure of a macromolecule correlates with a particular function. The unbranched, extended nature of the cellulose molecule endows it with resistance to pulling forces, an important property of plant cell walls. The hydrophobic character of lipids underlies many of their biological roles, explaining, for example, how waxes are able to provide plants with a waterproof covering. Protein function is correlated with protein shape. Just as a key is shaped to open a specific lock, a protein is shaped for a particular molecular interaction. For example, the shape of each polypeptide chain of hemoglobin enables a molecule of oxygen to fit perfectly into its binding site. A single alteration in the amino acid sequence of a hemoglobin chain can drastically reduce the molecule's oxygen-carrying capacity.

Biological Order, Regulation, and Homeostasis

Both blood sugar levels and body weight in humans are controlled by complex homeostatic mechanisms. The level of glucose in your blood is regulated by factors acting on the liver, which stimulate either glycogen breakdown (which increases blood sugar) or glycogen formation (which decreases blood sugar). Your body weight is, at least partly, determined by factors emanating from fat cells which either increase metabolic rate (which tends to decrease body weight) or slow down metabolic rate (which tends to increase body weight).

Acquiring and Utilizing Energy

The chemical energy that fuels biological activities is stored primarily in two types of macromolecules: polysaccharides and fats. Polysaccharides, including starch in plants and glycogen in animals, function primarily in the short-term storage of chemical energy. These polysaccharides can be rapidly broken down to sugars, such as glucose, which are readily metabolized to release energy. Gram-for-gram, fats contain even more energy than polysaccharides and function primarily as a long-term storage of chemical energy.

Unity within Diversity

All organisms, from bacteria to humans, are composed of the same four families of macromolecules, illustrating the unity of life—even at the biochemical level. The precise nature of these macromolecules and the ways they are organized into higher structures differ from organism to organism, thereby building diversity. Plants, for example, polymerize glucose into starch and cellulose, while animals polymerize glucose into glycogen. Similarly, many proteins (such as hemoglobin) are present in a variety of organisms, but the precise amino acid sequence of the protein varies from one species to the next.

Evolution and Adaptation

Evolution becomes very apparent at the molecular level when we compare the structure of macromolecules among diverse organisms. Analysis of the amino acid sequences of proteins and the nucleotide sequences of nucleic acids reveals a gradual change over time in the structure of macromolecules. Organisms that are closely related have proteins and nucleic acids whose sequences are more similar than are those of distantly related organisms. To a large degree, the differences observed among diverse organisms derives from the evolutionary differences in their nucleic acid and protein sequences.

SYNOPSIS

Carbon is the central element in all organic compounds. Carbon's versatile properties include its ability to form four covalent bonds; to link up with other carbons to form straight, branching, or ring-shaped chains; to attach itself to various functional chemical groups; and to form single, double, or triple bonds. Its versatility makes the extraordinarily complex molecules on which life depends possible.

Some biochemicals are small molecules; others are polymers (macromolecules) made up of smaller building blocks. Macromolecules constitute the fabric of an organism and conduct its needed functions. Most macromolecules are polymers constructed by linking together the same class of subunits—monomers—into long chains. Macromolecules are disassembled by hydrolysis.

There are four families of biochemicals. Carbohydrates include simple sugars and larger molecules (polysaccharides) that are formed by linking together many sugars. Some polysaccharides store energy (e.g., glycogen in animals and starch in plants), others are structural polysaccharides (e.g., cellulose and chitin). Glycogen, starch, and cellulose are all polymers of glucose but, because of differ-ences in the way the sugars are linked, these polysaccharides have different properties. Lipids are a diverse array of hydrophobic molecules, including fats, which are composed of fatty acids and glycerol; phospholipids, which contain a phosphate group; steroids, which are constructed of four rings linked together; and waxes, which contain a number of fatty acids. Fats are primarily energy stores, phospholipids are major components of membranes, steroids are found in animal membranes and hormones, and waxes form waterproof coverings. Proteins consist of chains of 20 or so different amino acids, which differ by the structure of their R group. The properties of the R groups of particular amino acids give the protein its particular activity. The sequence of amino acids in a protein determines the shape of the protein, which in turn determines the protein's biological role. Some proteins have a structural role; others, such as hormones, antibodies, hemoglobin, and enzymes, have functional roles. Nucleic acids are informational molecules that consist of strands of nucleotide monomers. Both types of nucleic acids, DNA and RNA, consist of four different types of nucleotides. The precise sequence of nucleotides in a chain determines the information content of the molecule. Changes in the sequence of nucleotides in DNA, which occur during evolution, lead to alterations in the sequence of amino acids in the corresponding protein.

Key Terms

organic (p. 68)
biochemical (p. 68)
functional group (p. 69)
macromolecule (p. 69)
condensation (p. 69)
monomer (p. 69)
polymer (p. 69)
hydrolysis (p. 69)
carbohydrate (p. 70)
polysaccharide (p. 73)

lipid (p. 75)
fat (p. 75)
fatty acid (p. 75)
phospholipid (p. 75)
steroid (p. 75)
wax (p. 75)
set-point (p. 76)
protein (p. 79)
R group (p. 79)

polypeptide chain (p.80)
peptide bond (p. 80)
conformational change (p. 81)
denaturation (p. 81)
deoxyribonucleic acid (DNA) (p. 82)
ribonucleic acid (RNA) (p. 82)
nucleic acid (p. 82)
nucleotide (p. 82)
adenosine triphosphate (ATP) (p. 83)

Review Questions

1. Why are macromolecules described as polymers?

2. Why is a potato more fattening than an equivalent weight of spinach, when both are composed largely of carbohydrates?

3. Discuss how carbon's ability to form four chemical bonds is critical to life.

4. A popular myth states that camels store water in their humps. In actuality, a camel's hump functions in storing fat. From what you know about hydrolysis, would utilizing this stored fat produce more water or use it up?

5. Complete the following table:

Compound	Class of Compound	Monomer(s)	Function
——	disaccharide	——	——
ribonuclease	——	——	——
——	fat	——	——
DNA	——	——	——

Critical Thinking Questions

1. Frederick Sanger worked out the amino acid sequence of beef insulin. How would you predict this sequence would compare to the insulin found in buffaloes? To that found in fish? Why?

2. Silicon is similar to carbon in that it has four electrons in its outer shell; however, it is less suitable for forming biological molecules than is carbon. Research other properties of silicon and discuss why this is so.

3. Discuss the ways in which the formation of macromolecules from smaller molecular subunits is analogous to the manufacture of automobiles on an assembly line. In each case, what are the advantages of constructing large, complex units from smaller subunits?

4. Knowing that diabetes is characterized by a high level of glucose in the blood and that diabetics are treated by injections of insulin, what effect would you expect insulin to have on glycogen levels in the liver?

5. Using their chemical composition, structure, and functions, develop a classification system for the following "species" of organic molecules: monosaccharides, disaccharides, polysaccharides, fats, phospholipids, waxes, steroids, DNA, RNA, and proteins. For example, you might begin by placing the nitrogen-containing molecules (DNA, RNA, proteins) in one group, and those without nitrogen (all of the others) in a second group. You would then subdivide each group until you had classified each of the molecules.

6. Termites harbor colonies of microorganisms in their digestive tracts. Why do you suppose this is adaptive? What would happen if you were to treat termites with drugs that killed all of these microorganisms?

7. Explain why macromolecules, but not small organic molecules, can be used to trace evolutionary changes in organisms.

8. Bacteria are known to change the kinds of fatty acids they produce as the temperature in which they are living changes. What types of changes in the fatty acids would you expect as the temperature drops, and why would this be adaptive?

Additional Readings

Burrows, G. D. , et al., eds. 1988. *Handbook of eating disorders*. New York: Elsevier. (Intermediate)

Field, H. L., ed. 1987. *Eating disorders throughout the life span*. Chicago: Greenwood. (Intermediate)

Fruton, J.S. 1972. *Molecules and life*. New York: Wiley. (Advanced)

Schulz, G. E., and R. H. Schirmer. 1990. *Principles of protein structure*. New York: Springer-Verlag. (Advanced)

Judson, H.F. 1980. *The eighth day of creation*. New York: Simon & Schuster. (Intermediate)

Lehninger, A. L. 1982. *Principles of biochemistry*. New York: Worth. (Intermediate)

Richards, F. M. 1991. The protein folding problem. *Sci. Am.* Jan: 54–63. (Intermediate)

Stryer, L. 1988. *Biochemistry*. 3d ed., New York: Freeman. (Advanced)

Voet, D., and J. G. Voet. 1990. *Biochemistry*. New York: Wiley. (Advanced)

Scientific American October 1985. "The Molecules of Life." (Special Edition) (Intermediate)

CHAPTER

◂ 5 ▸

Cell Structure and Function

STEPS
TO
DISCOVERY
The Nature of the Plasma Membrane

DISCOVERING THE CELL

BASIC CHARACTERISTICS OF CELLS

TWO FUNDAMENTALLY
DIFFERENT TYPES OF CELLS

THE PLASMA MEMBRANE:
SPECIALIZED FOR INTERACTION
WITH THE ENVIRONMENT

THE EUKARYOTIC CELL:
ORGANELLE STRUCTURE AND
FUNCTION

The Nucleus: Genetic Control Center of the Cell

Membranous Organelles of the Cytoplasm:
A Dynamic Interacting Network

Mitochondria and Chloroplasts: Acquiring and
Using Energy in Cells

The Cytoskeleton: Providing Support and
Motility

JUST OUTSIDE THE CELL

Cell Walls

Intercellular Junctions

EVOLUTION OF EUKARYOTIC
CELLS: GRADUAL CHANGE OR AN
EVOLUTIONARY LEAP?

THE HUMAN PERSPECTIVE
Lysosome-Related Diseases

Observing the parts of a cell through a light microscope is like looking for cars and people through the window of an airplane at 35,000 feet. In both situations, the objects are simply too small to be seen. Before the advent of the electron microscope, information on cell structure depended largely on indirect techniques and subsequent interpretations. This was particularly true of the plasma membrane —the delicate structure at the outer edge of all cells; it is so thin that its presence is not revealed by the light microscope.

In the 1890s, a German physiologist, Ernst Overton, thought he could obtain information about the structure of a cell's outer boundary layer by analyzing the types of substances that passed from the outside environment through the "invisible" barrier into the cell. Overton placed living plant cells into solutions containing various types of solutes. He found that the more nonpolar the solute, the better it was able to penetrate the cell boundary and enter the cell. Overton concluded that the dissolving power of a cell's outer layer matched that of a fatty oil. He hypothesized that the cell possessed a "lipid-containing membrane" that separated its living contents from the outer world.

In 1925, two Dutch scientists, E. Gorter and F. Grendel, designed an experiment to answer two questions: (1) Does the plasma membrane contain lipid? (2) If so, how much? Gorter and Grendel extracted the lipid present in human red blood cells and concluded that the amount of lipid in each cell was just enough to form a layer two mole-

By fusing the membranes of a mouse and human cell, scientists discovered the fluid state of the lipid bilayer which allows

cules thick—a **lipid *bilayer*.** This conclusion assumed that all of the lipid of the cell was part of the plasma membrane. This is not an unreasonable assumption to make when it comes to human red blood cells since they are essentially a "bag of dissolved hemoglobin" with virtually no internal structure, not even a nucleus.

The lipid bilayer was soon shown to be composed of phospholipids (page 75). Phospholipids have a "split personality"—a hydrophilic end (the phosphate and polar R group) that can form bonds with water, and a hydrophobic end (the two nonpolar fatty acid tails) that shuns the watery medium. As a result of their structure, phospholipids become aligned into a bilayered sheet, with the hydrophilic ends of the lipid molecules facing out toward the water, and the fatty acid tails facing inward toward each other and away from the water (see Figure 4-10). Gorter and Grendel concluded that cells were surrounded by a lipid bilayer that acts as a barrier to protect the cell's internal contents. A plasma membrane is more than just lipid, however; it also contains protein. In 1935, James Danielli of Princeton University and Hugh Davson of University College in London proposed a model for membrane structure that became the focal point of experimentation for 30 years. In the Davson–Danielli model, as it became known, the protein was present as a layer of globular molecules on both sides of the lipid bilayer, not unlike two slices of bread surrounding a double layer of cheese in a sandwich.

Nearly 40 years later, in 1972, S. Jonathan Singer and Garth Nicolson of the University of California proposed a new model of plasma membrane structure, which they named the *fluid-mosaic model* (see Figure 5-4). According to this model, the phospholipid bilayer of a membrane exists in a liquid (*fluid*) phase, having a viscosity similar to that of light machine oil. In the fluid-mosaic model, proteins are embedded in the membrane, penetrating into or passing completely through the lipid bilayer. What led Singer and Nicolson to view the membrane so differently from the earlier models? The answer relates back to the late 1960s, when a new procedure, termed *freeze-fracturing*, was developed. In this procedure, tissue that had been frozen solid was struck with a knife blade and fractured into two pieces, and the exposed surfaces then examined with an electron microscope. Many of the cracks produced by the knife edge ran directly through the interior of plasma membranes. Examination of the interior of these membranes revealed proteins that were deeply embedded in the lipid bilayer or even extended completely through the membrane (see Figure 5-4). The scattered protein particles seen in the interior of the plasma membrane gave rise to the "mosaic" component of the fluid-mosaic model.

The "fluid" component of the model was based on an earlier experiment employing the technique of *cell fusion*. In these experiments, cells taken from humans and mice were fused to form single cells containing both human and mouse cell nuclei surrounded by a common cytoplasm and a continuous human-mouse plasma membrane (depicted in the illustration that accompanies this essay). Locations of human and mouse membrane proteins were mapped at various times after fusion. At the instant of fusion, the two types of proteins were located in their respective separate portions of the membrane. Within 40 minutes, however, the membrane proteins from the two cells were completely intermixed. This experiment was the first to suggest that membrane proteins were not necessarily fixed in place but were capable of diffusing laterally within the membrane. For this to happen, the lipid bilayer of the membrane must be in a fluid state. Since its initial proposal, the fluid-mosaic model has been confirmed for virtually all cell membranes, regardless of their location in the cell or the type of organism in which they are found, once again emphasizing the unity of life.

The Singer–Nicolson model of membrane structure has ramifications far beyond the structure of the plasma membrane. Just as the discoveries of atomic structure led to an understanding of the way atoms interact (Chapter 3), the Singer–Nicolson model led to an appreciation of the dynamic quality of membranes, which, as we will see, participate in nearly all cellular activities.

proteins to move from one membrane to another.

*I*n these times of burgeoning computer technology, we have grown accustomed to hearing how smaller and smaller microchips are able to store and process greater and greater amounts of information. A 75-volume encyclopedia, for instance, can now be stored on a single chip and "read" by a computer in about 1 second. As remarkable as this may be, there is another microscopic package that can accomplish this feat and even more—the cell. In addition to storing an enormous amount of information—from the contours of a thumb print to the rate of hair growth—the cell is the center of life itself and is responsible for screening, sheltering, organizing, and coordinating a multitude of life-sustaining chemical reactions.

▼ ▼ ▼

DISCOVERING THE CELL

The invention of the microscope around 1600 sparked one of the greatest revolutions in the history of science.[1] By the mid-1600s, a handful of pioneering scientists had uncovered a new world that would have never been revealed had they relied on observations available to the naked eye. The discovery of cells is generally credited to Robert Hooke, an English microscopist who, at age 27, was awarded the position of curator of the Royal Society, England's foremost scientific academy. Three years later, in 1665, Hooke published *Micrographia*, a book in which he described his microscopic observations on such far-ranging subjects as fleas, feathers, and cork. Hooke wondered why stoppers made of cork were so well suited to holding air in a bottle. While viewing thin slices of cork that he cut with his pen knife, Hooke saw rows of tiny compartments resembling a honeycomb (Figure 5-1). He called these compartments "cells" because they reminded him of the cells that housed monks living in a monastery.

Meanwhile, Anton van Leeuwenhoek, a Dutchman who earned a living selling clothes and buttons, was spending his spare time grinding lenses and constructing *simple* (single-lensed) microscopes of remarkable quality. For 50 years, Leeuwenhoek sent letters to the Royal Society of London describing his microscopic observations—along with a rambling discourse on his daily habits and the state of his health. Leeuwenhoek was the first to examine a drop of pond water and, to his amazement, to observe the teeming microscopic "beasties" that darted back and forth before his

FIGURE 5-1

The discovery of the cell. Microscope used by Robert Hooke, with lamp and condensor for illumination of the object. (Inset) Hooke's drawing of a thin slice of cork, showing the honeycomb-like network of "cells."

eyes. He was also the first to describe various forms of bacteria which he obtained from water in which pepper had been soaked and from scrapings of his own teeth. His initial letters to the Royal Society describing this unseen world were met with such skepticism that the Society dispatched its curator, Robert Hooke, to confirm the observations. Hooke did just that, and Leeuwenhoek, an untrained amateur scientist, was soon a worldwide celebrity, receiving visits in Holland from Peter the Great of Russia and the Queen of England.

As microscopes gradually improved, biologists began to see a consistent pattern: Cells were present in all types of plants and animals. In 1838, Matthias Schleiden, a German lawyer-turned-botanist, proposed that all plants were constructed of cells. The following year, Theodor Schwann, a German zoologist, extended this generalization to include animals. Summarizing the new observations gained from studying the microscopic structure of life, Schwann proposed the first two tenets of the **cell theory.**

---

[1] The properties and uses of microscopes are discussed in Appendix B at the back of the book.

1. *All organisms are composed of one or more cells.*
2. *The cell is the basic organizational unit of life.*

    More than 10 years later, after studying cell reproduction, Rudolf Virchow, a German physician, proposed the third tenet of the cell theory

3. *All cells arise from preexisting cells.*

The cell theory is one of the greatest unifying concepts in biology; no matter how diverse organisms appear, all are made up of one or more cells. The cell is the most fundamental structure to harbor and sustain life as well as to give rise to new life. The evolution of the first primitive cells was one of the most crucial—and poorly understood—steps in the entire course of biological evolution on this planet. For all of these reasons, understanding the cell is one of the cornerstones of biology.

Technological invention often triggers new scientific advancements. The invention of the light microscope empowered biologists with the ability to see a multitude of previously unseen life forms and enabled them to investigate the fundamental unit of life—the cell. (See CTQ #2.)

BASIC CHARACTERISTICS OF CELLS

Cells can be removed from a plant or animal and kept alive and healthy outside the organism. Certain human cells, for example, have been kept in culture (growing in laboratory dishes) for decades. These cultured cells have all the necessary regulatory mechanisms to maintain a homeostatic condition. They take up nutrients, digest them, and excrete waste products; they take up oxygen and release carbon dioxide; they maintain a particular water and salt content; they are capable of growth, reproduction, and movement; they respond to external stimulation; they expend energy to carry out their activities; they inherit a genetic program from their parents and pass it on to their offspring; and, finally, they die. These are the characteristics of life, and they are all exhibited by individual cells. Cells come in a variety of shapes and sizes, ranging from enormously extended nerve cells that connect your spinal cord to your big toe (while very long, these cells have a very small diameter), to minute bacteria so small that more than a thousand would be needed to fill the dot in the letter *i*.

Cells are the fundamental units of life, exhibiting all of the basic properties of life. (See CTQ #3.)

TWO FUNDAMENTALLY DIFFERENT CLASSES OF CELLS

⬢ All cells are either prokaryotic or eukaryotic (Figure 5-2), distinguishable by the types of internal structures, or **organelles,** they contain. The existence of two distinct classes of cells, without any known intermediates, repre-

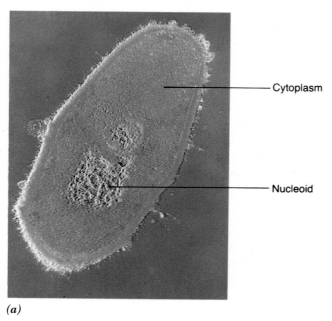

(a)

— Cytoplasm

— Nucleoid

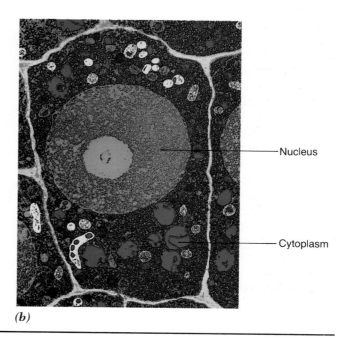

(b)

— Nucleus

— Cytoplasm

FIGURE 5-2

A comparison of prokaryotic *(a)* and eukaryotic *(b)* cells, the two fundamental cell types. *(a)* A bacterial cell showing the cytoplasm devoid of visible organelles and the DNA housed within an ill-defined nucleoid. *(b)* A plant root tip cell showing the varied cytoplasmic organelles and the membrane-bound nucleus, that houses the cell's DNA.

sents one of the most fundamental evolutionary discontinuities that exists in the biological world. Both types of cells —prokaryotic and eukaryotic—are bounded by a similar plasma membrane, and both may be surrounded by a nonliving cell wall. In other respects, the two classes of cells are very different. **Prokaryotic cells** (*pro* = before, *karyotic* = nucleus) are the simpler type; they contain much less genetic information and house it in a *nucleoid*, a rather poorly demarcated region of the cell that lacks a membrane to separate it from the surrounding **cytoplasm.** Prokaryotic cells are essentially devoid of organelles (other than ribosomes) and are capable of much less complex activities than are eukaryotic cells. Prokaryotic cells are found in prokaryotic organisms, all of which are bacteria (members of the kingdom Monera; see Chapter 1). Prokaryotic cells are thought to represent the vestiges of an early stage in the evolution of life which was present before the more complex eukaryotes appeared.

Eukaryotic cells (*eu* = true) are larger and much more structurally and functionally complex than are prokaryotic cells. As the name suggests, the genetic information is housed in a true **nucleus**—one surrounded by a complex membranous envelope. The cytoplasm outside the nucleus is filled with a variety of membranous and nonmembranous organelles that are specialized for various activities. Eukaryotic cells make up eukaryotic organisms, including fungi, protists, plants, and animals.

Nearly all cells—prokaryotic and eukaryotic—are microscopic; that is, they are too small to be seen without a microscope. Prokaryotic cells seldom reach diameters greater than just a few micrometers (μm), while most eukaryotic cells range in diameter from only 10 μm to 100 μm (Figure 5-3). Each micrometer is one-millionth of a meter. In general, any organism larger than 100 μm in diameter is composed of more than one cell. There are a number of reasons for the small size of cells. The larger a cell's volume,

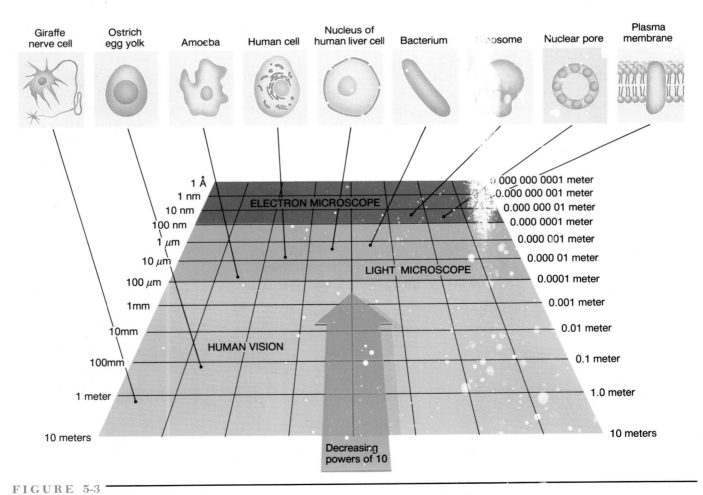

FIGURE 5-3

Relative sizes of cells and cell components. Each unit of measurement is one-tenth as large as the preceding unit. While the huge ostrich egg is technically a single cell, the living portion is present only as a microscopically thin disk located on one edge of the large inert yolk mass.

the more difficult it becomes for that cell to receive the oxygen and nutrients necessary to sustain its metabolic needs. This problem arises because larger cells have *relatively* less plasma membrane surface across which nutrients and oxygen can enter the cell. This problem of surface and volume is dealt with in detail in Chapter 22. Another limitation on cell size stems from the fact that a single nucleus can only support the activities of a limited volume of active cytoplasm. Cells that grow to large sizes, such as those of many protists (see Figure 1-6*b*), typically have more than one nucleus.

Two distinct classes of cells have evolved: a structurally simpler prokaryotic type characteristic of all bacteria, and a more complex eukaryotic type that constitutes all other living organisms. The simpler plan limits function, restricting life to the unicellular level. (See CTQ #4.)

THE PLASMA MEMBRANE: SPECIALIZED FOR INTERACTION WITH THE ENVIRONMENT

All cells are surrounded by an exceedingly thin **plasma membrane** of remarkably similar construction in all organisms—from bacteria to mammals. Among its numerous functions, the plasma membrane (1) forms a protective outer barrier for the living cell and (2) helps maintain the internal environment of the cell by regulating the exchange of substances between the cell and the outside world (Chapter 7). As we discussed earlier, the plasma membrane consists of a fluid, phospholipid bilayer that is pierced by proteins (Figure 5-4). In addition, the plasma membranes of animal cells contain cholesterol, a steroid that generally increases the fluid nature of the lipid bilayer.

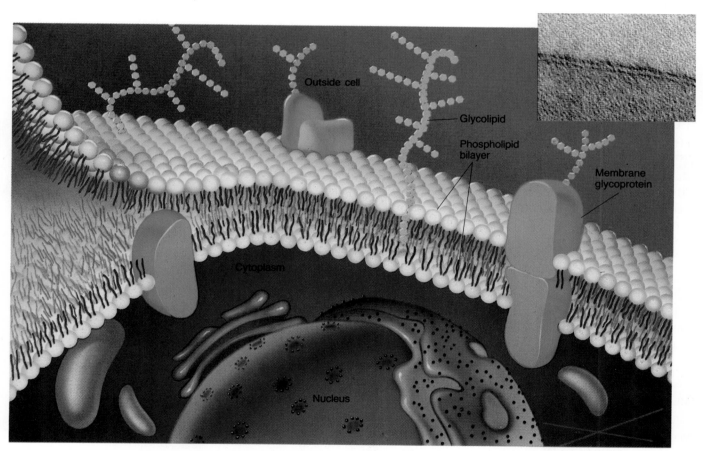

FIGURE 5-4

The structure of the plasma membrane. All cells have a plasma membrane of similar structure. Proteins penetrate into or through the lipid bilayer, with their hydrophobic surfaces forced into proximity with the fatty acid tails of the phospholipids. The hydrophilic portions of the membrane proteins project out into the external environment and inward into the cytoplasm. Chains of sugars are covalently bonded to the outer surface of most, if not all, of the proteins, and to a small percentage of the phospholipids. Membranes appear in the electron micrograph (inset) as three-layered structures having two dense outer layers and a less dense middle layer.

Membrane proteins provide a good example of the relationship between amino acid structure and protein function. The outer surface of membrane proteins consists predominantly of nonpolar amino acids (see Figure 4-14), which form hydrophobic interactions (page 59) with the nonpolar tails of the membrane's phospholipids (Figure 5-4). This facilitates the insertion of these proteins into the hydrophobic interior of the membrane, much like pegs can be inserted into the holes of a pegboard. Those portions of the membrane proteins that protrude beyond the lipid bilayer generally contain attached carbohydrates, making them *glycoproteins.* The carbohydrate groups are present as short, branched hydrophilic chains (Figure 5-4). In contrast to the carbohydrates discussed in Chapter 4 (glycogen, starch, etc.), which are polymers of a single sugar, the carbohydrates of the membrane are composed of a number of different sugars. The sequences of sugars and the branching patterns of the carbohydrate chains vary from one glycoprotein to another. Because of their variable structure, the carbohydrate chains of membrane glycoproteins can engage in *specific* interactions with other molecules. The basis of a person's blood type (A, B, AB, or O), for example, is determined by the particular sugars in the carbohydrate chains present on the glycoproteins at the surface of the person's blood cells.

The plasma membrane is able to recognize and interact with certain substances in the cell's environment. These specific interactions are mediated by those parts of the glycoproteins that protrude from the outer surface of the membrane. Membrane proteins having this role are termed **receptors.** In eukaryotes, each type of specialized cell has its own set of receptors, which allows the cell to respond to particular ions, hormones, antibodies, and other circulating molecules.

> **All cells are bounded by a plasma membrane which separates the living and nonliving worlds. In this capacity, the plasma membrane regulates the exchange of materials between the cell and its environment. (See CTQ #5.)**

TABLE 5-1
THE PRIMARY FUNCTIONS OF CELL COMPONENTS OF EUKARYOTIC CELLS

Component	Primary Functions
Plasma membrane	Boundary of cell, exchange with environment
Nucleus	Storage of hereditary information, control of cell activities
Nuclear envelope	Exchange between nucleus and cytoplasm
Nucleolus	Ribosome synthesis
Nuclear matrix	Support, synthetic machinery
Endoplasmic reticulum (ER)	Synthesis of protein, steroids, lipids; Storage of Ca^{2+}; detoxification
Ribosomes	Sites of protein synthesis
Mitochondria	Chemical energy conversions for cell metabolism
Chloroplasts (plant cells only)	Conversion of light energy into chemical energy, storage of food and pigments
Golgi complex	Synthesis, packaging, and distribution of materials
Lysosomes	Digestion, waste removal, discharge
Central vacuoles (plant cells only)	Storage, excretion
Microfilaments, microtubules, intermediate filaments	Cellular structure, movement of internal cell parts, cell movement
Cilia and flagella	Locomotion, production of currents
Vesicles	Storage and transport of materials
Cell wall (plant cells only)	Protection, fluid pressure, support
Intercellular junctions	Cell-to-cell adhesion, occlusion, communication

THE EUKARYOTIC CELL: ORGANELLE STRUCTURE AND FUNCTION

The unity and diversity of life is revealed at the cellular level, just as it was at the atomic and biochemical levels (Chapters 3 and 4). You have several hundred different types of cells in your body, each recognizably different from the others. Yet virtually all of these different cells are composed of the same types of organelles. The primary functions of all the components found in and around eukaryotic cells are summarized in Table 5-1. Figure 5-5 shows a "generalized" plant and animal cell, each containing a combination of organelles typically found in eukaryotic cells. Keep in mind that there is really no such thing as a "generalized" or "typical" cell. The appearance and distribution of cellular organelles vary greatly from a nerve cell, to a bone cell, to a gland cell. The analogy might be made to a variety of orchestral pieces: All are composed of the same notes, but varying arrangement gives each its unique character and beauty.

Organelles are specialized compartments inside the cell, in which specific activities take place without interference from other events. Viewed in this way, organelles maintain order within a cell, preventing biochemical chaos. A eukaryotic cell could no more operate without its organelles than a restaurant could operate without a kitchen, dining room, restrooms, and garbage bin. The dimensions of some cellular organelles are shown in Figure 5-3.

THE NUCLEUS: GENETIC CONTROL CENTER OF THE CELL

Typically, the most prominent structure in a eukaryotic cell is the nucleus (Figure 5-6). The nucleus is the residence of DNA, the cell's genetic material. Genetic instructions leave the nucleus via a form of RNA and enter the cytoplasm, where they direct the synthesis of specific proteins. Because of its role as genetic headquarters, the nucleus is often thought of as the "control center" of the cell. The nucleus contains chromosomes, nucleoli, and the nucleoplasm, and is surrounded by the nuclear envelope.

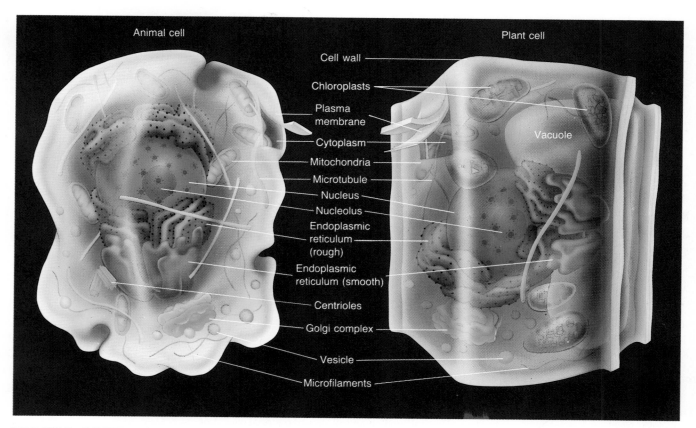

Animal cell

Plant cell

Cell wall
Chloroplasts
Plasma membrane
Cytoplasm
Vacuole
Mitochondria
Microtubule
Nucleus
Nucleolus
Endoplasmic reticulum (rough)
Endoplasmic reticulum (smooth)
Centrioles
Golgi complex
Vesicle
Microfilaments

FIGURE 5-5

Generalized structure of eukaryotic cells. Both plant and animal cells have a plasma membrane, nucleus, and cytoplasmic organelles. They differ in some types of organelles and by the presence of cell walls just outside of plant cells, which animal cells do not have.

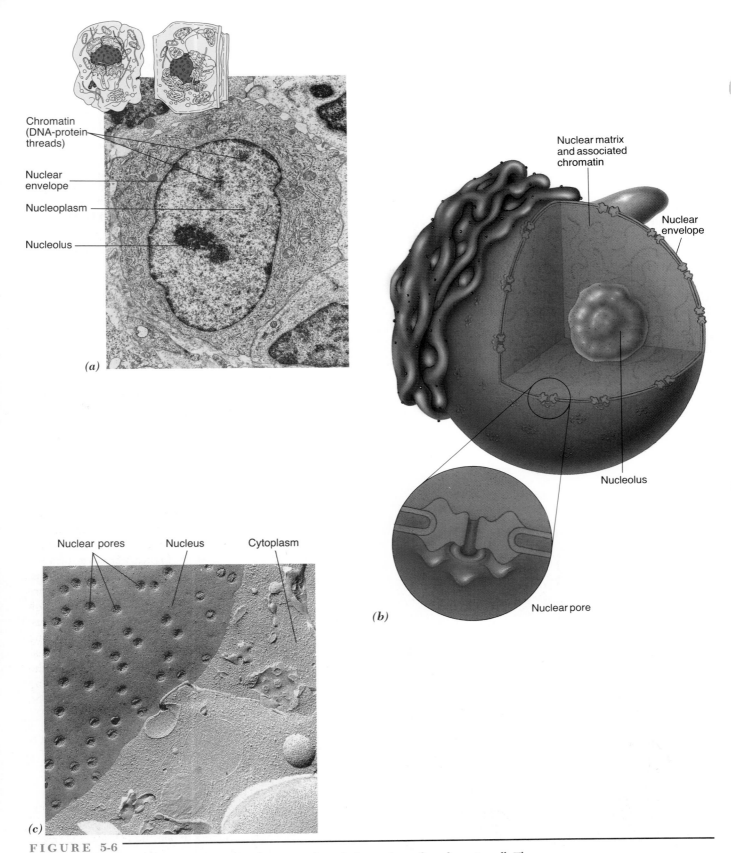

Chromatin
(DNA-protein
threads)

Nuclear
envelope

Nucleoplasm

Nucleolus

(a)

Nuclear matrix
and associated
chromatin

Nuclear
envelope

Nucleolus

Nuclear pore

(b)

Nuclear pores   Nucleus   Cytoplasm

(c)

FIGURE 5-6

The nucleus. *(a)* An electron micrograph of a section through the nucleus of a eukaryotic cell. The nucleolus is visible as a dense internal structure. The scattered clumps of stained material consist of DNA-protein fibers (chromatin) that make up the chromosomes. *(b)* Cut-away diagram showing the nuclear envelope that bounds the nuclear compartment and the nucleolus. *(c)* Using the technique of freeze-fracture (discussed in the chapter opener), the interior of cell membranes can be examined. In this freeze-fractured preparation, the pores in the nuclear envelope are readily visible.

Chromosomes

An organism's genes are located in its **chromosomes,** structures composed of DNA molecules and bound proteins. The number of chromosomes in the nucleus varies according to species; for example, a parasitic roundworm (*Ascaris*) has only four chromosomes, whereas the common sunflower has 34, and humans have 46. During most of the life of a cell, the material that makes up the chromosomes is unraveled to form highly elongated threads, called **chromatin.** It is only when the cell gets ready to divide that the threads become packaged into microscopically visible chromosomes (Figure 5-7).

Nucleoli

The nucleus often contains one or more dense-looking structures called **nucleoli** (Figure 5-6a). If you look closely at an electron micrograph of a single nucleolus, you will see tiny granules. These granules are the precursors of **ribosomes,** particles consisting of RNA and protein that serve as sites of protein synthesis in the cytoplasm. The two subunits that make up each ribosome are formed in the nucleoli and then squeezed through the nuclear pores into the cytoplasm, where they combine to form functioning ribosomes.

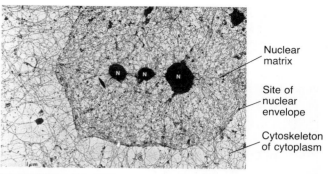

Nuclear matrix

Site of nuclear envelope

Cytoskeleton of cytoplasm

FIGURE 5-8

The nuclear matrix. The interconnecting fibers of the nuclear matrix help maintain the shape of the nucleus and provide sites for the attachment of enzymes involved in nuclear activities. Several nucleoli (N) are also shown.

Nucleoplasm

The **nucleoplasm** is the semifluid substance in the nucleus that contains proteins, granules, and an interconnecting network of fibers, termed the **nuclear matrix** (Figure 5-8). In addition to maintaining the shape of the nucleus, the nuclear matrix is thought to organize the contents of the nucleus and to provide attachment sites for enzymes involved in the duplication and expression of the DNA.

Nuclear Envelope

The **nuclear envelope** is a complex structure containing a double membrane that separates the nuclear contents from the rest of the cell (Figure 5-6). The nuclear envelope is studded with complex pores (Figure 5-6c) that form the passageway for materials entering or exiting the nucleus. For instance, proteins needed for making new DNA are synthesized in the cytoplasm and then transported through the nuclear pores to DNA assembly sites in the nucleus. Conversely, the genetic messages needed for synthesizing those proteins are formed in the nucleus and then transported through nuclear pores to the cytoplasm. Thus, the nuclear envelope is an important gateway for regulating traffic and maintaining essential differences between the two major regions of the cell.

MEMBRANOUS ORGANELLES OF THE CYTOPLASM: A DYNAMIC INTERACTING NETWORK

The cytoplasm of most eukaryotic cells is filled with membranous structures that extend to every nook and cranny of the cell's interior. Included among the cytoplasmic membranes are small spherical containers—or **vesicles**—of varying diameter, long interconnected channels, and flattened membranous sacs. These membranes illustrate the unity and diversity of cellular structure. While the membranes of the cytoplasm have the same basic structure as the plasma membrane, the particular proteins embedded in the

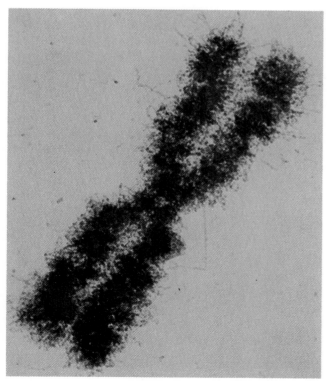

FIGURE 5-7

A chromosome. During cell division in a eukaryotic cell, each chromosome becomes greatly compacted to form the type of structure seen in this electron micrograph. The chromatin fibers that make up the chromosome are visible at its edges.

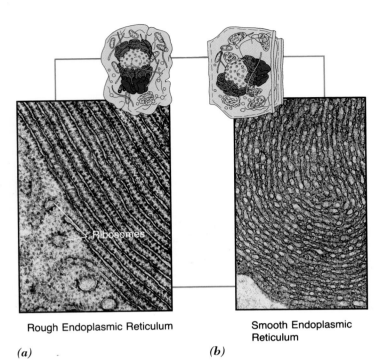

Rough Endoplasmic Reticulum

(a)

Smooth Endoplasmic Reticulum

(b)

FIGURE 5-9

Endoplasmic reticulum. *(a)* Portion of a pancreatic cell showing the rough endoplasmic reticulum (RER) where digestive enzymes are assembled. *(b)* Portion of a cell from the testis showing the smooth endoplasmic reticulum (SER) where steroid hormones are synthesized.

lipid bilayer vary from one part of the cell to another. These membrane proteins give each organelle many of its specialized functions.

Although the membranes of the cytoplasm may appear disconnected in an electron micrograph, in reality they form a highly interdependent membranous network composed of the endoplasmic reticulum, the Golgi complex, lysosomes, and vesicles.

Endoplasmic Reticulum

Throughout the cytoplasm is an elaborate system of folded, stacked, and tubular membranes known as the **endoplasmic reticulum,** or simply the **ER.** The ER occurs in two different forms: rough ER and smooth ER. **Rough ER (RER)** appears bumpy (rough) in electron micrographs because of its many attached ribosomes (Figure 5-9*a*). ER lacking ribosomes is called **smooth ER (SER)** (Figure 5-9*b*). SER is generally more tubular, whereas RER is usually composed of flattened sacs of membranes.

The membranes of the ER divide the cytoplasm into two compartments: one within the confines of the ER membranes, and one outside the ER membranes. As a result, materials confined within the ER space are segregated from the remainder of the cytoplasm. Moreover, the continuous membranes of the endoplasmic reticulum create a system of passageways that allows materials to be channeled to different locations within the cell.

Functions of the RER. The ribosomes located on the outer surface of the RER membranes are sites where pro-

teins are assembled, one amino acid at a time. The newly synthesized protein passes through the ER membrane and is then segregated within the ER compartment. These proteins have specific destinations: Some are exported out of the cell, some end up as membrane proteins, and some become part of other cytoplasmic organelles. Interest is currently focused on how these proteins are sorted and targeted to their specific destinations. The RER is most highly developed in cells that export (*secrete*) large quantities of protein, including the cells of the pancreas (which produce digestive enzymes) and the salivary glands (which produce salivary proteins). The RER is also the primary site of the synthesis of membrane phospholipids.

Functions of the SER. Depending on the particular type of cell, the SER may participate in various functions. The SER may be responsible for the synthesis of steroids; sex hormones of the gonad, for example, are SER products. The SER may also destroy toxic materials in the liver. For example, ingestion of a barbiturate, such as phenobarbitol, is followed by a rapid increase in the SER of a person's liver cells, leading to the destruction of these potentially toxic drug molecules. The SER also regulates calcium ion concentration. In many cells, the controlled release of calcium ions from the SER acts to trigger physiologic responses, including muscle contraction (Chapter 26).

Golgi Complex

In 1898, an Italian biologist, Camillo Golgi, was working with a new type of metallic stain when he discovered a dark

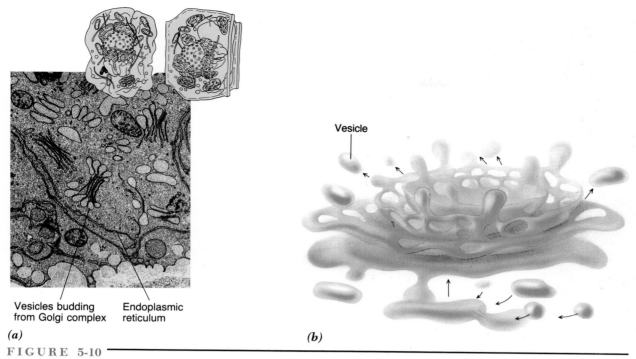

Vesicle

(a)   **(b)**

Vesicles budding   Endoplasmic
from Golgi complex   reticulum

F I G U R E   5 - 1 0

The Golgi complex. **(a)** Electron micrograph of the Golgi complex of a plant cell. Vesicles are seen budding from the lateral edges of the Golgi sacs. **(b)** Diagrammatic view of a Golgi complex. As with most organelles, the number of Golgi complexes varies from cell to cell. For example, dividing plant cells contain many Golgi complexes clustered near the area where a new cell wall is being manufactured. A single cell of the human pancreas, churning out enzymes needed to digest three meals a day, may contain thousands of Golgi complexes.

yellow network near the nuclei of stained nerve cells. This network was later named the **Golgi complex** and helped earn its discoverer the 1906 Nobel Prize. The Golgi complex remained a center of controversy between those who believed that the structure existed in living cells and those who believed it was an *artifact*—an artificial structure formed during preparation for microscopy. The electron microscope finally verified the existence of the Golgi complex as a true organelle—one composed of flattened, membranous sacs and associated vesicles (Figure 5-10). The sacs are arranged in an orderly pile, resembling a stack of "hollow" pancakes. The Golgi complex is a way station in the movement of materials from the ER through the cell (Figure 5-11). The Golgi membranes themselves are thought to form by the fusion of small vesicles that arrive from the RER. The contents of these vesicles, including the proteins destined for export, are chemically modified in the Golgi complex. For example, the insulin that controls the level of sugar in your blood is originally synthesized in the RER as a much larger protein than that ultimately secreted; the larger, precursor protein is cut down to length in the Golgi complex. Similarly, the sugars that determine your blood type are added as the glycoproteins pass through the Golgi complex. The Golgi complex is also the site of

synthesis of many of the complex polysaccharides that make up the plant cell wall, which we will discuss later in the chapter.

Materials processed in the Golgi complex are packaged into vesicles that bud from the lateral edges of the Golgi sacs (Figure 5-10 and 5-11). Some vesicles remain in the cell, storing important chemicals until they are needed; others move to the cell surface and discharge their contents to the outside by a process called **exocytosis.** Exocytosis occurs when the membrane of a vesicle fuses at some point with the overlying plasma membrane, thus opening the vesicle and allowing its contents to be released into the medium.

The Secretory Pathway. We have seen in this discussion that the Golgi complex forms a crucial link in a *secretory pathway* that carries materials from their site of synthesis in the ER to the cell's exterior (Figure 5-11). In addition to carrying enclosed materials, the secretory pathway provides a mechanism for building plasma membrane since the membrane of a cytoplasmic vesicle becomes *incorporated* into the plasma membrane during exocytosis (Figure 5-11). This is, in fact, how the plasma membrane is formed—from vesicles whose membranes were originally produced in the ER.

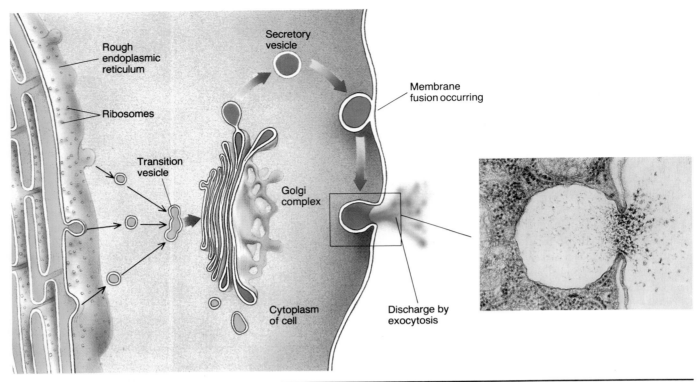

FIGURE 5-11

The secretory pathway. Proteins destined for secretion (such as the digestive enzymes of the pancreas or the proteins of salivary mucus) are synthesized on ribosomes situated on the outer surface of the ER membranes. The proteins are then passed into the ER compartment, packaged in a transition vesicle, and transferred to the Golgi complex, where the proteins are concentrated and often modified before being sent on their way in cytoplasmic vesicles. The vesicles may be stored in the cytoplasm before fusing with the plasma membrane (exocytosis), allowing the vesicle contents to be discharged into the external medium (shown in inset).

The studies that identified the various steps along the secretory pathway were carried out primarily in the 1960s in the laboratory of George Palade of Yale University, one of the corecipients of the 1974 Nobel Prize in Medicine. Palade and his colleagues were working with pancreatic cells that manufacture and export large quantities of digestive enzymes destined for service in the small intestine. They prepared thin slices of pancreatic tissue, placed the slices into a medium containing radioactive amino acids, and then waited for varying amounts of time before the cells were killed and examined. Palade and his co-workers found that the pancreatic cells took up the labeled amino acids, which were rapidly incorporated by the ribosomes of the RER into newly synthesized digestive enzymes. As a result, radioactivity rapidly appeared within the RER (Figure 5-12). If they waited for about 20 minutes before killing the cells, however, much of the radioactivity had moved away from the RER and into the Golgi complex. After an additional 20 minutes or so, radioactivity appeared in material discharged from the cell. Based on these experiments, the

researchers concluded that secretory proteins were synthesized in the RER and passed through the Golgi complex on their way out of the cell.

Lysosomes

Some of the vesicles that bud from the Golgi complex remain in the cytoplasm as **lysosomes**—storage vesicles that contain dozens of dangerously powerful hydrolytic enzymes capable of digesting virtually every type of macromolecule in the cell. The membrane surrounding the lysosomes keeps these lethal enzymes safely sequestered (see The Human Perspective: Lysosome-Related Diseases).

Why do cells house such potentially injurious materials? Lysosomes play a key role in maintaining constancy and order within the cell; they function as digestion chambers for destroying dangerous bacteria and foreign debris, as well as sites for disposing of impaired or worn-out organelles. In some organisms, including protists and sponges, lysosomes digest trapped food particles. In a few cases, an organism's own cells are deliberately digested by

◁ THE HUMAN PERSPECTIVE ▷
Lysosome-Related Diseases

Almost from the time of the discovery of lysosomes it was proposed that their malfunction might be a major cause of various diseases. For example, a miners' disease known as *silicosis* results from the uptake of silica fibers by wandering scavenger cells in the lungs. The fibers become enclosed within lysosomes but cannot be digested; instead, they cause the lysosomal membrane to leak, spilling the contents of digestive enzymes into the cell and damaging the tissue of the lungs. A similar result occurs when asbestos fibers are taken up by scavenging cells, resulting in the disease *asbestosis*. Both conditions can be debilitating and may even be fatal. Certain types of inflammatory diseases, such as rheumatoid arthritis, are believed to result, in part, from the release of lysosomal enzymes from the cell into the extracellular space, causing damage to materials in the joints.

One of the functions of anti-inflammatory drugs, such as steroids, is to stabilize the membrane of lysosomes, thus preventing their rupture.

In 1992, several papers were published that implicated lysosomes in the development of Alzheimer's disease. This type of senility is characterized by the formation of plaques, located outside of nerve cells, which contain a protein called *beta-amyloid*, a small fragment of a larger protein. Researchers have now found that the conversion of the larger protein to the small, apparently harmful, fragment occurs in the lysosomes of the nerve cells. This finding suggests that Alzheimer's disease may occur as the result of some defect in the lysosomal processing of the large precursor protein. While this matter is still highly speculative, it opens up the possibility that treatment of the disease might be

achieved by inhibiting the activity of certain lysosomal enzymes.

Just as there are diseases resulting from excessive lysosomal activity, there are also serious consequences associated with a lack of lysosomal enzymes. These rare genetic disorders are called *lysosomal storage diseases* (of which about 30 are known) and are characterized by the buildup within the tissues of the particular macromolecule that the missing enzyme would normally hydrolyze. Tay-Sachs disease, the most familiar lysosomal storage disease, is characterized by severe mental retardation and death by about age 5. Damage to the brain results from an accumulation of certain plasma membrane lipids in the child's nerve cells. The lethal nature of lysosomal storage diseases illustrates the importance of these hydrolytic enzymes in cell metabolism.

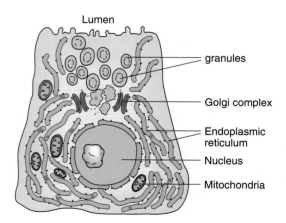
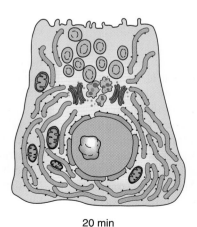
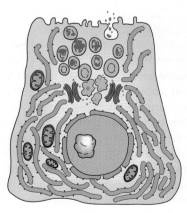

Lumen

granules

Golgi complex

Endoplasmic reticulum

Nucleus

Mitochondria

5 min   20 min   40 min

FIGURE 5-12

Experimental demonstration of the secretory pathway. When a secretory cell is given radioactively labeled amino acids, the amino acids are immediately incorporated into proteins in the rough ER. After about 20 minutes, the radioactively labeled proteins have moved onto the Golgi complex, and within about 40 minutes the label is seen discharged into the external medium. Radioactivity is indicated by the red spots.

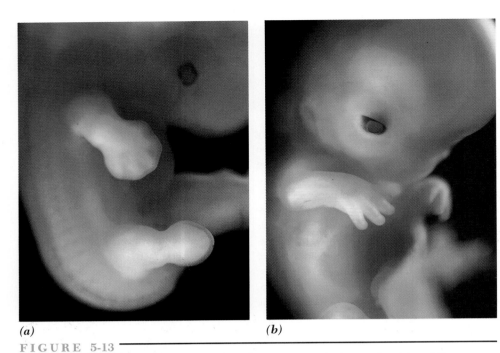

(a)   *(b)*

F I G U R E 5-13

Carving human fingers. *(a)* At an early embryonic stage, the human hand is paddle-shaped. *(b)* The fingers are "carved" out of the paddle by the death of the intervening cells. Cell death results from the release of lysosomal enzymes.

lysosomal enzymes. Your hand, for example, began as a paddle-shaped structure. When you were an early embryo, your fingers were carved out of the "paddle" due to the death of the intervening cells by digestive enzymes released from lysosomes (Figure 5-13). Lysosomes also play a key role in digesting "unwanted" tissues in a tadpole as it undergoes metamorphosis into a frog, or in a caterpillar as it transforms into a butterfly.

Vacuoles

Vacuoles are essentially large vesicles; that is, large, membrane-bound, fluid-containing sacs. Vacuoles are particularly prominent in mature plant cells, where they may occupy more than 90 percent of the cell's volume (see Figure 5-5). The plant vacuole is surrounded by a single membrane that governs which materials are exchanged between the cytoplasm and the fluid inside the vacuole. In addition to containing water, the fluid in the plant vacuole may contain gases (oxygen, nitrogen, and/or carbon dioxide), acids, salts, sugars, crystals, pigments that account for some of the colors of flowers and leaves, and even toxic wastes. Unlike animals, who have elaborate excretory systems for expelling toxic wastes, plants apparently "excrete" such substances into their own vacuoles, safely partitioning the toxins from the plant's cytoplasm. Plant vacuoles also maintain high internal water pressure, which aids in the support of the plant (discussed in Chapter 7).

MITOCHONDRIA AND CHLOROPLASTS: ACQUIRING AND USING ENERGY IN THE CELL

Mitochondria can be likened to miniature "power plants" located within the cytoplasm of eukaryotic cells. Whereas power plants convert the energy stored in energy-rich fuels (such as oil and coal) into electricity—a form usable by the consumer, **mitochondria** convert the energy stored in energy-rich macromolecules (such as fats and polysaccharides) into adenosine triphosphate (ATP)—a form usable by the cell in running virtually all of its immediate activities (Chapter 6).

Mitochondria are typically sausage-shaped (Figure 5-14*a*), although their contours change as they flow through the cytoplasm. Regardless of shape, each mitochondrion is constructed of two membranes—an *outer membrane* surrounding an elaborately folded *inner membrane* (Figure 5-14*b,c*). Just as the plasma membrane is specialized for interacting with external substances, and the RER membrane is specialized for synthesizing and packaging proteins, the inner mitochondrial membrane is specialized for energy transfer. The inner mitochondrial membane illustrates another important feature of membranes: their ability to organize molecules into a stable, ordered array. The inner mitochondrial membrane contains dozens of different components arranged in the precise spatial order required for the formation of a cell's ATP (Chapter 9). The

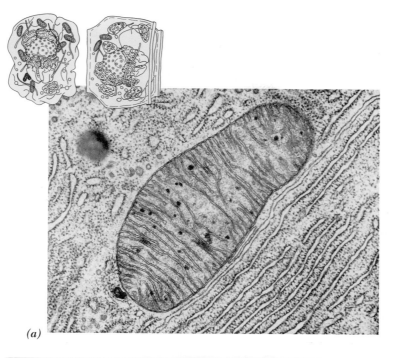

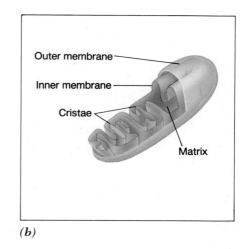

(b)

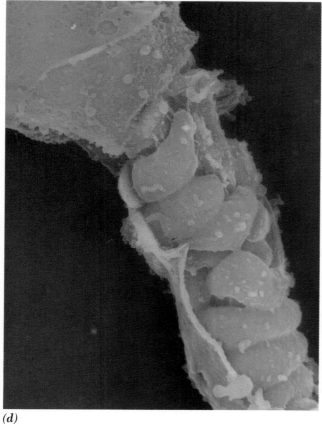

(a)

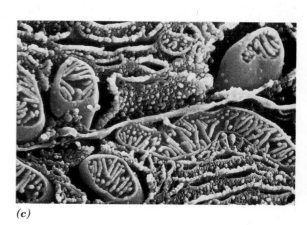

(c)

(d)

FIGURE 5-14

Mitochondria. Electron micrographs *(a,c,d)* and drawing *(b)* of mitochondria showing the outer and inner membranes. The labyrinth of convolutions created by the inner membrane forms the cristae, which project into an internal, semifluid compartment, the matrix. The matrix also contains DNA and ribosomes. The electron micrograph in part *(d)* shows the way in which mitochondria (orange structures) are packaged in the middle portion of a mammalian sperm.

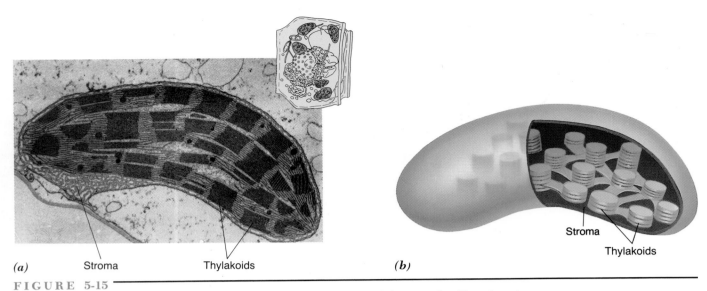

(a)　　　　Stroma　　　　　　Thylakoids　　　　　　*(b)*　　　　　Stroma

Thylakoids

FIGURE 5-15

Chloroplasts. *(a)* An electron micrograph, and *(b)* a three-dimensional diagram of a chloroplast. A chloroplast contains an outer and inner membrane as well as stacks of membranous thylakoids that are contained in the central stroma. The stroma also contains DNA and ribosomes.

labyrinth of convolutions created by the inner membrane forms the **cristae,** which project into the **matrix,** an internal, semifluid compartment (Figure 5-14*b,c*)

As with other organelles, the number and location of mitochondria in a cell depend on the cell's activities. Mitochondria are particularly numerous in muscle cells, which reflects the high energy requirements of muscle contraction. Often, mitochondria associate with fat-containing oil droplets, from which they derive the raw materials for energy production. A particularly striking arrangement of mitochondria is seen in many sperm cells, where the mitochondria are distributed in a spiral in the middle of the cell (Figure 5-14*d*). The movements of the sperm are powered by the ATP produced in these mitochondria. Since the mitochondria in a human sperm remain outside the fertilized egg, all the mitochondria of the offspring are derived from the mother. This is not an inconsequential matter since mitochondria contain a small amount of genetic material, which is thus inherited maternally.

Chloroplasts (Figure 5-15) are found only in plants and certain photosynthetic protists and are sites of photosynthesis, a complex process during which light energy is captured and used to construct complex biochemicals (Chapter 8). Virtually all life on earth depends on the energy captured by photosynthesis. Each chloroplast is bounded by a double-membrane envelope and contains an elaborate internal system of flattened membranous discs called **thylakoids,** in which light-capturing pigments, such as chlorophyll, are contained. The stacks of thylakoids are located in the central **stroma.** It is the green-colored chlorophyll of the chloroplasts that give plants their color.

Both mitochondria and chloroplasts contain small circular DNA molecules and ribosomes, which function as sites for the synthesis of certain proteins of these organelles. The presence of this system for storing and translating genetic information has sparked an interesting hypothesis concerning the origin of these organelles, which we will discuss later in the chapter.

THE CYTOSKELETON: PROVIDING SUPPORT AND MOTILITY

The human skeleton consists of hardened parts of the body which support the soft tissues and play a key role in mediating body movements. The cell also has a "skeletal system" —a **cytoskeleton** (*cyto* = cell), with some analogous functions. The cytoskeleton serves as an internal framework that supports the shape of the cell, organizes its contents, and provides the machinery for moving cells and their organelles. The cytoskeleton includes three components—microtubules, microfilaments, and intermediate filaments—which are typically interconnected to form an elaborate interactive network of fibers (Figures 5-8 and 5-16*a*). The elements of the cytoskeleton are a dynamic group of structures capable of rapid and dramatic reorganization. Unfortunately, this property cannot be appreciated by viewing images in a photograph which are fixed in time.

Microtubules are hollow, cylindrical structures whose wall is composed of subunits made of the protein *tubulin* (Figure 5-16*b*). **Microfilaments** (Figure 5-16*c*) are solid, thinner and more flexible than are microtubules and are polymers of the contractile protein *actin*. **Intermediate**

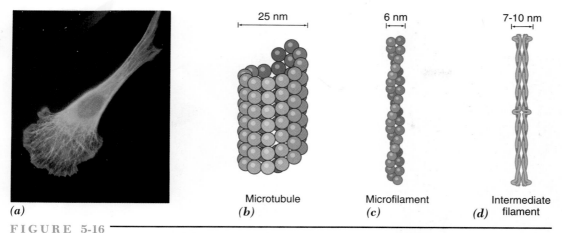

25 nm

6 nm

7-10 nm

Microtubule
(b)

Microfilament
(c)

Intermediate
filament
(d)

(a)

FIGURE 5-16

The cytoskeleton. *(a)* This fibroblast cell was caught in the act of moving over the surface of a culture dish. The cell has been stained to reveal the distribution of microfilaments and microtubules. The rounded edge of the cell is leading the way; the clusters of microfilaments at the leading surface are sites where the forces required for movement are generated. *(b-d)* Microtubules, microfilaments, and intermediate filaments are composed of different types of protein subunits, that become arranged in characteristic patterns.

filaments (Figure 5-16*d*) are tough, ropelike fibers composed of a variety of proteins. One type of intermediate filament, which is present only in nerve tissue, accumulates in deteriorating regions of the brain of Alzheimer's disease patients.

The structures that make up the cytoskeleton function in two interrelated activities:

1. as a scaffold, providing structural support, maintaining the shape of the cell, and organizing the internal contents of the cytoplasm;

2. as part of the machinery required for the movement of materials within the cell or for the movement of the cell itself.

An example of the first activity, structural support, is seen in the long "tentacles" that project from the "body" of certain unicellular organisms (Figure 5-17*a*). These fragile processes are supported in their extended position by an internal skeleton of microtubules (Figure 5-17*b*). We will now take a closer look at the role of cytoskeletal elements in cellular movements.

Microtubules

(b)

FIGURE 5-17

Microtubules as structural supporting rods. *(a)* This protist has tentacles projecting from its cell body. *(b)* The tentacles are supported by an elaborate array of microtubules.

(a)

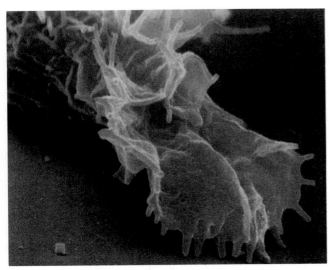

FIGURE 5-18
The leading edge of a moving cell may contain ruffles, which are formed by the contractile activity of microfilaments located just beneath the plasma membrane.

The Movement of Cells from One Place to Another

At one time, when you were an early embryo, cells that went on to form the pigmentation in the skin of your arm and the nerve cells in your brain were very close neighbors. These two types of cells end up in very different locations in the adult because pigment cells migrate away from the developing brain, across the entire width of the embryo. Even as you read this book, cells are leaving your bloodstream by squeezing through the walls of blood capillaries and making their way through the surrounding tissue. We know relatively little about how cells are able to carry out these remarkable movements. This is particularly unfortunate because the movement of single cancer cells away from the site of a tumor is one of the primary causes of the spread of malignant diseases.

One way to learn about the function of a cell structure is to destroy it selectively and study the consequences. If migrating cells are treated with drugs that cause microfilaments to fall apart into their subunits, cell locomotion halts. Studies of this type have drawn attention to microfilaments as important structures in cell locomotion. The microfilaments that drive cell movement tend to be concentrated in a thin layer just beneath the plasma membrane. In fact, the plasma membrane at the leading edge of a moving cell undergoes a dramatic ruffling activity (Figure 5-18) due to the contractile activity of the underlying microfilaments.

Cilia and Flagella. Microscopic organisms typically dart through their aqueous environment, powered by tiny locomotory organelles called cilia and flagella, which are really two versions of the same structure. **Cilia** are short, hairlike organelles that project from the surface of many small eu-

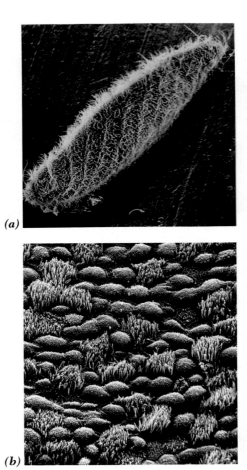

(a)

(b)

FIGURE 5-19
Cilia. *(a)* The numerous cilia that cover the external surface of this protist, *Paramecium*, propel the cell through the water while creating currents that channel food into its "mouth." *(b)* The coordinated beating of the cilia lining the mammalian trachea moves mucus and its trapped dust particles out of the respiratory tract into the throat.

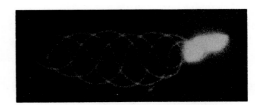

FIGURE 5-20
The undulating beat of the flagellum of a sea urchin sperm. Although this cell has only a single flagellum, the multiple-exposure photograph shows it in several different positions.

karyotic organisms; they act like oars to propel the organism through the water (Figure 5-19*a*). In larger organisms, such as mammals, cilia often line the surfaces of various tracts (Figure 5-19*b*), where they help propel materials through the channel. Generally, cilia occur in large numbers and are densely packed.

Flagella are longer than cilia and are present in fewer numbers. Flagella beat with an undulating motion that pushes or pulls the cell through the medium. In addition to powering microscopic organisms, including flagellated protozoa and algae, flagella provide the propulsive force for the movement of sperm in most animals (Figure 5-20).

◫▶ Cilia and flagella are powered by microtubules. The microtubules are arranged in a specific pattern called a *9 + 2 array*—nine pairs of microtubules encircling two central microtubules (Figure 5-21). This same 9 + 2 array is found in the cilia and flagella of virtually every eukaryotic organism, from fungi to humans, another of many reminders of the unity of life. We must presume that the 9 + 2 array is critical to ciliary and flagellar function, otherwise it would surely have undergone some evolutionary modification over the 2 billion years that eukaryotic cells have existed on earth.

Cilia and flagella move in a whiplike motion that requires that these fibers bend along their length. Bending results from the sliding of microtubules past one another (Figure 5-21). This sliding is actively accomplished by the movement of small arms that connect each pair of microtubules with its neighbor. The importance of the movable arms became apparent when it was discovered that people suffering from certain types of sterility produced sperm devoid of these arms. Further examination indicated that these people also suffered from respiratory ailments as a result of the absence of arms on their respiratory tract cilia.

Eukaryotic cells contain a variety of organelles that are specialized for particular activities. Many of these organelles are bound by cellular membranes; other organelles are nonmembranous. Eukaryotic cell organelles separate complex activities that could interfere, compete, or somehow disrupt the orderly sequence of reactions essential to each biological task. Compartmentalized organelles also enable cells to run many complex processes simultaneously. (See CTQ #6.)

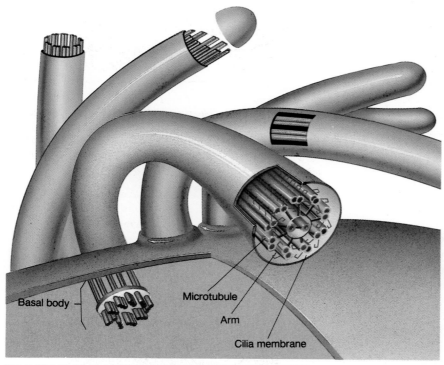
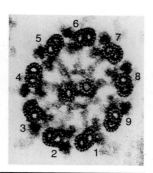

Basal body
Microtubule
Arm
Cilia membrane

FIGURE 5-21

Cilia structure and movement. The 9 + 2 array of microtubules is shown in the cutaway sketch and accompanying electron micrograph. The bending of a cilium results from the sliding of one pair of microtubules over its neighboring pair. Those microtubules at one edge of the cilium (such as numbers 5 and 6 in the inset) move outward, while microtubules at the opposite edge (such as numbers 9 and 1), move inward. Sliding is powered by the movement of small arms that connect adjacent microtubules. A cilium moves when the arms projecting from one pair of microtubules "walks" along the neighboring pair of microtubules. The base of each cilium (or flagellum) contains a specialized structure called a *basal body*.

JUST OUTSIDE THE CELL

Even though the living portion of a cell ends at the plasma membrane, most cells produce extracellular materials which, as the name suggests, remain just outside the cell. Most animal cells secrete proteins and polysaccharides that form a thin **extracellular matrix** around their outer surface, which mediates interactions between different cells, and between cells and their nonliving environment. A few cell types, such as those found in bone and cartilage, are surrounded by an extensive extracellular matrix that gives these tissues their hardness and durability. Collagen—the most abundant protein in the human body—is a major constituent of the extracellular matrix.

CELL WALLS

In the beginning of this chapter we described how Robert Hooke was the first person to have observed a cell. In actual fact, the compartments he described were the empty cell walls of dead cork tissue which had originally been pro-

duced by the living cells they once surrounded. Cell walls, of one type or another, are present in bacteria, fungi, and plants but are absent in animals. These walls form a rigid outer casing that provides support, slows dehydration, and prevents a cell from bursting when internal pressure builds due to an influx of water. We will focus our discussion on the cell walls of plant cells (Figure 5-22).

You might not think that plant cell walls and reinforced concrete have much in common, but they are both built of materials having a similar basic function. Concrete is an amorphous (nonstructured) material designed to resist crushing (compression forces). The steel rods used to reinforce the concrete are designed to help the concrete resist being pulled apart (tension forces). Plant cell walls have a similar structure. The steel rods of reinforced concrete are analogous to the cellulose molecules of a plant cell, which become bundled together to form thicker cables, called *microfibrils* (Figure 5-22). Microfibrils enable the cell wall to resist forces that might pull it apart. The microfibrils are embedded in an amorphous polysaccharide matrix that is formed in the Golgi complex and secreted outside the cell.

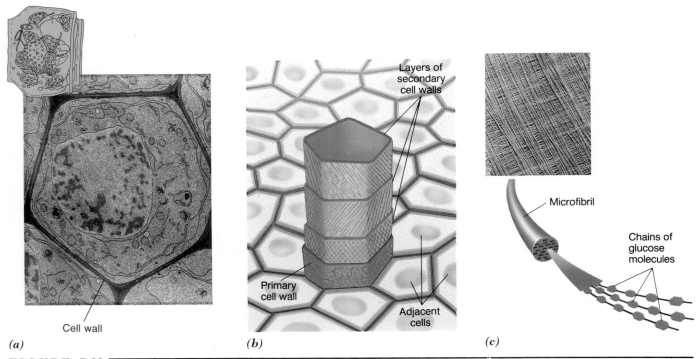

(a)   *(b)*   *(c)*

FIGURE 5-22

Plant cell walls. *(a)* Electron micrograph of a plant cell surrounded by its cell wall. *(b)* A diagrammatic plant cell wall, telescoped to reveal the primary cell wall, layers of the secondary cell wall, and the arrangement of cellulose microfibrils. *(c)* A surface view of the layers of parallel microfibrils in a secondary cell wall. Each microfibril is a complex of cellulose chains, held together by hydrogen bonds between adjacent glucose molecules to form flat, sheetlike strips. Cellulose molecules are assembled from glucose monomers by proteins embedded in the plasma membrane. Plant cell walls provide support and resist external forces.

The matrix of the cell wall holds the fibrils together; at the same time, it resists forces that might crush the enclosed cell. The various polysaccharides of the wall form a porous matrix that allows passage of dissolved gases and nutrients into the cell.

When a "young" plant cell is formed at the end of cell division it becomes surrounded by a thin (0.1 μm) **primary cell wall** composed largely of cellulose. The primary cell wall provides a firm, protective casing for the fragile internal cell, but it is not absolutely confining. As the young cell grows, the pressure exerted against the surrounding wall increases, causing the microfibrils to slide over one another, allowing the cell to increase in volume. When cell growth slows or ceases, the cell wall may become modified and thickened by the addition of other materials. This strengthened **secondary cell wall** greatly increases the cell's resiliency. Plant cells with secondary cell walls are found clustered in stems, as well as in the stalks that support leaves, flowers, and fruits. Secondary walls are often impregnated

with a hardening substance called *lignin*. Lignified secondary wall layers are responsible for the strength of wood and natural fibers, providing the means by which trees can grow hundreds of meters taller than the tallest animal.

INTERCELLULAR JUNCTIONS

Animal cells often interact with one another, forming **intercellular junctions** that can be seen in the electron microscope. The function of each type of junction is correlated with its structure (Figure 5-23).

- *Anchoring junctions or adhering junctions contain adhesive materials between the plasma membranes* The cells of tissues subjected to pulling or stretching forces, such as the uterine cervix or the skin, are typically "welded" together by anchoring junctions. A rare and potentially fatal disease called *pemphigus* results from deterioration of the junctions that anchor the

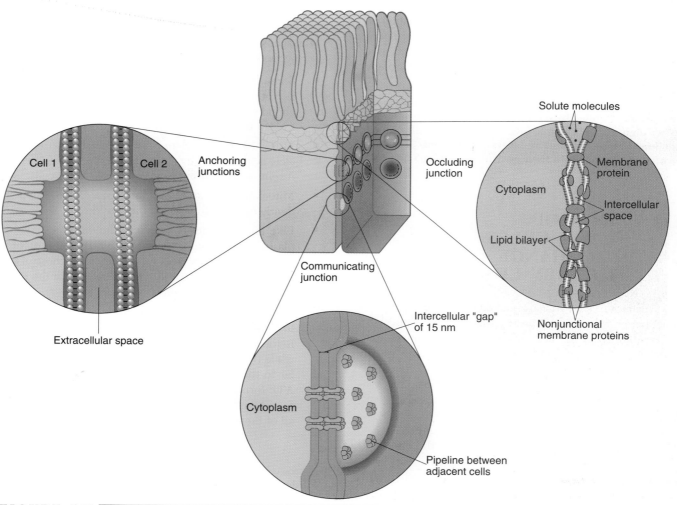

FIGURE 5-23

Intercellular junctions. This epithelial cell contains anchoring, occluding, and communicating junctions.

cells of the skin. The disease is characterized by severe blistering due to fluid leaking into the loosened skin tissue.

- ***Occluding junctions*** *are sites where adjacent plasma membranes make direct contact* Occluding junctions prevent materials from moving between the cells. Junctions of this type are found between the cells lining the capillaries in the brain, for example. Because of these junctions, many drugs are unable to pass from the bloodstream into the tissues of the brain. Pharmaceutical companies are working on drugs that are better able to penetrate this so-called blood–brain barrier.

- ***Communicating junctions*** *contain pipelines between adjoining plasma membranes* Communicating junctions join cells in such a way that materials are able to move directly through a channel from the cytoplasm of one cell into the cytoplasm of the adjoining cell. Communicating junctions often function to coordinate activities occurring in a sheet or mass of cells, as evidenced by the muscle tissue of your heart. Each heartbeat is initiated by direct electrical stimulation of only a small part of the heart (known as the *pacemaker*). The muscular wall of the heart contracts as a result of an ionic current that spreads rapidly through the muscle cells via communicating junctions.

Cells are surrounded by secreted materials that form structures ranging from a simple layer to a complex cell wall or junction. These surrounding layers protect cells and may govern interactions with their environment. (See CTQ #8.)

EVOLUTION OF EUKARYOTIC CELLS: GRADUAL CHANGE OR AN EVOLUTIONARY LEAP?

▮▶ Eukaryotic cells evolved about 2 billion years ago, presumably from simpler, prokaryotic ancestors. With specialized cytoplasmic organelles and a nucleus that sequestered genetic material, eukaryotic cells were more highly organized than were prokaryotes and were capable of conducting more complex functions. One of the questions that has intrigued biologists for decades is how the transition from the prokaryotic to eukaryotic state took place. Since there are no organisms living today that are intermediate in complexity between prokaryotes and eukaryotes, the question has been the subject of considerable speculation. Two modern hypotheses—the membrane invagination hypothesis and the endosymbiosis hypothesis—offer possible explanations.

The **membrane invagination hypothesis** (Figure 5-24*a*) proposes that the plasma membrane of the ancestral prokaryotic cell gradually folded in on itself, forming pockets, which pinched off from the surface membrane to form organelles in which particular enzymes or cellular materials could be concentrated. Over long periods of time, these simple organelles became highly specialized, forming the complex organelles found in modern eukaryotic cells.

It is easy to see how membrane invagination could have formed the endoplasmic reticulum, Golgi apparatus, and even the nucleus. However, the evolution of complex DNA-containing organelles, such as mitochondria and chloroplasts, is usually explained by another theory, the **endosymbiosis hypothesis** (Figure 5-24*b*), which proposes that the mitochondria and chloroplasts of eukaryotic cells were originally derived from small prokaryotic cells that took up residence *inside* a larger cell. (The name of this hypothesis uses the word *symbiosis,* a term that refers to a close association between different kinds of organisms.) For example, chloroplasts would have derived from a small photosynthetic bacterium, and mitochondria from a non-photosynthetic bacterium that had evolved the machinery to use oxygen in the formation of ATP (Chapter 9). Convincing evidence in behalf of the endosymbiosis hypothesis has been assembled by Lynn Margulis of the University of Massachusetts. Among the evidence she cites are the following facts:

- Some bacteria form symbiotic partnerships with eukaryotic cells today. These bacteria resemble mitochondria in structure and carry out the same steps in oxygen utilization as do mitochondria.

- Mitochondria and chloroplasts contain their own nucleic acids, including a circular strand of DNA which resembles the chromosomes of prokaryotic cells.

- Mitochondria and chloroplasts can divide independently of the cell in which they reside.

- Like an independent cell, mitochondria and chloroplasts contain their own ribosomes whose structure and drug sensitivity more closely resemble those of the ribosomes of prokaryotic cells than those in the cytoplasm of eukaryotic cells.

- The thylakoid membranes of chloroplasts contain photosynthetic pigments and proteins that are very similar to those in the membranes of photosynthetic bacteria, while the inner mitochondrial membrane is very similar to the plasma membrane of most bacteria where ATP formation takes place. The enzymes responsible for ATP synthesis, for example, are virtually identical in all of these membranes.

Evidence indicates that prokaryotic cells evolved before —and later gave rise to—eukaryotic cells. How this transition took place is currently the focus of vigorous scientific investigation and debate. (See CTQ #9.)

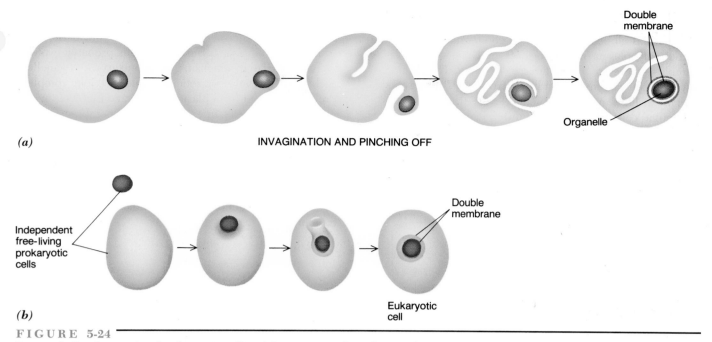

INVAGINATION AND PINCHING OFF

(a)

(b)

F I G U R E 5-24

Proposals for the origin of eukaryotic cells. *(a)* Invagination hypothesis. *(b)* Endosymbiosis hypothesis.

R E E X A M I N I N G T H E T H E M E S

The Relationship between Form and Function

The structure of every cell component is well-suited to its function. For example, a plant cell wall provides protection and support for the fragile cell it surrounds. The cellulose microfibrils within the wall provide resistance to pulling (tension) forces, while the amorphous matrix in which the microfibrils are embedded provides resistance to crushing (compression) force. This type of composite structure is analogous to that of reinforced concrete; the steel cables resist tension, while the concrete resists compression. Reinforced concrete and cell walls allow office buildings and trees to tower over the horizon.

Biologic Order, Regulation, and Homeosatasis

Like an entire organism, a cell remains alive only as long as the composition of its internal medium is maintained within a tolerable range. Each of a cell's organelles plays some role in maintaining this internal stability. For example, the plasma membrane prevents a cell from losing essential ingredients; the nuclear envelope helps maintain essential differences between the composition of the nucleus and cytoplasm; mitochondria provide the chemical energy necessary to fuel the cell's activities and to prevent its downhill slide to greater disorder; lysosomes digest various materials, including invading bacteria whose presence would threaten the cell; plant cell walls help maintain a cell's volume; and a plant cell's central vacuole serves as a storage site that removes toxic substances from the cell itself.

Acquiring and Using Energy

Two cellular organelles—the mitochondrion and the chloroplast—are the principal players in life's energy balance. Chloroplasts trap light energy from the sun and convert it to chemical energy stored in carbohydrate. Mitochondria convert the energy stored in fats and polysaccharides into ATP, which is used to run cellular activities. In both organelles, the inner membranes hold the components necessary for energy transfer.

Unity within Diversity

⚠ Whereas all cells share certain common features—a surrounding plasma membrane of similar structure, for example—there are also notable differences. The dichotomy between prokaryotic and eukaryotic cells represents diversity at life's most fundamental level—the cell. Plant and animal cells also have basic differences; chloroplasts, cell walls, and large central vacuoles are all present in plants, for example, but absent in animals. Even within an individual plant or animal, cellular diversity is evident from the various cell types specialized for particular functions. Even though these various cells contain the same kinds of organelles, their form and function can be very different.

Evolution and Adaptation

▶ The appearance of eukaryotic cells was one of the great evolutionary advances. Eukaryotic cells probably evolved by two separate mechanisms. Some eukaryotic organelles, such as the endoplasmic reticulum and the Golgi complex, probably evolved from plasma membrane that had invaginated into the cell's interior, whereas membrane-bound mitochodria and chloroplasts were probably derived from prokaryotic organisms that took up residence inside a larger host cell. Eukaryotic organelles remain relatively similar in structure over long periods of evolution. For example, the 9 + 2 array of mirotubules in cilia and flagella is seen across the entire spectrum of eukaryotic organisms, from protists to mammals.

SYNOPSIS

The cell theory has three parts. 1) All organisms are composed of one or more cells; 2) the cell is the basic organizational unit of life; and 3) all cells arise from preexisting cells.

Cells are either prokaryotic or eukaryotic. Prokaryotic cells are found only among bacteria (kingdom Monera); all other organisms are composed of eukaryotic cells. Both types of cells are surrounded by a fluid-mosaic plasma membrane consisting of a lipid bilayer and embedded proteins. Unlike prokaryotic cells, eukaryotic cells contain a true membrane-bound nucleus and a variety of distinct cytoplasmic organelles.

Eukaryotic cells contain similar organelles, but their number, form, and distribution vary from one cell type to another. The plasma membrane serves as a barrier between the cell and the outside world and regulates exchanges between the two. The plasma membrane contains receptors that recognize and interact with specific substances in the external medium. The nucleus is bound by a complex nuclear envelope and houses the chromosomes, which contain highly elongated DNA molecules and bound protein. Genetic messages move into the cytoplasm through pores in the nuclear envelope. The cytoplasm contains an interconnected, interfunctional network of membranous organelles. Membrane proteins, secretory proteins, and the proteins of some organelles are synthesized on ribosomes bound to the outer surface of the ER membranes, passed through the ER membrane to the internal space, and then sent off in vesicles to the Golgi complex. Once in the Golgi complex, materials are often modified

and then packaged for transport either to specific vesicles, such as lysosomes, or for discharge outside the cell during exocytosis. Lysosomes contain a variety of hydrolytic enzymes that are utilized in digesting macromolecules and aging cytoplasmic organelles. Plant cells often contain large vacuoles that store various substances. Mitochondria are specialized for the transfer of energy from stored macromolecular fuels to ATP for immediate use in the cell. Chloroplasts, which are not found in animal cells, are organelles in which light energy is captured and used to manufacture biochemicals during photosynthesis. The cytoskeleton of microtubules, microfilaments, and intermediate filaments maintains the shape of a cell, organizes its internal components, and provides the machinery necessary for cell movement. Some eukaryotic cells have cilia or flagella that are employed in the movement of single cells as well as the movement of materials along the surface of various tracts.

Cells secrete materials that form various types of extracellular coverings. The most complex extracellular layers are the walls that surround plant cells. These walls contain bundles of cellulose fibers embedded in an amorphous polysaccharide matrix. Animal cells may form specialized intercellular junctions that function in cell–cell attachment, occlusion of the space between cells, and intercellular communication.

The evolution of eukaryotic cells from prokaryotic ancestors is explained by two hypotheses. Many of the cytoplasmic organelles of a eukaryotic cell may derive from invaginated membrane, while mitochondria and chloroplasts are probably derived from prokaryotic symbionts.

Key Terms

lipid bilayer (p. 89)
cell theory (p. 90)
organelle (p. 91)
prokaryotic cell (p. 92)
eukaryotic cell (p. 92)
nucleus (p. 92)
cytoplasm (p. 92)
plasma membrane (p. 93)
receptor (p. 94)
chromosome (p. 97)
chromatin (p. 97)
nuclear envelope (p. 97)
nucleoli (p. 97)
ribosome (p. 97)
nucleoplasm (p. 97)
nuclear matrix (p. 97)

vesicle (p. 97)
endoplasmic reticulum (ER) (p. 98)
rough ER (RER) (p. 98)
smooth ER (SER) (p. 98)
Golgi complex (p. 99)
exocytosis (p. 99)
lysosome (p. 100)
vacuole (p. 102)
mitochondria (p. 102)
cristae (p. 104)
matrix (p. 104)
chloroplast (p. 104)
thylakoid (p. 104)
stroma (p. 104)
cytoskeleton (p. 104)

microtubule (p. 104)
microfilament (p. 104)
intermediate filament (p. 104)
cilia (p. 106)
flagella (p. 107)
extracellular matrix (p. 108)
primary cell wall (p. 109)
secondary cell wall (p. 109)
intercellular junction (p. 109)
anchoring junction (p. 109)
occluding junction (p. 110)
communicating junction (p. 110)
membrane invagination hypothesis
(p. 110)
endosymbiosis hypothesis (p. 110)

Review Questions

1. List the similarities and differences between prokaryotic and eukaryotic cells.

2. Prepare a chart comparing the similarities and differences between plant and animal cells.

3. Why do the surfaces of membrane proteins contain both hydrophobic and hydrophilic regions?

4. Which cytoplasmic organelles synthesize steroid hormones? Which synthesize proteins to be secreted? Which store calcium? Which store hydrolytic enzymes? Which synthesize plasma membrane proteins? Which store toxic waste products in plant cells? Which synthesize polysaccharides for plant cell walls?

5. Try to convince a friend that the mitochondria in their cells are derived from bacteria.

Critical Thinking Questions

1. Gorter and Grendel used human red blood cells for their determination of a cell's lipid content. Would their results have differed if they had used a salivary gland cell instead? In what way? What do you think they would have been able to conclude about plasma membrane structure?

2. In 1674, Anton van Leeuwenhoek used a single glass lens (a "simple" microscope) to look at a drop of lake water. He discovered a new world, one of tiny organisms he called animalcules ("little animals"). Imagine you are van Leeuwenhoek. Write a letter to the Royal Society of London in which you to try to convey your feelings about your momentous discovery.

3. When Leeuwenhoek looked at what he saw through his microscopes, he concluded that they were living organisms. Most of what he saw we now know are single-celled organisms. List the evidence Leeuwenhoek could have used in drawing the conclusion that the things he saw were indeed alive.

4. Given that both single-celled prokaryotes and single-celled eukaryotes carry out all of the life functions, why do you think biologists describe the former as "simple" and the latter as "complex?" Compare your answer to that of a professional biologist.

5. What changes would occur in the interaction of cells with external substances if enzymes are used to remove: (1) the carbohydrate chains that project from the outer surface of the plasma membrane; (2) the polar amino acids that project from the outer surface of the plasma membrane; (3) the phospholipids that make up much of the plasma membrane?

6. The individual cells found in living organisms exhibit the characteristics that distinguish living from nonliving things (see Chapter 1). Complete the chart below by listing the appropriate cellular components (organelles) in the right-hand column. You may use a component more than once.

Characteristic	Cellular Components
maintaining order	
acquiring and using energy	
reproduction/ heredity	
growth/ development	
responsiveness to environment	
metabolism	
homeostasis	

7. Suppose you wanted to test the hypothesis that cellulose fibers were synthesized in the Golgi complex. (They are actually synthesized at the plasma membrane.) How might you go about testing this hypothesis?

8. How do the materials secreted by cells in multicellular organisms contribute to the life functions of homeostasis and responsiveness for those organisms?

9. Fossil evidence indicates that prokaryotic cells evolved before eukaryotic cells. If you did not know this, what evidence from the structure of these two main types of cells would lead you to the same conclusion?

Additional Readings

Alberts, B., et al. 1989. *Molecular biology of the cell,* 2d ed. New York: Garland. (Intermediate)

Becker, W. and D. W. Deamer. 1991 *The world of the cell,* 2d ed. San Francisco: Benjamin-Cummings. (Intermediate)

Darnell, J., H. Lodish, and D. Baltimore. 1990. *Molecular cell biology,* 2d ed. New York: Freeman. (Advanced)

Karp, G. 1984. *Cell biology.* New York: McGraw-Hill. (Intermediate)

Kleinsmith, L. J., and V. M. Kish. 1988. *Principles of cell biology.* New York: Harper & Row. (Intermediate)

Loewy, A. G., et al. 1991. *Cell structure and function,* 3d ed. Philadelphia: Saunders. (Advanced)

Marx, J. 1992. Boring in on β-Amyloid's Role in Alzheimer's. *Science* 255:688–689. (Intermediate).

Prescott, D. M. 1988. *Cells.* Boston: Jones and Bartlett. (Intermediate)

Sharon, N. and Lis, H. 1993. Carbohydrates in cell recognition. *Sci. Amer.* Jan:82–89. (Intermediate)

Sheeler, P. and D. Bianchi. 1987. *Cell and molecular biology.* New York: Wiley. (Intermediate)

Wolfe, S. L. 1993. *Molecular and cell biology.* Belmont, Ca: Wadsworth. (Advanced)

CHAPTER

◄ 6 ►

Energy, Enzymes, and Metabolic Pathways

STEPS
TO
DISCOVERY
The Chemical Nature of Enzymes

ACQUIRING AND USING ENERGY

The Laws of Thermodynamics

Energy is Transferred During Chemical Reactions

ENZYMES

Why Enzymes are Such Effective Catalysts

The Effect of Temperature and pH on Enzyme Activity

METABOLIC PATHWAYS

Oxidation and Reduction: A Matter of Electrons

ATP: The Energy Currency of Life

Directing Metabolic Traffic

MACROMOLECULAR EVOLUTION AND ADAPTATION OF ENZYMES

BIOLINE

Saving Lives by Inhibiting Enzymes

THE HUMAN PERSPECTIVE

The Manipulation of Human Enzymes

Hexokinase

Alcohol
dehydrogenase

Yeast

The Chemical Nature of Enzymes

*I*n 1779, the French Academy of Science offered 1 kilogram (2.2 pounds) of gold to anyone who could unravel the mystery by which sugars present in grape juice are converted to alcohol during the formation of wine. The prize was never collected.

A hundred years later, two conflicting explanations for the nature of alcohol formation (a process known as *fermentation*) prevailed. On one side of the argument were the biologists, led by Louis Pasteur, a biologist working on behalf of the French wine industry. Pasteur correctly believed that the conversion of sugars to alcohol was a process carried out by living yeast cells. In fact, he applied this scientific fact to the faltering French wine industry, thereby restoring it to its previous station as the best winemaker in the world, simply by manipulating the organisms used for fermentation. In adopting this position, however, Pasteur hypothesized that fermentation required a "vital force" that could only be supplied by an intact, highly organized, living organism. Pasteur rejected the notion that life processes, such as fermentation, could be reduced to simple chemical reactions.

On the other side of the argument were the organic

Yeast cells provide the enzymes that ferment the sugar in grapes to form the alcohol found in wine.

chemists of the period, including the aggressive debater Justus von Liebig, who ridiculed the suggestion that yeast cells were responsible for fermentation. Liebig and his fellow chemists proposed that fermentation was simply an organic reaction that occurred on its own, no different from those reactions they had been studying in the test tube. The battle lines were drawn, biologists versus chemists.

Then came Eduard Buchner. Buchner was both a chemist—teaching chemistry at a German university—and a biologist—working in his brother's bacteriology laboratory. In 1897, 2 years after Pasteur's death, Buchner was preparing "yeast-juice,"—an extract made by grinding yeast cells with sand grains and then filtering the mixture through filter paper. Buchner was planning to use this "yeast-juice" in a series of medical studies, but he wanted to preserve it for later work. After trying to preserve the extract with antiseptics and failing, Buchner attempted to protect his cell juice from spoilage by adding sugar, the same procedure used to preserve jams and jellies. Instead of preserving the solution, the sugar produced gas, which bubbled continuously for days. Buchner's scientific curiosity was aroused by this unexpected occurrence, and he kept the solution instead of discarding it. Analysis revealed that fermentation was occurring, producing alcohol and bubbles of carbon dioxide. All this was taking place without a single living yeast cell in the "soup," a severe blow to Pasteur's hypothesis that fermentation required living organisms.

Buchner had accidentally discovered that chemical agents (later identified as *enzymes*) promote biological reactions, and can do so *outside* of the cells in which they were originally produced. Investigators could now purify enzymes and study their activities without interference from other cellular ingredients. The door to modern biochemistry had swung wide open, raising the question: What was the nature of these remarkable molecular mediators?

One prominent scientist to turn his attention to the study of enzymes was the German chemist Richard Willstater, who had earlier won a Nobel Prize for his work on the structure of chlorophyll, the central plant pigment in photosynthesis. Willstater set out to purify plant enzymes but found that his most active preparations contained so little material they could not be chemically identified. Today, we understand that enzymes are so efficient that they are active at extremely low concentrations—too low to be characterized using 1920s' methods. To Willstater, it was bewildering. Because he was unable to detect any protein in the purified enzyme preparations, Willstater erroneously concluded that enzymes could not be made of protein.

Then in 1926, James Sumner, an American biochemist, prepared the first crystals of an enzyme (urease), which he had purified from the seeds of a tropical plant; he determined that the crystals were indeed composed of protein. He also demonstrated that protein-denaturing treatments destroyed the preparation's enzymatic activity and concluded that the enzyme was a protein. Sumner's finding was not greeted with much acclaim. Willstater argued that the crystals may have been partly composed of protein but that the enzyme was only a minor contaminant of the preparation and it was not being detected in the chemical analysis. A few years later, however, another American, John Northrop, crystallized a number of different enzymes (including pepsin, a digestive enzyme of the stomach) and showed conclusively that all were made of protein. As a result of their work, Buchner won the Nobel Prize in chemistry in 1907, while Sumner and Northrop shared the same prize in 1946.

*E*very organism is a precisely coordinated collection of chemicals, each of which is created through biochemical reactions. Every detail of your body, from the hair on your head down to the nails on your toes, is the product of such reactions. Your eyes, for example, owe their color to pigments manufactured by metabolic activity. The chemical changes that construct these pigments don't happen automatically or haphazardly; they require the tight supervision provided by enzymes, which may be thought of as "metabolic traffic directors." Enzymes act on prepigment chemicals, changing them step by step until they become the molecules that ultimately produce eye color. Most of your inherited characteristics developed because your cells produced enzymes that encouraged the biochemical construction of each particular trait. In this sense, an organism is the product of its unique combination of enzymes (Figure 6-1). Most series of reactions—whether they lead to the production of pigment or of some other material—also require a source of chemical energy. The topic of energy is a good starting point in a discussion of metabolism.

▼ ▼ ▼

ACQUIRING AND USING ENERGY

A living cell bustles with activity. Materials constantly enter and exit the cell; genetic instructions flow from the nucleus to the cytoplasm, where proteins and other substances are synthesized or degraded. To maintain such a high level of activity a cell needs energy. For most of us, the word "energy" conjures up a rather vague concept of being energetic, such as playing baseball, feeling enthusiastic, or making an effort. For scientists, **energy** has a more precise definition: It is the capacity to do work; that is, the capacity to change or move something. There are several different forms of energy. Chemical energy can change the structure of molecules, mechanical energy can move objects, light energy can boost electrons to an outer shell, thermal energy (heat) can increase the motion of molecules, and electrical energy can move electrically charged particles.

Energy exists in two general states: potential and kinetic. **Potential energy** is stored energy; it has the potential to perform work if allowed. **Kinetic energy,** which literally means "energy in motion," is energy expended in the process of performing work. Each form of energy (chemical, mechanical, light, thermal, and electrical) can exist in either state. The glycogen stored in your muscle and liver cells, for example, contains abundant potential energy. When the glycogen molecules are broken down and metab-

(a)

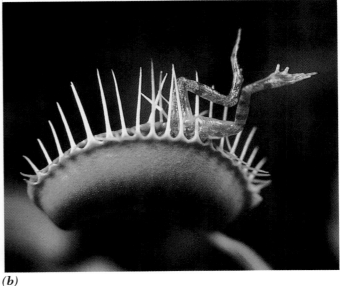

(b)

FIGURE 6-1

A sampler of enzyme performance. The properties and activities of every organism depend on enzymes. *(a)* The unusual color of this blue frog is due to a pigment manufactured by enzymes, a set different from that of his green counterparts. *(b)* Digestive enzymes secreted by the Venus-flytrap will disassemble the body of this imprisoned frog. Other enzymes will reassemble the molecules released from the frog's flesh into plant tissue.

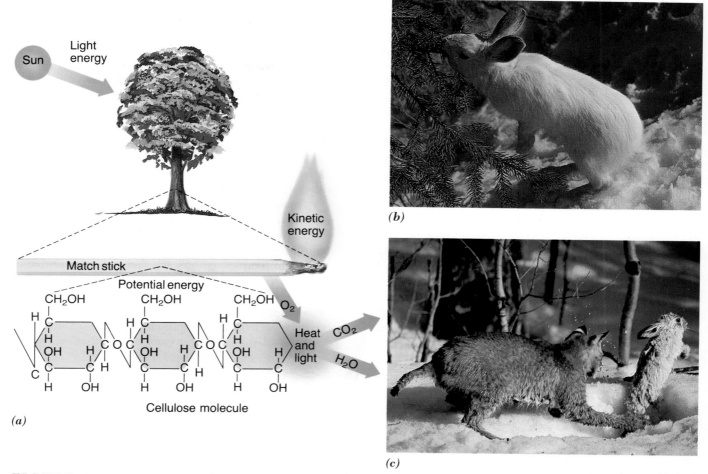

FIGURE 6-2

Biological transfers of energy. *(a)* Energy is captured for the biosphere by photosynthesis. Plant leaves transform sunlight into chemical energy stored in the plant's tissues primarily in the form of cellulose and starch. In this illustration, the potential energy stored in the cellulose (structure shown) that makes up a match becomes kinetic energy when the cellulose molecules are "burned." Animals that feed on plants, such as a hare *(b),* utilize the chemical energy in starch to fuel their diverse energy-requiring activities. Plant-eating organisms (herbivores), such as the hare, serve as energy sources for other animals, this lynx for example *(c).* In each case, however, most of the potential energy of the food matter escapes back into the environment as unusable energy.

olized, the energy appears as kinetic energy and is used by the body to contract muscles, to move fluid through various tracts and vessels, to produce body heat, and myriad other activities.

THE LAWS OF THERMODYNAMICS

Much of what we know about energy is summarized in two basic laws of nature—the laws of thermodynamics:

- **The first law of thermodynamics** is the *law of energy conservation.* Simply put, it states that energy can

neither be created nor destroyed. Energy can be converted from one form to another, however. For example, electrical energy is converted to mechanical energy when we plug in a clock or turn on the blender; chemical energy is converted to heat and light energy when fuel is burned in an oil heater. Living organisms are also capable of energy conversion. Green plants perform *photosynthesis,* a process by which solar energy is converted into chemical energy that is stored in organic chemicals, such as starch or cellulose (Figure 6-2a). The next time you eat a piece of fruit or meat, remember that the chemical energy you are receiving was originally derived as light energy from the sun.

- **The second law of thermodynamics** expresses the concept that events in the universe proceed in a predictable "downhill" direction, from a higher energy level to a lower energy level. Rocks fall off cliffs to the ground, but they never spontaneously lift themselves back to the higher energy level at the top of the cliff. The second law of thermodynamics carries with it the notion that nature is "wasteful." Any time energy is exchanged, some *usable* energy is inevitably lost. In other words, the energy *available to perform additional work* decreases. For example, when a hare browses on the leaves of a tree (Figure 6-2*b*), or a lynx preys on the hare (Figure 6-2*c*), most of the chemical energy in the food is inevitably lost to the animal having the meal. What happens to this additional energy? A significant portion is "wasted" by increasing the random movements of atoms and molecules. This so-called unusable energy has a special name—entropy.

The Concept of Entropy

Entropy can be measured in terms of an increased state of randomness or disorder in the universe, which is often a result of the release of heat. Entropy can be illustrated by innumerable familiar activities. Suppose you were to walk into a library and search for books by three of your favorite authors. You find the books, take them to a desk, browse through them, and leave them on a table. You have just increased the entropy of the universe. Prior to your visit, the books were in an ordered location; there is only one proper place for each book in the library. In contrast, after you left the library, the books were left on a table that was selected essentially at random; you could just as well have left them in any one of a hundred or more places. As a result of your activities, the library has become more disordered; that is, its entropy has increased. Another example of entropy is shown in Figure 6-3.

One of the themes we continue to stress throughout this book is the complexity and order of living organisms. If every event increases the disorder in the universe, how is it possible for organisms to maintain their state of order and, even more perplexing, how did complex organisms ever evolve in the first place? The fact that entropy in the universe always increases does not mean that *every part* of the universe has to become more and more disordered. On the contrary, each organism represents a temporary departure from the relentless march toward disorganization.

FIGURE 6-3

A victim of entropy. A house left on its own for a number of years will spontaneously fall into a state of disorder, increasing the entropy of the universe. All organized structures require a constant input of energy to maintain their complexity and order; otherwise, they slowly deteriorate. The same is true for organisms; without the input of energy, they die and decompose.

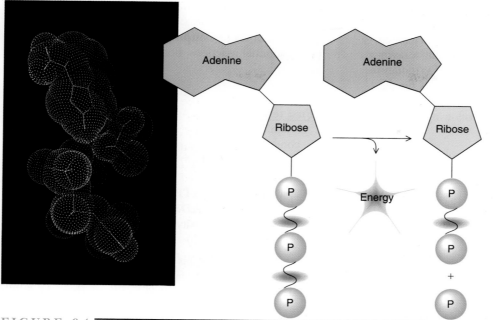

FIGURE 6-4

ATP. ATP is a nucleotide consisting of a sugar (ribose), a nitrogenous base (adenine), and three phosphate groups joined to each other. When hydrolyzed, the molecule is split into two products: ADP and inorganic phosphate (P_i). The squiggly lines connecting the second and third phosphate groups represent the bonds that can be broken when ATP is hydrolyzed. Space-filling model of ATP generated by a computer. Each ball represents an atom (white = hydrogen, green = carbon, blue = nitrogen, yellow = phosphorus).

Maintaining a state of low entropy requires the input of energy, which, in keeping with the first law of thermodynamics, lowers the energy content of the remainder of the universe. Consider just one molecule of DNA located in one cell in your liver. That cell has dozens of different enzymes whose sole job it is to patrol the DNA, looking for damage and repairing it. Without this expenditure of energy, the ordered arrangement of nucleotides in the cells of your body would literally disappear overnight. Energy is expended to maintain order at every level of biological organization—from molecules to ecosystems.

The evolution of complex organisms from simpler forms has also been driven by the input of energy. Energy is required to build and maintain the organisms on which natural selection operates. As organisms became larger and more complex, they required more energy to develop, survive, and reproduce; this energy is ultimately derived from the sun.

ENERGY IS TRANSFERRED DURING CHEMICAL REACTIONS

It is easy to understand how a moving sledgehammer releases mechanical energy as it slams through a cement sidewalk or how a pot of water absorbs thermal energy as it becomes hotter and hotter. Understanding how energy is released or absorbed during the transformation of one type of chemical compound into another is a bit more complex. Let us begin with one of the most important compounds found in all living organisms, the key compound of energy metabolism, *adenosine triphosphate* or *ATP* (Figure 6-4). ATP is a nucleotide (Chapter 4); it is made up of a sugar (ribose), a nitrogenous base (adenine), and three phosphate groups linked to one another. ATP is the molecule in which all cells temporarily store the chemical energy used to run virtually all of a cell's activities. These small, usable amounts of energy are stored in a form that is instantly accessible as the energy is needed. For this reason, ATP is often described as the "energy currency" of the cell. As we will see, organisms spend their ATP on various energy-requiring tasks, such as the assembly of macromolecules, the contraction of muscles, and the buildup of ions on opposite sides of the plasma membrane.

The amount of ATP present in a cell at a given moment is surprisingly small. The average bacterial cell, for example, has fewer than 5 million ATP molecules—only enough to sustain the cell's activities for a second or two. Similarly, the human body has only enough ATP "on hand" to last about 20 seconds. Consequently, ATP supplies must be continually replenished from the energy stored in its large organic molecules, particularly the polysaccharides and fats, which maintain the energy reserves of the cell. In keeping with the comparison of ATP to currency, a cell's energy-rich macromolecules can be likened to money banked in a savings account, while its ATP supply can be thought of as money in the cell's "pocket," available to be "spent" on its needs at that very instant.

ATP is broken down in a cell by the following hydrolytic reaction

$$ATP + H_2O \rightarrow ADP + P_i(\text{inorganic phosphate}) + \text{energy}$$

It is a highly *favored* reaction, that is, a reaction that will proceed spontaneously toward the formation of products, in this case ADP and P_i. The reason this reaction is so highly favored is that ATP has considerably more energy than do ADP and P_i; therefore, when ATP is hydrolyzed the contents of the test tube attain a lower energy level. A similar principle underlies the use of common explosives, such as nitroglycerin $[C_3H_5(NO_3)_3]$ and trinitrotoluene (TNT) $[C_7H_5(NO_2)_3]$. The explosive reaction of these high-energy molecules is driven by the production of one of the most stable, lowest energy molecules in the universe — N_2 gas — the most common component of the atmosphere.

Highly favored chemical reactions, such as ATP hydrolysis or the explosion of TNT, are referred to as **exergonic** because they occur spontaneously with the release of energy. Whereas the energy released by a TNT explosion is simply released into the environment, the energy released by ATP hydrolysis is utilized by the cell to accomplish useful work. This will be illustrated in the following discussion.

Unlike ATP hydrolysis, many chemical reactions that occur in a cell are not favored at all; they are described as thermodynamically *unfavorable*. An example of a thermodynamically unfavorable reaction is the synthesis of the amino acid glutamine

$$\text{glutamic acid} + \text{ammonia} \rightarrow \text{glutamine}$$

Such reactions, which lead to the formation of products with more energy than the original reactants, are referred to as **endergonic.** The most important endergonic reaction in biology is the splitting of water (H_2O) into its component elements, hydrogen and oxygen; this is the fundamental reaction that occurs during photosynthesis. In contrast to exergonic reactions, endergonic reactions, such as glutamine formation or the splitting of water, do not occur spontaneously but require an input of external energy. As we will see later in the chapter, this is a crucial factor of metabolism. First, let us examine how endergonic reactions can be made to occur.

The Use of Energy-Releasing Reactions to Drive Energy-Requiring Reactions

If the conversion of glutamic acid into glutamine cannot occur spontaneously, how does a cell produce this essential amino acid? The answer was revealed in 1947, when it was shown that the formation of glutamine is *coupled* to the hydrolysis of ATP. **Coupling** is accomplished by utilizing two sequential reactions, both of which are favored.

First reaction:
  glutamic acid + ATP → glutamyl phosphate + ADP
Second reaction:
  glutamyl phosphate + ammonia → glutamine + P_i

Net reaction:
  glutamic acid + **ATP** + ammonia → glutamine
    + **ADP** + **P_i**

The situation is analogous to the scene depicted in Figure 6-5, where a mallet driven down onto a platform (an energy-releasing reaction) is used to lift another object (an energy-requiring reaction). ATP hydrolysis is used to drive

FIGURE 6-5

A mechanical analogy for the use of ATP in driving endergonic reactions. The energy released as the mallet strikes the base is used to drive the uphill reaction as the weight is moved upward against the force of gravity to ring the bell.

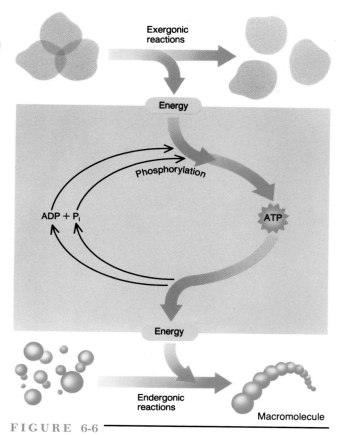

FIGURE 6-6 ——————————————

Producing and "spending" the cell's energy currency.
Endergonic and exergonic reactions are coupled by ATP, the
common denominator in energy production and utilization. Exergonic reactions generate ATP, whereas endergonic reactions,
such as macromolecular assembly, utilize ATP.

virtually every endergonic process within the cell, including
chemical reactions such as the formation of glutamine, the
assembly of macromolecules (Figure 6-6), the concentration of ions across a membrane (Chapter 7), and the movement of filaments in a muscle cell (Chapter 26). In most
coupled reactions, the phosphate group is transferred from
ATP to an acceptor—such as glutamic acid, a sugar, or
often a protein—and is subsequently removed in a second
step. The molecules that make all of this possible are
enzymes.

—————————————————————————

**Every time energy is exchanged, as occurs during chemical
reactions, a portion of that energy is "lost" for further use.
As a result, events in the universe tend to proceed toward a
state of lower energy. Life is a highly organized state. Maintaining such a high degree of organization demands a constant input of energy. (See CTQ #2.)**

ENZYMES

Even though a biochemical reaction is thermodynamically
favored, it usually takes place at extremely slow rates.
Herein lies the value of enzymes. Enzymes are biological
catalysts—substances that greatly increase the rate of
particular chemical reactions. An enzyme may cause a reaction to proceed billions of times faster than it would
otherwise occur. Consider, for example, what would happen if you were to add a teaspoon of glucose to a test tube,
seal the tube, and sterilize the solution. The glucose would
remain essentially intact for years. If you were to add bacteria to the solution, however, the sugar molecules would be
taken into the cells and degraded in seconds by the enzymes
of the bacteria.

🔄 Because of their powerful catalytic activity, enzymes
are effective in small numbers. Even a tiny bacterial cell has
space for hundreds of different enzymes, all catalyzing different reactions. Just as importantly, enzymes help maintain biological order by converting reactants into specific
products needed by the cell. In contrast, when these same
reactions are attempted in the laboratory by organic chemists without the benefit of enzymes, many different byproducts are usually produced. In a cell, formation of unwanted
byproducts would disrupt the progress of other reactions
and kill the cell. The importance of enzymes can be appreciated when one considers that such devastating diseases as
phenylketonuria (PKU) and Tay-Sachs, which lead to severe mental retardation and infant death, result from the
body's deficiency of a single enzyme.

WHY ENZYMES ARE SUCH EFFECTIVE CATALYSTS

We have seen how certain reactions, such as glucose breakdown or ATP hydrolysis, are thermodynamically favored
because they lead to the formation of lower-energy products. But why don't these reactions occur at significant rates
in the absence of enzymes? The reason is that the reactant
molecules—the starting substances—must contain a certain amount of energy to become transformed into product
molecules—the new substances. This required energy is
called **activation energy**. In Figure 6-7, the activation
energy is represented by the height of the barrier. Enzymes
act by lowering a reaction's activation energy—the amount
of energy needed by molecules to undergo a reaction. We
will return to the matter of activation energy below.

The question still remains: Why are enzymes such effective catalysts? How is it that an enzyme can cause a
reaction to occur ten times a second when that same reaction might occur only once every hundred years in the
enzyme's absence? A simple analogy may clarify the answer.

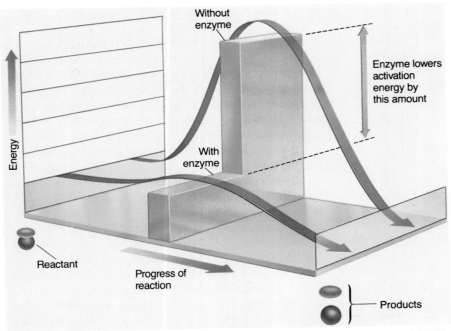

FIGURE 6-7

Activation energy and enzymes. Activation energy is the energy required to initiate a chemical reaction—to overcome the reactants' tendency to resist change. This energy barrier must be hurdled before a chemical reaction can occur. Enzymes reduce activation energy, so a reaction proceeds at a much faster rate than would occur in their absence.

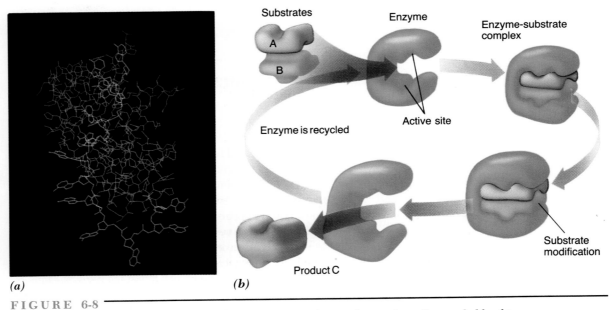

(a) *(b)*

FIGURE 6-8

An enzyme in action. *(a)* The closeness of the enzyme–substrate fit is realistically revealed by this computer-generated version showing the RNA substrate (in green) being attacked and disassembled by the enzyme ribonuclease (in purple). *(b)* Substrates A and B are enzymatically altered to form product C. The reaction occurs following the binding of the substrates within the enzyme's active site. Enzymes are recycled—following completion of a reaction and the dispersal of the product, the enzyme binds a new pair of substrates and catalyzes another reaction.

Suppose you were to place a handful of nuts and bolts into a bag and shake the bag for 15 minutes. It is very unlikely that any of the bolts would have a nut firmly attached to its end when you stopped shaking the bag. In contrast, if you were to pick up a bolt in one hand and a nut in the other, you could guide the bolt into the nut very rapidly. In this analogy, enzymes are the guiding hands, and reactants are the nuts and bolts. Enzymes bind reactants—termed **substrates** when an enzyme is involved—increasing the likelihood that they will be converted to products.

The area on the enzyme that binds the substrate(s) is called the **active site,** and it is often situated within a groove or cleft of the enzyme. The active site and the substrate(s) have complementary shape enabling them to bind together with a high degree of precision, like pieces of a jigsaw puzzle (Figure 6-8a). In addition to binding the substrate(s), the active site contains a particular array of amino acids whose presence lowers the activation energy the substrates need to undergo a reaction. There are a number of ways this can happen (Figure 6-9). These include:

- *Substrate orientation* Like the connection between nuts and bolts, many reactions require that two different substrate molecules collide with each other in just the right way. Enzymes hold their substrates in a particular orientation that forces them together in the proper relationship.

- *Physical stress* Once a substrate molecule is in the "grip" of an enzyme, certain bonds within the substrate molecule may be placed under physical stress, increasing the likelihood that the bond will rupture.

- *Changes in substrate reactivity* Refer back to Figure 4-14 and notice the variety in the structures of amino acid R groups. When these R groups on the enzyme's amino acids come into close proximity with a part of the substrate, they can change the charge of the substrate, alter the distribution of electrons within the substrate's bonds, tie up surrounding water molecules, and cause other changes that increase the reactivity of the substrate.

The interaction between enzyme and substrate is not a rigid one, as would occur, for example, between a lock and key. Rather, enzymes are built with an internal flexibility that allows them to undergo **conformational changes,** that is, changes in shape. As a result of conformational changes in the enzyme, the fit between an enzyme and substrate often improves after the substrate initially binds to the active site. In addition, conformational changes invariably occur within an enzyme while a substrate is being chemically transformed into a product. Once the reaction has taken place, the product(s) leaves the enzyme, which is then ready to bind a new set of substrates. This cycle of substrate binding, chemical modification, and product release is illustrated in Figure 6-8b.

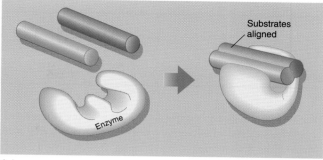

(a)

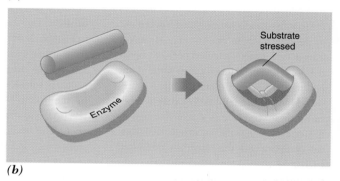

(b)

(c)

FIGURE 6-9

Three processes by which enzymes modify substrates: *(a)* maintaining precise substrate orientation, *(b)* physical stress, and *(c)* change in substrate reactivity.

The complementarity in the shapes of the active site of the enzyme and substrate accounts for the high degree of *specificity* of enzyme activity. Specificity is a term used to describe the high degree of selectivity and precision with which biological molecules interact; it is a key ingredient in maintaining biological order. Regarding enzyme activity, for example, only the proper substrate fits into the active site of an enzyme. As a result, the course of an enzymatic reaction is not affected by the hundreds of other types of molecules present in the cell at the time. However, the activity of an enzyme can be affected by adding a substance whose structure is very similar to that of a substrate (see Bioline: Saving Lives by Inhibiting Enzymes).

◁ B I O L I N E ▷
Saving Lives by Inhibiting Enzymes

You may be alive today because of the ability of chemicals to inhibit enzyme activity. Many of the diseases that ravaged our ancestors are treatable today with chemicals that selectively block the action of enzymes present in disease-causing bacteria but not in people. Such inhibitors may be safely introduced into an infected person's body, to inhibit bacterial metabolism without impeding the person's metabolism.

Many of these inhibitors function because their structure is similar enough to the enzyme's normal substrate that they can compete with the substrate in binding to the enzyme's active site. Unlike the substrate, however, the inhibitor cannot be converted to products and, while in the active site, prevents normal substrate bind-

ing (Figure 1a). In this state, the enzyme ceases to function.

Sulfa drugs—compounds that have saved countless human lives—provide an example of this type of **competitive inhibitor.** These agents are given to people to fight bacteria that cause diseases ranging from urinary bladder infections to pneumonia. The drugs block the ability of bacteria to transform a particular chemical in their cells—*para-aminobenzoic acid* (PABA)—to the essential coenzyme *folic acid*. The structural similarity between sulfa drugs and PABA (Figure 1b) creates metabolic confusion in the bacteria. Although many critical reactions in the human body also depend on folic acid, we cannot naturally manufacture this coen-

zyme. Instead, all of our folic acid is obtained as a vitamin in our diet. For this reason, we are not affected by the sulfa drugs in the same way as the bacteria, which must make their own folic acid. As a result, we survive the treatment unimpaired, but the bacteria do not.

Penicillin is another drug that owes its lifesaving antibacterial activity to its ability to compete with the substrate of a bacterial enzyme that directs formation of the cell wall of the bacterium. Penicillin is even more effective than the sulfa drugs: Once penicillin enters the active site of the enzyme instead of the normal substrate, it forms a covalent bond with the enzyme, permanently inactivating it. Without their cell walls, bacteria responsible for such diseases as pneumonia and syphilis are readily destroyed.

Enzymes may also be inactivated by heavy metals, such as lead, silver, mercury, and arsenic. In contrast to competitive inhibitors, these poisonous substances bear no resemblance to the enzyme's substrates but bind nonspecifically to many proteins. Some of these inhibitors inactivate enzymes by altering their shape; others bind in place of required cofactors. Since these metals act as inhibitors of essential human enzymes, as well as those of bacterial cells, they are not particularly useful for treating human diseases. Before the widespread use of penicillin during the 1940s, however, arsenic-containing solutions were commonly employed as "home remedies" and may have been responsible for the gradual poisoning of large numbers of people who used these drugs. The bacteria responsible for syphilis are particularly sensitive to an arsenic compound developed by the microbiologist Paul Ehrlich in 1910. The drug—which was the first compound discovered that would kill bacteria without also killing the human taking the drug—was the 606th arsenic compound that Ehrlich had methodically synthesized in the laboratory and tested.

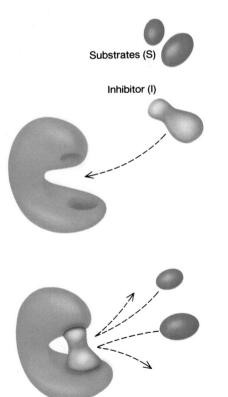

Substrates (S)

Inhibitor (I)

Enzyme-inhibitor complex
(active site blocked)

(a)

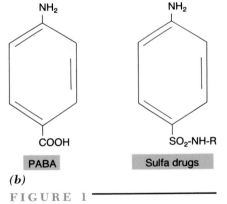

PABA

Sulfa drugs

(b)

FIGURE 1

Competitive inhibition. *(a)* Competitive inhibition of enzyme activity is due to the structural similarity between the inhibitor and the substrate(s). The enzyme's active site recognizes both the inhibitor (I) and the substrate (S), but the inhibitor cannot be converted to products. *(b)* Due to its similar structure, the sulfa drug acts as a competitive inhibitor of PABA, the enzyme's normal substrate.

Cofactors: Nonprotein Helpers

Although enzymes are proteins, many enzymes utilize non-protein helpers, called **cofactors,** which are required for the enzyme to carry out its function. Depending on the enzyme, the cofactor may be an organic molecule, termed a **coenzyme,** or a metal atom. Cofactors typically carry out chemical activities for which amino acids are not well suited. Thus, they add to the repertoire of chemical reactions that enzymes can catalyze. The importance of coenzymes is illustrated by your daily need for vitamins. For the most part, these nutritional supplements act as essential coenzymes that you cannot synthesize for yourself.

THE EFFECT OF TEMPERATURE AND pH ON ENZYME ACTIVITY

Increases in temperature boost enzyme activity for the same reason that heat accelerates chemical reactions—they increase molecular motion. As a result, more reactant molecules possess the activation energy needed to be converted to products. Above a critical temperature, however, the rate of an enzymatic reaction rapidly decreases because enzymes suffer heat damage. Excessive heat disrupts hydrogen bonding and other forces that stabilize the shape of a protein. Enzymes change shape, lose solubility, and coagulate (thicken), as illustrated by the protein in a cooked egg (Figure 6-10). Proper cooking destroys (*denatures*) the proteins of many disease-causing microbes residing in food. This is one of the reasons why it is safer to eat cooked meats and eggs than raw ones. Pasteurization (controlled heating) of milk and other dairy foods has dramatically reduced the spread of serious milk-borne diseases, such as tuberculosis. High temperatures can also destroy human enzymes, however, which is one reason why high, prolonged fevers can be life-threatening and should be taken seriously.

Enzyme activity is also altered by changes in pH. A rise or fall in H^+ concentration can change the charge of many of the amino acids, thus affecting the structure of the active site. An enzyme that functions optimally at neutral pH (7.0) will usually be inactivated when its environment becomes either too basic or too acidic. Not all enzymes have their optimum near pH 7, however. Enzymes that act in an acidic environment, such as within a lysosome or the stomach of a mammal, have a structure conducive to functioning at very low pH.

Now that we have described the structure and properties of enzymes, we can examine how these remarkable biological catalysts are functionally linked into pathways to form molecules needed in the construction of cellular materials.

Life demands precisely orchestrated chemical activity. Cells govern metabolic activities by producing specific enzymes that vastly increase the rate of specific chemical reactions. Reactions promoted by enzymes generate only the desired products, which prevents metabolic chaos. (See CTQ #3.)

METABOLIC PATHWAYS

The formation of complex biochemicals within a cell occurs in stepwise fashion, similar to the way an automobile or appliance is built on an assembly line. Each biochemical is either assembled or dismantled by an orderly sequence of chemical reactions which comprise a **metabolic pathway.** Each pathway is catalyzed by a series of specific enzymes in which the product(s) of one reaction become the substrate(s) of the following reaction. The substances formed along the pathway (**b**, **c**, and **d** below) are called **metabolic intermediates.**

FIGURE 6-10
Denaturation. The protein in the bacon and eggs has been denatured by heat.

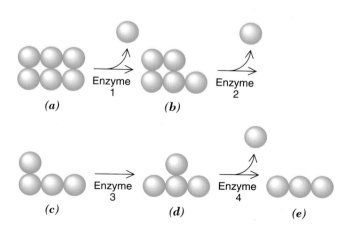

Many of the same metabolic pathways are present in virtually all modern organisms—from bacteria to mammals to trees—indicating that they arose at an early stage in biological evolution and have been retained for billions of years.

Metabolic pathways fall into two categories, depending on whether the products are more or less complex than the original substrates. **Catabolic pathways** (illustrated by the diagram above) degrade complex compounds into simpler molecules, releasing building materials and chemical energy. Therefore, catabolic pathways are exergonic (energy releasing). In contrast, **anabolic pathways** lead to the biosynthesis of complex molecules from simpler components. Since molecular construction requires energy, anabolic reactions are said to be endergonic (energy absorbing). But there is more to the differences between anabolism and catabolism than just molecular complexity and energy transfers. A key feature of both types of pathways is the transfer of electrons. In order to understand these electron transactions, we must first consider the processes of oxidation and reduction.

OXIDATION AND REDUCTION: A MATTER OF ELECTRONS

Oxidation and reduction describe the relationship between atoms and electrons. For example, we saw in Chapter 4 how the electrons of a covalent bond may be shared unequally. When a carbon atom is covalently bonded to a hydrogen atom, the carbon atom has the strongest pull on the shared electrons. Because of its greater "possession" of electrons, carbon atoms bonded to hydrogen atoms are said to be in a *reduced* state. In contrast, if a carbon atom is bonded to a more electron-attracting atom, such as an oxygen or a nitrogen atom, the electrons are pulled *away* from the carbon atom; the carbon is said to be in an *oxidized* state. But the state of **reduction** or **oxidation** of a carbon atom is not an all-or-nothing condition. Since carbon has four outer-shell electrons it can share with other atoms, it can exist in various *oxidation states*. This is illustrated by a series of one-carbon molecules (Figure 6-11), ranging from the fully reduced methane (CH_4) to the fully oxidized carbon dioxide (CO_2). The oxidation state of an organic molecule is a measure of its energy content. The compounds that we use as chemical fuels to run our furnaces and automobiles are highly reduced organic molecules, such as natural gas (CH_4) and petroleum derivatives. Energy is released when these molecules are burned in the presence of oxygen, converting the carbons to more oxidized forms, primarily carbon dioxide gas. The energy-rich fuel molecules in a living organism are its carbohydrates and fats. Carbohydrates are rich in chemical energy because they contain strings of ($H-\overset{|}{\underset{|}{C}}-OH$) units. Fats contain even greater energy per unit weight because they contain strings of more reduced ($H-\overset{|}{\underset{|}{C}}-H$) units. Oxidation of these cellular fuels provides organisms with their usable energy.

Transferring Electrons during Chemical Reactions

We used the terms reduced and oxidized in the previous sections to describe the electronic state of a carbon atom. These terms are also used to describe a *change* in that state.

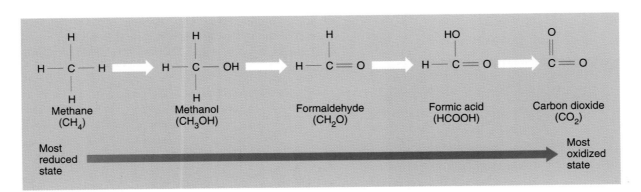

H	H	H	HO	O
H—C—H	H—C—OH	H—C=O	H—C=O	C=O
H	H			
Methane	Methanol	Formaldehyde	Formic acid	Carbon dioxide
(CH_4)	(CH_3OH)	(CH_2O)	($HCOOH$)	(CO_2)

Most reduced state ⟶ Most oxidized state

—— Covalent bond in which carbon atom has greater share of electron pair
—— Covalent bond in which oxygen atom has greater share of electron pair

FIGURE 6-11

The oxidation state of a carbon atom. The oxidation state of a given carbon atom depends on the other atoms to which it is bonded. Each carbon atom can form a maximum of four bonds with other atoms. This series of simple, one-carbon molecules illustrates the various oxidation states in which the carbon atom can exist. In its most reduced state, the carbon is bonded to four hydrogens (forming methane); in its most oxidized state it is bonded to two oxygens (forming carbon dioxide). The relative oxidation state of an organic molecule can be roughly determined by counting the number of hydrogen versus oxygen (or nitrogen) atoms per carbon atom.

A molecule is said to be *oxidized* when it loses one or more electrons, whereas it is said to be *reduced* when it gains one or more electrons. Electrons lost during oxidation are not simply released free into the medium, however; instead, they are immediately captured by another molecule, which is therefore reduced in the process. Consequently, whenever one substance is oxidized, another must be reduced.

We can illustrate these features by examining one of the key reactions of the anabolic pathway of photosynthesis. Don't be concerned with the chemical structures of the molecules in these reactions. Your attention is best directed to the red carbons and the transfer of the electrons indicated by the circles.

In this reaction, a pair of electrons (together with a proton) are transferred from NADPH (nicotinamide adenine dinucleotide phosphate) to the substrate DPG. NADPH is an important molecule in many anabolic reactions that occur in all types of organisms, from bacteria to humans. NADPH functions primarily as an electron donor, raising the energy level and state of reduction of the substrate. A cell's reserve of NADPH is referred to as its **reducing power,** which is one measure of a cell's energy supply. As a result of the transference of electrons in the above reaction, the NADPH becomes oxidized to $NADP^+$, while the DPG becomes reduced to PGAL. This reduction is evidenced by the red carbon; before the reaction, the carbon was bonded to an oxygen, and after the reaction, it is bonded to a hydrogen. In this reaction, energy has been transferred in the form of electrons from the coenzyme NADPH to the substrate, leaving the substrate with a higher energy content.

DIRECTING METABOLIC TRAFFIC

Just as organisms must adapt to changes in external conditions, cells must respond to changes in internal conditions. One of the mechanisms by which cells maintain an ordered internal environment is by directing materials into the metabolic pathways that best satisfy the cell's needs at that particular moment. Glucose, for example, can be directed along several different metabolic routes. It can be dismantled into carbon dioxide and water to generate energy; it can be linked to other glucose molecules to form a polysaccharide; it can be partially disassembled into fragments used to build lipids; or it can be modified to form amino acids or nucleotides. Even the simplest cell can accurately "evaluate" these needs and direct the fate of glucose (and the thousands of other molecules for which multiple options exist). How can cells "decide" which of these options best satisfies its needs? The answer is they can't, but there is no need for them to because a cell has built-in mechanisms that regulate enzyme activities according to the cell's needs.

Changing an enzyme's activity is usually accomplished by modifying the enzyme in a way that alters the shape of its active site. Two of the most common mechanisms for doing so are covalent modification and feedback inhibition.

Altering Enzyme Activity by Covalent Modification

During the mid 1950s, Edmond Fischer and Edwin Krebs of the University of Washington were studying *phosphorylase,* an enzyme found in muscle cells, which disassembled glycogen into its glucose subunits. The enzyme could exist in either an inactive or an active form. Fischer and Krebs prepared a crude extract of muscle cells and found that inactive enzyme molecules in the extract could be converted to active ones simply by adding ATP to the test tube. Further analysis revealed a second enzyme in the extract—a "converting enzyme," as they called it—that transferred a phosphate group from ATP to one of the 841 amino acids that make up the glycogen-disassembling enzyme. The presence of the phosphate group altered the shape of the active site of the enzyme molecule and increased its catalytic activity. Subsequent research has shown that covalent modification of enzymes, as illustrated by the addition of phosphates, is a general mechanism for activating (or inactivating) enzymes. Enzymes that transfer phosphate groups to other proteins are called **protein kinases,** and are involved in regulating such diverse activities as hormone action, cell division, and gene expression. The discovery of protein kinases earned Krebs and Fischer the Nobel Prize in 1992.

Altering Enzyme Activity by Feedback Inhibition

Feedback inhibition is a mechanism whereby a key enzyme in a metabolic pathway is temporarily inactivated when the concentration of the end product of that pathway —an amino acid, for example—becomes elevated. This is illustrated by the simple pathway depicted in Figure 6-12 in which two substrates A and B are converted to the end product E. As the concentration of the product E—an amino acid, for example—rises, it binds to the enzyme BC, causing a conformational change in the enzyme that decreases the enzyme's activity.

Feedback inhibition is analogous to a refrigerator or air conditioner whose function is to generate cold air rather than a chemical product. When the environment becomes too cold, the cold air acts on the thermostat to switch off production of additional cold air. In the case of feedback inhibition, the end product E does not bind to the enzyme's active site but to a separate "feedback site" on the large enzyme molecule. Like the sequential collapse of a row of dominoes, the binding of the end product to the feedback site sends a "ripple" through the protein, producing a de-

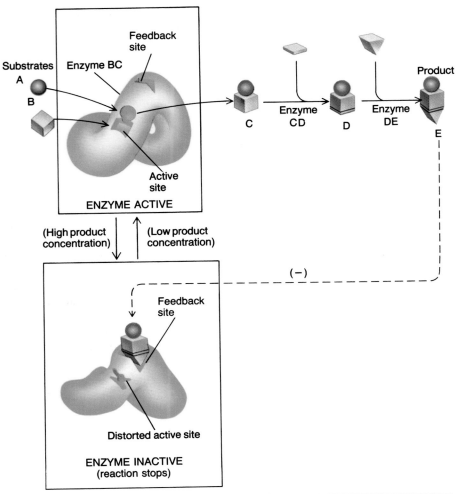

FIGURE 6-12

FIGURE 6-12

Metabolic control by feedback inhibition. When concentrations of the product E are low, the first enzyme (BC) is active, and the pathway proceeds to completion. As the end product (E) accumulates, it binds to the enzyme's feedback (or *allosteric*) site, changing the shape of the enzyme's active site, decreasing further production of the end product (E).

fined change in the shape of the active site, thereby decreasing the enzyme's activity. Feedback inhibition prevents an organism from wasting resources by continuing to produce compounds that are not needed at the time and may even become toxic if allowed to accumulate to high levels.

Control of Enzyme Synthesis

Although covalent modification and feedback inhibition of enzyme activity may provide immediate inactivation of a metabolic pathway, these activities still allow energy to be wasted on synthesizing unneeded enzymes. Organisms possess additional control mechanisms that prevent such metabolic extravagance by inhibiting the continued production of unneeded enzymes. For example, the enzymes

for digesting the sugar lactose are not manufactured when there is no lactose to digest. The mechanism by which cells avoid squandering their resources, making unneeded enzymes, is described in Chapter 15.

The molecules of biological importance are usually quite different from one another structurally and cannot be converted from one to another in a single step. Rather, such conversions require a series of chemical reactions, each catalyzed by a different enzyme, forming a metabolic pathway. Certain key reactions in these pathways include the transfer of electrons from one substrate to another, a process that changes the energy level of the substrates and products. (See CTQ #4.)

◁ THE HUMAN PERSPECTIVE ▷
The Manipulation of Human Enzymes

Recent advances in recombinant DNA technology (Chapter 16) have allowed investigators to isolate an individual gene from human chromosomes, to alter its information content, and to synthesize the modified protein with its altered amino acid sequence. One protein that has been the subject of such manipulation is tissue plasminogen activator (TPA), a key enzyme in the complex process by which blood clots are naturally dissolved in the human body. Human TPA is one of the first proteins to be produced commercially by the biotechnology industry, and its synthesis is finding widespread application in hospital emergency rooms around the world.

Most heart attacks result from blood clots that become lodged in the coronary arteries—the vessels that carry fresh oxygenated blood to the heart. TPA is a "clot-buster," an enzyme that, if injected very soon after a heart attack, dissolves the blood clot that is blocking the flow of blood to the heart tissues. The enzyme presently being marketed is identical to that produced naturally within the human body. There are certain problems with the natural form of the enzyme, however. For example, the blood contains a TPA inhibitor that binds to an inhibitor site on the enzyme, decreasing activity at the active site. In response to this problem, a synthetic version of the enzyme has been produced which lacks a number of amino acids in the inhibitor-binding site, thus preventing the body's natural TPA inhibitor from binding to the synthetic enzyme.

A much more drastically modified version of TPA has been produced by combining the catalytic portion of the TPA molecule with a totally different protein (an antibody molecule). The antibody portion of this "hybrid enzyme" acts to target the molecule directly to the blood clot with which it binds very tightly. This brings the TPA portion of the hybrid enzyme into very close proximity with its intended substrate. In the laboratory, this hybrid enzyme is a very effective clot dissolver, but whether or not such modified enzymes are safe for human use remains to be determined.

MACROMOLECULAR EVOLUTION AND ADAPTATION OF ENZYMES

▐▐▶ Enzymes subject to feedback inhibition probably began as simple catalytic proteins without a separate feedback site. Over time, natural selection favored the survival of organisms whose enzymes were inhibited by metabolic products because these organisms made more efficient use of available nutrients than did their counterparts lacking such regulatory mechanisms. This, in turn, led to the appearance of enzymes with sites to which metabolic end products could bind.

"Fingerprints" left by evolution can also be revealed by comparing the same enzyme in similar organisms that live under very different environmental conditions. Consider two species of bacteria, one living in hot springs in Yellowstone Park at temperatures above 90°C, and another living in a nearby pond fed by cool spring water. The enzymes present in these two types of bacteria have their optimal activity at drastically different temperatures. For example, the purified, DNA-synthesizing enzyme from the bacteria that lives in the hot springs is active in the test tube at 90°C, while the same type of enzyme from the other bacteria is totally inactive at this temperature. Differences in primary, secondary, and tertiary structure provide added stability to the enzyme having the higher temperature optimum. Biotechnologists have taken advantage of the temperature stability of the DNA-synthesizing enzyme from hot springs bacteria to develop a revolutionary new technique that allows them to produce large amounts of DNA from a miniscule amount of starting material (Chapter 16). Another example of the scientific manipulation of enzymes is discussed in the accompanying Human Perspective box.

Enzymes are adaptations, just as are fingers in humans and feathers in birds. Just as natural selection leads to the survival of species with more efficient anatomic traits, it also leads to survival of species with more efficient enzymes. (See CTQ #6.)

REEXAMINING THE THEMES

Relationship between Form and Function

The ability of an enzyme to catalyze a specific reaction derives from its structure, most notably from the position of particular amino acids within the active site. The active site of an enzyme has a precise molecular shape that selectively binds to a specific substrate and converts the molecule to a product. If the shape of the active site should be altered, even slightly (as occurs, for example, when the feedback site of an enzyme is occupied by an end product), the activity of the enzyme may be greatly decreased.

Biological Order, Regulation, and Homeostasis

Just as organisms must maintain a stable, ordered internal environment to remain alive, so too must cells. Cells accomplish this, in part, by regulating the activity of enzymes that govern critical steps in metabolic pathways. Covalent modification and feedback inhibition provide two of the best examples of biological regulation at the molecular level. During covalent modification, the activity of an enzyme is changed by the attachment of a small chemical group, such as a phosphate, by another enzyme. Feedback inhibition operates when a particular end product of a metabolic pathway reaches an elevated concentration, which promotes its binding to a key enzyme in the pathway, preventing further production of the product.

Acquiring and Using Energy

Energy is the central topic of this chapter. Here, we have defined energy, described how it is converted from one form to another, discussed how some of it inevitably raises the level of disorder in the universe, and seen how the direction taken by chemical reactions can be explained by a shift from a higher to a lower energy state. Most of the body's energy is stored in energy-rich macromolecules. Energy released by the disassembly of these macromolecules is stored temporarily in ATP, whose subsequent hydrolysis drives reactions that would otherwise be unfavored.

Unity within Diversity

No matter how different their form and function, all cells use ATP as energy currency. The same enzymes and metabolic pathways use ATP to drive endergonic reactions in organisms as diverse as bacteria and humans, illustrating that not only are the types of biochemicals remarkably similar among diverse organisms (Chapter 4), so too are the enzymes and the metabolic pathways that form these biochemicals.

Evolution and Adaptation

Enzymes are adaptations and, therefore, are the result of natural selection. For example, organisms that live at markedly different temperatures have enzymes that are active at temperatures that correspond to that of their normal environment. The DNA-synthesizing enzyme from bacteria living in certain hot springs, for example, is active at temperatures above 90°C, while the DNA-synthesizing enzyme from most bacteria is totally inactive at such high temperatures.

SYNOPSIS

Energy is the capacity to do work. Energy can occur in various forms, including chemical, mechanical, light, electrical, and thermal, which are interconvertible. Whenever an exchange of energy occurs, the total amount of energy in the universe remains constant, but there is a loss of usable energy, that is, energy available to do additional work. This loss of usable energy, called entropy, results from an increase in the randomness and disorder of the universe. Living organisms are systems of low entropy, maintained by the constant input of external energy—energy ultimately derived from the sun. All spontaneous (favorable) reactions proceed to a lower state of energy. The hydrolysis of ATP is an example of a favorable reaction. Many unfavorable reactions that would normally fail to occur in a cell are driven by coupling them to ATP hydrolysis.

Enzymes are proteins that vastly accelerate the rate of specific chemical reactions by binding to the reactant(s) and increasing the likelihood that they will be converted to products. Enzymes act by lowering the activation energy—the energy required by reactants to undergo a reaction. To do this, reactants may be held in the proper orientation, subjected to physical stress, and made more reactive by interacting with amino acids in the protein. The specificity and catalytic activity of enzymes are due to the complementary structure of the active site that binds the substrate(s), enabling the enzyme to exert its influence. Many enzymes also contain nonprotein "helpers," which may be organic coenzymes or metals. Enzyme structure—and thus function—is affected by temperature, pH, and the presence of inhibitors.

Enzymatic reactions are organized into metabolic pathways in which the product(s) of one reaction becomes the substrate(s) of a subsequent reaction. Catabolic pathways lead to less complex products, with the accompanying release of energy. In contrast, anabolic pathways build more complex products at the expense of cellular energy. Many reactions include the transfer of electrons from one molecule to another. When a molecule receives one or more electrons it becomes reduced; when a molecule gives up one or more electrons it becomes oxidized. Electrons carry with them chemical energy; thus, the more reduced an organic molecule, the higher its energy content. The activity of many key enzymes is under cellular control and can be altered by interaction with certain metabolic end products or by covalent modification, such as the addition of a phosphate group.

Key Terms

energy (p. 118)
potential energy (p. 118)
kinetic energy (p. 118)
first law of thermodynamics (p. 119)
second law of thermodynamics (p 120)
entropy (p. 120)
exergonic (p. 122)
endergonic (p. 122)
coupling (p. 122)

catalyst (p. 123)
activation energy (p. 123)
substrate (p. 125)
active site (p. 125)
conformational change (p. 125)
competitive inhibitor (p. 126)
cofactor (p. 127)
coenzyme (p. 127)
metabolic pathway (p. 127)

metabolic intermediate (p. 127)
catabolic pathway (p. 128)
anabolic pathway (p. 128)
reduction (p. 128)
oxidation (p. 128)
reducing power (p. 129)
protein kinase (p. 129)
feedback inhibition (p. 129)

Review Questions

1. Contrast potential and kinetic energy, using both a biological and a nonbiological example.

2. Describe how enzymes decrease activation energies; how they are able to bind only specific substrates; why they may need cofactors; and why their activity is sensitive to temperature and pH.

3. What is the relationship between enzyme activity and each of the following:

   a. *increasing temperature (two answers)*
   b. *cold*
   c. *the presence of a structural analog of the substrate (an analog is a molecule with a similar shape)*
   d. *lead poisoning*
   e. *extreme acidity*

4. Why does NADPH provide a cell with reducing power?

Critical Thinking Questions

1. Suppose that Eduard Buchner had added sugar to his yeast-juice preparation and failed to observe the occurrence of fermentation. Would this have proven that fermentation could only occur in a living yeast cell. Why or why not?

2. Using the energy principles discussed in the beginning of this chapter, explain why it is a more efficient use of agricultural land for people to eat grains rather than meat from grain-fed cattle.

3. Lead was formerly used in the manufacture of house paints. Young children who live in homes where lead paint still exists unintentionally ingest flakes of it, which may result in serious and permanent brain damage. Using your knowledge of enzymes, explain how even minute quantities of lead can have such disastrous consequences.

4. Study the diagram *(right)* and identify which compounds (A–K) would *not* be produced under each of the following conditions. Explain why.
   (1) Compound E binds to and inhibits enzyme 2 when E is present in high concentrations.
   (2) A compound is added that competes with compound F for enzyme 7 but cannot be converted to compound G.
   (3) Enzyme 5 is inactivated by a change in pH.
   (4) Mercury is added and contacts enzymes 1–10.

5. How is it that feedback inhibition can block the formation of a particular product, such as an amino acid, even though it binds to only one enzyme in the entire pathway?

6. The enzyme lactase digests lactose, a disaccharide found in milk, to glucose and galactose, two monosaccharides. When lactase is not present, lactose is not digested and can cause intestinal disturbances in humans. Lactase is normally present in young children and in adults who continue to eat a diet that includes milk products, but it is not present in adults who do not consume milk. What do you think was the adaptive value of this enzyme for primitive humans?

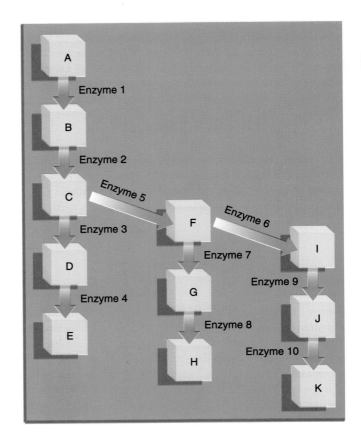

Additional Readings

Cutnel, J. D., and K. W. Johnson, 1992. *Physics.* New York: Wiley. (Introductory)

Kornberg, A. 1989. Never a dull enzyme. *Ann. Revs. Biochem.* 58:1–30. (Intermediate)

Kornberg, A. 1991. *For the love of enzymes: The odyssey of a biochemist.* Cambridge: Harvard. (Intermediate)

Laszlo, P. 1986. *Molecular correlates of biological concepts: A history of biochemistry.* Comprehensive Biochemistry,

Vol. 34A. New York: Elsevier. (Advanced)

Lehninger, A. L. 1982. *Principles of biochemistry.* New York: Worth. (Intermediate)

Stryer, L. 1988. *Biochemistry.* 3d ed., New York: Freeman. (Advanced)

Voet, D., and J. G. Voet. 1990. *Biochemistry.* New York: Wiley. (Advanced)

Movement of Materials Across Membranes

STEPS
TO
DISCOVERY
Getting Large Molecules Across Membrane Barriers

MEMBRANE PERMEABILITY

DIFFUSION: DEPENDING ON THE RANDOM MOVEMENT OF MOLECULES

Conditions For Diffusion Across Membranes

Osmosis

Facilitated Diffusion

ACTIVE TRANSPORT

Generating Ionic Gradients And Storing Energy

ENDOCYTOSIS

Phagocytosis

Pinocytosis

THE HUMAN PERSPECTIVE

LDL Cholesterol, Endocytosis, and Heart Disease

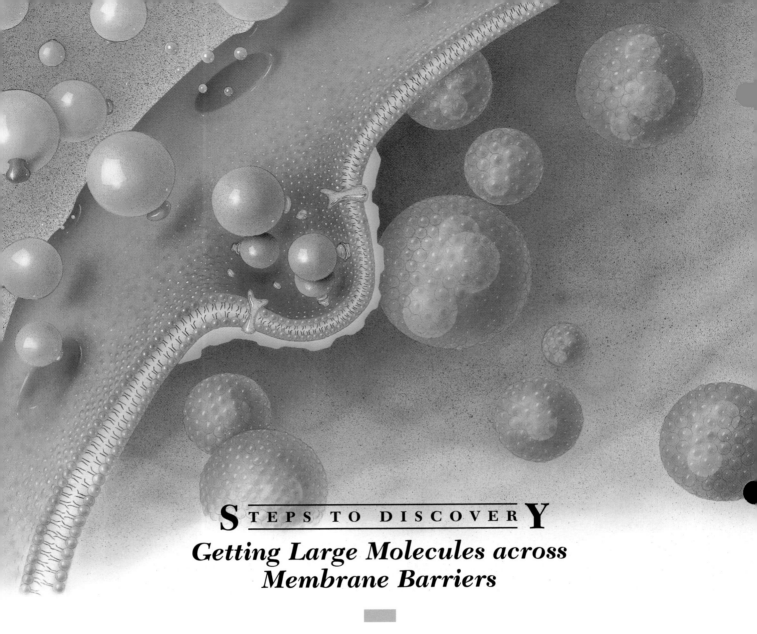

Getting Large Molecules across Membrane Barriers

For many years, a question that puzzled cell biologists was how large molecules were able to get across an intact plasma membrane. Your blood, for example, delivers a variety of large molecules to all your cells. These molecules pass from the bloodstream across the plasma membrane, and into the cytoplasm of the cells in which they are used. It was not until the 1970s that this process came to be understood at the molecular level, primarily as a result of the work of Michael Brown and Joseph Goldstein at the University of Texas.

Brown and Goldstein stumbled into the area of mem-

brane transport as a result of their interest in a genetic disease called *familial hypercholesterolemia* in which people suffer from extremely high levels of blood cholesterol. Cholesterol is a hydrophobic molecule. Because it is virtually insoluble in water, cholesterol is transported through the circulation in special packages called low-density lipoproteins (LDLs). LDLs are spheres that contain about 1,500 cholesterol molecules packaged in a sac of phospholipids and protein that keeps the enclosed cholesterol suspended in the blood. Cells take up these large LDLs and

Cholesterol-containing particles enter a cell after binding to receptors situated in coated pits on the cell surface. The particles

liberate the cholesterol molecules, which — depending on the type of cell — may be used for various functions, including assembling cell membranes, forming steroid hormones, or producing bile in the liver.

Brown and Goldstein began their studies on cholesterol by examining cells growing in cultures, thus avoiding the complications of working with whole animals. The investigators used cells called *fibroblasts,* which remove cholesterol-containing LDLs from the culture medium in which they are growing. Brown and Goldstein incubated the fibroblasts with a preparation of LDL that contained radioactive atoms (the radioisotope acted as a marker to reveal the location of the labeled LDLs). They found that the radioactively labeled LDLs bound tightly to the surface of the fibroblast, suggesting that the surface of these cells contained *receptor* molecules with binding sites that specifically fit the LDLs. Within 10 minutes of attachment, virtually all of the bound LDL had entered the cell and was being converted to cholesterol for use in cellular metabolism. The LDL receptors then returned to the plasma membrane where they could bind additional LDLs.

These early findings raised an important question: How do these giant LDL particles get into the cell? To answer this question, Brown and Goldstein, together with Richard Anderson, prepared LDLs that were chemically bonded to iron particles (which are easily seen in the electron microscope). When they examined the cells under the microscope, they were surprised to find that the iron-containing LDL particles were not spread uniformly over the fibroblast surface, but were concentrated into special sites. These sites are called *coated pits* because the membrane is indented to form a pit that is coated on its inner surface by a

layer of bristly looking protein. This finding suggested that the coated pits were sites where LDL receptors became concentrated. The investigators envisioned the LDL receptors as proteins that spanned the membrane. As these proteins moved around within the membrane, they would inevitably bump into a coated pit. When this happened, the receptors would bind to the bristly coating and remain in the pit. Viewed in this way, the coated pits were acting as "molecular traps" that ensnared the LDL receptors, concentrating them within small regions of the membrane. When they waited longer periods of time before killing and examining the cells, the scientists found that the coated pits eventually bulged inward, forming pouchlike structures, which then pinched off from the remainder of the membrane, forming bristly coated cytoplasmic vesicles (see Figure 7-11). The bristly protein and the LDL receptors then returned to the membrane, while the LDL was broken down by lysosomal enzymes, releasing free cholesterol molecules, which passed into the cytoplasm for use by the cell.

Within a few years, a dozen or so different receptors were discovered which acted in a similar manner to the LDL receptor, suggesting that this was a general mechanism for the specific ingestion of a variety of large, blood-borne molecules or particles. For their work on the role of the LDL receptor, as well as their important discoveries concerning cholesterol metabolism, Brown and Goldstein were awarded the 1985 Nobel Prize for Medicine.

are enclosed in vesicles pinched off from the plasma membrane.

As we have seen in previous chapters, a living cell is a bastion of complexity and order—a microscopic package that contains thousands of different molecules whose concentrations are maintained by carefully controlled reactions. A cell's internal order is protected by the plasma membrane—a structure so thin it would take approximately 40,000 of them stacked one on top of the other to equal the thickness of one page of this book. The plasma membrane keeps essential substances inside the cell and regulates the passage of materials between a cell and its environment. The basis for these functions becomes apparent when we consider the plasma membrane's structure.

▼ ▼ ▼

MEMBRANE PERMEABILITY

We described in Chapter 5 how the plasma membrane is a mosaic of glycoproteins embedded in a fluid, bilayered sheet of phospholipids (Figure 7-1; see also Figure 5-4). The two major components of the plasma membrane—its proteins and lipid bilayer—affect whether substances enter or leave a cell. On the one hand, the lipid bilayer acts primarily as a barrier, blocking the passage of most molecules based primarily on their insolubility in lipid. On the other hand, membrane proteins provide an alternate pathway across the membrane for a substance that would be unable to penetrate a hydrophobic lipid sheet. Unlike the lipid bilayer, which allows any lipid-soluble material to pass, membrane proteins are able to *control* the entrance and exit of "selected" substances. Just as enzymes bind specific substrates and catalyze their reaction, these membrane proteins bind specific solutes and promote their movement across the membrane.

Together, the phospholipid bilayer and embedded proteins make the plasma membrane **selectively permeable;** that is, it allows certain substances to cross, while restricting the passage of others. Selective permeability enables a cell to import and accumulate essential molecules to concentrations high enough for normal metabolism. It also enables the cell to export wastes and other substances that might interfere with metabolism. The appearance of a lipid-containing membrane that allowed a primitive "cell" to retain its nutrients and other essential materials must have been one of the most critical steps in the early evolution of life on earth.

If we are to understand the selective permeability of the plasma membrane, we need to consider the means by which individual molecules can traverse the membrane. There are basically two means for such movement: passively, by diffusion, or actively, by a transport process requiring an energy boost from the cell.

Normal cell function requires a dynamic barrier between the cell and its surroundings in order to prevent dilution of cell contents, to safeguard contents against external conditions, and to maintain a concentration of essential molecules and ions inside the cell different from that in the surrounding fluid. The plasma membrane accomplishes these functions by selectively controlling the materials entering and exiting the cell. (See CTQ #2.)

DIFFUSION: DEPENDING ON THE RANDOM MOVEMENT OF MOLECULES

Diffusion is the passive movement of molecules in a gas or liquid from a region where they are present at higher concentration to a region where they are present at lower concentration; in other words, *down a concentration gradient.* Diffusion is readily demonstrated by dropping colored dye into a glass of water. The dye molecules are localized at first in one region, which appears dark in color. After time, however, the color spreads until the dye molecules are distributed rather evenly throughout the glass. Let's examine why this happens.

All molecules exist in a state of continuous, random (that is, unpredictable) movement. If molecular move-

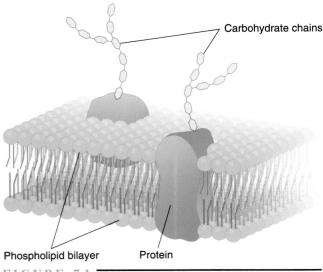

Carbohydrate chains

Phospholipid bilayer   Protein

FIGURE 7-1

The structure of the plasma membrane.

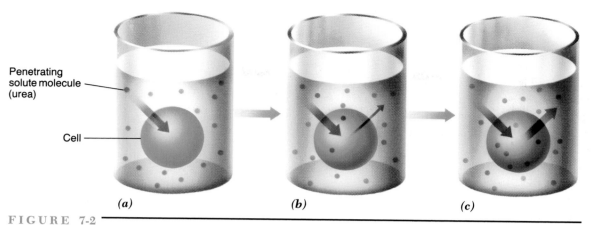

Penetrating
solute molecule
(urea)

Cell

(a)   *(b)*   *(c)*

FIGURE 7-2

Diffusion of a solute through the plasma membrane. In this example, a cell is dropped into a solution containing a solute (urea) capable of penetrating the membrane. Initially, the concentration of solute is much greater outside the cell than inside, resulting in the rapid movement of solute into the cell *(a)*. After the solute enters the cell, some of the molecules will diffuse back across the membrane to the outside, but the *net* movement (as indicated by the larger arrow) will still be into the cell *(b)*. Eventually, the concentration on the two sides of the membrane will be equal, and the movement of solute will occur at the same rate in both directions *(c)*.

ments are random, why is there a predictable diffusion of molecules from a region of high concentration to one of low concentration? The answer can best be visualized using an analogy. Consider what would happen if two people were to go into a room with a line painted down the center, and each person were to stand at equal distances from the line on opposite sides. Then, one person were to drop ten bouncing rubber balls on his side of the line and the other were to drop only a single ball on her side. Even though the movement of each ball occurred at random, the chances would be very high that, after the balls had come to rest, there would be more than one ball on the second person's side and fewer than ten balls on the first person's side. In other words, a *net* movement of balls from the side of initial high concentration to the side of initial low concentration is likely to occur. Thus, even though the movement of an individual ball (or molecule) is not predictable, the collective action of a large number of them is quite predictable. In the case of molecules (Figure 7-2), which, unlike rubber balls, never come to rest, the ultimate outcome will be a state of equilibrium in which the concentration of molecules will be equal on both sides of the membrane. When this occurs, molecules will be crossing back and forth across the membrane at equal rates; there will no longer be any *net* movement of molecules. This process is known as *simple diffusion*. Diffusion plays a key role in the movement of many substances within the body. When you breathe, for example, oxygen moves by diffusion from the lungs into the bloodstream. Then, when the oxygen-rich blood moves through the tis-

sues, oxygen again moves by diffusion out of the blood into the cells where it is utilized.

CONDITIONS FOR DIFFUSION ACROSS MEMBRANES

Two qualifications must be met before a substance can diffuse across a plasma membrane. First, the substance must be present at a higher concentration on one side of the membrane than on the other, and second, the membrane must be permeable to (able to be penetrated by) the substance. The *rate* of diffusion of a given substance across the plasma membrane—that is, the number of molecules that will diffuse through a given section of membrane per unit time—depends on a number of conditions. In general, the lipid bilayer is much more permeable to nonpolar molecules and to small molecules than to ions, polar molecules, or large molecules. Some cells are quite permeable to specific ions (Na^+, K^+, or Ca^{2+}) because they contain special channels that allow the ions to pass unimpeded across the lipid bilayer. These ion channels consist of clusters of membrane proteins that form a "doughnut-shaped" complex that spans the membrane (Figure 7-3). The ions pass through the center of the "doughnut." Most ion channels are equipped with *gates* that may be either open or closed. The opening and closing of these gates play a key role in the movement of an impulse along the membrane of a nerve cell (discussed in Chapter 23). Two types of diffusion of particular importance in biology are osmosis and facilitated diffusion.

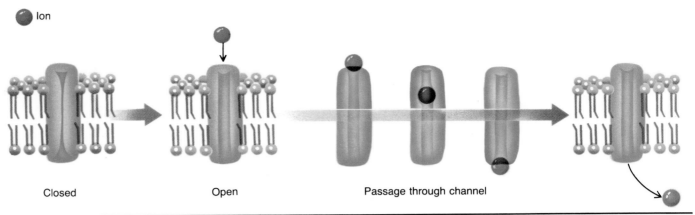

Ion

Closed   Open   Passage through channel

FIGURE 7-3

Ion channels consist of pores situated within the center of membrane proteins or within clusters of such proteins. Most ion channels possess a "gate" that can exist in either an open or a closed conformation, depending on conditions in the cell.

OSMOSIS

Even though water is highly polar, it diffuses readily across the lipid bilayer of the plasma membrane because it is a very small molecule and is able to squeeze between neighboring phospholipids. The diffusion of water through a membrane has a special name—**osmosis.** In understanding osmosis, let us consider what might happen to a sea urchin egg under different environmental conditions. Sea urchin eggs (a favorite of both embryologists and gourmet diners) are spawned into sea water, where they are fertilized and develop into microscopic swimming larvae. The concentration of dissolved substances (solutes) in a sea urchin egg is virtually identical, or **isotonic,** to that of the surrounding sea water. In such isotonic (*iso* = equal) solutions, the movement of water into and out of a cell is equal; therefore, the egg neither gains nor loses water by osmosis (Figure 7-4*a*).

Now let us consider what might happen if the same egg should be caught in a shallow tidepool on a hot day at a very low tide. As the water in the tidepool evaporates, its solute concentration rises and, conversely, its water concentration decreases. As a consequence of the difference in water concentration between the egg and its surroundings, water will move out of the cell by osmosis, causing the cell to shrink. In other words, cells shrink when surrounded by **hypertonic** solutions—solutions with a higher solute concentration (*hyper* = more) than that found inside the cell (Figure 7-4*b*).

Finally, let us consider what might happen if the sea urchin egg drifts into an *estuary,* a site along the coastline where a freshwater river empties into the sea. Under these conditions, the egg would find itself in a medium of diluted sea water, which would have a lower solute concentration

—and, thus, a higher water concentration—than that of the egg. In this case, the external medium is said to be **hypotonic** to that of the egg (*hypo* = less), and water will move by osmosis into the egg, causing the egg to swell (Figure 7-4*c*).

To summarize, water always moves across a membrane by osmosis from the hypotonic compartment (one with a lower solute concentration) to the hypertonic compartment (one with a higher solute concentration). It makes no difference if the types of solute molecules on the two sides of the membrane are completely different—the only important factor is the total solute concentration on each side.

▮▶ Osmosis is a fact of life with which cells—and entire organisms—have to live. In some circumstances, osmosis is a challenge that natural selection has been able to overcome. For example, as organisms moved from the sea to inhabit fresh waters, they generally moved from an environment that was isotonic to their bodily fluids into one that was very hypotonic. These organisms evolved adaptations that allowed them to expel the excess water that flooded into their bodies (Chapter 28). In other circumstances, osmosis has facilitated certain bodily functions. Your digestive tract, for example, produces several liters of fluid, which is reabsorbed osmotically by the cells that line your intestine. If this reabsorption process did not occur, as happens in cases of extreme diarrhea, you would face the prospect of rapid dehydration.

Plants utilize osmosis in other ways. The movement of water is affected not only by solute concentration, but also by pressure. For example, when a plant cell is placed in a hypotonic solution, water diffuses into the cell by osmosis. Because the cell has a rigid wall that resists expanding, however, the plant cell will not swell to the point of burst-

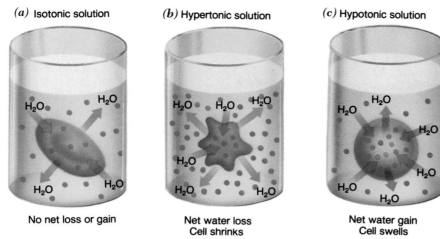

(a) Isotonic solution **(b) Hypertonic solution** **(c) Hypotonic solution**

No net loss or gain

Net water loss
Cell shrinks

Net water gain
Cell swells

FIGURE 7-4

The effects of isotonic, hypertonic, and hypotonic solutions on osmosis. *(a)* A cell placed in an isotonic solution (one containing the same solute concentration as the cell) neither swells nor shrinks because it gains and loses equal amounts of water. *(b)* A cell in a hypertonic solution (one containing a higher solute concentration than the cell) soon shrinks because of a net loss of water by osmosis. *(c)* A cell placed in a hypotonic solution (one having a lower solute concentration than the cell) swells because of a net gain of water by osmosis.

ing, as would an animal cell. Instead, the internal water pressure builds as water continues to diffuse into the plant cell. Eventually, the water pressure builds high enough inside the cell to stop the influx of more water by osmosis, even though the solute concentration inside the cell remains higher than the surrounding solution. Plant cells utilize the water pressure that develops due to osmosis to maintain their rigidity. This internal pressure, called **turgor pressure,** provides support for nonwoody plants or for the nonwoody parts of plants, such as the leaves. When plant cells lose water, as occurs when they are placed in a hypertonic solution, the cell shrinks away from its cell wall, a process termed **plasmolysis** (Figure 7-5). The loss of water due to osmosis (or evaporation in a dry, terrestrial habitat) causes plants to lose their support and wilt.

FACILITATED DIFFUSION

As we mentioned earlier, the diffusion of a substance across a membrane always occurs from a region of higher concentration to a region of lower concentration, but the penetrating molecules do not always move through the membrane unaided. Many plasma membranes contain **carrier proteins** that bind specific substances and facilitate their diffusion across the membrane. This process, termed **facilitated diffusion** (Figure 7-6), might be likened to a ferry

moving people across a river. Like the lipid bilayer, the river is a barrier to traffic, while the ferry provides a specialized avenue across the barrier. The ferry can move its cargo in both directions but, as long as there are more people waiting on the north side than on the south side of the river, there will be a net flow of traffic from north to south. Similarly, in facilitated diffusion, as long as there are more molecules of a substance on the outside of the cell, there will be a net flow of those molecules into the cytoplasm. Facilitated diffusion is particularly common in the inward movement of sugars and amino acids, substances that would be unable to penetrate the hydrophobic lipid bilayer directly. Just as enzymes are specific for particular substrates, carrier proteins are specific for particular solutes; a protein that facilitates diffusion of glucose, for example, will not carry amino acids.

If a specific solute is present at higher concentration on the outside of the plasma membrane than in the cytoplasm, and the membrane is permeable to that particular solute, then there will be a net movement of the solute into the cell without the need for the cell to expend energy. If the membrane is *not* permeable to the solute, then water will move in the direction of higher solute concentration. Such osmotic movements of water are critical to life. (See CTQ #3.)

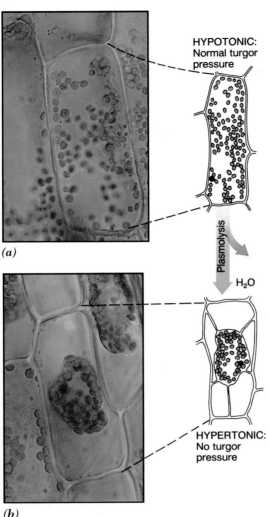

(a)

(b)

HYPOTONIC:
Normal turgor
pressure

Plasmolysis

H₂O

HYPERTONIC:
No turgor
pressure

FIGURE 7-5

The effects of osmosis in plants. Aquatic plants living in fresh water are surrounded by a hypotonic environment. Water therefore tends to flow into the cells, creating high water (*turgor*) pressure *(a).* If the plant is placed in a hypertonic solution, such as seawater, the cell loses water, and the plasma membrane pulls away from the cell wall, a process known as *plasmolysis* *(b).* If the cells of land plants evaporate large amounts of water to dry air, water pressure plunges, and the plant wilts.

ACTIVE TRANSPORT

↻ A frog sitting in a shallow freshwater pond has internal fluids with a solute concentration 40 to 50 times greater than its aqueous environment, which is not much different from pure water. How is this frog able to maintain a salt concentration that is so much higher than its surrounding medium? It can do so because its skin contains proteins capable of "pumping" salts from the environment into the body *against* a very large concentration gradient (Figure 7-7). Movement of substances against a concentration gradient—that is, from a region of lower concentration to one of higher concentration—is called **active transport.** This process requires the input of considerable chemical energy, which is either directly or indirectly supplied by the hydrolysis of ATP.

GENERATING IONIC GRADIENTS AND STORING ENERGY

One of the most widely occurring active transport proteins is the *sodium–potassium pump,* a protein found in all types of eukaryotes. This protein transports sodium ions out of cells and potassium ions into cells, both against steep concentration gradients (Figure 7-8). Human nerve cells, for example, have internal concentrations of potassium that are 30 times greater than that of the extracellular fluid, due to the pumping activities of this protein. The sodium–potassium pump is more than just a transport protein, it is also an enzyme—an *ATPase.* As an ATPase, the protein hydrolyzes ATP, utilizing the energy to move ions against their concentration gradients. During this reaction, the phosphate group released by ATP hydrolysis becomes transiently linked to the transport protein, changing the shape

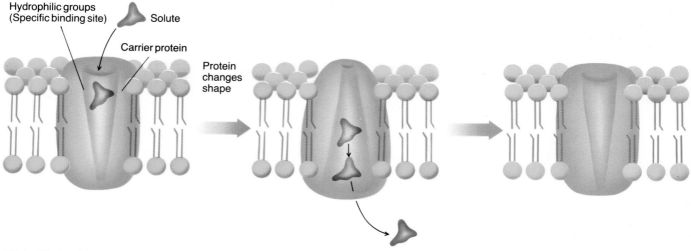

Hydrophilic groups
(Specific binding site)

Solute

Carrier protein

Protein
changes
shape

FIGURE 7-6

Facilitated diffusion. Facilitated diffusion is responsible for the uptake of many small hydrophilic molecules, such as sugars and amino acids. Facilitated diffusion is initiated when a solute molecule binds to a carrier protein at the outer surface of the membrane. The binding of the solute is thought to trigger a conformational change within the carrier protein, exposing the solute molecule to the inner surface of the membrane. The solute molecule is then free to diffuse into the cytoplasm down its concentration gradient. Once the solute diffuses into the cytoplasm, the carrier protein snaps back to its original conformation, ready to bind another solute molecule. Like simple (unaided) diffusion, facilitated diffusion can only occur in a direction from higher concentration to lower concentration and does not require the input of energy.

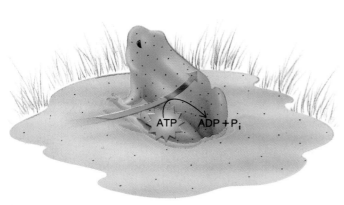

ATP   ADP + P_i

FIGURE 7-7

An example of active transport. Animals that live in freshwater habitats tend to lose salts from their body to their environment by diffusion. Some of this salt is regained in their diet, but most freshwater animals, including this frog, possess mechanisms to take back salts from their environment against a concentration gradient. This process requires the input of energy and is called *active transport*. Active transport proteins in this frog are located in the plasma membranes of its skin.

of the protein and forcing the ions through the membrane (Figure 7-8). With the exception of maintaining an elevated temperature, a resting person utilizes more ATP in pumping ions than in any other activity.

Recall from Chapter 6 that energy can be stored in several forms. ATP serves as a storage of chemical energy, some of which is expended in forming ionic gradients, such as those of sodium and potassium. Ionic gradients are themselves a form of stored (potential) energy since, given the opportunity, these ions will spontaneously diffuse back across the membrane, thereby eliminating the concentration difference. Like the flow of water over a dam, the directed movement of ions back across the membrane is a form of kinetic energy that can be used to accomplish work. For example, when you eat a food that is rich in starch, such as a potato, the glucose molecules derived from the starch are transported from the intestine into the bloodstream — across the intervening intestinal cell — *against a concentration gradient* (Figure 7-9). In this case, glucose molecules are transported by a membrane protein that also carries sodium ions. The "downhill" movement of sodium ions into the cell provides the energy used to *cotransport* the "uphill" movement of glucose molecules. The movement of nerve impulses back and forth between your brain and the other

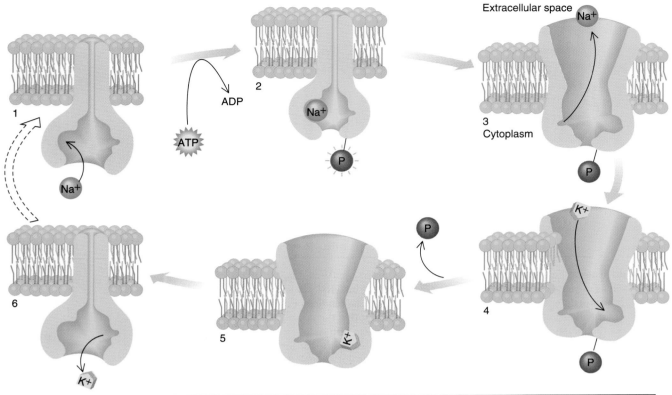

FIGURE 7-8

Schematic concept of the mechanism of the sodium–potassium pump. Sodium ions (1) bind to the protein on the inside of the membrane; ATP is hydrolyzed and the phosphate produced is linked to the protein (2), causing a change in the shape (conformation) of the protein (3), allowing sodium ions to be expelled to the external space. Potassium ions then bind to the protein (4), followed by the removal of the phosphate group (5) which causes the protein to "snap" back to its original conformation, moving the potassium ions to the inside of the cell (6). Unlike facilitated diffusion, the changes in the shape of the protein are driven by energy from ATP hydrolysis, which allows the transport system to move these ions against their concentration gradient.

parts of your body also depends on the movement of sodium ions across a membrane (Chapter 23). As we will see in the following chapters, the formation of ATP in mitochondria and chloroplasts also results from energy stored in ionic gradients.

> **The plasma membrane utilizes energy to transport solutes into regions of higher concentration, a process vital to the survival and normal functioning of a cell. This ability enables cells to set up concentration gradients necessary for generating the energy to drive many energy-requiring activities. (See CTQ #5.)**

ENDOCYTOSIS

Both diffusion and active transport can move smaller-sized solutes directly through the plasma membrane, but what of the uptake of materials that are too large to penetrate a membrane, regardless of its nature? We saw in the last

chapter how materials stored in cytoplasmic vesicles can be released from cells by exocytosis (page 99). As described in the chapter opener, cells also carry out a reverse process, called **endocytosis,** in which large molecular weight materials are enclosed within invaginations of the plasma membrane, which subsequently pinch off to form cytoplasmic vesicles. The ability of the plasma membrane to perform this remarkable activity depends on the dynamic, fluid nature of the membrane. There are two forms of endocytosis:

1. **phagocytosis** (*phago* = eating), in which a cell ingests large particles, such as bacteria or pieces of debris, and
2. **pinocytosis** (*pino* = drinking), in which a cell ingests liquid and/or dissolved solutes and small suspended particles.

PHAGOCYTOSIS

An example of phagocytosis is pictured in Figure 7-10*a*. In Figure 7-10*b*, a unicellular amoeba engulfs its microscopic

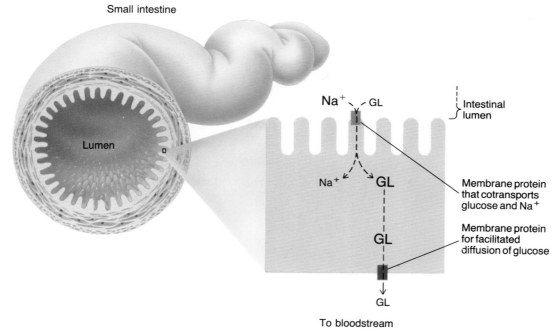

Small intestine

Lumen

Na$^+$ — GL

Na$^+$ GL

Intestinal lumen

Membrane protein that cotransports glucose and Na$^+$

GL

Membrane protein for facilitated diffusion of glucose

GL

To bloodstream

FIGURE 7-9

An example of the use of energy stored in an ionic gradient. The sodium–potassium pump establishes a steep sodium gradient by pumping sodium ions out of the cell. This gradient represents a storehouse of potential energy that can be put to use in various ways. In this case, the gradient is employed to move glucose molecules into an intestinal cell against a concentration gradient. The glucose molecules exit the opposite end of the cell by facilitated diffusion. The relative size of the "Gl" and "Na$^+$" letters indicates the directions of the concentration gradients.

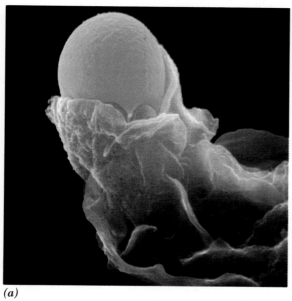

(a)

FIGURE 7-10

Cell uptake by phagocytosis. *(a)* Cell ingesting a synthetic particle. *(b)* A food particle becomes enclosed within an invagination of the plasma membrane, which subsequently pinches off to form a large membranous vesicle (called a *food vacuole*). Food matter is digested by enzymes discharged from lysosomes that fuse with the vesicle membrane. The digested food is then absorbed into the amoeba's cytoplasm.

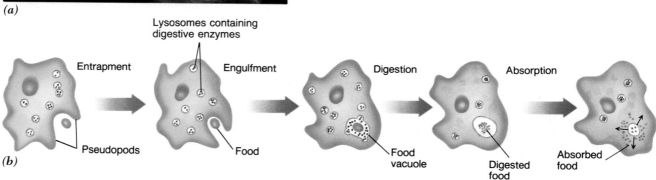

Lysosomes containing digestive enzymes

Entrapment   Engulfment   Digestion   Absorption

Pseudopods   Food   Food vacuole   Digested food   Absorbed food

(b)

◁ THE HUMAN PERSPECTIVE ▷
LDL Cholesterol, Endocytosis, and Heart Disease

We saw in the beginning of the chapter how cholesterol is transported in protein-containing particles called LDLs, which are taken into cells by receptor-mediated endocytosis. This is a critical process since excess LDLs—those that are not ingested by cells and thus remain in the bloodstream—can build up as plaques on the walls of blood vessels and eventually cause *atherosclerosis* (narrowing of the arteries) or heart attacks (Figure 1).

In its severe form, the rare genetic disease *familial hypercholesterolemia* strikes about 1 in every million individuals. Children with this disease have six to ten times the normal level of blood cholesterol and often suffer heart attacks before age 15. Goldstein and Brown, the scientists depicted in the chapter opener, discovered the reason for this inherited condition: These individuals lack LDL receptors; consequently, their cells are not able to remove cholesterol from the blood. This genetic condition has provided the strongest evidence that elevated blood cholesterol levels *alone* are sufficient to predispose a person to cardiovascular disease.

It is evident why these genetically deficient individuals would suffer from atherosclerosis and heart attacks, but why do some people who possess *normal* LDL receptors develop these same diseases, albeit at an older age? Evidence strongly suggests that the root of the problem still lies in elevated blood cholesterol levels which, in many people can be traced to diet (along with genetic factors, smoking, and a lack of exercise).

⟳ Cholesterol is required by all animal cells as a component of their cell membranes. When a cell's cholesterol needs are

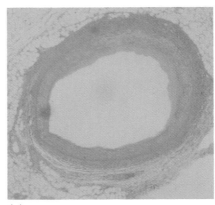

(a)

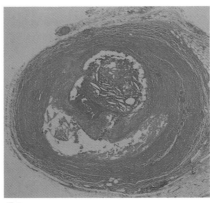

(b)

FIGURE 1 ─────────

Comparison of a normal coronary artery *(a)* and an artery whose channel is almost completely closed as the result of atherosclorosis *(b).*

met, the production of LDL receptors is shut down, and the receptors present are gradually removed from the cell's surface. Evidence suggests that people with a diet high in cholesterol (or saturated fats, which lead to the formation of cholesterol) quickly provide their cells with their needed allotment of cholesterol. As a result, the cells stop producing LDL receptors and stop removing cholesterol from the blood, leading to a rise in blood cholesterol levels and an increased risk of atherosclerosis.

These findings have produced an experimental strategy for lowering blood cholesterol in individuals with particularly high levels. The key to the strategy is to inhibit the synthesis of cholesterol by the cells of the body, particularly those of the liver. This can be accomplished by the prescription drug mevinolin (trade name *Lovastatin*), a compound derived from a fungus that inhibits a key enzyme in cholesterol synthesis. As the cells produce less cholesterol, they must take up more cholesterol from the blood, which is accomplished by an increased production of LDL receptors. The more LDL receptors, the more cholesterol that is taken up, and the less cholesterol that remains in the blood to become deposited on the walls of arteries. Recent evidence suggests mevinolin may actually reverse arterial disease.

The breakthroughs achieved by Goldstein and Brown have paved the way to understanding—and hopefully treating—other receptor-related diseases, including diabetes. Don't be surprised if you soon hear that the medical treatment for a disease is aimed at the plasma membrane's receptors.

meal, a bacterium, gradually enclosing the food particle in a large vesicle. Once the bacterium is enclosed, the vesicle fuses with a lysosome, allowing the hydrolytic enzymes of the lysosome to digest the bacterium. After the macromolecules have been hydrolyzed, the products (sugars, amino acids, etc.) of the digested bacterium are transported though the vesicle membrane and utilized by the amoeba. A similar mechanism is employed by your white blood cells when defending your body against invading microorganisms; the cells engulf the microorganisms in a vesicle and destroy them.

PINOCYTOSIS

Most cells are not capable of phagocytosis, but virtually all eukaryotic cells are capable of pinocytosis. This form of endocytosis is often initiated by the interaction of dissolved or suspended molecules with specific receptors on the plasma membrane; it is known as **receptor-mediated endocytosis,** which we introduced at the beginning of the chapter. Included within this group are receptors for hormones, enzymes, and blood materials, such as LDL (see The Human Perspective: LDL Cholesterol, Endocytosis,

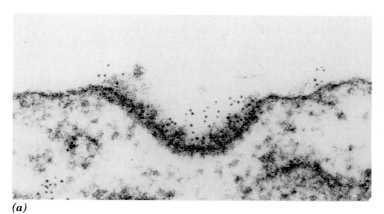

(a)

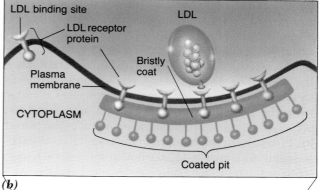

(b)

FIGURE 7-11

Binding and uptake of LDL. (*a*) LDL particles (indicated by black dots on photo, left) bind to receptors situated in coated pits in the plasma membrane. (*b*) The receptors are held in the coated pit by their attachment to a bristly protein on the inner surface of the membrane. Following binding to the LDL receptor, the LDL particles are taken into the cell as part of a bristly-coated vesicle. (*c*) Events that occur following LDL uptake. LDL receptors are returned to the plasma membrane while cholesterol molecules are released from the LDL by lysosomal enzymes.

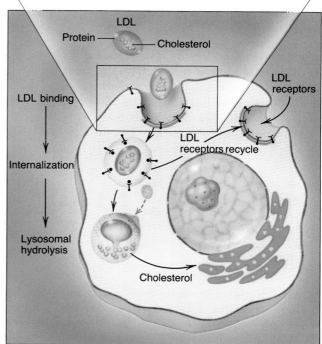

(c)

and Heart Disease). Certain viruses and bacteria take advantage of this cellular activity by binding to a surface receptor and gaining entry to the cell by receptor-mediated endocytosis (Figure 7-11). The cytoplasmic vesicle formed during endocytosis normally fuses with a lysosome whose hydrolytic enzymes degrade the vesicle's contents. The digested products of the vesicle then move across the surrounding membrane into the fluid of the cytoplasm.

Cells must sometimes take in molecules or particles that are too large to pass through the membrane by diffusion or active transport. The fluid nature of the plasma membrane enables cells to surround large molecules or particles completely and to fuse with itself to engulf the enclosed substances. (See CTQ #6.)

REEXAMINING THE THEMES

Relationship between Form and Function

The structure of the plasma membrane—consisting of a fluid lipid bilayer and a mosaic of embedded proteins—accounts for many of the cell's life-sustaining functions. For example, the lipid bilayer acts as a barrier, preventing essential molecules from leaking out of the cell. The fluidity of the lipid bilayer gives the membrane the necessary flexibility it needs to facilitate such dynamic processes as endocytosis, which requires the membrane to invaginate and pinch off vesicles. Proteins provide the membrane with selectivity. Protein receptors bind specific extracellular substances, including hormones and LDL cholesterol, or act as passageways for the transmembrane movement of specific ions or dissolved solutes.

Biological Order, Regulation, and Homeostasis

The concentration of many substances within a cell is maintained by transport activities of the plasma membrane. Sodium ions, for example, are maintained at a low concentration inside a cell by the expulsion of these ions by membrane transport. Cholesterol levels, both inside the cell and in the blood, are also determined by the activity of membrane proteins. When cholesterol levels in a cell are low, the cell produces additional LDL receptors, which then import more cholesterol from the blood. When cellular cholesterol levels are high, LDL receptors are removed from the plasma membrane, causing less cholesterol to be taken up.

Acquiring and Using Energy

One of the essential uses of energy in living organisms is the active transport of materials across cell membranes. Active transport of sodium and potassium ions is accomplished by conformational changes brought about by the transfer of a phosphate group from ATP to one of the amino acids in the transport protein. This is an example of one of the common mechanisms by which the chemical energy in ATP is used to power an otherwise energetically unfavorable reaction. Although energy is used to establish these ionic gradients, the energy is not "wasted" since the gradient itself is a form of potential energy. The energy stored in the gradient, for example, is used to move other substances across a membrane against their concentration gradient. This is illustrated by the movement of sugar molecules in your intestine.

SYNOPSIS

The cell's plasma membrane is a dynamic boundary that protects the internal environment of the cell while directing the exchange of materials between the cell and the extracellular medium. The barrier properties of the membrane stem largely from the lipid bilayer, which is impermeable, except to very small polar molecules (such as water and carbon dioxide) or to lipid-soluble substances, such as steroids. Proteins promote the passage of selected molecules. Together, the lipids and proteins make the membrane selectively permeable.

Diffusion is the passive movement of a substance from a region of higher concentration to a region of lower concentration. Diffusion depends solely on the random movements of molecules and does not require the input of energy. The rate of diffusion depends on the polarity and size of the molecule and the magnitude of the concentration gradient. The diffusion of water across a membrane, called osmosis, is of great importance in the lives of diverse organisms. During osmosis, water moves from a compartment of lower solute concentration to one of higher solute concentration. Some membranes have special channels that allow ions to diffuse across the membrane without having to traverse the lipid bilayer directly. Membranes may also possess carrier proteins that bind to specific solutes and facilitate the diffusion process.

Active transport is a process in which substances are moved against a concentration gradient by mem- brane proteins at the expense of cellular energy. The establishment of gradients by active transport provides a storage of energy the cell can use to drive various energy-requiring activities.

Endocytosis is the uptake of materials in the extracellular environment by invaginations of the plasma membrane. The material becomes enclosed within a cytoplasmic vesicle, which generally fuses with a lysosome, allowing digestive enzymes to disassemble the material. Endocytosis is categorized as either phagocytosis —the uptake of relatively large particles, such as bacteria or debris—or pinocytosis—the uptake of smaller materials or fluid. Pinocytosis is often initiated by the interaction of material (such as LDL) with specific membrane receptors.

Key Terms

selectively permeable (p. 138)
diffusion (p. 138)
osmosis (p. 140)
isotonic (p. 140)
hypertonic (p. 140)

hypotonic (p. 140)
turgor pressure (p. 141)
plasmolysis (p. 141)
carrier protein (p. 141)
facilitated diffusion (p. 141)

active transport (p. 142)
endocytosis (p. 144)
phagocytosis (p. 144)
pinocytosis (p. 144)
receptor-mediated endocytosis (p. 147)

Review Questions

1. Hexane (C_6H_6), a commercial organic solvent, and glucose ($C_6H_{12}O_6$) contain six carbon atoms each. Which one would move more rapidly across the membrane by simple diffusion? Why?

2. Why does water stop moving into a plant cell even if the cell remains hypertonic to its environment?

3. In comparing simple diffusion, facilitated diffusion, and active transport, which of these processes is being described by the following phrases:

   a. penetrating molecules move in only one direction across the membrane;
   b. requires the use of ATP hydrolysis;
   c. rate of penetration is related to lipid solubility;
   d. involves a membrane protein.

4. Sea urchins are isotonic to the surrounding sea water, while most fishes are extremely hypotonic. What type of osmotic challenge, if any, would each of these animals face?

Critical Thinking Questions

1. If you were studying viruses and discovered one whose protein coat resembled the protein found in an LDL particle, what conclusion might you draw about the way the virus entered the cell?

2. If you were designing instruments to look for signs of life on Mars, would you include a search for a selectively permeable membrane, such as that found in living cells

on Earth? Adopt a position on this question and defend it.

3. How does osmosis explain each of the following?
   (1) a red blood cell swells and bursts when placed in pure water
   (2) a red blood cell shrinks when placed in a concentrated salt solution

4. Amino acids are found to enter most cells rapidly, even if the cells are treated with drugs that block ATP formation. Can you conclude that amino acids are capable of rapid diffusion across the lipid bilayer? Why or why not?

5. For each case given below, indicate whether it illustrates simple diffusion, facilitated diffusion, or active transport.
   (1) Some marine algae contain much higher concentrations of iodine than does the surrounding sea water.
   (2) A single-celled freshwater organism uses a special organelle to pump excess water out of its cytoplasm.
   (3) A single-celled freshwater organism is hypertonic to the surrounding water, and water moves into the organism.
   (4) Urea, a small molecule that is relatively soluble in lipids is produced inside of cells. It moves into the surrounding medium, which contains little or no urea.

6. Use the terms below to prepare a concept map of the various means by which materials are transported across the cell membrane. [To make a concept map, the concepts (terms listed) are arranged graphically to represent relationships among the concepts. Concepts are enclosed by boxes, circles, or ovals and are connected by lines indicative of relationships. Connecting words are written on the lines.]

   passive transport; active transport; diffusion; osmosis; facilitated diffusion; endocytosis; phagocytosis; pinocytosis.

7. Recall from Chapter 6 that events in the universe tend to proceed spontaneously in the direction that increases entropy (disorder) in the universe. Do you think that diffusion is an example of this principle? Why or why not?

Additional Readings

Alberts, B., et al. 1989. *Molecular biology of the cell.* New York: Garland. (Intermediate)

Brown, M. S., and J. L. Goldstein. 1984. How LDL receptors influence cholesterol and atherosclerosis. *Scientific American* Nov: 58–66. (Intermediate)

Darnell, J., H. Lodish, and D. Baltimore. 1990. *Molecular cell biology*, 2d ed. New York: Freeman. (Advanced)

Lienhard, G. E., et al. 1992. How cells absorb glucose. *Scientific American* Jan: 86–91. (Intermediate)

Myant, N. B. 1990. *Cholesterol metabolism, LDL and the LDL receptor.* New York: Academic Press. (Advanced)

Stein, W. D. 1990. *Channels, carriers, and pumps: An introduction to membrane transport.* San Diego: Academic Press. (Advanced)

CHAPTER

◄ 8 ►

Processing Energy: Photosynthesis and Chemosynthesis

STEPS TO DISCOVERY
Turning Inorganic Molecules Into Complex Sugars

AUTOTROPHS AND HETEROTROPHS: PRODUCERS AND CONSUMERS

AN OVERVIEW OF PHOTOSYNTHESIS

THE LIGHT-DEPENDENT REACTIONS: CONVERTING LIGHT ENERGY INTO CHEMICAL ENERGY

Photosynthetic Pigments: Capturing Light Energy

Organization of Photosynthetic Pigments

Obtaining Electrons By Splitting Water

Transporting Electrons and Forming NADPH

Making ATP

Cyclic Photophosphorylation: An Alternate Pathway For Energy Conversion

THE LIGHT-INDEPENDENT REACTIONS

C_3 Plants

C_4 Plants: Adaptations To A Hot, Dry Environment

CAM Plants: Further Adaptations To A Hot, Dry Environment

CHEMOSYNTHESIS: AN ALTERNATE FORM OF AUTOTROPHY

BIOLINE
Living on the Fringe of the Biosphere

THE HUMAN PERSPECTIVE
Producing Crop Plants Better Suited to Environmental Conditions

Turning Inorganic Molecules into Complex Sugars

*I*n the late 1930s, Martin Kamen, having recently received his doctorate in nuclear physics, arrived at the University of California at Berkeley, the site where the original cyclotron had recently been constructed. A cyclotron is an instrument that accelerates subatomic particles to high speeds and directs them at the nuclei of atoms. The subsequent crash changes the target atom's structure. Kamen's goal was to use the cyclotron to produce new isotopes, such as ^{11}C, a radioactive isotope of carbon. Soon after his arrival, Kamen met Sam Ruben, a chemist at Berkeley, who pointed out the possibilities for using a radioactive carbon atom for working out the steps of metabolic pathways. One of the least understood, and most important, metabolic pathways was that of photosynthesis. At the time, scientists knew that the process of photosynthesis converted carbon dioxide and water to carbohydrates, such as glucose, but they knew very little about the steps by which such a remarkable conversion took place. How are these inorganic molecules converted to the "food" substances on which virtually all life on earth depends? Kamen and Ruben set out to provide the answer.

The scientists planned to grow either barley plants or cultures of photosynthetic algae in the presence of radioactive carbon dioxide ($^{11}CO_2$). After a short period of time, they would kill the cells with acidified, boiling water, which would extract the small polar organic molecules that form the intermediate compounds in the photosynthetic path-

Water from the soil and carbon dioxide from the air combine during photosynthesis to produce carbohydrates and oxygen,

way. By determining the chemical identity of the radioactive compounds, Kamen and Ruben felt they could assemble a diagram of the metabolic pathway. The approach was sound, but the technical problems were overwhelming.

The first problem was presented by the isotope itself. ^{11}C has a half-life of only 21 minutes, meaning that within 21 minutes of the onset of the experiment, only half the starting amount was left; the other half had already disintegrated. Before the experiment was an hour old, only 15 percent of the original radioactivity remained. Because of the isotope's short life span, the entire procedure—from producing the isotope at the cyclotron, to carrying out the experiment, to measuring the radiation with a counter—had to be performed impossibly fast. Kamen and Ruben found themselves running across campus in the middle of the night. The second problem was the lack of a reliable method for separating radioactive organic molecules from one another. After 3 years of research, no substantial progress on the project had been made.

Then, in 1940, Kamen and Ruben discovered a longer-lived isotope of carbon. They found that bombarding nitrogen atoms with neutrons produced a new isotope of carbon—^{14}C—which has a half-life of 5,700 years. Unfortunately, just as they were gearing up for an onslaught on photosynthesis using the new isotope, World War II broke out. Kamen was assigned to work on war-related isotope research; Ruben was put to work on the development of chemical weapons and was killed in a laboratory accident.

By the end of the war, the second problem that had plagued Kamen and Ruben—how to separate organic molecules from one another—was solved with a surprisingly simple technique. Archer Martin and Richard Synge, a pair of British biochemists, discovered that they could separate closely related organic compounds with a technique they called *paper chromatography*. In this procedure, a sample containing a mixture of compounds is applied to a piece of filter paper, and one end of the paper is submerged in a solvent. The solvent then migrates through the paper in much the same way a paper towel "sucks up" water. As the solvent moves through the applied sample, it dissolves the organic molecules and carries them along at different speeds—the smaller, more soluble compounds moving fastest. At the end of the "run," all the compounds lie in different locations on the paper strip. The isolated spots can then be cut out of the strip and dissolved in a solvent to obtain purified preparations of each compound in a sample.

Following the war, another group of scientists at Berkeley picked up where Ruben and Kamen had been forced to stop. Headed by the biochemist Melvin Calvin, this team used the approach of Ruben and Kamen. Armed with ^{14}C and paper chromatography, however, their work was much more fruitful. Calvin and his colleagues, James Bassham and Andrew Benson, grew algae in the presence of $^{14}CO_2$, killed the cells after a time, and extracted the labeled compounds with boiling alcohol. They reasoned that a very brief exposure to the radioactive carbon dioxide would allow radioactive carbon to become incorporated into the initial compounds formed during photosynthesis. Exposing the cells to ^{14}C a little longer allowed the next compounds in the pathway to become radioactive, and so on, until all the intermediates in the photosynthetic pathway were labeled. Using this approach, the researchers attempted to trace the route by which carbon travels from carbon dioxide to carbohydrate.

In their initial studies, they found that even short exposures of about 5 seconds to labeled carbon dioxide produced a number of radioactive compounds (see Figure 8-10). When they shortened the exposure to 2 seconds, one radioactive spot appeared, which corresponded to the first compound in the pathway formed after carbon dioxide was "fixed" to another molecule within the cells. The exposures were lengthened until the team had identified all the compounds in what Calvin concluded was a *circular* pathway—the route from carbon dioxide to carbohydrates. For his work in unraveling the pathway of the photosynthetic conversion of carbon dioxide to carbohydrate, Calvin received the Nobel Prize in Chemistry in 1961.

which is released into the air.

*L*ife is powered by the sun. The energy that gives a cheetah its incredible speed or a bird its power to fly does not originate within the planet, but travels 93 million miles from the sun. The energy of sunlight, however, becomes energy for life only after it has been transformed into chemical energy by the process of **photosynthesis.** The cheetah, for example, gets its food by eating an antelope, which in turn ate grasses and other plants that had photosynthesized. Virtually all organisms depend on photosynthesis for supplies of energy-rich food. For this reason, photosynthesis might be considered the most important set of chemical reactions in biology.

▼ ▼ ▼

AUTOTROPHS AND HETEROTROPHS: PRODUCERS AND CONSUMERS

◉ Life on earth can be divided into two groups: autotrophs and heterotrophs. **Autotrophs** (*auto* = self, *troph* = feeding) are organisms that use carbon dioxide as their sole source of carbon. They convert simple inorganic molecules, such as carbon dioxide and water, into all of the complex, energy-rich organic materials needed to sustain their life. This feat requires the input of external energy derived from a nonliving source. Most autotrophs harness the energy from the sun; these are called *photosynthetic autotrophs.* A number of autotrophic bacteria, however, have evolved an alternate mechanism for capturing external energy. They acquire the energy from inorganic substances through *chemosynthesis,* a process discussed in more detail at the end of the chapter.

Although all organisms depend on the metabolic activity of autotrophs, only plants, some protists, and some bac-

(a)

FIGURE 8-1

Life is centered around photosynthesis. *(a)* Plants supply energy for virtually all organisms by transforming light energy into chemical energy. Every year, plants convert about 600 billion tons of carbon dioxide into sugars, a weight equivalent to about 1,000 times the combined weight of all people living on earth today. *(b)* Tea fields growing near Mt. Fuji, Japan. Photosynthetic plants are essential to human existence as well as to the existence of virtually all other animals. Although about 3,000 species of plants have been cultivated throughout human history, today only 15 plant species supply most of the world's human population with food.

(b)

teria are categorized as autotrophs (Figure 8-1). The remainder of organisms—more than 80 percent of living species—are unable to carry out either photosynthesis or chemosynthesis. These organisms are called **heterotrophs** (*hetero* = other). Since heterotrophs cannot manufacture complex organic molecules from inorganic precursors, they must consume preformed organic materials initially produced by the autotrophs. Thus, autotrophs not only provide themselves with food, but they provide food to a vast array of heterotrophs as well—you included. It is estimated that each year plants convert about 600 billion tons of carbon dioxide into organic nutrients and release about 400 billion tons of oxygen into the environment. Much of this activity is accomplished by single-celled algae living in the thin upper layer of the world's sensitive—and increasingly more polluted—oceans.

Organisms depend on their environment to provide them with the raw materials and energy necessary for life. Autotrophs derive their energy and materials from inorganic sources and provide the resources on which heterotrophs depend. This process is not always a one-way street, however, because heterotrophs help provide the inorganic resources (carbon dioxide) needed by the autotrophs. (See CTQ #2.)

AN OVERVIEW OF PHOTOSYNTHESIS

In the early 1600s, it was generally accepted that plants, which obviously did not prey on other organisms, absorbed all of their food and nutrients from the soil alone. This was the common belief until Jan Baptista van Helmont, a Belgian physician, conducted a simple experiment that disproved this hypothesis of "soil-eating" plants. After planting a 5-pound willow tree in 200 pounds of dry soil, van Helmont added only rainwater to the soil for a period of 5 years. When he reweighed the soil and the willow tree, the tree had gained 164 pounds, yet the soil had lost only 2 ounces—not 164 pounds, as would be expected of a soil-eating plant. Van Helmont concluded that the willow's weight gain must have come from the added water, a conclusion we now know is only partly correct.

Willows, like most other plants, gather the carbon needed to build their tissues from the air, not the soil. Through photosynthesis, the carbon in atmospheric carbon dioxide is used to form the backbone of new organic compounds that are fashioned into the cellular material required for growth. The soil provides only a few minerals needed for growth, which accounts for the 2-ounce weight loss measured by van Helmont.

Photosynthesis in eukaryotes takes place inside chloroplasts (page 104), organelles whose structure is ideally suited to their function. Chloroplasts contain an elaborate array of membranes in which clusters of light-absorbing

pigments are embedded. These pigments capture solar energy, which is used to transform carbon dioxide into energy-rich sugars. These sugars not only provide a store of chemical energy but serve as the raw materials for producing thousands of other molecules needed by the plant. Oxygen, which is vital to energy utilization by both plants and animals, is released as a byproduct of photosynthesis.

Expressing the overall process in a chemical equation makes photosynthesis appear rather simple:

$$CO_2 + H_2O \xrightarrow[\text{chlorophyll}]{\text{light energy}} (CH_2O) + O_2$$

$$\underset{\substack{\text{Unit of}\\\text{carbohydrate}}}{}$$

You might conclude from this equation that the energy in light is used to split carbon dioxide, releasing molecular oxygen (O_2) and transferring the carbon atom to a molecule of water to form a unit of carbohydrate (CH_2O). In fact, this was the prevailing line of thought as late as 1941, the year Ruben and Kamen reported on an experiment using a specially labeled isotope of oxygen (^{18}O) as a replacement for the common isotope (^{16}O). In this experiment, one group of plants was given labeled $C^{18}O_2$ and "regular" water, while the other group was given "regular" carbon dioxide and labeled $H_2{}^{18}O$. The investigators asked a simple question: Which of these two groups of plants released labeled $^{18}O_2$ gas? The results showed that those plants given the labeled water produced labeled oxygen, while the plants given the labeled carbon dioxide produced "regular" oxygen. Con-

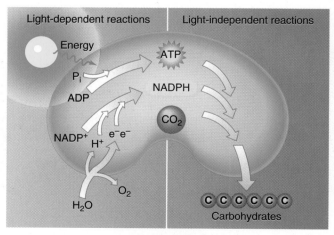

FIGURE 8-2 ——————

Photosynthesis: A two-stage process. Photosynthesis is divided into light-dependent reactions and light-independent (dark) reactions. During the light-dependent reactions, the energy of sunlight provides the power to generate ATP and NADPH from ADP, P_i, NADP$^+$, and water. Oxygen gas, derived from water, is given off as a byproduct. In the light-independent reactions, ATP and NADPH from the light-dependent reactions provide the energy and electrons to convert low-energy carbon dioxide molecules into energy-rich carbohydrates.

trary to popular belief, it wasn't carbon dioxide that was being split into its two atomic components, but water.

Botanists—biologists who specialize in the study of plant life—group the reactions of photosynthesis into two general stages: light-dependent reactions and light-independent reactions (Figure 8-2). During the **light-dependent reactions,** energy from sunlight is absorbed and converted to chemical energy, which is stored in two energy-rich molecules: ATP and NADPH. This process involves the splitting of water. During the second stage of photosynthesis, the **light-independent reactions** (or "dark reactions"), light is not required; carbohydrates are synthesized from carbon dioxide using the energy stored in the ATPs and NADPHs formed in the light-dependent reactions.

> **Life depends on the nonliving environment. Photosynthesis brings together light energy and simple inorganic molecules (H_2O and CO_2) to produce living material by a complex series of light-dependent and light-independent reactions. This new living material is the source of nutrients that support virtually every organism on earth. (See CTQ #3.)**

THE LIGHT-DEPENDENT REACTIONS: CONVERTING LIGHT ENERGY INTO CHEMICAL ENERGY

Chloroplasts are large organelles, by organelle standards. Their role in photosynthesis was revealed in 1881 in an ingenious demonstration by German biologist Theodor Engelmann. Engelmann showed that when the cells of an aquatic plant were illuminated, actively moving bacteria would collect outside the cell near the site of the large ribbonlike chloroplast. The bacteria were utilizing the minute quantity of oxygen released by photosynthesis in the chloroplast to stimulate their energy metabolism.

Recall from Chapter 5 that chloroplasts contain stacks of membrane discs, or thylakoids, surrounded by a semifluid stroma. The light-dependent reactions take place within the thylakoid membranes. They begin when the energy in sunlight is captured by light-absorbing pigments.

PHOTOSYNTHETIC PIGMENTS: CAPTURING LIGHT ENERGY

Before we can understand how photosynthetic pigments capture light energy, we need to consider a few basic properties of light. Light energy travels in packets called **photons.** The energy content of a photon depends on the wavelength of the light: the shorter the wavelength, the higher the energy content. Since a photon cannot be divided into smaller packets of energy, a molecule that absorbs light must absorb an entire photon. Consequently, each pigment molecule absorbs only those wavelengths whose energy matches that needed to boost an electron to a higher orbital (page 54). As we will see below, the fate of these "photoexcited" electrons is a key aspect of photosynthesis.

Plants are able to absorb a wide range of wavelengths because they contain a variety of pigment molecules that have different structures and absorption properties. As a group, these pigments absorb light energy with wavelengths between 400 nanometers (or 400 billionths of a meter), violet light, and 700 nanometers, red light, which repre-

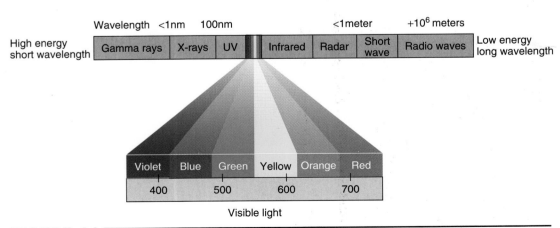

FIGURE 8-3

Solar energy for life. The wavelengths of radiation reaching the earth range from short-wavelength (high-energy) gamma rays, to extremely long-wavelength (low-energy) radio waves. Our eyes can detect only a small segment of the radiation spectrum, the portion we call *visible light* (from 380 to 750 nanometers). Photosynthetic pigments are also sensitive to visible light. They selectively absorb wavelengths between 400 and 700 nanometers.

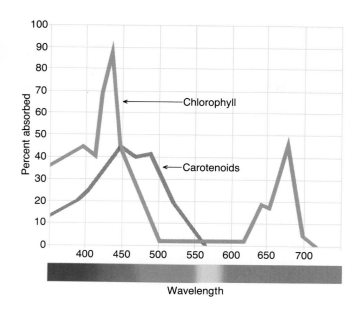

FIGURE 8-4
Absorption spectrum of photosynthetic pigments. Most absorption of light energy by photosynthetic pigments takes place at the peaks of each graph; the valleys indicate wavelengths that are reflected by the pigments. All plants use chlorophylls and carotenoids as photosynthetic pigments. Chlorophylls absorb primarily violet-blue and orange-red wavelengths, while carotenoids absorb wavelengths only in the violet-blue range of the spectrum. Differences in absorption from one chlorophyll molecule to the next depend largely on differences in the proteins to which the pigments are bound.

sents only a small fraction of the electromagnetic spectrum that reaches the earth (Figure 8-3).

There are two major groups of light-capturing pigments in plant chloroplasts: chlorophylls and carotenoids. **Chlorophylls** are complex, magnesium-containing compounds that absorb light of red and blue wavelengths (Figure 8-4). Plants are typically green because chlorophyll, which is the predominant plant pigment, does not absorb light of green wavelengths. Instead, this light is reflected to our eyes, which is why we see the green color. **Carotenoids** absorb light of blue and green wavelengths (Figure 8-4). Yellow, orange, and red wavelengths are reflected by carotenoids, producing the characteristic orange and red colors of carrots, oranges, tomatoes, and the leaves of some plants during the fall (Figure 8-5). As the leaves stop producing chlorophyll in the cooler weather, the gold and red carotenoids become more visible, which accounts for the brightly colored landscapes of autumn.

FIGURE 8-5
Fall colors. As chlorophyll is broken down in autumn, the underlying red, orange, and yellow carotenoids in the leaf chloroplasts are exposed.

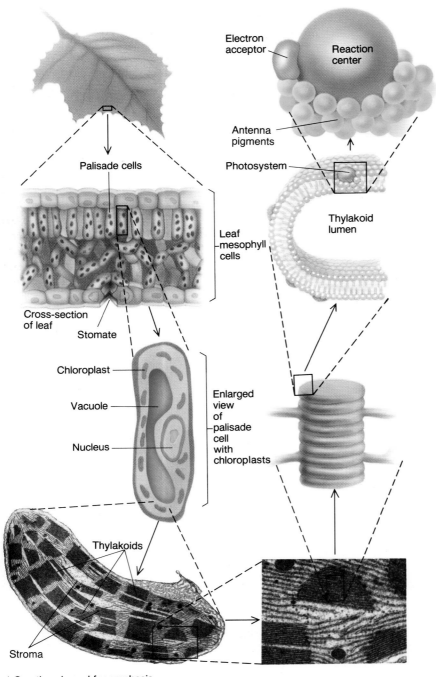

Electron acceptor

Reaction center

Antenna pigments

Photosystem

Thylakoid lumen

Palisade cells

Leaf mesophyll cells

Cross-section of leaf

Stomate

Chloroplast

Vacuole

Nucleus

Enlarged view of palisade cell with chloroplasts

Thylakoids

Stroma

* Greatly enlarged for emphasis

FIGURE 8-6

Organization for photosynthesis is evident at many different levels. Photosynthesis takes place in the chloroplasts of plant cells. A leaf cell may contain as many as 60 chloroplasts. The thylakoid system of the chloroplast is the site of the light-dependent reactions, whereas the stroma is the site of the light-independent reactions. Photosynthetic pigments are precisely arranged in the thylakoid membranes to form light-harvesting photosystems. Within a photosystem, light is absorbed by antenna pigments, and the energy is transferred to the reaction-center pigment, boosting an electron to a higher energy level from which it is passed to an electron-acceptor molecule. This starts a chain of chemical reactions that leads to the formation of ATP and NADPH and, ultimately, energy-rich carbohydrates.

ORGANIZATION OF PHOTOSYNTHETIC PIGMENTS

Capturing light energy and converting it into usable chemical energy requires an ordered biological structure, which can be seen at all levels of organization from the arrangement of individual pigment molecules to the entire leaf (Figure 8-6). Photosynthetic pigments are embedded in the thylakoid membranes of chloroplasts in clusters called **photosystems.** A single chloroplast contains several thousand photosystems. Each photosystem has an ordered structure in which a special chlorophyll molecule, called the **reaction center,** is surrounded by 250 to 350 **antenna pigments** including both chlorophyll and carotenoid molecules. Antenna pigments are so named because they gather light energy of wavelengths varying between 400 and 700 nanometers, and then channel this energy to the reaction center, much as an antenna receives and relays radio and television waves. As a result, the reaction center always ends up with the absorbed energy.

The membranes of the chloroplast contain two types of photosystems, **Photosystem I** and **Photosystem II**—distinguishable by their reaction-center pigments (Figure 8-7). The reaction center of Photosystem I is referred to as **P700**—"P" standing for "pigment," and "700" standing for the wavelength of light that this particular chlorophyll molecule absorbs most strongly. The reaction center of Photosystem II is referred to as **P680,** for comparable reasons.

When sunlight strikes the thylakoid membrane, the energy is absorbed simultaneously by the antenna pigments of both Photosystems I and II and passed to the reaction centers of both photosystems. Electrons of both reaction-center pigments are boosted to an outer orbital, and each photoexcited electron is then passed to an "open" **electron acceptor;** that is, a molecule that will receive the electron. The transfer of electrons out of the photosystems leaves the two reaction-center pigments missing an electron and, thus, are positively charged. After losing their electrons, the reaction centers of Photosystem I and II can be denoted as

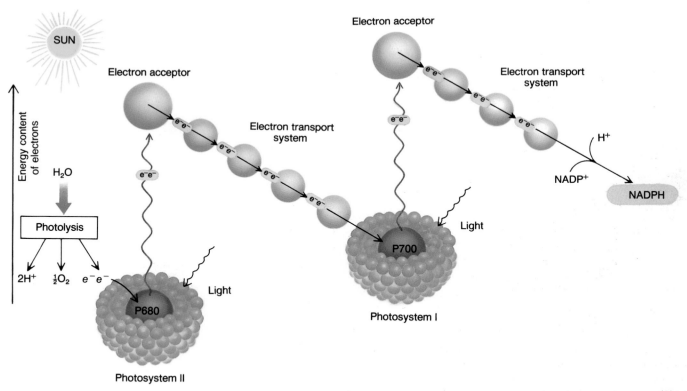

FIGURE 8-7

Electron flow from water to NADP+. Photoexcited electrons from the P700 and P680 reaction centers are passed to electron acceptors, which in turn transfer electrons to an electron transport system. The electron vacancy created by the absorption of a photon by P680 is filled with an electron released during photolysis. The electron vacancy in P700+ is filled with electrons that originated in P680. The photoexcited electrons from P700 pass down a second electron transport system until they react with NADP+ and H+ to form NADPH. The overall consequence of the process is to remove low-energy electrons from water, boost them to a higher energy level, and transfer the energized electrons to NADP+, providing the cell with reducing power.

P700⁺ and P680⁺, respectively. Let's look at the subsequent events occurring at each photosystem separately.

OBTAINING ELECTRONS BY SPLITTING WATER

The positively charged P680⁺ has a very strong attraction for electrons — strong enough to pull tightly held (low-energy) electrons from water (Figure 8-7), splitting the molecule. The magnitude of this statement becomes evident when you consider that the splitting of water in a laboratory requires the use of a strong electric current or temperatures approaching 2,000°C. Yet, a plant cell can accomplish this feat on a snowy mountainside using the small amount of energy of visible light.

The splitting of water is termed **photolysis** (*photo* = light, *lysis* = splitting) and can be expressed as

$$\text{energy} + H_2O \rightarrow \tfrac{1}{2}O_2 + 2e^- + 2H^+$$

This reaction generates three different products: (1) oxygen produced by photolysis is released into the atmosphere; (2) electrons that travel to the P680⁺ reaction center of

Photosystem II, returning the system to the uncharged (P680) state; and (3) protons that are released into the thylakoid lumen (see Figure 8-8). We will return to the fate of these protons shortly. But first, let's consider the fate of the electrons that had been transferred earlier from P680, producing the electron-deficient P680⁺ reaction center.

TRANSPORTING ELECTRONS AND FORMING NADPH

Each electron lost by P680 following absorption of a photon is passed through the "hands" of a number of electron carriers that together form an **electron transport system** (Figure 8-8). The components of the electron transport system are precisely arranged within the thylakoid membrane to facilitate the passage of electrons through the system. Electron transport systems provide one of the best examples of the role of membranes in organizing components (proteins, lipids, pigments, etc.) in spatially precise arrays. As a result of its position in the membrane, each carrier in the system is able to receive electrons from the previous member and to transfer them to the next member,

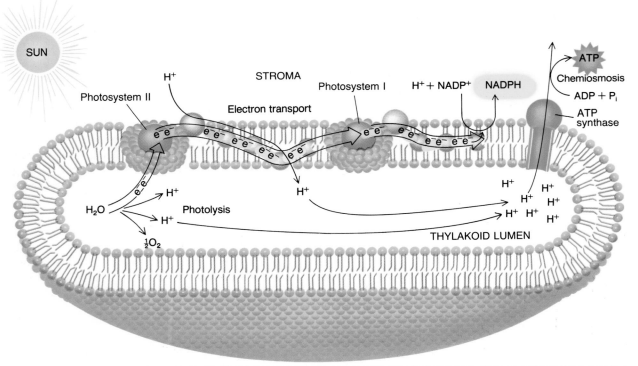

FIGURE 8-8

Photosynthesis and the thylakoid membrane. The carriers that transport electrons during the light-dependent reactions of photosynthesis are embedded within the thylakoid membrane. The transport of energized electrons "downhill" along an electron transport system and the splitting of water during photolysis both lead to the concentration of protons inside the thylakoid lumen. This produces a concentration gradient of protons across the thylakoid membrane with sufficient energy content to drive the phosphorylation of ADP to ATP as protons diffuse across the membrane through special channels in the ATP synthase. This process is called *noncyclic photophosphorylation*. Measurements indicate that approximately four ATPs are formed per molecule of oxygen released.

releasing energy in the process. This process of shuttling electrons along an electron transport chain is analogous to the passage of buckets from one hand to another along a bucket brigade.

At the end of the electron transport chain lies P700$^+$, the electron-deficient reaction center of Photosystem I, which accepts the electron from the last carrier in the chain. When P700$^+$ accepts an electron, it returns to the uncharged state (P700), at which time it is ready to absorb another photon. But what happens to the electrons that had been previously removed from P700 to create the electron-deficient P700$^+$ of Photosystem I? These electrons are passed through another set of electron carriers and on to NADP$^+$ to form NADPH (Figure 8-7, 8-8).

We have now followed the process in which electrons flow from water to Photosystem II to Photosystem I to NADP$^+$. The NADPH formed in this process provides the cell with reducing power, which is a form of stored energy (page 129). This process is called **noncyclic photophosphorylation** because electrons move in one direction, from water to NADP$^+$. But the flow of electrons and the storage of energy in NADPH is only half the story. The term "photophosphorylation" describes the fact that light energy is also used to drive a reaction in which ATP is formed, the subject of the next section.

MAKING ATP

Recall from Chapter 6 that spontaneous reactions proceed to a state of lower energy, releasing energy as they occur. Electron transport is a spontaneous process in which electrons drop to lower and lower energy levels, releasing energy that is used to manufacture ATP (discussed at greater length in Chapter 9). For decades, a question of intense interest was: How is the "downhill" process of electron transport coupled to the "uphill" process of ATP formation? In 1961, the British physiologist Peter Mitchell formulated what seemed like a radical hypothesis to explain this phenomenon. In essence, Mitchell proposed that the energy released as the electrons fell to lower and lower energy levels during electron transport produced a *proton gradient* across the thylakoid membrane. A proton gradient is one in which hydrogen ions (H$^+$) are present at higher concentration on one side of the membrane than on the other side. Recall from Chapter 7 that an ionic gradient, whether Na$^+$ or H$^+$, represents a store of potential energy. In chloroplasts, the H$^+$ gradient is used to drive the energy-requiring (endergonic) reaction in which ADP reacts with P$_i$ to form ATP. Mitchell called his mechanism for ATP formation **chemiosmosis.** Initially, Mitchell's hypothesis met with skepticism because it lacked direct evidence. However, experimental support for the proposal accumulated over the next decade to an overwhelming degree, culminating in Mitchell's receiving the Nobel Prize in 1978.

There are two matters to consider in this description of ATP formation: (1) How is a proton gradient formed across the thylakoid membranes of chloroplasts? and (2) How does this gradient lead to the formation of ATP?

Storing Energy in a Proton Gradient

We partially answered the question of how a proton gradient is established earlier in the chapter. Recall that the splitting of water (photolysis) releases protons into the thylakoid lumen (Figure 8-8). Since the thylakoid membrane is highly impermeable to protons, the protons remain at a higher concentration within the thylakoid lumen than in the stroma, thus forming a proton gradient across the membrane.

Protons are also released into the thylakoid lumen as a result of electron transport. If you look closely at the electron transport chain between Photosystem II and Photosystem I in Figure 8-8, you will notice that protons move into the thylakoid membrane at its outer surface and are discharged into the thylakoid lumen at the inner surface of the membrane. This movement of protons from the stroma to the thylakoid lumen accompanies electron transport, contributing to the proton gradient across the thylakoid membrane.

Using the Stored Energy in the Formation of ATP

The second question remains: How does the energy stored in a proton gradient lead to the formation of ATP? ATP synthesis requires a large enzyme complex called **ATP synthase,** whose shape resembles a ball sitting on a stalk (Figure 8-8). Each part of this complex has a role to play. The stalked portion of the enzyme crosses through the thylakoid membrane and contains an internal channel or "proton pore" that allows the protons massed in the lumen to flow through the thylakoid membrane toward the region of lower concentration in the stroma. The flow of protons through these channels is analogous to water flowing through a bathtub drain: Water trapped in the tub escapes through the drain, flowing to a lower, more stable energy level. The active site of the enzyme—where ADP and P$_i$ come together to form ATP—is located in the ball-shaped portion of the complex at the end of the proton channel. The "downhill" movement of protons through the channel drives the "uphill" synthesis of ATP at the enzyme's active site. The mechanism by which this happens remains one of the central questions in cell biology.

CYCLIC PHOTOPHOSPHORYLATION: AN ALTERNATE PATHWAY FOR ENERGY CONVERSION

In addition to noncyclic photophosphorylation, plants have a second mechanism to form ATP by chemiosmosis. This mechanism is called **cyclic photophosphorylation** (Figure 8-9). Only Photosystem I participates in cyclic photophosphorylation, a process that involves neither photolysis nor the formation of NADPH.

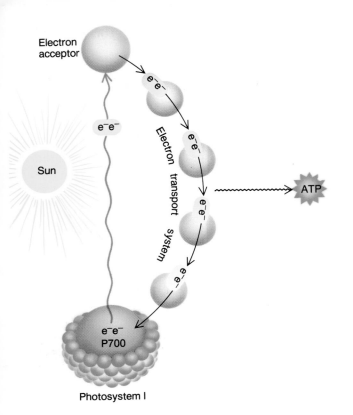

Electron acceptor

Sun

Electron transport system

ATP

e⁻e⁻
P700

Photosystem I

FIGURE 8-9

Electron flow during cyclic photophosphorylation. During cyclic photophosphorylation, light energy is absorbed by Photosystem I. Photoexcited electrons from the P700 reaction center are transferred to an electron acceptor and passed down an electron transport system which returns the electron to the P700⁺ pigment. During this cyclic flow of electrons, protons are shuttled across the membrane, generating a proton gradient used to form ATP. The electron carriers that participate in cyclic photophosphorylation include some of the same molecules used in noncyclic electron flow. Cyclic photophosphorylation does not involve Photosystem II and does not generate NADPH.

During cyclic photophosphorylation, light energy is absorbed by antenna pigments and funneled into the P700 reaction center. Electrons from P700 are passed to an electron acceptor and then relayed through a number of electron carriers, which return the electrons back to the P700⁺ reaction center. As the electrons move along this circular electron transport pathway, a pair of protons are picked up on the outer surface of the thylakoid membrane and deposited into the thylakoid lumen, forming the proton gradient

necessary to generate ATP by chemiosmosis. The differences between cyclic and noncyclic photophosphorylation are summarized in Table 8-1.

Cyclic photophosporylation is likely the means by which the earth's first autotrophic bacteria converted light energy into chemical energy. Even today, the only means of making ATP for some prokaryotic bacteria is through cyclic photophosphorylation. However, the failure of the cyclic pathway to generate NADPH limits its ability to provide

TABLE 8-1

SUMMARY OF CYCLIC AND NONCYCLIC PHOTOPHOSPHORYLATION

Cyclic	Noncyclic
1. Light is absorbed by pigments in Photosystem I.	1. Light is absorbed by pigments in Photosystems I and II.
2. Energized electrons from P700 pass to an electron acceptor.	2. Energized electrons from P680 of Photosystem II and energized electrons from P700 of Photosystem I pass to separate electron acceptors.
3. Energized electrons are relayed down an electron transport system and returned to the original P700⁺ reaction center.	3. Splitting of water fills the electron vacancy in P680⁺, releasing protons and oxygen.
4. ATPs are generated a proton gradient as a result of the cyclic flow of electrons.	4. P680's electron acceptor passes energized electrons to an electron transport system that relays them to the P700⁺ reaction center of Photosystem I.
	5. The electron acceptor for P700 passes energized electrons to another electron transport chain that shuttles them to NADP⁺, forming NADPH.
	6. ATPs are generated by a proton gradient as a result of a noncyclic flow of electrons and the release of protons as water is split.

chemical energy for larger autotrophic organisms. This was probably the major selective pressure that led to the evolution of the noncyclic pathway.

> **When light is absorbed by chlorophyll molecules, water molecules are split, and electrons are boosted to a higher energy state. These high-energy electrons are used to form NADPH and ATP, which provide the energy required to convert carbon dioxide to carbohydrate. The capture of light energy and its conversion to chemical energy require the precise arrangement of chlorophyll and other components within the thylakoid membrane. (See CTQ #4.)**

THE LIGHT-INDEPENDENT REACTIONS

Since both ATP and NADPH are forms of chemical energy, why do photosynthesizers go on to manufacture another form of chemical energy—energy-rich sugars—during the synthesis reactions? The answer is that neither ATP nor NADPH can be stored or moved from one part of a plant to another, but sugars can. Transport and storage of energy-rich compounds are critical to the survival of a multicellular plant. For example, glucose produced in leaves is converted to the disaccharide sucrose and moved to growing regions of roots and stems where photosynthesis does not take place. Sucrose is an ideal transport molecule since it is highly stable and can be rapidly disassembled and used to construct other organic compounds. Thus, the function of the light-independent reactions is to use the readily available energy of ATP and the energy-rich electrons of NADPH to manufacture glucose. Carbon dioxide is gathered from air to supply carbon for glucose construction.

To date, three variations of light-independent reactions have been identified in plants—C_3, C_4, and *CAM synthesis*. Table 8-2 lists common plants that utilize each of these

TABLE 8-2
PLANTS WITH C_3, C_4, AND CAM SYNTHESIS REACTIONS

C_3	C_4	CAM
Legumes (bean, pea)[a]	Sugar cane[a]	Pineapple[a]
Wheat[a]	Corn[a]	Cactus
Oats[a]	Sorghum[a]	Jade plant
Barley[a]	Crabgrass	Lillies
Rice[a]	Russian thistle (tumbleweed)	Agave (tequila)[a]
Bluegrass	Bermuda grass	Some orchids

[a] Agricultural crops.

variations. All three modes of carbohydrate synthesis share the same pathway for producing glucose; they differ in the way carbon dioxide is first combined with a carbon-acceptor molecule. As we will see, in certain environments, this seemingly small difference has given some plants an adaptive edge over others.

C_3 PLANTS

C_3 synthesis—and indeed all three types of synthetic pathways—begins with **carbon dioxide fixation,** the combining of carbon dioxide with a carbon-acceptor molecule. When he began his studies on the incorporation of $^{14}CO_2$ by photosynthetic algae, one of Melvin Calvin's first objectives was to determine the nature of the carbon dioxide acceptor—a goal that proved to be rather elusive.

Calvin and his colleagues noticed that when the cells were exposed to $^{14}CO_2$ for very short periods of time—5 seconds or less—one spot appeared that contained most of the radioactivity (Figure 8-10). This spot contained the three-carbon compound, phosphoglyceric acid (PGA). Calvin concluded that the one-carbon compound carbon dioxide reacted with a two-carbon acceptor to form the three-

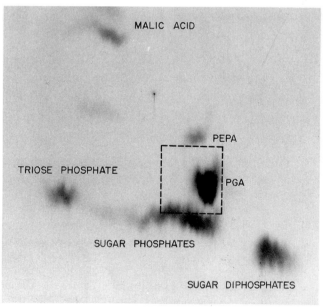

FIGURE 8-10

Unraveling the metabolic pathway for photosynthesis. A photograph of a chromatogram obtained after incubating algal cells with $^{14}CO_2$ for 5 seconds. Each of the spots corresponds to a different organic molecule containing one or more radioactive carbon atoms derived from the labeled carbon dioxide. During short incubations, the spot (boxed) corresponding to the compound phosphoglyceric acid (PGA) was most heavily labeled. PGA is the first stable compound into which carbon dioxide becomes incorporated.

carbon PGA, hence the designation **C₃ pathway,** also known as the *Calvin–Benson cycle.* A search for the mysterious two-carbon compound began. Further studies revealed, however, that carbon dioxide is not fixed to a two-carbon molecule but to a five-carbon molecule—**ribulose biphosphate (RuBP)**—by the enzyme **RuBP carboxylase,** nicknamed "Rubisco" (Figure 8-11). However, the six-carbon molecule formed by this union is so unstable that it "immediately" splits into two molecules of PGA, thus accounting for the early PGA spot on the chromatogram.

PGA is not a high-energy molecule. In fact, two molecules of PGA contain less energy than does a single molecule of RuBP, the actual carbon dioxide acceptor. This explains why carbon dioxide fixation is spontaneous and does not require energy from the light-dependent reactions. However, producing high-energy molecules from PGA requires the energy from ATP and electrons from NADPH formed by the light-dependent reactions. These compounds are used to reduce each PGA to a more energy-rich molecule, phosphoglyceraldehyde (PGAL), in a two-

step reaction, the first step using ATP, and the second using NADPH (Figure 8-11). (The second of these reactions—DPG to PGAL—was used as an example of an oxidation–reduction reaction on page 129.) Two PGAL molecules can ultimately give rise to one glucose by the C₃ pathway depicted in Figure 8-11. The vast majority of all plants use the C₃ pathway exclusively for constructing glucose, making it the predominant synthetic pathway among the earth's photosynthesizers.

C₄ PLANTS: ADAPTATIONS TO A HOT, DRY ENVIRONMENT

An alternate means by which plants can fix carbon dioxide was uncovered by Hugo Kortschak in the early 1960s. While working with sugar cane in Hawaii, Kortschak was surprised to find that when these plants were provided ¹⁴C-labeled carbon dioxide, radioactivity first appeared in organic compounds containing four carbons rather than three, hence the name **C₄ synthesis.** It soon became apparent that these

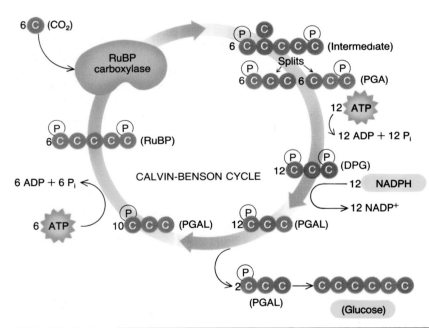

FIGURE 8-11

The Calvin-Benson Cycle (C₃ pathway). During the formation of glucose from carbon dioxide in C₃ plants, six carbon dioxide molecules are combined with six RuBPs to form six unstable, six-carbon carbohydrates. Each carbohydrate immediately splits into two, producing a total of 12 molecules of PGA. The PGAs are converted to DPGs by the transfer of phosphate groups from ATP. This is one of the places where the ATPs produced by the light-dependent reactions are utilized. NADPHs formed by the light-dependent reactions are used in the following step in which the 12 DPGs are reduced to 12 PGALs by the transfer of electrons. Two of the PGALs are "drained" away to form a molecule of glucose, whereas the remaining 10 PGALs (representing a total of 30 carbon atoms) recombine to regenerate six molecules of RuBP. The regeneration of six RuBPs is the same number of acceptor molecules with which we began.

◁ THE HUMAN PERSPECTIVE ▷

Producing Crop Plants Better Suited to Environmental Conditions

Biologists are currently investigating whether some plants are naturally able to shift among C_3, C_4, and/or CAM pathways, depending on conditions, and whether it is possible to genetically engineer "flexible" plants that can shift between two, or possibly all three, of these synthetic pathways. Shifting between different pathways could give plants a selective advantage over those with only one pathway. For example, a "flexible" plant could be in the CAM pathway when conditions are very hot and dry

and then shift into C_3 or C_4 following irrigation or a rainfall, enabling it to grow many times faster than it could if it remained in CAM.

So far, scientists have not found a plant capable of switching among all three synthetic pathways. But a few species have been described as natural C_3–C_4 intermediates, and preliminary results indicate that some intermediates show higher C_4 photosynthetic rates. Other investigations reveal natural C_3–CAM intermediates.

As for genetically engineered "flexible" plants, scientists have engineered C_3–C_4 hybrid plants. Unfortunately, none of the hybrids have shown increased rates of photosynthesis over the C_3 level as of yet. But investigations will continue, for if biologists can engineer biochemically flexible crop plants, much of the world's hunger problems could be solved—assuming there is no corresponding increase in human population.

plants utilize a different mechanism for carbon dioxide fixation, giving them an adaptive advantage over C_3 plants in hot, dry habitats. At high temperatures (45°C, 103°F or higher), plants that utilize C_4 synthesis produce as much as six times more carbohydrate than do C_3 plants. C_4 plants also survive better than do C_3 plants in environments in which water is limited and soil nutrients are scarce. This is

why crabgrass, a C_4 plant, tends to take over a lawn, displacing the domestic C_3 grasses originally planted.

The adaptive advantage of C_4 plants can be traced to the way they fix carbon dioxide which, in turn, is related to the structure of their leaf (Figure 8-12). RuBP carboxylase, the enzyme that fixes carbon dioxide in C_3 plants, requires relatively high concentrations of carbon dioxide before it

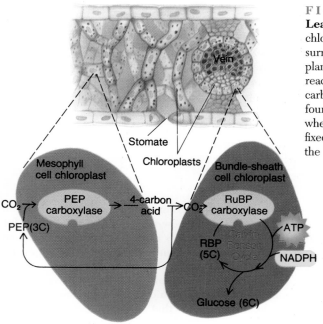

FIGURE 8-12

Leaf anatomy and C_4 synthesis. Plants that undergo C_4 synthesis have chloroplasts in their *mesophyll* cells and in large *bundle-sheath* cells surrounding the leaf veins. This unique chloroplast distribution enables C_4 plants to photosynthesize up to six times faster than can C_3 plants. The C_4 reactions take place in different leaf cells. Carbon dioxide fixation by PEP carboxylase takes place in the chloroplasts of mesophyll cells. The resulting four-carbon products are moved to the chloroplasts in bundle-sheath cells, where the carbon dioxide is released. The released carbon dioxide is then fixed to RuBP by RuBP carboxylase and carbohydrates are synthesized by the C_3 pathway.

can bind this substrate. This is important since carbon dioxide constitutes only about 0.03 percent of the gas content of the atmosphere. C_4 plants have a different carbon dioxide-fixing enzyme—**PEP carboxylase**—which is able to bind carbon dioxide at very low concentrations. Consequently, carbon dioxide levels can be very low in a leaf of a C_4 plant, even much lower than normal air concentrations, and carbon dioxide will still be linked to its acceptor.

How does this information fit with our earlier statement that all plants use the same cyclic pathway to produce glucose? The answer is explained by a C_4 plant's structure and function. In C_4 plants, PEP carboxylase fixes carbon dioxide in the leaf's *mesophyll cells* (Figure 8-12). The four-carbon products are then shipped to *bundle-sheath cells* lying deeper in the leaf. Once in the chloroplast of the bundle-sheath cells, the carbon dioxide is split from the four-carbon carrier, generating high levels of dissolved carbon dioxide—up to 100 times that in the mesophyll cell. This elevated carbon dioxide level is now sufficient to drive fixation by the less efficient RuBP carboxylase in the bundle sheath, and glucose is produced through the same reactions that operate in C_3 plants (Figure 8-11). It appears, therefore, that the C_4 pathway evolved as a mechanism to "get around" the problems resulting from the inability of RuBP carboxylase to bind carbon dioxide at low concentrations.

In hot, dry environments where water loss is critical, C_4 plants possess a major survival advantage. Carbon dioxide enters the leaves of all plants through small pores, called **stomates.** When a stomate opens for carbon dioxide uptake, however, water vapor escapes through the same pore. Since C_4 plants are able to fix carbon dioxide even at extremely low carbon dioxide concentrations, stomates can be kept very narrow or closed altogether, reducing water loss,

without affecting their rate of carbon dioxide fixation. In contrast, when C_3 plants save water in hot, dry environments by closing their stomates, they drastically reduce their rate of carbon dioxide fixation and subsequent growth.

Another survival advantage of C_4 plants stems from their resistance to **photorespiration**—a phenomenon whereby high levels of oxygen actually interfere with photosynthesis by binding to RuBP carboxylase and inhibiting its carbon dioxide-fixing activity. Photorespiration can cut glucose production in C_3 plants by as much as 50 percent during peak sunlight hours. C_4 plants are less susceptible to photorespiration because they are able to photosynthesize under lower carbon dioxide and oxygen levels.

So why haven't vigorously growing C_4 plants taken over all the earth's habitats? When C_4 plants are introduced by people into some environments, the consequences to the native flora have indeed been swift. But C_4 plants, such as sugar cane and kudzu, do better than C_3s only where it is warm or dry (Figure 8-13). In cold environments, C_3s do better than C_4s because they are less sensitive to low temperature.

CAM PLANTS: FURTHER ADAPTATIONS TO A HOT, DRY ENVIRONMENT

Less than 5 percent of plants have another biochemical adaptation that allows them to survive in very hot, dry desert habitats. These plants, called **CAM plants,** utilize the same enzyme for carbon dioxide fixation as do C_4 plants, but conduct the light-dependent reactions and CO_2 fixation at different times rather than in different cells. (CAM is an acronym for *crassulacean acid metabolism,* named for plants of the family Crassulaceae in which it was first discovered.) While C_3 and C_4 plants open their stomates and fix carbon dioxide during the daytime, CAM plants, such as cactus and pineapple, open their stomates at night, when the rate of water vapor loss is greatly reduced. Throughout the night, more and more carbon dioxide is fixed and the products stored in the cell's vacuole. At sunrise, stomates close (conserving water), and sunlight powers ATP and NADPH production via the light-dependent reactions. At the same time, the products of carbon dioxide fixation are gradually removed from the vacuole and converted to carbohydrates. Attempts to manipulate C_3, C_4, and CAM plants in the laboratory are discussed in "The Human Perspective: Producing Crop Plants Better Suited to Environmental Conditions."

New life is built mostly from air and water. The original source of carbon to build new organic molecules needed for growth and reproduction is carbon dioxide gas obtained from the nonliving environment. Carbon becomes incorporated into organic molecules during the light-independent reactions of photosynthesis, forming the life-sustaining bridge across which carbon passes from the realm of the nonliving into that of the living. (See CTQ #6.)

FIGURE 8-13

Kudzu, a C_4 plant, is a familiar sight growing along the highways in the southeastern United States. Introduced from Asia, kudzu is able to overgrow many indigenous species.

◁ B I O L I N E ▷
Living on the Fringe of the Biosphere

In 1977, Robert Ballard and fellow scientists of the Woods Hole Oceanographic Institute guided a small research submarine through the pitch-black depths of the Pacific Ocean near the Galapagos Islands. Ballard's group was looking for cracks in the ocean floor that had appeared in photographs taken by cameras suspended at the end of long cables. There, at a depth of more than 2,500 meters (8,000 feet), the researchers discovered a flourishing community of diverse organisms (Figure 1a), including sea anemones, clams, mussels, blind crabs, and giant-sized red tube worms that (Figure 1b) had no mouths and were many times larger than their closest known relatives. All of these strange marine creatures were crowded near "chimneys" that spewed scalding, black water. The scientists crowded into the small research vessel were amazed at having discovered one of the most extraordinary and productive ecosystems on earth—lying at the very fringe of the biosphere. Animals living on the deep ocean bottom generally must depend on a trickle of organic material that filters through the water from the surface. How is it possible, then, that such an abundant fauna could live in such a hostile environment?

The cracks in the ocean floor are termed **hydrothermal vents.** These are sites where frigid sea water seeps into red-hot fissures and becomes heated to more than 350°C (662°F). The scalding water

(a)   *(b)*

FIGURE 1

mixes with hydrogen sulfide released from the earth's core, producing a habitat capable of supporting the growth of dense clouds of chemosynthetic bacteria that depend on hydrogen sulfide as an energy source. These bacteria form the diet of barnacles, clams, mussels, and small worms, which filter the microscopic organisms from the sea water. These filter

feeders, in turn, are eaten by larger crabs and fishes. A number of animals, including the giant red tube worm, house the chemosynthetic bacteria *inside their tissues,* thereby receiving a direct supply of food. Remarkably, the organisms of the hydrothermal vents form a self-sufficient community independent of life elsewhere on earth.

CHEMOSYNTHESIS: AN ALTERNATE FORM OF AUTOTROPHY

◉ Although the predominant means of generating energy-rich organic molecules is through photosynthesis, a number of bacteria depend on chemosynthesis for a supply of chemical energy. In photosynthesis, electrons and protons come from the splitting of water, while in **chemosynthesis,** they are stripped from reduced inorganic substances, such as ammonia (NH_3) and hydrogen sulfide (H_2S). The chemical equation for chemosynthesis using hydrogen sulfide is:

$$CO_2 + 2 H_2S \rightarrow (CH_2O) + H_2O + 2 S$$

The removal of electrons (oxidation) provides the energy needed for the formation of NADPH and ATP. Until cently, chemosynthetic bacteria did not receive much attention from scientists, but this changed in 1977 when a whole new type of community was discovered that depended entirely on chemosynthesis of microscopic prokaryotes. This community is discussed in the Bioline entitled "Living on the Fringe of the Biosphere."

Light is not the only source of energy for autotrophs. The energy contained in some inorganic molecules is harvested by a few types of autotrophs. In some habitats devoid of light, these autotrophs supply the energy that supports an entire community of organisms. (See CTQ #8.)

REEXAMINING THE THEMES

Relationship between Form and Function

Differences in the growth of C_3 and C_4 plants in different environmental conditions are based on differences in the structure of the leaves and the carbon dioxide-fixing enzymes of these plants. C_4 plants are able to flourish in warm, dry habitats partly because their leaves separate parts of photosynthesis into two specialized types of cells, each with a different carbon dioxide-fixing enzyme. PEP carboxylase, whose active site is able to bind carbon dioxide at low levels, is present in the outer mesophyll cells, while RuBP carboxylase, a less efficient enzyme but an important component of the pathway leading to carbohydrate synthesis, is present in bundle-sheath cells lying deeper in the leaf. C_3 plants do not use PEP carboxylase, nor do they separate reactions in different cells.

Biological Order, Regulation, and Homeostasis

Chloroplast thylakoids probably contain the most highly ordered components of any cellular membrane. Within these thin membrane sheets is an array of proteins and pigments capable of harvesting light energy and converting it to chemical energy of ATP and NADPH. Each of the components required for the light-dependent reactions —from the elements of the two photosystems to those of the electron transport chains—must be held in a precise orientation within the thylakoid membrane; otherwise, electrons could not pass along a defined sequence of carriers, and protons could not be transported across specific sites of the thylakoid membrane.

Acquiring and Using Energy

Photosynthesis (and chemosynthesis) are the only means by which the energy used and lost by living organisms is replaced, allowing life to continue. The energy present in sunlight drives endergonic reactions in the chloroplast, which lead to the synthesis of ATP and NADPH. The energy stored in these compounds is subsequently used in the synthesis of carbohydrate and other organic materials.

Unity within Diversity

Organisms can be divided into two groups based on their ability or inability to manufacture their own organic materials from "scratch," that is, from inorganic precursors such as carbon dioxide and water. Organisms with this ability are autotrophs—life's "producers"—while organisms lacking this ability are heterotrophs—life's "consumers." Organisms as different as the chemosynthetic bacteria living in the cracks in the ocean floor, the flagellated protists that form a scum over ponds during the summer, and the trees in the forest are all autotrophs. Conversely, organisms as different as disease-causing bacteria, mushrooms, baker's yeast, and mammals are all heterotrophs.

Evolution and Adaptation

Plants living in hot, dry environments are frequently threatened with severe water loss. Saving water requires that the plants close or reduce the size of their stomates during the day, which reduces the carbon dioxide supply to the chloroplasts. C_4 and CAM plants have evolved adaptations that get around these problems. Both groups possess a carbon dioxide-fixing enzyme that can operate at low carbon dioxide levels. The C_4 plant separates fixation of atmospheric carbon dioxide and carbohydrate synthesis into different cells of the leaf, whereas CAM plants carry out these two processes at different times of the day.

SYNOPSIS

Organisms are categorized as autotrophs if they manufacture organic nutrients from carbon dioxide and other inorganic precursors; they are categorized as heterotrophs if they depend on organic materials synthesized by others. Most autotrophs are photosynthesizers, converting light energy into chemical energy, which is then used to manufacture carbohydrates from carbon dioxide and water.

During the light-dependent reactions, photons of light are absorbed by pigments embedded in the thylakoid membranes. The pigments are associated into clusters, called photosystems, each of which includes several hundred antenna pigments that absorb varying wavelengths of light and a single reaction-center pigment that ultimately receives the energy. There are two types of photosystems, I and II, which operate in conjunction with each other. When energy is absorbed by the reaction-center pigments, electrons are boosted to an outer orbital from which they are transferred to an electron acceptor, leaving the reaction-center pigments positively charged. The reaction center of Photosystem II ($P680^+$) is sufficiently electron attractive to pull electrons from water, splitting the water molecules into protons, electrons, and oxygen atoms. The protons pass into the inner space of the thylakoid; the oxygen atoms form molecular oxygen, which is released as a byproduct; and the electrons pass to $P680^+$, filling the electron vacancy. The electrons previously lost by P680 pass along a chain of electron carriers to the electron-deficient reaction center ($P700^+$) of Photosystem I, filling its electron vacancy. The electrons previously lost by P700 pass along a chain of electron carriers to $NADP^+$, forming NADPH. During the process of electron transport, protons are moved across the thylakoid membrane into the lumen.

The protons that accumulate in the thylakoid lumen as the result of the splitting of water and electron transport produce a proton gradient—a form of stored energy that is used to drive the reaction in which ADP is converted to ATP. ATP formation occurs as a result of the movement of protons back across the thylakoid membrane through a channel in an ATP-synthesizing complex. This entire process in which electrons pass from water to NADPH and ATP is formed by a proton gradient is called noncyclic photophosphorylation.

During the light-independent reactions, the chemical energy stored in NADPH and ATP is used in the synthesis of carbohydrates from carbon dioxide. There are three variations in these synthetic pathways. The majority of plants utilize C_3 synthesis in which carbon dioxide is fixed by RuBP carboxylase to a five-carbon compound, RuBP, forming an unstable six-carbon compound, which immediately splits into two molecules of PGA. NADPH and ATP are used to convert PGA molecules to glucose, regenerating RuBP to participate in another round of the cycle. In C_4 and CAM plants, carbon dioxide is initially fixed by a different enzyme (PEP carboxylase) to a three-carbon compound. PEP carboxylase has the ability to bind carbon dioxide at much lower carbon dioxide concentrations than can RuBP carboxylase. This allows C_4 and CAM plants to flourish under hot, dry conditions where they can keep their stomates closed, which greatly lowers the carbon dioxide concentration but prevents water loss.

Chemosynthesis practiced by certain bacteria utilizes energy obtained by oxidizing inorganic substrates, such as ammonia and hydrogen sulfide to form NADPH and ATP needed in the synthesis of carbohydrates.

Key Terms

photosynthesis (p. 154)
autotroph (p. 154)
heterotroph (p. 155)
light-dependent reaction (p. 156)
light-independent reaction (p. 156)
photon (p. 156)
chlorophyll (p. 157)
carotenoid (p. 157)
photosystem (p. 159)
reaction center (p. 159)
antenna pigment (p. 159)

Photosystem I (p. 159)
Photosystem II (p. 159)
P700 (p. 159)
P680 (p. 159)
electron acceptor (p. 159)
photolysis (p. 160)
electron transport system (p. 160)
noncyclic photophosphorylation (p. 161)
chemiosmosis (p. 161)
ATP synthase (p. 161)
cyclic photophosphorylation (p. 161)

carbon dioxide fixation (p. 163)
C_3 pathway (p. 164)
ribulose biphosphate (RuBP) (p. 164)
RuBP carboxylase (p. 164)
C_4 synthesis (p. 164)
PEP carboxylase (p. 166)
stomate (p. 166)
photorespiration (p. 166)
CAM plant (p. 166)
chemosynthesis (p. 167)
hydrothermal vent (p. 167)

Review Questions

1. Rearrange the order of the following terms to match the correct sequence of reactions during photosynthesis:

   glucose formation
   splitting of water
   carbon dioxide fixation
   noncyclic flow
   PGA production
   chemiosmosis
   absorption of light
   PGAL production

2. Contrast the functions of the light-dependent and light-independent reactions. Why do the light-independent reactions depend on the light-dependent reactions having already occurred?

3. How does light absorption by the Photosystem II reaction center contribute to the splitting of water? How does the splitting of water contribute to the formation of ATP?

4. Contrast the basic roles of Photosystems I and II in generating NADPH.

5. Describe the adaptations in C_4 and CAM plants which allow them to flourish in hot, dry environments.

Critical Thinking Questions

1. If Melvin Calvin had been working with a C_4 plant, what type of compound would he have found most heavily labeled after administering $C^{18}O_2$ for 2 seconds? What if he had been working with a CAM plant?

2. Considering the second law of thermodynamics (Chapter 6), which states that usable energy becomes lost whenever energy is exchanged, would you expect there to be a greater number of heterotrophic or autotrophic individuals in the world? How does your conclusion fit with the statement that approximately 80 percent of all species are heterotrophs?

3. Technological advances and scientific discoveries often go hand in hand. List the technological advances that made it possible for scientists to unravel the mysteries of photosynthesis, and explain how each advance contributed to our knowledge of photosynthesis.

4. Suppose you discovered a mutant plant having thylakoids lacking Photosystem II. How would this affect carbohydrate production? Why? Do you suppose this plant would be able to produce any ATP from converted light energy?

5. Organization is one of the characteristics of all living things. Explain how organization is reflected in the structure of plant cells and the biochemical processes that constitute photosynthesis.

6. Construct a diagram that illustrates the flow of carbon from carbon dioxide in the atmosphere to heterotrophic organisms and back to the atmosphere.

7. If you were trying to engineer C_3 plants genetically to make them more efficient and able to survive better in hot, dry climates, on which enzyme would you focus? What changes would you want to develop in this enzyme?

8. Suppose somebody were to tell you that all life is dependent on energy from the sun and that the "sudden death" of the sun would quickly extinguish all life forms. Would you agree with their conclusion? Why or why not?

Additional Readings

Alberts, B., et al. 1989. *Molecular biology of the cell.* New York: Garland. (Intermediate)

Cone, J. 1991. *Fire Under the Sea: The Discovery of the Most Extraordinary Environment on Earth — Volcanic Hot Springs on the Ocean Floor.* New York: Morrow. (Intermediate)

Govindjee, and W. J. Coleman. 1990. How cells make oxygen. *Sci. Amer.* Feb: 50–58. (Intermediate)

Gregory, R. P. 1989. *Photosynthesis.* London: Chapman and Hall. (Intermediate)

Kaharl, V. A. 1991. *Water baby: The story of Alvin.* Oxford. (Alvin is the name of the submersible vessel that discovered life in the hydrothermal vents) (Introductory)

Kamen, M. D. 1986. A cupful of luck, a pinch of sagacity. *Ann Rev. Biochem.* 55:1–34. (Introductory)

Prebble, J. N. 1981. *Mitochondria, chloroplasts, and bacterial membranes.* New York: Longman. (Advanced)

Stryer, L. 1988. *Biochemistry.* 3d ed., New York: Freeman. (Advanced)

Youvan, D. C., and B. L. Marrs. 1987. Molecular mechanisms of photosynthesis. *Sci. Amer.* June: 42–50. (Intermediate)

Processing Energy: Fermentation and Respiration

STEPS
TO
DISCOVERY
The Machinery Responsible For ATP Synthesis

FERMENTATION AND AEROBIC
RESPIRATION: A PREVIEW OF THE
STRATEGIES

GLYCOLYSIS

The Reactions of Glycolysis

FERMENTATION

AEROBIC RESPIRATION

The Krebs Cycle

The Electron Transport System:
The Value of Oxygen

BALANCING THE METABOLIC
BOOKS: GENERATING AND
YIELDING ENERGY

COUPLING GLUCOSE OXIDATION TO
OTHER PATHWAYS

BIOLINE

The Fruits of Fermentation

THE HUMAN PERSPECTIVE

The Role of Anaerobic and Aerobic Metabolism in Exercise

The Machinery Responsible
for ATP Synthesis

With the improvement of light microscopes in the latter part of the nineteenth century, biologists observed thread-like particles within the cytoplasm of various types of cells. They named the organelle "mitochondrion," from the Greek meaning "thread granule." Despite their prevalence in cells, the function of mitochondria remained a mystery until the 1940s, when scientists perfected techniques to remove and isolate various organelles from cells. The isolation of mitochondria depended on the development of a new instrument — the "ultracentrifuge." This device could spin tubes in a circular path at speeds high enough to gen-

erate centrifugal forces (forces that pull an object outward when it is rotating around a center) over 100,000 times the force of gravity. Cells were first broken open to release their contents, which were placed in a tube and spun at various speeds. Larger organelles, such as cell nuclei, would settle at the bottom of the tube at slower speeds. The supernatant (the liquid remaining in the upper portion of the tube) containing smaller particles was then removed and recentrifuged *at higher speeds*, causing the mitochondria to settle at the bottom. The supernatant was then discarded, and the purified mitochondria were resuspended.

When mitochondria are broken, the inner membranes form vesicles (red spheres) and particles (yellow dots).

In 1948, using these techniques, Eugene Kennedy and Albert Lehninger, then at the University of Chicago, purified mitochondria and demonstrated that these organelles were the cells' "chemical power plants." These purified mitochondria were able to oxidize organic compounds (such as fatty acids) and to use the released energy to make ATP.

The mitochondrion is a complex organelle composed of many parts. Which part was responsible for synthesizing ATP? Efraim Racker and his colleagues in New York attempted to answer this difficult question. According to Racker, "We had no new ideas, so we did what I call 'instrumental research': You have no ideas, use a new instrument. At about that time, Dr. Nossal in Australia had designed a mechanical shaker for breaking up yeast cells with glass beads. So we used this machine to break up mitochondria."

When Racker's team subjected the suspension of broken mitochondria to high-speed centrifugation, they divided the material into two portions: one portion consisted of membrane vesicles that went to the bottom of the centrifuge tube, the other consisted of material that remained in the liquid supernatant. The isolated vesicles at the bottom could oxidize various organic substrates, but they were unable to make ATP. When mixed with the liquid supernatant, however, these same vesicles acquired full ATP-generating capacity. It seemed that the supernatant contained some factor needed by the vesicles for ATP production. Racker called this factor F_1.

Meanwhile, in Boston, Humberto Fernandez-Moran of Massachusetts General Hospital was examining mitochondrial membranes using a new electron-microscopic technique called negative staining, whereby objects are brightly outlined against a dark background. This makes very small objects consisting of macromolecules much easier to see. In this case, the technique revealed never-before-seen rows of particles protruding from the inner mitochondrial membrane. Each particle (see Figure 9-9) was attached to the membrane by a thin stalk.

Racker then used the electron microscope to examine the supernatant formed from disrupted mitochondria to see if his F_1 factors could be visualized. Racker observed particles in the supernatant that appeared identical to the stalked particles Fernandez-Moran had discovered attached to the mitochondrion's inner membrane. Furthermore, when Racker added the F_1 particles to the membrane vesicles formed from disrupted mitochondria, the particles became attached to the membranes by "little stalks." Racker concluded that the F_1 particles released during mitochondrial disruption were the same ones as those seen on the inner wall of mitochondria. After a series of biochemical studies, Racker discovered that these particles were in fact the sites in mitochondria where ATP was synthesized. He called the entire complex (sphere, stalk, and membrane-base piece) *ATP synthase*.

Similar types of stalked particles have been found attached to the thylakoid membranes of chloroplasts and the plasma membrane of bacteria (which have no mitochondria). In all of these organisms, the stalked particles are the sites where energy from substrate oxidation drives the synthesis of ATP. Once again, the unity of life reveals the common evolutionary origins of all organisms. Bacteria, plants, and animals are all descended from the same ancestor, one from which they inherited their energy-generating mechanisms.

When mixed, particles reattach to the vesicles to form complexes capable of ATP formation.

*I*t may have been the most catastrophic single occurrence in the history of life on earth. It was presumably responsible for the deaths of billions of individual organisms and spelled extinction for countless species. Creatures that had prevailed for millions of years disappeared forever, banished by one of the most universally poisonous substances the world had ever known. This unwelcome toxic intruder is still around today; in fact, it saturates today's biosphere. We are referring to molecular oxygen (O_2).

Molecular oxygen gradually appeared on earth beginning about 3 billion years ago. It was produced by a new type of microorganism—an evolutionary innovator that changed the face of the planet even more dramatically than would the eventual reign of human beings billions of years later. These devastating microbes were *cyanobacteria* (formerly known as "blue-green algae"), the prokaryotes that "invented" oxygen-evolving photosynthesis.

Up to that time, none of the earth's organisms required molecular oxygen, and the gas was not present in the atmosphere. The world was populated by **anaerobes** —organisms that utilized *oxygen-free (anaerobic)* metabolic processes to disassemble nutrients and obtain energy. This was the ancient order on the early oxygen-devoid earth. Following the rise of the cyanobacteria, however, the oceans and the atmosphere were infused with so much molecular oxygen that the old order was overthrown, forever altering the history of life on earth. The success of the cyanobacteria's energy-acquiring strategy catapulted these microbes to the top of the evolutionary hierarchy, where they remained for more than 2 billion years.

When it first appeared, molecular oxygen fatally oxidized the cytoplasm of all but a few types of microorganisms. Eventually, however, species evolved that not only withstood the poisonous effects of molecular oxygen but actually became dependent on it, using oxygen to increase their metabolic efficiency enormously. Organisms dependent on oxygen are called **aerobes.** These evolutionary innovators—the first users of oxygen—were the pioneers of an evolutionary line that ultimately produced all oxygen-using organisms, including humans.

Living with oxygen is still a dangerous business, however. As we saw on page 56, molecular oxygen can pick up an extra electron, becoming transformed into the highly destructive superoxide radical. The survival of aerobes depends on protective enzymes that disarm these free radicals by converting them to water the instant they are formed. Without this protection you inherited from early aerobic prokaryotes, the oxygen in your next breath would bring certain death instead of sustaining your life.

Perhaps after reading this chapter, you will have a new appreciation of breathing (Figure 9-1). Breathing draws in life-supporting oxygen and expels the waste product, carbon dioxide. Molecular oxygen is needed for you to release the energy stored in the chemical fuel you consume as food. Without oxygen, the metabolic furnace that powers your life would be extinguished.

FERMENTATION AND AEROBIC RESPIRATION: A PREVIEW OF THE STRATEGIES

One of the foundations of biology is the evolutionary unity of organisms. Even such distantly related organisms as a bacterium, a garden weed, and a human being share many common biological characteristics. We now come to another of these biological "universals." Recall from previous chapters that reduced organic molecules represent a storehouse of chemical energy. All organisms harvest chemical energy by oxidizing these organic molecules, generating

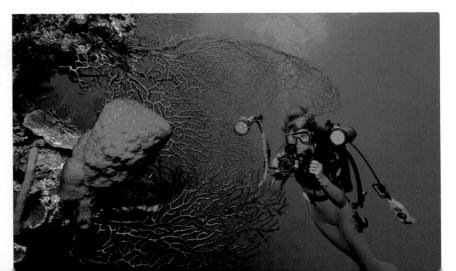

FIGURE 9-1
Oxygen—a precious commodity. Without a supply of oxygen from the tank, this scuba diver would only be able to remain submerged for a minute or two.

raw materials, and releasing energy that is trapped momentarily in the form of ATP. Moreover, they accomplish this shuffling of chemical energy from one compound to another by virtually the same set of metabolic reactions that arose at a very early stage of evolution. The starting point for most of these metabolic reactions is glucose.

Glucose is an energy-rich molecule, but it cannot directly power biological activities. In other words, a glucose molecule in a cell is like a person carrying around a $500 bill and trying to use it to purchase a few gallons of gasoline: Both the cell and the gasoline purchaser require smaller denominations. Therefore, cells convert energy-rich molecules, such as glucose, to other compounds containing more usable quantities of energy, most often ATP.

◐ The *complete* oxidation of a gram of glucose releases 3,811 calories. Not all organisms can take full advantage of glucose's energy content, however. The extent to which glucose is dismantled to release its energy is the basis for distinguishing between two fundamental processes: fermentation and aerobic respiration. As we will see shortly, **fermentation** is an incomplete breakdown of glucose, which occurs in the absence of oxygen and extracts only a small portion of the sugar's energy content. In contrast, during **aerobic respiration,** each glucose molecule is *completely* disassembled, step by step, with pairs of high-energy electrons being stripped from the substrate and transferred to molecular oxygen.[1] Electrons removed from the substrate are passed through a series of membrane-embedded electron carriers that comprise an electron transport system similar to that utilized during photosynthesis, which we discussed on page 160. The energy released during electron transport is used in the formation of ATP.

Regardless of the strategy (fermentation or aerobic respiration), tapping the energy stored in glucose begins with **glycolysis** (*glyco* = sugar, *lysis* = split), a universal pathway in which glucose is split into two fragments, both of which are converted to pyruvic acid, which contains a storehouse of readily available energy. We will now take a closer look at the process of glycolysis.

If all the energy in an energy-rich molecule were released in one step, most of the energy would be lost as heat and could not be used to power biochemical reactions. A multistep, energy-harvesting pathway enables cells to release the energy gradually to form molecules capable of delivering smaller, usable amounts of energy, most often as ATP. (see CTQ #2.)

---

[1] Cellular respiration should be distinguished from the term "respiration" as it applies to inhaling and exhaling. Nearly all organisms, even those that have never taken a breath, rely on aerobic respiration as their primary energy-harvesting strategy. These organisms include animals, plants, protists, and bacteria—every organism on earth, with the exception of a few prokaryotes and animals (such as intestinal parasites) that live in conditions where oxygen is not available. Clearly, "respiration" implies more than just breathing.

GLYCOLYSIS

In 1905, two British chemists, Arthur Harden and W. J. Young, were studying glucose breakdown by yeast cells, a process that generates bubbles of carbon dioxide. Harden and Young noted that the carbon dioxide bubbling eventually slowed and stopped, even though there was plenty of glucose left to metabolize. Apparently, some component of the broth was being exhausted. After experimenting with a number of substances, the chemists found that adding inorganic phosphates started the reaction going again. They concluded that the reaction was exhausting phosphate, the first clue that phosphate played a role in metabolic pathways. It would be several decades before biochemists would understand that the inorganic phosphate was being used to form ATP, which was subsequently used in glucose disassembly. This is illustrated by the first reaction of glycolysis.

THE REACTIONS OF GLYCOLYSIS

Glycolysis begins with the linkage of two phosphate groups to glucose at the expense of two molecules of ATP (Figure 9-2). This process of phosphate addition, called *phosphorylation*, "activates" the sugar—in this case, making it reactive enough to be split into two fragments, each containing three carbons and a phosphate. (Keep in mind that, because of the splitting of glucose, two molecules of each reaction product are formed from a single molecule of glucose. This is indicated in Figure 9-2.) In this case, the loss of a pair of ATPs can be considered the cost of getting into the glucose-oxidation business.

The next two reactions are particularly important because they have the potential to generate ATP. The first of these reactions is the conversion of phosphoglyceraldehyde (PGAL) to 1,3-diphosphoglycerate (DPG). You needn't be concerned with the chemical structures of the molecules in these reactions; your attention is best directed to the red carbons and the transfer of the electrons, indicated by the circle.

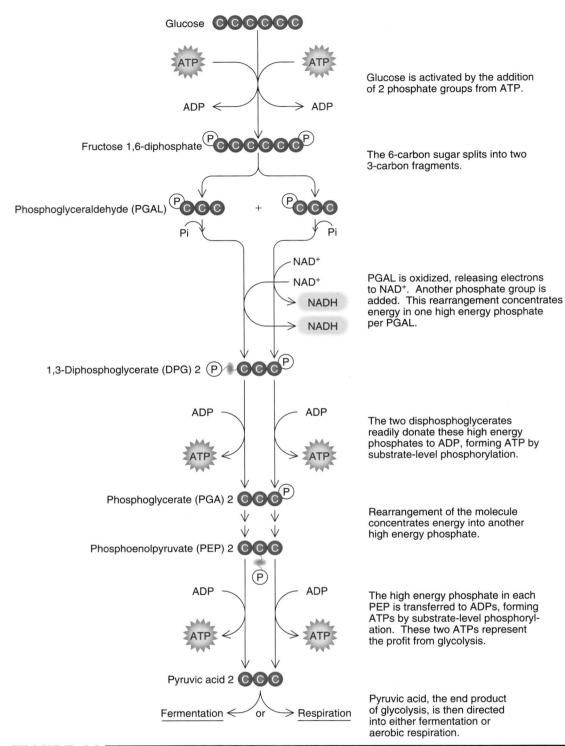

Glucose is activated by the addition of 2 phosphate groups from ATP.

The 6-carbon sugar splits into two 3-carbon fragments.

PGAL is oxidized, releasing electrons to NAD⁺. Another phosphate group is added. This rearrangement concentrates energy in one high energy phosphate per PGAL.

The two disphosphoglycerates readily donate these high energy phosphates to ADP, forming ATP by substrate-level phosphorylation.

Rearrangement of the molecule concentrates energy into another high energy phosphate.

The high energy phosphate in each PEP is transferred to ADPs, forming ATPs by substrate-level phosphoryl-ation. These two ATPs represent the profit from glycolysis.

Pyruvic acid, the end product of glycolysis, is then directed into either fermentation or aerobic respiration.

FIGURE 9-2 ━━━━━━━━━━━━━━━━━━━━━━━━━━━━━━━━━━━━━━

A condensed version of glycolysis. Although some of the ten reactions have been omitted for clarity, the overall activities of the pathway are evident—the oxidation of glucose to two molecules of pyruvic acid, with the release of two NADHs and a profit of two ATPs. The overall reaction can be written: Glucose + 2 ATPs + 2 NAD⁺ → 2 pyruvic acid + 4 ATPs + 2 NADH. Note: each ball represents a carbon atom.

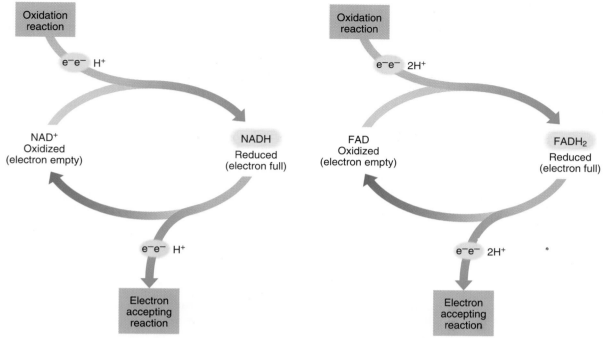

FIGURE 9-3

Electron carriers in action. A number of enzymes oxidize substrates by removing a pair of electrons, which is transferred to NAD^+, forming NADH. The electrons are only stored temporarily in NADH. As we will see, during fermentation, the electrons stored in NADH are transferred to another substrate, thereby regenerating NAD^+. During aerobic respiration, the electrons are passed along an electron transport system and used to generate ATP. (NADH may also be converted to NADPH, whose electrons are used in anabolic pathways rather than to form ATP.) As we will see later in the chapter, FAD (flavin adenine dinucleotide) is another coenzyme that accepts electrons from substrates and passes the electrons along an electron transport system, thereby generating ATP.

NAD^+ *(nicotinamide adenine dinucleotide)* is a coenzyme for a number of enzymes that oxidize substrates. This key compound of metabolism is derived from the vitamin niacin, a substance readily obtained from various types of meats and leafy vegetables. In the reaction shown above, the enzyme removes "high-energy" electrons (and a proton) from the PGAL molecule and transfers them to NAD^+, forming NADH (Figure 9-3), which is now free to leave the enzyme. In respiring organisms, these high-energy electrons stored in NADH will be "cashed in" for ATP at a later stage in the process.

In the next reaction of glycolysis, a phosphate group is transferred from DPG to ADP to form a molecule of ATP. Once again, you needn't be concerned with the chemical structures shown in this reaction, for it is the transfer of the phosphate group (indicated by the circle) that is important here:

This "direct" route of ATP formation is referred to as **substrate-level phosphorylation,** because a phosphate group is transferred directly from a substrate molecule to ADP *and does not require an electron transport chain.* A second substrate-phosphorylation step occurs later in glycolysis during the formation of pyruvic acid (Figure 9-2).

We can now look back over glycolysis and total up the energy profits. The two substrate-level phosphorylation steps produce four ATPs for each glucose entering the pathway. Because two ATPs must be spent just to get glycolysis "rolling," however, the *net* ATP yield is only two. In addition to a net gain of two ATPs, glycolysis also generates two pairs of energized electrons (carried in two NADHs).

The molecular remains of glycolysis are two molecules of pyruvic acid, each containing three carbons. This is the point where anaerobic (oxygen-independent) fermentation and aerobic (oxygen-dependent) respiration diverge. They differ in the way they solve a common problem: how to regenerate the NAD$^+$ molecules that were converted to NADH during the oxidation of PGAL. In all organisms, the reserves of available NAD$^+$ are replenished by transferring electrons from NADH to another molecule, regenerating NAD$^+$. If NADH donates its electrons directly to an organic substrate, which is simply excreted as a waste product, the process is called *fermentation.* If, instead, NADH passes its electrons to an electron transport chain in which oxygen is the final electron acceptor, the process is called *aerobic respiration.* If the regeneration of NAD$^+$ could not occur, the cell would rapidly run out of its small supply of this coenzyme, and neither glycolysis nor the subsequent energy-generating process could proceed.

Glucose contains a large amount of chemical energy. Before its energy can be extracted, however, glucose molecules must first be activated by phosphorylation. Activation requires adding energy. *Glycolysis* accomplishes this activation, and extracts a small amount of usable energy in the process. (See CTQ #3.)

FERMENTATION

During fermentation, the electrons from NADH are transferred either to pyruvic acid, the end product of glycolysis, or to a compound formed from pyruvic acid. The end products of fermentation (Figure 9-4) vary according to the organism but always lead to the regeneration of NAD$^+$. In the most familiar type of fermentation—**alcoholic fermentation**—yeast cells convert pyruvic acid to ethyl alcohol, the alcohol consumed in alcoholic beverages. Alcoholic beverages differ primarily in the raw materials used in the fermentation process. Wines are produced by fermenting fruits, particularly grapes, whereas beers are produced by fermenting malted cereals, such as barley. Both wine and beer production are ancient arts; a recipe for making beer, for example, was found inscribed on stone tablets from Mesopotamia that dated back 9,000 years. By itself, fermentation is only able to produce a liquid containing about

15 percent alcohol. Those alcoholic products containing higher quantities of alcohol, such as whiskey or brandy, are produced by *distillation*, whereby the alcohol is evaporated from the fermentation liquid and then recondensed at higher concentration. The role of microorganisms in producing alcohol and other commercial products of fermentation is discussed in the Bioline: The Fruits of Fermentation.

In 1907, the British biochemist Frederick Hopkins demonstrated that when isolated frog muscles were caused to contract in an anaerobic environment, the muscles produced large amounts of lactic acid. Lactic acid was being produced by fermentation of pyruvic acid. **Lactic acid fermentation** also occurs in human muscle cells when they undergo such strenuous activity that they cannot obtain enough oxygen to oxidize pyruvic acid fully (see "The Human Perspective: The Role of Anaerobic and Aerobic Metabolism in Exercise," at the end of the chapter).

To reiterate, all types of fermentation are anaerobic processes. During the early stages of life on earth, when oxygen had not yet appeared, glycolysis and fermentation were probably the primary metabolic pathways by which energy was extracted from sugars by primitive prokaryotic cells. Today, many organisms still live in environments lacking oxygen, as in deep soil or inside the body of another animal, and glycolysis and fermentation remain the only

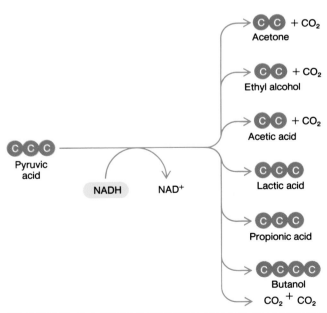

FIGURE 9-4

Some metabolic products of fermentation. Pyruvic acid may be converted to several end products, depending on the organism and the presence or absence of oxygen. In all cases, the final compound contains electrons donated by NADH. This recycles the coenzyme NAD$^+$, allowing it to continue its essential role as electron acceptor in glycolysis.

◁ B I O L I N E ▷
The Fruits of Fermentation

FIGURE 1

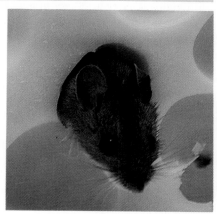

FIGURE 2

Biological events that take place in the absence of oxygen have been creating fortunes and livelihoods for people since the dawn of recorded history. The metabolic byproducts of fermentation, for example, make our diets more interesting, protect us from diseases, retard food spoilage, and provide us with alcoholic beverages (Figure 1). Alcoholic drinks are the products of converting pyruvic acid to ethyl alcohol by brewer's yeast, *Saccharomyces cerevisiae*. Alcoholic fermentation also releases carbon dioxide, which, if not allowed to escape, becomes trapped in the liquid,

producing the natural carbonation of beer and champagne. The same process, alcoholic fermentation by yeast cells, elevates dough during the leavening of bread and pastries. In fact, until the mid-1800s, commercial bakeries used yeast left over from the brewing of beer to leaven their pastries. The dried baker's yeast familiar to modern cooks contains living cells that quickly convert the sugar in dough to ethyl alcohol and carbon dioxide. The carbon dioxide forms the expanding bubbles that cause the dough to rise; the dough literally inflates with the gas. The alcohol is driven off by baking, which explains why you can eat bread without fear (or hope) of inebriation.

Bacteria generate a wide variety of valuable fermentation products (see Figure 9-4), of which lactic acid is perhaps the most common. This slightly sour acid imparts flavor to yogurt, rye bread, and some cheeses (Figure 2). It also lowers the pH of food below a level tolerable by many spoilage microbes. Dairy products that contain lactic acid are therefore more resistant to spoilage than is the milk from which they were made. Swiss cheese is produced by other types of bacteria that convert pyruvic acid to propionic acid and carbon dioxide. The acid imparts a nutty flavor to the cheese, whereas the gas creates the large bubbles that form its characteristic holes. Vinegar is a product of acetic-acid-

producing bacteria that grow in apple cider or grape mash. The tangy taste of pickles, sauerkraut, and olives reflects the presence of lactic and acetic acids, microbial byproducts that provide a flavorful twist to these foods. The process of making soy sauce requires an 8- to 12-month fermentation period before the mash of soy beans acquires its characteristic flavor.

Acetone and other industrially produced solvents are also byproducts of fermentation, as is isopropyl (rubbing) alcohol. Fermentation may be an inefficient means for an organism to harvest biological energy, but it has proven a very efficient source of valuable products.

---

methods of obtaining energy.[2] The total ATP yield of such organisms is a mere two molecules per glucose oxidation. Most of the potential energy of glucose is left in the waste products that are simply discarded, such as ethyl alcohol or

lactic acid. In fact, more than 90 percent of glucose's chemical energy is untapped by glycolysis and fermentation. (The flammability of ethyl alcohol, which can be used as a fuel in automobiles, testifies to the high energy content of this fermentation product.)

---

[2] A few bacteria in these oxygen-devoid environments use anaerobic respiration rather than fermentation. Anaerobic respiration resembles a less efficient form of aerobic respiration that uses a terminal electron acceptor such as nitrate or sulfate rather than oxygen.

In oxygen-poor environments, cells utilize a less efficient metabolic pathway that allows them to continue to harvest a limited amount of energy. (See CTQ #4.)

AEROBIC RESPIRATION

As the atmosphere became infused with molecular oxygen, organisms evolved a new metabolic strategy of *aerobic respiration,* which could completely oxidize the two pyruvic acids generated by glycolysis (instead of fermenting them) and obtain more than 30 additional ATPs. The complete oxidation of glucose can be likened to a three-act drama, as illustrated in the three steps of Figure 9-5. These same three acts are played out in all types of organisms that live in an aerobic environment.

The first act, glycolysis, which we have already described, takes place in the cytoplasmic fluid of all cells. The second act is called the **Krebs cycle,** after the British biochemist Hans Krebs who worked out the cyclic metabolic pathway in the 1930s (Figure 9-6). In eukaryotic cells, the Krebs cycle occurs in the innermost compartment (*matrix*) of the mitochondrion, while in prokaryotic cells, the same cyclical pathway occurs in the fluid of the cytoplasm. Fortunately for biochemists, the roles of the actors in the first two acts are played by so-called soluble enzymes; that is, enzymes more or less dissolved in the cell's fluids. This made

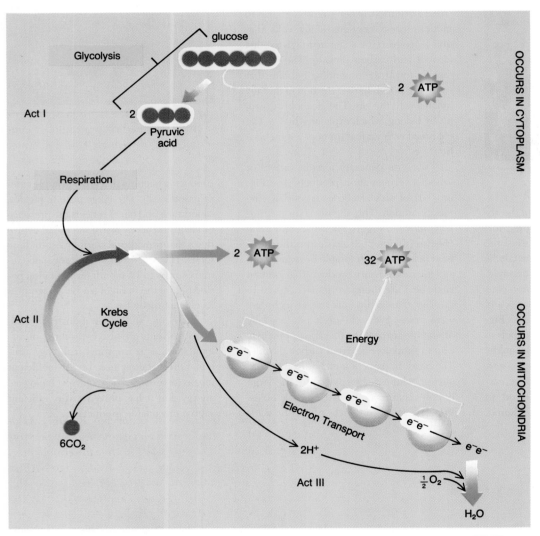

FIGURE 9-5

Respiration in eukaryotic cells: An overview. Respiration is a three-part process. (Act I) glycolysis; (Act II) the Krebs cycle, and (Act III) electron transport. During glycolysis, glucose is converted into two molecules of pyruvic acid with a net formation of two ATPs. As substrates pass through the Krebs cycle, carbons are stripped away and released as carbon dioxide, and high-energy electrons are stripped away and transferred to the coenzymes NAD^+ and FAD. The electron transport system allows the energy of the electrons to be used in the formation of large numbers of ATP. Electron transport depends on the availability of oxygen. Oxygen combines with electrons and protons to form water. Without oxygen, only glycolysis can continue, the pyruvate converted to fermentation byproducts such as lactic acid or ethyl alcohol.

their study relatively easy because these enzymes were readily purified and were active in a test tube solution. The script for the first two acts was well in hand by the 1940s. The third act, which includes both electron transport and ATP formation, has proven much harder to study because these processes depend on electron carriers' maintaining their precise positions within a membrane—either the mitochondrial membrane of a eukaryotic cell or the plasma membrane of a bacterium. If the structure of the membrane is disrupted, as was described in the chapter opener, so is its function. It wasn't until techniques were developed to isolate intact mitochondrial membranes that biologists began to understand the complexity of the third act.

THE KREBS CYCLE

As noted earlier, very little oxidation occurs during glycolysis. The oxidation of the substrate increases rapidly during the Krebs cycle, however, as (1) hydrogens (each consisting of a proton and two electrons) are stripped from several of the carbons and transferred to electron-carrying coenzymes (either NAD^+ or FAD), and (2) the carbons are split from the rest of the molecule in the form of carbon dioxide, the most oxidized state of a carbon atom. These two activities are illustrated in the very first reaction (step 1 in Figure 9-7), which connects glycolysis with the Krebs cycle.

During this reaction, the three-carbon substrate, pyruvic acid, is split into a two-carbon fragment and a one-carbon fragment. The one-carbon fragment loses a pair of electrons to NAD^+ and is released as carbon dioxide. The two-carbon *acetyl* fragment is temporarily coupled with a carrier molecule, coenzyme A (CoA), forming the complex *acetyl coenzyme A* (acetyl CoA). Coenzyme A is derived from the vitamin pantothenic acid, a substance present in virtually all foods, particularly meats. The discovery of acetyl CoA by

F I G U R E 9-6

Hans Krebs at work in the laboratory.

Fritz Lipmann in 1951 was the last piece in the puzzle of glucose oxidation.

The Krebs cycle itself begins when the two carbons of acetyl CoA are joined with a four-carbon molecule, *oxaloacetic acid* (step 2 in Figure 9-7), releasing CoA for reuse. The product of step 2 is a six-carbon compound, *citric acid,* the substance that gives the Krebs cycle its alternate name, the "citric-acid cycle" and also gives citric fruits (lemons, grapefruits, oranges) their tart flavor.

During the next reactions (steps 3 and 4 in Figure 9-7), two carbons are fully oxidized to carbon dioxide, with the electrons being transferred to two NAD^+ molecules, forming two NADHs. One ATP (formed from GTP) is also generated by substrate-level phosphorylation. This happens as the final carbon dioxide is removed in step 4, leaving a four-carbon compound *(succinic acid)*. The last of the high-energy electrons are removed in the final steps of the Krebs cycle, at which time succinic acid is converted to oxaloacetic acid. In the first of these steps (step 5), electrons (and two protons) are transferred to a different coenzyme FAD *(flavin adenine dinucleotide,* which is derived from the vitamin riboflavin, found in meats, fruits, and leafy vegetables), forming the reduced coenzyme $FADH_2$. Like NADH, $FADH_2$ is able to feed electrons into the electron transport system to generate ATP. In the subsequent step (step 6), electrons (and a proton) are transferred to NAD^+, forming NADH. Oxaloacetic acid, the last product of the Krebs cycle, is then free to react with another acetyl CoA to form another citric acid, starting the cycle all over again.

For their work in unraveling the pathway by which pyruvic acid is oxidized, Krebs and Lipmann, both early refugees from Nazi Germany, shared the 1953 Nobel Prize in physiology and medicine (Figure 9-6). Ironically, when Krebs first obtained sufficient evidence to support the idea of a metabolic cycle, he submitted the research article to the British journal *Nature.* The paper was returned several days later accompanied by a letter of rejection. The editor had concluded it wasn't important enough to be published in the journal.

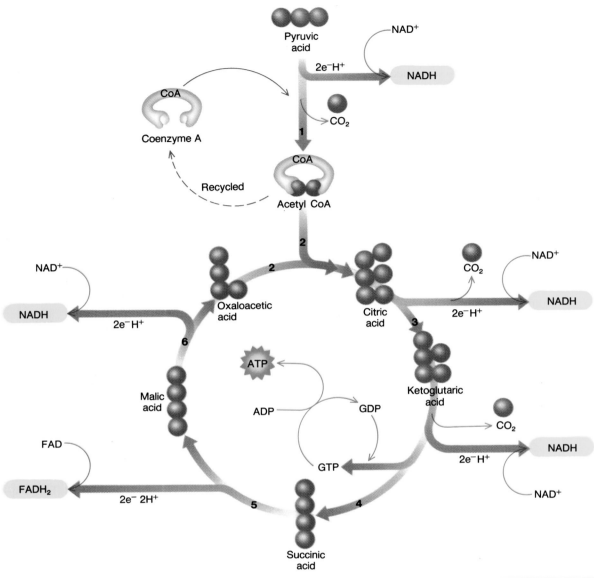

FIGURE 9-7

An abbreviated version of the Krebs cycle. As a result of the reactions shown in this figure, one molecule of pyruvic acid is completely oxidized to three molecules of carbon dioxide. In the process, four NADHs, one FADH₂, and one GTP are formed. GTP (guanosine triphosphate) is a high-energy compound that is readily converted to ATP and is thus considered equivalent to ATP. The energy in the NADH and FADH₂ coenzymes can be cashed in for ATPs via the electron transport system and chemiosmosis. Because glycolysis generates *two* pyruvic acids, oxidation of a single glucose molecule produces *twice* the amounts shown here.

In summary, each "turn" of the Krebs cycle oxidizes pyruvic acid to its simplest, most oxidized form, carbon dioxide, gradually releasing energy stored in the original glucose molecule. Two turns of the cycle are required for each glucose molecule. Most of the energy is transferred to NADH and FADH₂. In most organisms, the key that unlocks this treasury of energy is a small, electron-attracting molecule called oxygen.

THE ELECTRON TRANSPORT SYSTEM: THE VALUE OF OXYGEN

Fermentation taps less than 3 percent of the energy stored in glucose, whereas aerobic respiration captures about 40 percent—enough to form nearly 40 molecules of ATP per molecule of glucose oxidized. Only a small number of ATPs are produced directly by substrate-level phosphorylation (a

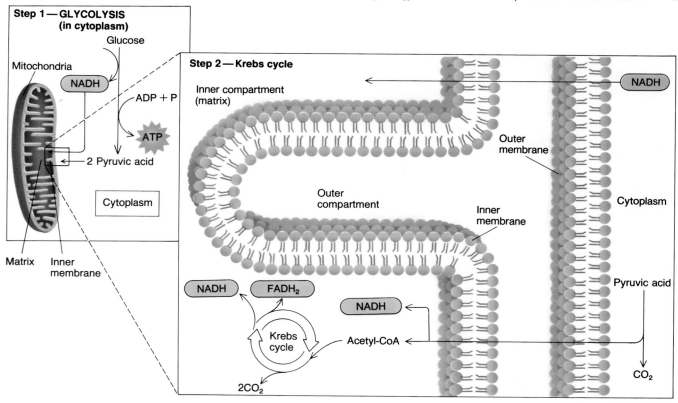

Step 1 — GLYCOLYSIS (in cytoplasm)

Glucose

Mitochondria

NADH

ADP + P

ATP

2 Pyruvic acid

Cytoplasm

Matrix

Inner membrane

Step 2 — Krebs cycle

Inner compartment (matrix)

Outer membrane

Outer compartment

Inner membrane

Cytoplasm

NADH

FADH$_2$

NADH

Krebs cycle

Acetyl-CoA

Pyruvic acid

2CO$_2$

CO$_2$

NADH

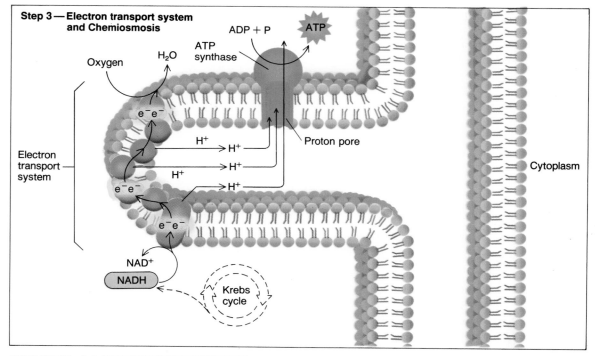

Step 3 — Electron transport system and Chemiosmosis

H$_2$O

ADP + P

ATP

ATP synthase

Oxygen

e$^-$ e$^-$

Proton pore

Electron transport system

H$^+$ → H$^+$

H$^+$ → H$^+$

H$^+$ → H$^+$

Cytoplasm

e$^-$ e$^-$

e$^-$ e$^-$

NAD$^+$

NADH

Krebs cycle

FIGURE 9-8

A closer look at the steps of glucose oxidation in a eukaryotic cell. Step 1: Glycolysis converts glucose to pyruvic acid in the soluble portion of the cytoplasm, generating pyruvic acid, NADH, and a small amount of ATP. Step 2: Pyruvic acid is moved into the inner compartment of the mitochondrion, where it is converted to acetyl CoA and then completely oxidized via the Krebs cycle to carbon dioxide, generating several molecules of NADH and one FADH$_2$. Step 3: The energized electrons carried by NADH and FADH$_2$ are transported from carrier to carrier along the electron transport system and, in the process, generate a proton gradient with a higher concentration of protons on the outer side of the inner mitochondrial membrane. As the protons pass down their concentration gradient through the channel of the ATP synthase, they drive the phosphorylation of ADP to form ATP.

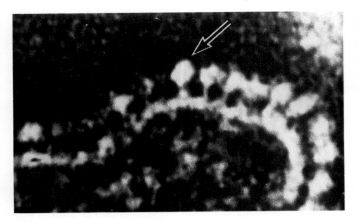

FIGURE 9-9
The stalked particles projecting from the inner mitochondrial membrane house the machinery for ATP manufacture. The entire structure, which consists of a sphere (upper arrow), stalk, and baseplate embedded within the inner mitochondrial membrane, is called the ATP synthase.

net of two during glycolysis and two during the Krebs cycle), the remainder are formed from the energized electrons transferred to NAD$^+$ and FAD and are processed by the electron transport system (Figures 9-5 and 9-8)—the third and final "act" in glucose oxidation.

Recall from Chapter 8 that an electron transport system consists of a chain of electron carriers embedded within a membrane. The electron carriers in the mitochondria are organized within the inner membrane in a precise, ordered arrangement so that electrons are passed from carriers of lesser electron attraction to carriers of greater electron attraction. The final acceptor of electrons is molecular oxygen, which, because of its atomic structure, has a very high attraction for electrons.

The electrons removed from the organic intermediates of the Krebs cycle are high-energy electrons. As they are transferred to the coenzymes NADH and FADH$_2$ and then passed down the electron transport chain to oxygen, these electrons give up nearly all of their energy, which is utilized to drive the energy-requiring reaction in which ADP is converted to ATP. The formation of ATP follows the same mechanism of chemiosmosis that accompanies photosynthesis. You will recall from Chapter 8 that chemiosmosis is the process by which a proton gradient drives the phosphorylation of ADP. The electron carriers of aerobic respiration are embedded in the inner mitochondrial membrane. So too is the enzyme complex (the ATP synthase) responsible for ATP synthesis (Figure 9-9). The energy released by electrons as they move down the electron transport chain is used to move protons across the membrane from the inner-most compartment of the mitochondrion (the matrix) to the outer compartment (Figure 9-8), forming a proton-concentration gradient across the membrane. The structure of the ATP synthase is well suited for its function. A central channel in the stalk of the ATP synthase provides a pore for protons to stream across the membrane into the matrix, powered by the force of the proton gradient. This proton flow through the active site of the enzyme located in the spherical (F$_1$) portion of the ATP synthase provides the

necessary energy to drive the phosphorylation of ADP to ATP.

The importance of the proton gradient in ATP formation is readily demonstrated by treating cells with chemicals that abolish the gradient. Dinitrophenol, for example, eliminates the gradient by making the inner mitochondrial membrane freely permeable to protons. Without a gradient, ATP formation ceases. During the 1920s, a few physicians prescribed dinitrophenol as a "diet pill." When taking the drug, obese patients would continue to oxidize the body's stores of fats and carbohydrates in a vain attempt to maintain high ATP levels. The practice ceased with the deaths of a number of patients taking this medication.

Now you know why you breathe—to provide a powerful electron-attracting agent capable of removing the electrons from the last carrier of the electron transport chain. This last electron carrier is a huge complex of proteins, called **cytochrome oxidase,** which spans the mitochondrial membrane and facilitates the final transfer of electrons to oxygen. If electrons were not removed by oxygen, electron flow along the electron transport chain would stop and ATP production would cease. A common poison, cyanide, causes just that to happen. This poison exerts its toxic effect by binding to cytochrome oxidase, blocking the transfer of electrons to oxygen. The lethal effects of this poison again reemphasize the essential role of cellular respiration to all aerobic organisms, from bacteria to humans. Without cellular respiration, ATP production drops to practically nothing. This quickly leads to energy starvation and, consequently, to fatal deterioration of biological order, unless the organism can switch to fermentation to provide the ATP needed to sustain a much slower pace of biological activity. Although such metabolic dexterity is common among microorganisms, fermentation does not provide enough energy to sustain the vast energy demands of larger multicellular organisms, such as humans. In these organisms, the efficiency of respiration has created a biological dependency: We either respire or we die.

In oxygen-rich environments, cells are able to utilize the products of glycolysis to extract much greater amounts of usable energy. This is possible because of the very strong electron-attracting properties of oxygen. High-energy electrons that are removed from substrates are transported to oxygen via a mechanism that energizes a membrane by forming a proton gradient and leads to the formation of ATP. (See CTQ #6.)

BALANCING THE METABOLIC BOOKS: GENERATING AND YIELDING ENERGY

As indicated in the metabolic ledger of Figure 9-10, the complete oxidation of one pyruvic acid via the Krebs cycle yields four NADHs, one $FADH_2$, and one GTP (= ATP) by direct substrate-level phosphorylation. Because each glu-

cose produces two pyruvic acids, we double these numbers and get eight NADHs, two $FADH_2$s, and two ATPs. The reduced coenzymes generated by the Krebs cycle in mitochondria can be "cashed in" at the rate of three ATPs for each NADH, and two for each $FADH_2$. Thus, the total profit from oxidizing both pyruvic acids is 30 ATPs (24 or (3 × 8) from NADH; 4 or (2 × 2) from $FADH_2$, and 2 that are formed directly).

In addition to these 30 ATPs, there are still the profits of glycolysis. In prokaryotes, electron transport yields an additional six ATPs from the two NADH molecules produced during glycolysis. In eukaryotic cells, however, these two NADH molecules usually yield only four ATPs as a result of the transfer of the electrons from the fluid of the cytoplasm, where glycolysis occurs, into the mitochondria, where electron transport occurs. When we add the other two ATPs from substrate-level phosphorylation during glycolysis, we find that the aerobic oxidation of one glucose molecule releases enough energy to produce 36 (in eukaryotes) or 38 (in prokaryotes) molecules of ATP. Although

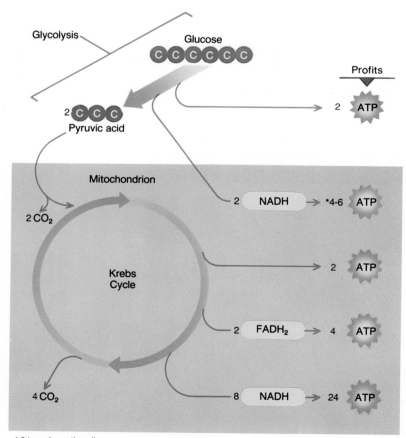

FIGURE 9-10

A metabolic ledger reveals the energy profits from oxygen-dependent respiratory oxidation of one glucose molecule. More than 90 percent of the ATPs are collected via the electron transport system. Strictly fermentative organisms lack not only an electron transport system but also the Krebs cycle. They therefore obtain only two ATPs per glucose fermented.

◁ THE HUMAN PERSPECTIVE ▷
The Role of Anaerobic and Aerobic Metabolism in Exercise

Most of you have tried lifting a barbell or doing "push-ups." You may have noticed that the more times you repeat the exercise, the more difficult it becomes until you are no longer able to perform the activity. The failure of your muscles to continue to work can be explained by oxidative metabolism. Skeletal muscles (the muscles that move the bones of the skeleton) consist of at least two types of fibers (insets): "fast-twitch" fibers—which can contract very rapidly—and "slow-twitch" fibers—which contract more slowly. Fast-twitch fibers are nearly devoid of mitochondria, which indicates that these cells are unable to produce much ATP by aerobic respiration. In contrast, slow-twitch fibers contain large numbers of mitochondria—sites of aerobic ATP production. These two types of skeletal muscle fibers are suited for different types of activities. Lifting weights or doing push-ups depends primarily on fast-twitch fibers, which are able to generate more force than are their slow-twitch counterparts. Fast-twitch fibers produce nearly all of their ATP anaerobically as a result of glycolysis. The problem with producing ATP by glycolysis is the rapid use of the fiber's available glucose (stored in the form of glycogen) and the production of an undesirable end product, lactic acid. Let's consider this latter aspect further.

FIGURE 1

Skeletal muscles (*insets*) contain a mix of fast-twitch fibers (darkly stained) and slow-twitch fibers (lightly stained). The muscles of world-class weightlifters (*top photo*) typically have a higher-than-average percentage of anaerobic, fast-twitch fibers, while world-class marathon runners (*bottom photo*) tend to have a higher-than-average percentage of aerobic, slow-twitch fibers.

Recall that the continuation of glycolysis requires the regeneration of NAD$^+$, which occurs by fermentation. Muscle cells regenerate NAD$^+$ by reducing pyruvate—the end product of glycolysis—to lactic acid. Most of the lactic acid diffuses out of the active muscle cells into the blood, where it is carried to the liver and converted back to glucose. Glucose produced in the liver is released into the blood, where it can be returned to the active muscles to continue to fuel the high levels of glycolysis. Not all of the lactic acid is transported to the liver, however; some of it remains behind in the muscles. The buildup of lactic acid and the associated drop in pH within the muscle tissue may produce the pain and cramps that accompany vigorous exercise and, together with the depletion of glycogen stores, accounts for the sensation of muscle fatigue.

If, instead of trying to use your muscles to lift weights or do push-ups, you were to engage in an "aerobic" exercise, such as jumping jacks or fast walking, you would be able to continue to perform the activity for much longer periods of time without feeling muscle pain or fatigue. Aerobic exercises, as their name implies, are exercises designed to allow your muscles to continue to perform aerobically; that is, to continue to produce the necessary ATP by electron transport. Aerobic exercises depend largely on the contraction of the slow-twitch fibers of your skeletal muscles; these muscles are able to generate less force but can continue to function for long periods of time, due to the continuing aerobic production of ATP without the corresponding buildup of lactic acid. Aerobic exercise is initially fueled by the glucose molecules stored as glycogen in the muscles themselves, but after a few minutes, the muscles depend on glucose and fatty acids released into the blood by the liver. The longer the period of exercise, the greater the dependency on fatty acids. After 20 minutes of vigorous aerobic exercise, it is estimated that about 50 percent of the calories being consumed by the muscles are derived from fat. Aerobic exercise, such as jogging, fast walking, swimming, or bicycling, is one of the best

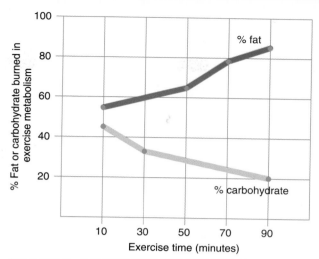

FIGURE 2

ways of reducing the body's fat content (Figure 2).

The ratio of fast-twitch to slow-twitch fibers varies from one particular muscle to another. For example, postural muscles in the back that are needed to allow a person to stand consist of a higher proportion of slow-twitch fibers than do arm muscles used to throw or lift an object. The ratio of fast-twitch to slow-twitch fibers in a particular muscle is genetically determined—a factor that may play a role in allowing a particular individual to excel in certain types of physical activities. For example, among world-class athletes, those who excel in activities that require short bursts of exertion, such as weight lifters (top photo) or sprinters, tend to have a higher proportion of fast-twitch fibers in their muscles than do long-distance runners (bottom photo), who excel in events that require endurance (Table 9-1).

TABLE 9-1

TYPICAL MUSCLE FIBER COMPOSITION IN ELITE ATHLETES REPRESENTING DIFFERENT SPORTS AND IN NONATHLETES

Sport	Slow-Twitch Fibers (%)	Fast-Twitch Fibers (%)
Distance running	60–90	10–40
Track sprinters	25–45	55–75
Shot-putters	25–40	60–75
Nonathletes	47–53	47–53

From Scott K. Powers and Edward T. Howley, *Exercise Physiology: Theory and Application to Fitness and Performance.* Copyright © 1990 Wm. C. Brown Communications, Inc., Dubuque, Iowa. All Rights Reserved. Reprinted by permission.

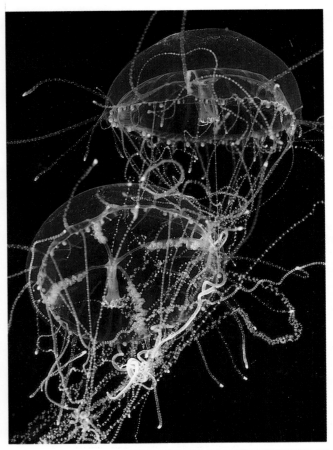

FIGURE 9-11
A luminescent jellyfish utilizes chemical energy stored in NADH to emit light.

some of the energy present in the original glucose molecule inevitably is lost, the formation of 36 to 38 ATPs represents an energy-capture efficiency of about 40 percent (the actual percentage varying with the conditions in the cell at the time). Compare this to a typical automobile engine that runs at less than 25 percent efficiency.

A metabolic ledger (Figure 9-10) pinpoints the source of each ATP. A balanced equation for the complete oxidation of glucose in a eukaryotic cell, for example, would be written

$$C_6H_{12}O_6 + 6\,O_2 + 36\,ADP + 36\,P_i \rightarrow 6\,CO_2 + 6\,H_2O + 36\,ATP$$

This equation represents the *potential* number of ATPs each molecule of glucose provides an organism. Electrons (and energy) are also needed for other purposes, however, such as biosynthesis of new molecules, movement, making sounds, or, in a more unusual example, light generation (Figure 9-11). Although these activities reduce the number of ATPs produced, their energy is not wasted because it is used to accomplish biological work.

The laws of thermodynamics apply to the living world. The chemical energy originally stored in glucose is converted by cells to the energy stored in proton gradients and then to the energy stored in ATP. Not all of the original available energy is used to form ATP, however; some of the energy is lost in the process. (See CTQ #7.)

COUPLING GLUCOSE OXIDATION TO OTHER PATHWAYS

Thus far, we have limited our discussion of energy to that which is obtained by glucose oxidation. But you have probably never sat down to a meal of pure glucose. What about the hamburger and french fries that may be fueling your biological processes right now? How is that energy utilized? The respiratory pathway we have described in this chapter (glycolysis, the Krebs cycle, and electron transport) not only extracts energy from glucose but is also the central pathway utilized for the breakdown of a diverse variety of materials (Figure 9-12). For example, the protein in your hamburger is converted to amino acids in your intestine and absorbed into the bloodstream. These amino acids are carried to the liver, where they are converted to molecules that are part of the pathways described in this chapter. The three-carbon amino acid alanine, for example, can be converted to pyruvic acid simply by removing its amino (NH_2) group. The pyruvic acid can then be fed into the Krebs cycle for complete oxidation. Similarly, the long fatty acid chains that make up fat molecules in your french fries are broken down into two-carbon units that enter the Krebs cycle as acetyl CoA molecules.

A pyruvic acid or acetyl CoA molecule is treated exactly the same by an enzyme, regardless of whether it came from glucose degradation or from some other carbohydrate, fat, or protein. Thus, any biochemical that can be degraded or converted to an intermediate of these central respiratory pathways can be completely oxidized for its energy.

We have just seen how the breakdown products of all types of materials are fed into glycolysis and the Krebs cycle. This principle also works in reverse. That is, virtually any compound that an organism must synthesize can be manufactured by diverting metabolic intermediates of glycolysis or the Krebs cycle into the appropriate biosynthetic pathway. For example, much of the surplus acetyl CoA produced in your body when glucose is abundant will be diverted to fat synthesis for energy storage.

The energy used to form more complex materials, such as proteins, fats, and nucleic acids, is derived largely from the ATP generated by electron transport. Many of these

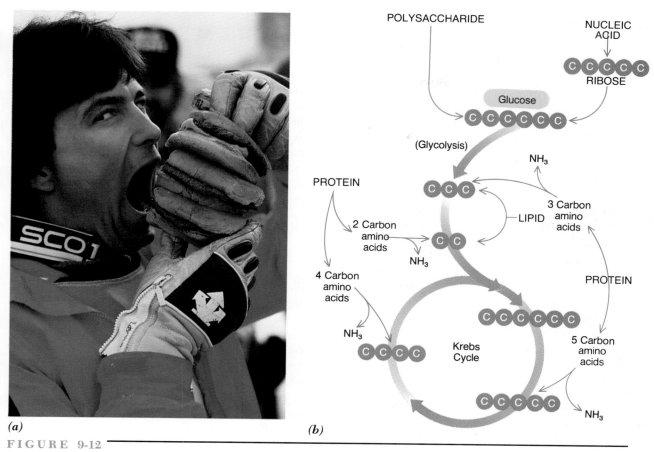

(a)   *(b)*

FIGURE 9-12

Some common denominators of metabolism. *(a)* Most meals contain a variety of macromolecules. The meat and cheese in this hamburger are rich in protein and fat, while the bun is rich in polysaccharide. *(b)* Each of the catabolic pathways by which proteins, lipids, carbohydrates, and nucleic acids are broken down produce metabolic intermediates that are channeled into glycolysis or the Krebs cycle, where they are completely oxidized. Alternatively, the intermediates of glycolysis and the Krebs cycle provide raw materials that are diverted into anabolic pathways, leading to the synthesis of macromolecules.

materials, particularly fats and other lipids, are more reduced than the intermediates from which they are built. Reduction of these intermediates is accomplished by electrons donated by NADPH. A cell's reservoir of NADPH represents its *reducing power* (page 129). Whereas NADPH is formed directly during photosynthesis (see Figure 8-7), humans and other nonphotosynthetic organisms must build a pool of reducing power by transferring electrons from NADH to NADP$^+$:

$$NADH + NADP^+ \rightarrow NAD^+ + NADPH$$

We can see, therefore, that not all of the NADH generated by glucose oxidation is used to make ATP. When energy is abundant, the production of NADPH is favored, providing an ample supply of electrons needed for biosynthesis of new macromolecules, such as proteins and lipids, which are essential for growth. When energy resources are scarce, however, most of the high-energy electrons of NADH are "cashed in" for ATP, whereas NADP$^+$ gets just

enough electrons for the minimal biosynthesis needed to maintain the status quo. As a result of this regulation, an organism integrates its metabolic processes into a "traffic pattern" that can be directed according to the demands of the cell.

Now that we have described how nutrients are dismantled and their energy content extracted, we can understand how this energy can be used to fuel human exercise. This is the subject of The Human Perspective.

Metabolic pathways, like roads, intersect one another. Glycolysis and the Krebs cycle are central pathways in all types of organisms. A diverse array of biochemicals are converted to compounds that can be fed into these pathways where they are dismantled and their energy content is put to use. Conversely, glycolysis and the Krebs cycle provide the cell with raw materials for the construction of other types of molecules. (See CTQ #8.)

REEXAMINING THE THEMES

Relationship between Form and Function

The structure of a mitochondrion is correlated with its role in energy metabolism. The inner mitochondrial membrane contains a highly ordered array of enzymes and electron carriers that must be kept in exact position for the removal of electrons from coenzymes and for their passage from carrier to carrier and, ultimately, to molecular oxygen. The impermeability of the inner mitochondrial membrane maintains the proton gradient. In contrast, each ATP synthase contains an internal channel that allows protons to move through the membrane and drive the reaction by which ATP is synthesized. The central compartment of the mitochondrion contains the soluble enzymes that catalyze the reactions of the Krebs cycle. The electrons removed during these reactions are then passed to the carriers in the surrounding inner mitochondrial membrane.

Biological Order, Regulation, and Homeostasis

The needs of a cell change from moment to moment. For example, the fat cells of a person ingesting a large number of calories are likely to be engaged in the synthesis of fat molecules, whereas these same cells may be engaged in the breakdown of fat during times when calorie intake is low. At certain times, the pathways of glycolysis and the Krebs cycle may run primarily in a direction that releases energy, while at other times they may be running primarily in a direction that produces intermediates for the construction of macromolecules. The rates of activity of the various metabolic pathways in a cell are closely regulated so that they meet the needs of the cell at the time.

Acquiring and Using Energy

Glycolysis, the Krebs cycle, and the electron transport system are the three primary pathways by which cells extract energy from organic molecules. Energy extraction is accomplished primarily through the removal of energized electrons. Lipids, proteins, carbohydrates, and nucleic acids are broken down either into glucose or into metabolic

intermediates that can be fed into these oxidative pathways. As the molecules are oxidized, the energy is conserved as usable packets in the form of ATP and/or NADPH. Cells that rely on fermentation are able to extract only a small percentage of the energy contained in their organic substrates. In contrast, aerobic respiration is much more "profitable" because high-energy electrons are allowed to drop to a much lower energy state by combining with molecular oxygen. Because no energy transfer can be 100 percent efficient, some useful energy is inevitably lost in the process.

Unity within Diversity

Remarkably, the chemical reactions and metabolic pathways described in this chapter are found in virtually every living cell, from the simplest bacterium to the most complex mammal. These pathways constitute a cell's most basic metabolic activity, providing usable intermediates, chemical energy (ATP), and reducing power (NADPH).

Evolution and Adaptation

The metabolic pathways described in this chapter did not appear "overnight"; they undoubtedly evolved step by step. Glycolysis, which is anaerobic, was probably one of the first metabolic pathways to appear, providing small amounts of ATP for use by primitive prokaryotic cells. Fermentation must have been an early accompaniment to glycolysis because it allowed the regeneration of NAD^+ needed for glycolysis to continue. Then, the cyanobacteria appeared and permeated the atmosphere with molecular oxygen. At first, cells must have evolved safeguards against this toxic substance, but they eventually acquired enzymes that could take advantage of oxygen's presence. Organisms that could utilize oxygen would have been able to obtain much greater quantities of energy from their nutrients. These aerobes would have been rapidly selected over their anaerobic competitors, forcing the anaerobes into habitats lacking oxygen, such as deep in the soil, where they can still be found today.

SYNOPSIS

During the early stages of life on earth, the atmosphere was devoid of oxygen and the earth was populated with anaerobic organisms. With the evolution of the cyanobacteria, the earth's atmosphere and bodies of

water became infused with oxygen. This paved the way for a new breed of organisms that not only withstood the toxic effects of the substance but also possessed metabolic pathways that took advantage of the ability of oxygen to attract electrons and, in the process, to extract energy from organic substrates.

The first stage in glucose disassembly is glycolysis. Located in the fluid phase of the cytoplasm, glycolysis converts glucose to two molecules of pyruvic acid (a three-carbon molecule), four ATPs (by substrate level phosphorylation), and two NADHs (which can be "cashed in" for up to six additional ATPs). On the negative side of the ledger, glycolysis costs the cell two ATPs to initiate the pathway by activating the glucose molecule before it is split into two fragments.

Under anaerobic conditions, cells carry out fermentation as a means to regenerate NAD+ from the NADH formed during glycolysis. At the end of fermentation, more than 90 percent of glucose's chemical energy remains in discarded end products, such as ethyl alcohol or lactic acid.

Aerobic respiration continues the disassembly of glucose in the presence of oxygen. The two pyruvic acid molecules generated from each glucose by glycolysis are completely oxidized to six carbon dioxide molecules. Located in the matrix of the mitochondria in eukaryotes (or the soluble cytoplasm of prokaryotes), the Krebs cycle generates reduced coenzymes for biosynthesis and energy production. It yields two ATPs by substrate-level phosphoryl-

ation, eight NADHs, and two $FADH_2$s, per glucose oxidized, or a potential yield of 30 ATPs. The high-energy electrons stored temporarily in NADH and $FADH_2$ are used to generate ATP (three ATPs per NADH, and two per $FADH_2$) by passage down an electron transport system embedded in the inner mitochondrial membrane. The electrons are passed to carriers having a greater affinity for electrons until eventually they are passed to molecular oxygen, leading to the formation of water. As the electrons pass down the electron transport system, the energy released is stored in the form of a proton gradient across the ion-impermeable inner mitochondrial membrane. As protons move down their concentration gradient through the channel in the ATP synthase and into the active site located in the spherical (F_1) portion of the complex, they somehow provide the energy necessary to drive the phosphorylation of ADP. NADH is also used in the formation of NADPH, providing the cell with reducing power.

Glycolysis, the Krebs cycle, and electron transport are central metabolic pathways that provide virtually all aerobic cells with energy and raw materials. A diverse array of materials, ranging from proteins to lipids, are broken down into compounds that can be fed into glycolysis or the Krebs cycle and further metabolized. Conversely, intermediates from these central pathways can be diverted into the formation of various types of biochemicals, depending on the needs of the cell at the time.

Key Terms

anaerobe (p. 174)
aerobe (p. 174)
fermentation (p. 175)
aerobic respiration (p. 175)

glycolysis (p. 175)
substrate-level phosphorylation
   (p. 177)
alcoholic fermentation (p. 178)

lactic acid fermentation (p. 178)
Krebs cycle (p. 180)
cytochrome oxidase (p. 184)

Review Questions

1. Under what conditions do muscle cells form lactic acid and why is this adaptive?

2. What is the role of NAD+ and FAD in the Krebs cycle reactions?

3. What is the role of the proton gradient and why would its function be disrupted by dinitrophenol?

4. Rank the following substances from the least electron attracting to the most electron attracting: NAD+, glucose, molecular oxygen, cytochrome oxidase.

5. Rank the following compounds in terms of energy content: pyruvic acid, glucose, carbon dioxide, lactic acid, PGAL.

Critical Thinking Questions

1. Why do you suppose the isolated vesicles in the supernatant of disrupted mitochondria in Racker's experiments were able to oxidize glucose?

2. Burning organic molecules (i.e. fuels) in an internal combustion engine is only about 25 percent efficient (about 25 percent of the potential chemical energy in the molecules is transformed into usable energy). In contrast, the release of energy from glucose molecules by cells in the process of respiration is about 40 percent efficient. What accounts for the greater efficiency of respiration, compared to combustion?

3. If you were to place some sugar in a small metal dish and expose it to the flame of a Bunsen burner, it would break down into carbon dioxide and water without being activated. Why is activation not necessary when sugar is burned, and why is it necessary when sugar is broken down during respiration?

4. When glucose is broken down during anaerobic respiration, only about 5 percent of the potential energy of glucose is transferred to ATP. Where is the remainder of the energy? (Remember: The first law of thermodynamics states that the total energy present when a process begins must equal the total energy present at the end of the process.)

5. Why would an organism with respiratory equipment ever resort to fermentation? Why can't you—as a human being—switch to fermentation to sustain yourself indefinitely in similar circumstances?

6. Compare fermentation and aerobic respiration by completing the table below:

	Fermentation	Aerobic Respiration
initial compound		
final products		
part of cell where it occurs		
net ATP molecules produced		
involves glycolysis (y/n)		
involves Krebs cycle (y/n)		
involves electron transport chain?		

7. Consistent with the laws of thermodynamics, not all of the energy stored in glucose is converted to ATP; some of the energy is "lost" in the process. Where does this "lost" energy go? Does it have any benefits, or is it entirely wasted?

8. How do you personally benefit from the fact that your metabolic pathways are interconnected? How does this fact benefit people who have had to subsist on starvation diets for long periods of time?

Additional Readings

Bursztyn, P. G. 1990. *Physiology for sportspeople: a serious user's guide to the body.* New York: Manchester University Press. (Intermediate)

Darnell, J., H. Lodish, and D. Baltimore. 1990. *Molecular cell biology.* 2d ed. New York: Freeman. (Advanced)

Holmes, F. L. 1991. *Hans Krebs: The formation of a scientific life.* New York: Oxford. (Introductory)

Kalckar, H. 1991. 50 years of biological research—from oxidative phosphorylation to energy requiring transport. *Ann. Rev. Biochem.* 60:1–37. (Intermediate)

Krebs, H. 1982. *Reminiscences and reflections.* Oxford. (Introductory)

Powers, S. K., and E. T. Howley. 1990. *Exercise physiology.* W. C. Brown. (Intermediate-Advanced)

Racker, E. 1976. *A new look at mechanisms in bioenergetics.* New York: Academic. (Intermediate)

Stryer, L. 1988. *Biochemistry.* 3d ed. New York: Freeman. (Advanced)

Cell Division: Mitosis

**STEPS
TO
DISCOVERY**
Controlling Cell Division

TYPES OF CELL DIVISION

Cell Division in Prokaryotes

Cell Division in Eukaryotes

THE CELL CYCLE

Regulating the Cell Cycle

THE PHASES OF MITOSIS

**Prophase: Preparing for Chromosome
Separation**

Metaphase: Lining Up the Chromosomes

Anaphase: Separating the Chromatids

Telophase: Producing Two Daughter Nuclei

**CYTOKINESIS: DIVIDING THE
CELL'S CYTOPLASM AND ORGANELLES**

THE HUMAN PERSPECTIVE
Cancer: The Cell Cycle Out of Control

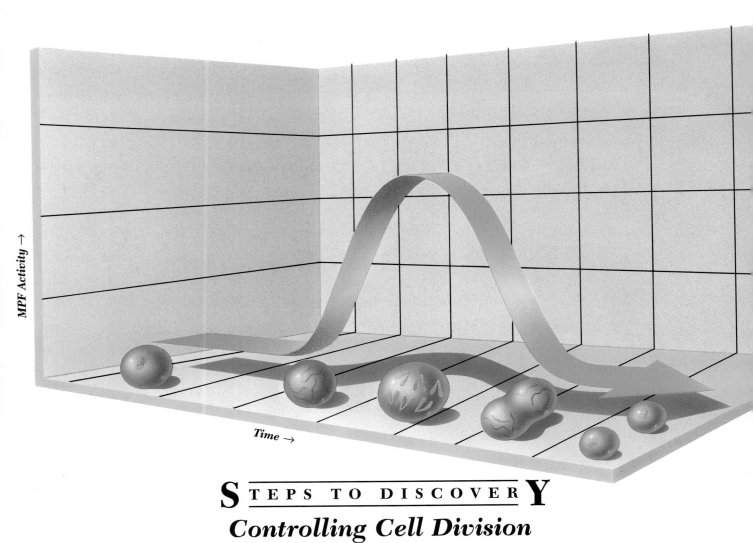

MPF Activity →

Time →

S TEPS TO DISCOVER Y
Controlling Cell Division

Cells reproduce—that is, they generate more of themselves—by dividing in two. Many of the events that occur during cell division have been known for decades, but the mechanism that triggers cell division has been shrouded in mystery until very recently. Research into the factors that control whether a cell is "resting" or dividing has enormous practical implications, particularly in combating cancer, a disease that results from a breakdown in a cell's ability to regulate its own division. Loss of this regulatory ability produces cells that divide out of control and form a continuously growing malignant tumor.

In 1970, Potu Rao and Robert Johnson of the University of Colorado devised an ingenious way of examining whether or not cells contain factors that trigger cell division. The scientists wanted to know what would happen if a cell that was just about to divide were to fuse with a cell that had divided recently, and would not divide again for many hours. To find out, they fused two cells to form a single "giant" cell that contained the nuclei and cytoplasm of both original cells. Consider the possible consequences of fusing these two types of cells: Two nuclei, one about to divide and the other totally unprepared to divide, are brought together in cytoplasm made up of a mixture of the two fused cells. Does the cytoplasm contain regulatory factors that affect

Cell division is triggered as MPF activity rises to its peak level.

the nuclei? If so, does the cytoplasm donated by the nondividing cell contain factors that can *block* the dividing nucleus from continuing its activities? Or, does the cytoplasm from the dividing cell contain factors that *stimulate* division activities in the nucleus derived from the nondividing cell?

Rao and Johnson found evidence for the latter alternative. As we will discuss later in the chapter, before a cell normally divides, its chromosomes become transformed from an "invisible" state to a condensed (packaged) state, in which they are readily visible under the microscope. In Rao and Johnson's experiment in which a dividing and a nondividing cell were fused, the chromosomes donated by the nondividing cell underwent condensation—a process that would not normally have occurred in that nucleus for many hours. The scientists concluded that the cytoplasm of the dividing cell contained one or more regulatory factors that triggered the nondividing nucleus to behave as if it were getting ready to divide.

Over the next few years, evidence accumulated that showed that the triggering agent for this behavior is a protein named *mitosis promoting factor* (MPF). This same protein was found in a wide variety of eukaryotic organisms, from yeast to humans. MPF purified from one species could be injected into the cells of a distant species, causing the injected cell to divide. Because of its widespread occurrence, MPF is thought to have appeared very early in the evolution of eukaryotic organisms and was retained as a key cellular component as more and more complex eukaryotes evolved.

If MPF triggers cell division, it seemed likely that it would be found in very small concentrations in nondividing cells, then jump to high levels in cells just before cell division, then return again to very low levels just after cell division is completed. The first evidence for such an oscillation in the level of MPF activity came in 1978 from a study conducted in the laboratory of Dennis Smith at Purdue

University. Smith and his colleagues had been studying the first few cell divisions in the life of a frog; that is, the first divisions after the egg had been fertilized. These investigators found that as cells got ready to divide, the level of MPF activity rose dramatically; it then fell to undetectable levels after division, only to rise once again as the next ensuing division approached. These results indicated that MPF was acting like an "alarm clock"—whenever its activity rose to a high enough level it would set off the events that led to cell division. The question was: How was MPF able to do this?

Much of the progress toward answering that question has come from recent studies by James Maller and his colleagues at the University of Colorado. One of the events that accompanies a cell's preparation for division is the addition of large numbers of phosphate groups to a variety of the cell's proteins, some of which are found in the chromosomes. MPF is a protein kinase—an enzyme that adds phosphates to proteins. Injecting MPF into a cell induces an immediate increase in protein phosphorylation.

Perhaps the most widely accepted hypothesis based on these and other observations proposes that nondividing cells contain the proteins that initiate cell division, but they are present in an inactive state. Before they can participate in cell division, they must first be activated by the addition of a phosphate by MPF. Thus, MPF provides the major activating stimulus by phosphorylating key proteins needed for cell division.

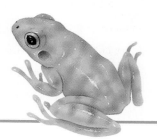

You were once just a fertilized egg—the product of the union of gametes: a sperm from your father and an egg from your mother. From this inauspicious beginning, you have grown into an organism consisting of trillions of cells. How did this remarkable transformation take place?

Recall the third tenet of the cell theory: New cells originate only from other living cells. The process by which this occurs is called **cell division.** For a multicellular organism like yourself, countless divisions of the fertilized egg result in an organism of astonishing cellular complexity and organization. Furthermore, cell division continues throughout life. Biologists estimate that more than 25 million cells are undergoing division each second of your life. This enormous output of cells is needed to replace those cells that have aged or died. Old, worn blood cells, for example, are removed and replaced by newcomers at the rate of about 100 million per minute. Not surprisingly, then, anything that blocks cell division, such as exposure to heavy doses of radiation, can have tragic effects. For example, many people who valiantly worked to seal the damaged nuclear reactor at Chernobyl, in the former Soviet Union, died because their bodies were unable to produce new, healthy blood cells (Figure 10-1). Ironically, the same type of destructive radiation is also used as a treatment for cancer because it selectively destroys rapidly dividing cells, such as those of a tumor.

Each dividing cell is called a **mother cell,** and its descendants are appropriately called **daughter cells.** There is a reason for using these "familial" terms. The mother cell transmits copies of its hereditary information (DNA) to its daughter cells, which represent the next cell generation. In turn, the daughter cells can become mother cells to their own daughter cells, passing along the same genes they inherited from their mother to yet another cellular generation. For this reason, cell division is often referred to as *cellular reproduction.*

Cell division is more than just a means of reproducing more cells; it is the basis for reproducing more *organisms.* Cell division, therefore, forms the link between a parent and its offspring; between living species and their extinct ancestors; and between humans and the earliest, most primitive cellular organisms.

▼ ▼ ▼

TYPES OF CELL DIVISION

Despite the great diversity of living organisms and the types of cells they contain, there are only three basic types of cell division, which are distinguished primarily by the way the genetic material is partitioned between the daughter cells. Prokaryotic cells partition their genetic material by a simple process involving membrane growth, whereas eu-

FIGURE 10-1

Testing for radiation in a village near the Chernobyl nuclear power plant in the former Soviet Union.

karyotic cells employ a complex process of nuclear division —either *mitosis* or *meiosis.* In all three of these processes, the cell first prepares for division by duplicating (*replicating*) its genetic material, a process discussed in detail in Chapter 14. The mother cell later splits itself in such a way that each daughter cell receives at least one complete set of hereditary instructions, as well as a portion of the cytoplasm.

CELL DIVISION IN PROKARYOTES

You will recall from earlier chapters that prokaryotes lack a nuclear membrane (page 92). The DNA of prokaryotic cells is attached directly to the cell's plasma membrane, providing a mechanism by which the replicated copies can be precisely distributed to the daughter cells (Figure 10-2). At the beginning of the division process, the replicated DNA molecules are attached to slightly different points on the plasma membrane. A new section of plasma membrane

grows between these attachment points, separating the DNA molecules. A partitioning plasma membrane and cell wall then develop between the DNA molecules to form two daughter cells. This process is called **prokaryotic fission.**

CELL DIVISION IN EUKARYOTES

The division of a eukaryotic cell into daughter cells occurs in two stages: First, its nuclear contents are divided by either *mitosis* or *meiosis.* This is followed by a second step, **cytokinesis,** in which the cell is actually split into two. Cell division in eukaryotes is much more complex than is prokaryotic fission for a number of reasons:

1. Eukaryotic cells are much larger than are prokaryotic cells and contain a diverse array of organelles, including a membrane-bound nucleus. The cytoplasmic organelles must be sorted out more or less equally between the daughter cells, and a new nuclear envelope must be reassembled for each daughter cell nucleus.

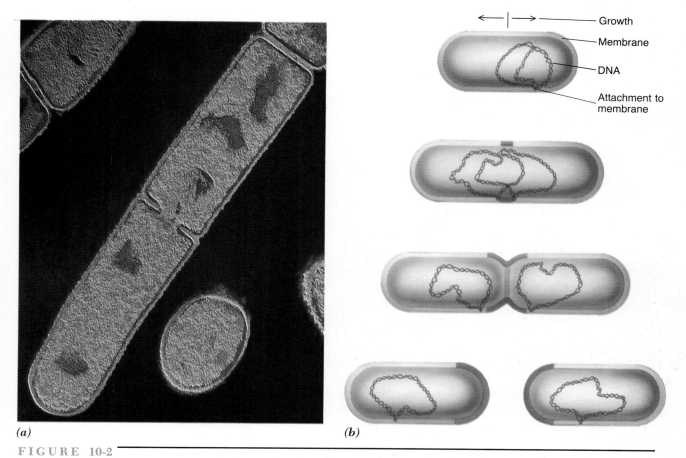

(a) *(b)*

F I G U R E 10-2
Cell division in prokaryotes. *(a)* Photograph of a bacterial cell in the final stages of cell division. The DNA has already been separated, and the crosswall that will separate the cells is partially manufactured. *(b)* Stages in the process of prokaryotic fission. The DNA is attached directly to the cell's plasma membrane. Separation of the duplicated DNA molecules is accomplished by growth of the plasma membrane between the points of attachment.

2. Eukaryotic cells house much greater amounts of DNA than do prokaryotic cells. The quantity of DNA in a human cell, for example, is roughly 700 times that found in the most complex prokaryote.

3. Eukaryotic cells contain anywhere from 2 to more than 1,000 chromosomes, each of which consists of a molecule of DNA associated with a complex variety of proteins. In contrast, prokaryotic cells contain a single "chromosome," consisting essentially of a "naked" DNA molecule; that is, one devoid of permanently associated proteins.

As we will see below, eukaryotic cells are able to deal with such large amounts of chromosomal material by organizing it into specialized "packages" that can be distributed to their daughter cells. During mitosis or meiosis, each "packaged" chromosome appears as two thick "rods," called **chromatids**, which are connected to one another in an indented region called the **centromere** (inset, Figure 10-3). The two "sister" chromatids that make up each chromosome are genetic duplicates of each other—exact copies that were constructed at an earlier stage, when the DNA was duplicated. Each chromosome of a dividing cell has a

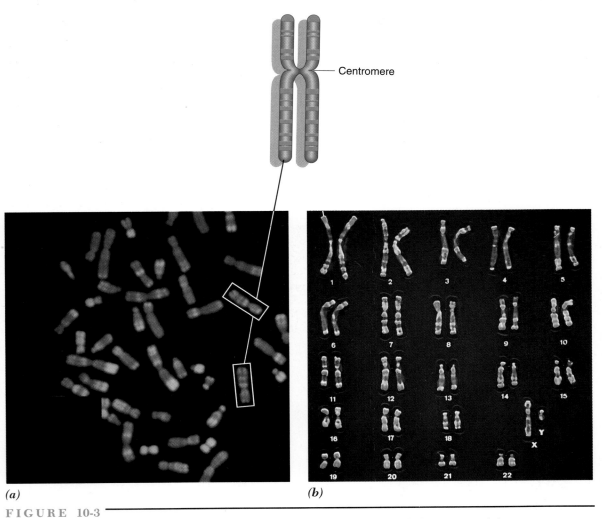

(a)   *(b)*

FIGURE 10-3

Human mitotic chromosomes. Top inset shows a stylized drawing of a stained chromosome. *(a)* Photograph of a cluster of mitotic chromosomes that spilled out of the nucleus of a single dividing human cell. This diploid set of 46 chromosomes has been stained with a dye that gives the chromosomes a banded pattern (see top inset). The two chromatids that make up each chromosome can be distinguished from each other. As discussed in the text, diploid cells contain pairs of homologous ("lookalike") chromosomes. One pair of homologous chromosomes is indicated by the boxes. *(b)* The chromosomes of a human male. In this figure, called a *karyotype,* homologous chromosomes are paired and arranged according to number (size). If the chromosome preparation had been made from the cells of a female, two X chromosomes would be seen, instead of an X and Y.

characteristic shape and size and can be identified from other chromosomes. Let's look more closely at the chromosomes of a dividing cell.

Chromosome Number: Haploid and Diploid

In eukaryotic cells, chromosomes occur in pairs (Figure 10-3*a,b*); each chromosome has a partner that is virtually identical in appearance. The two similar-shaped chromosomes are called **homologues** (*homo* = same, *log* = writing) because each one has the same sequence of genes along its length as the other. (You will recall from Chapter 1 that genes are the blueprints that determine our heritable traits, such as eye color and hair texture.) Homologous chromosomes can be recognized by their identical size, shape, and appearance (Figure 10-3*b*). Even though homologues have the same genes—in the sense that they determine the same trait—they may have different *versions* of those genes. For example, consider a gene for height in plants. One homologue may code for a tall plant, whereas the other may code for a short one. Both chromosomes have a gene for height at the same location, but each may produce a different version of the trait.

Most eukaryotic cells carry two complete sets of homologues—a **2N** number of chromosomes; such cells are said to be **diploid** (*di* = two, *ploid* = multiple). In diploid cells, one set of homologous chromosomes was originally contributed by each parent during sexual reproduction. Every cell in your brain, for example, has 46 chromosomes—a set of 23 from your father and a homologous set of 23 from your mother. Put differently, the cells in your body contain 23 pairs of homologous chromosomes. Each species has a characteristic diploid number, ranging from 2 in the horse roundworm and *Penicillium* fungus to a whopping 1,262 in Adder's tongue fern. Our closest relative, the chimpanzee, has 48 chromosomes, most of which are indistinguishable in appearance from our own (Chapter 13).

In contrast, **haploid** cells contain only one set of homologues—a **1N** number of chromosomes. The haploid number for humans is 23. Except for haploid sperm or eggs produced by your testes or ovaries, the cells in your body are diploid.

Mitosis: The Production of Two Identical Nuclei

The name "mitosis" comes from the Greek word *mitos,* meaning "thread." The name was first used in the 1870s to describe the threadlike chromosomes that appeared to "dance around" the cell just before it divided in two. **Mitosis** is a process of nuclear division in which duplicated chromosomes are separated from one another, producing two nuclei, each with one copy of each chromosome. Mitosis is usually accompanied by cytokinesis, resulting in *two* daughter cells with genetic potential identical to each other and to the mother cell from which they arose (Figure 10-4, left column). Mitosis, therefore, maintains the chromosome number and generates new cells for the growth, maintenance, and repair of an organism. Mitosis can take place in either diploid or haploid cells, the latter occurring in plants and a few animals (including male bees known as drones).

In some plants and animals, mitotic cell divisions are also a means of producing an entirely new organism through *asexual reproduction*—reproduction that does not involve the union of male and female gametes but involves the growth of offspring from the cells of a single parent (discussed further in Chapter 31). The offspring produced by asexual reproduction have exactly the same genes as does their parent.

Meiosis: The Production of Four Haploid Nuclei

Unlike mitosis, **meiosis** occurs only in diploid cells and produces *four* daughter nuclei, each containing a haploid (1N) number of chromosomes. A mother cell undergoing meiosis is able to produce four daughters by duplicating its own chromosomes and then proceeding through two consecutive nuclear divisions (Figure 10-4, right column). The nuclear divisions are usually accompanied by a corresponding division of the cytoplasm, resulting in four haploid cells. Haploid (1N) cells produced by meiosis form **gametes**—reproductive cells that fuse during fertilization, forming a diploid (2N) cell with a genetic potential different from that of either parent. The events that take place during meiosis, and their importance in sexual reproduction, are discussed in the next chapter.

For life to be perpetuated, organisms must reproduce. Since all organisms are composed of one or more cells, cell division is the foundation of reproduction. Prokaryotic cells have a simpler structure and divide by a simpler mechanism than do eukaryotic cells. In eukaryotic cells, there are two types of cell division, depending on whether the nucleus divides by mitosis or meiosis. Mitosis produces nuclei of equivalent genetic constitution to that of the original parent cell, whereas meiosis leads to the formation of nuclei with half the number of chromosomes of the original parental cell. (See CTQ #2.)

THE CELL CYCLE

The life of a cell begins with its formation from a mother cell by division and ends when the cell divides to form daughter cells of its own or when it dies. The stages through which a cell passes from one cell division to the next constitute the **cell cycle** (Figure 10-5).

Based on cellular activities visible with a light microscope, biologists divide the cell cycle into two phases: the "M phase" (*M* = mitotic) and the "interphase." The **M phase** includes (1) the process of mitosis, whereby chromosome separation occurs (which will be discussed in more detail later in the chapter); and (2) **cytokinesis,** whereby the entire cell is physically divided into two daughters. The M phase occupies only a small portion of the cell cycle since the separation of the chromosomes usually takes only 30 minutes to an hour, whereas the life of a cell may extend

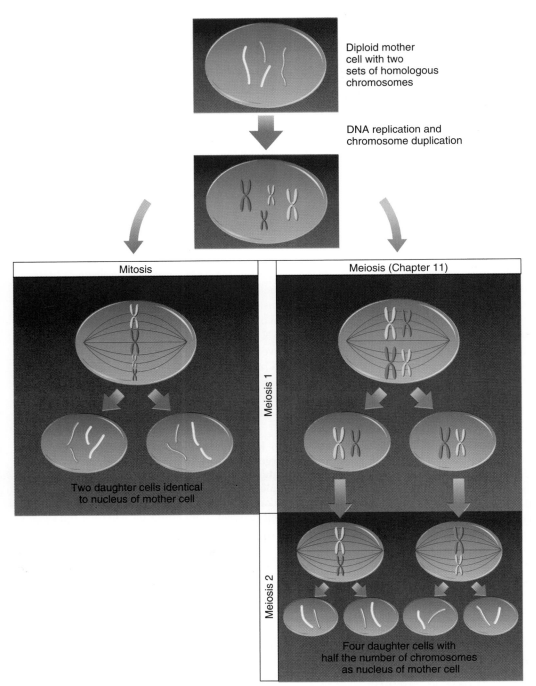

Diploid mother cell with two sets of homologous chromosomes

DNA replication and chromosome duplication

Mitosis

Meiosis (Chapter 11)

Meiosis 1

Meiosis 2

Two daughter cells identical to nucleus of mother cell

Four daughter cells with half the number of chromosomes as nucleus of mother cell

FIGURE 10-4

Schematic comparison of mitosis versus meiosis. DNA replication and the duplication of the chromosomes occur similarly prior to both mitosis and meiosis. When a cell enters either mitosis or meiosis, the chromosomes become condensed and visible in the light microscope as rodlike structures, each consisting of two chromatids. Mitosis produces two daughter nuclei that have exactly the same number of chromosomes as does the nucleus of the mother cell. In meiosis, however, the mother cell divides *twice,* producing four daughter nuclei, each with *half* the number of chromosomes of the nucleus of the mother cell. (Meiosis is discussed in detail in Chapter 11.)

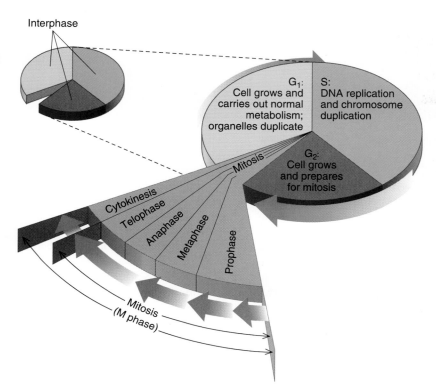

Interphase

G₁:
Cell grows and
carries out normal
metabolism;
organelles duplicate

S:
DNA replication
and chromosome
duplication

G₂:
Cell grows
and prepares
for mitosis

Mitosis

Prophase

Metaphase

Anaphase

Telophase

Cytokinesis

Mitosis
(M phase)

F I G U R E 10-5

The eukaryotic cell cycle. This diagram of the cell cycle indicates the stages through which a cell passes from one division to the next. The cell cycle is divided into two stages: M phase and interphase. M phase includes the successive stages of mitosis and cytokinesis. Interphase is divided into G₁, S, and G₂ phases—S phase being equivalent to the time of DNA synthesis. G₁ typically lasts 6 to 10 hours; S lasts 3 to 5 hours; and M less than 1 hour. G₁ is highly variable, lasting from a few minutes to months.

over many years. **Interphase** takes up the remainder of the cycle. During interphase, the cell grows in volume, and normal metabolic functions, such as glucose oxidation or the production of proteins for export, are carried out.

Although M phase is the period when the contents of a cell are actually divided, numerous preparations for an upcoming mitosis occur during interphase, including the important events involved in the replication of its DNA. The interphase period of the cell cycle is generally divided into three parts—G₁, S, and G₂ (Figure 10-5)—based on the timing of DNA replication. G₁ (Gap₁) is the period following mitosis and preceding the onset of DNA synthesis. S (Synthesis) is the period during which all the DNA in the chromosomes is replicated (Chapter 14). G₂ (Gap₂) is the period following DNA synthesis and preceding the subsequent mitosis. During G₂, the cell makes its final preparations for the upcoming division. Of the three phases, G₁ is the most variable. For example, in adult humans, the rapidly dividing cells that line the intestines remain in the G₁ phase for only about 2 hours, whereas it may take months for the very slowly dividing cells that make up the liver to move from mitosis through G₁ to S phase. Some types of cells, most notably nerve and muscle cells, do not divide under any circumstances. These cells can be considered to be "permanently" arrested in G₁ and, thus, will never enter S phase.

REGULATING THE CELL CYCLE

We described in Chapter 1 how organisms utilize homeostatic mechanisms to maintain stable internal conditions. Controlling the rates at which cells divide is an important element in maintaining homeostasis. Each type of tissue must maintain its own characteristic cellular makeup. Blood cells, for example, which are needed in great numbers and have relatively short life spans, must be produced at much more rapid rates than liver cells, which have long lives. The rate at which a particular cell divides is determined by the length of its cell cycle, which, in turn, is regulated by a number of substances, including hormones and growth factors that bind to the outer surface of the cell and either trigger or block cell division, as well as substances produced within the cell itself, such as MPF. For various reasons, these controls sometimes break down, causing some cells to divide independently. This malfunction can lead to one of the most dreaded types of diseases—cancer (see The Human Perspective: Cancer: The Cell Cycle Out of Control).

Maintaining homeostasis also requires that cells be able to change their rate of division depending on present conditions. For example, under ordinary conditions, a skin-producing cell takes about 20 hours to progress through interphase prior to producing two new skin cells by mitosis.

◁ THE HUMAN PERSPECTIVE ▷
Cancer: The Cell Cycle Out of Control

CANCER! The name alone strikes fear in all of us—and for good reason. Cancer is a leading cause of death in the Western world, second only to heart disease. Humans are not the only kind of organism that develops cancer. Humans, frogs, chickens, mice, and even plants—indeed all multicellular organisms—are candidates for cancer. And, contrary to what many believe, cancer is not a new disease. Cancerous lesions have been found in Egyptian mummies and even in dinosaur bones.

Cancer is a disease that results from uncontrolled cell divisions. Normal cells may divide very rapidly, as occurs, for example, among liver cells following the removal of a portion of the liver. These normal cells are closely regulated, however, and they stop dividing when the liver has been returned to its normal size. In contrast, cancer cells continue to grow and divide indefinitely. For some reason, they no longer respond to the normal metabolic checks and balances that would otherwise limit and coordinate their growth with other cells. To fuel this unbridled growth, cancer cells out-compete surrounding cells for energy and nutrients. Since cancerous cells invade and spread to other tissues, many organs of the body can become adversely affected.

Why would a perfectly normal cell begin dividing wildly? We don't yet know the answer to this question, but it is becoming increasingly apparent that the genes involved with the cell cycle play an important role. Normal cells become transformed into cancer cells when something happens to certain genes, converting them to **oncogenes,** which causes them either to change their level of activity or to produce proteins with altered amino acid sequences (that is, mutant proteins). This is the basis of the action of *carcinogens* (cancer-causing agents), such as cigarette smoke, ultraviolet radiation, X-rays, and more than 1,000 known chemicals, including numerous pesticides, household products, and food additives. Carcinogens act by causing alterations in the DNA.

The study of oncogenes has led to a better understanding of the genes that control normal cellular growth and division. The first oncogene discovered was subsequently shown to code for a protein kinase—an enzyme that phosphorylates other proteins (page 129). (For their discovery of this oncogene in the mid 1970s, J. Michael Bishop and Harold Varmus of the University of California were awarded the 1989 Nobel Prize for Medicine.) There are now several dozen known oncogenes, and the list continues to grow. Included on the list are genes that code for (1) growth factors—that is, substances that bind to cell receptors and stimulate the cell to divide; (2) receptors for growth factors—that is, the cell-surface protein that binds the growth factor and mediates its response; (3) regulatory proteins that bind to genes involved in cell growth; and (4) a number of protein kinases. These same genes and proteins are found not only in humans, or even mammals, but in virtually all eukaryotes. Since all of these proteins are normally involved in cell growth and division, it is easy to understand how changes in these proteins could make cells less responsive to the body's growth-control mechanisms.

Our understanding of the factors that control the cell cycle has received a boost in recent years with the purification of mitosis promoting factor, MPF—the subject of the chapter-opening essay. MPF is a protein kinase, as are the proteins encoded by some oncogenes. In fact, some of the proteins phosphorylated by MPF may be the same proteins that, when mutated, transform a normal cell into a cancerous one. A great deal of research is currently focused on how the activity of a single protein, MPF, can trigger such a complex process as mitosis. When we understand how mitosis and the cell cycle are regulated, we will be closer to winning one of the greatest medical battles ever waged.

If you cut yourself, however, the interphase period is reduced, allowing rapid cell replacement for healing.

Most cells play two critical roles: They conduct the various cellular functions that are characteristic of their cell type, and they divide. In general, a cell spends much more time conducting metabolic functions than it spends in cell division. However, cell division demands that virtually all cell components and activities be devoted to the process. As a result, cell division stands out as a distinct stage in the life of a cell. (See CTQ #3.)

THE PHASES OF MITOSIS

Even though interphase occupies nearly all of the cell cycle, it is the "dance of the chromosomes" during mitosis that has captured the attention of biologists for over a century. Mitosis is a continuous process. For the sake of discussion, however, we will divide mitosis into four sequential phases: prophase, metaphase, anaphase, and telophase (Figure 10-6). Each phase is defined by the behavior of the chromosomes.

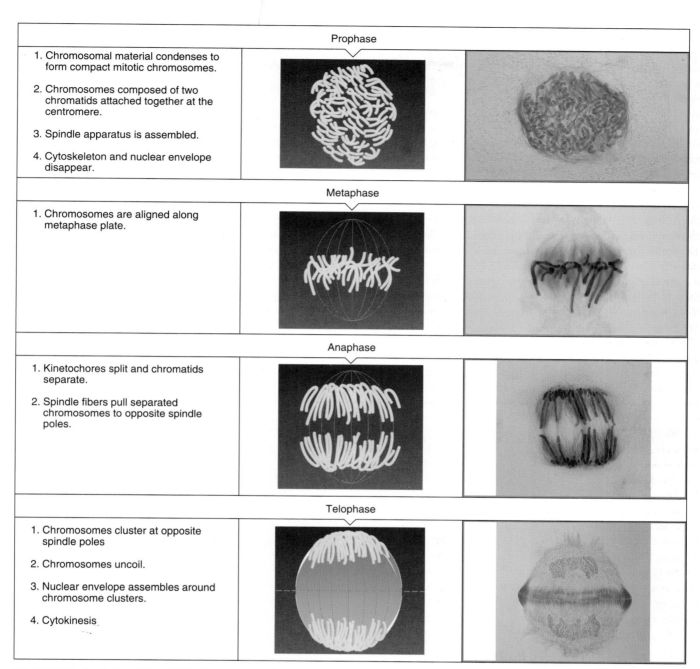

Prophase

1. Chromosomal material condenses to form compact mitotic chromosomes.

2. Chromosomes composed of two chromatids attached together at the centromere.

3. Spindle apparatus is assembled.

4. Cytoskeleton and nuclear envelope disappear.

Metaphase

1. Chromosomes are aligned along metaphase plate.

Anaphase

1. Kinetochores split and chromatids separate.

2. Spindle fibers pull separated chromosomes to opposite spindle poles.

Telophase

1. Chromosomes cluster at opposite spindle poles

2. Chromosomes uncoil.

3. Nuclear envelope assembles around chromosome clusters.

4. Cytokinesis.

FIGURE 10-6

The phases of mitosis.

PROPHASE: PREPARING FOR CHROMOSOME SEPARATION

The first phase of mitosis, **prophase** (*pro* = before), is the longest phase and includes a number of complex activities, including chromosome condensation and formation of the spindle apparatus (Figure 10-6).

Chromosome Condensation

Each chromosome contains a single molecule of DNA which, together with its associated protein, is spread throughout the nucleus of an interphase cell. In this extended state, DNA is well suited for interaction with various enzymes and regulatory proteins required for gene expression and replication. As cell division approaches, however, these remarkably long DNA-protein fibers undergo a coiling process, in which they are packaged into compact **mitotic chromosomes** (Figures 10-3, 10-6, 10-7)—structures that are ideally suited for the upcoming separation process.

The importance of chromosome coiling becomes clear when you consider that the DNA of a single human chromosome is 30,000 times longer in the extended state than when it is coiled. Imagine the molecular chaos that would reign inside a human cell during the separation of the duplicates of 46 *uncoiled* chromosomes. Inevitable tangling would pull the uncoiled DNA to pieces, forming a jumble of DNA fragments that would be unevenly sorted into the daughter cells. The result would be daughter cells with an incomplete set of genetic blueprints. No species could survive disorder of this magnitude.

As the condensation of the chromatin nears completion, it becomes evident that each mitotic chromosome consists of two identical, attached chromatids (Figure 10-7). These sister chromatids are visible evidence of chromosome duplication, a process that occurred earlier during interphase.

Formation of the Spindle Apparatus

In eukaryotes, the precise separation of duplicated mitotic chromosomes into two daughter cells requires the activity of a cellular "machine" called the **spindle apparatus,** which has no counterpart in prokaryotic cells. The spindle apparatus is constructed primarily of cytoskeletal materials. Recall from Chapter 5 that the cytoskeleton has the dual function of providing support and moving particles within a cell. The spindle apparatus consists of bundles of microtubules, called **spindle fibers,** which are organized to form both a supportive scaffolding and a chromosome-pulling machine (Figure 10-8). Remarkably, the microtubules that make up the spindle apparatus are formed from the same subunits that a few minutes earlier might have been part of a different microtubule having an entirely different function. This is analogous to demolishing a brick building and then using the same bricks to construct a new office building—all within a matter of minutes. The rapid assembly of the spindle apparatus during mitosis, and its even more rapid disassembly following mitosis, illustrates the dynamic nature of cytoskeletal elements.

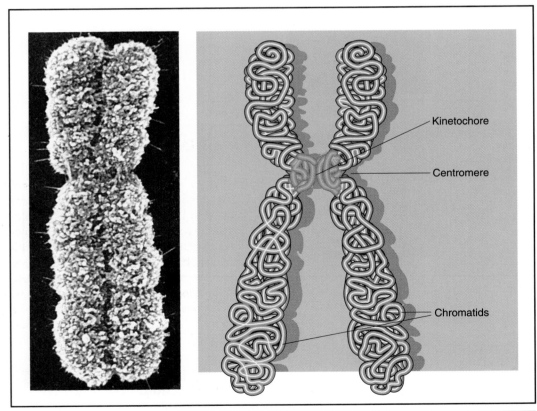

FIGURE 10-7

The structure of a mitotic chromosome.   During mitosis, the DNA-protein fibers coil, as is indicated in both the drawing and the electron micrograph. Each chromatid remains distinct but is joined to the other in the indented region of the chromosome (the *centromere*). When the centromere is examined under the electron microscope, it is seen to contain a dense, protein-containing structure called the *kinetochore.*

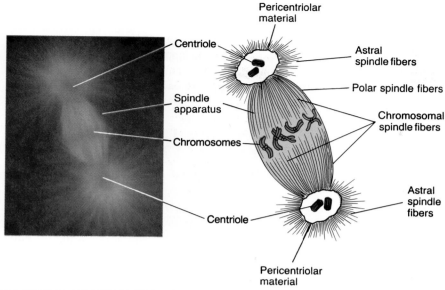

Pericentriolar material

Centriole

Astral spindle fibers

Polar spindle fibers

Spindle apparatus

Chromosomal spindle fibers

Chromosomes

Centriole

Astral spindle fibers

Pericentriolar material

FIGURE 10-8

Spindle apparatus of an animal cell in metaphase of mitosis. Each spindle pole contains a pair of centrioles surrounded by a shapeless pericentriolar material from which the microtubules which make up the spindle fibers radiate. Spindle fibers can be categorized as astral fibers, chromosomal fibers, or polar fibers, each having its own distinct functions. *Astral spindle fibers* radiate like a starburst around the centrioles; they probably help position the spindle apparatus in the cell. *Polar spindle fibers* stretch completely across the cell from pole to pole, forming "guidewires" that support the spindle apparatus during its operation. Shorter *chromosomal spindle fibers* attach to the kinetochores of each mitotic chromosome. When the kinetochores split at the start of anaphase, the sister chromatids are pulled to opposite poles by the action of the chromosomal spindle fibers.

Spindle fibers radiate out from the two poles (ends) of the spindle apparatus. In animal cells, the poles contain a pair of peculiar, pinwheel-shaped structures called **centrioles,** surrounded by dense, unstructured **pericentriolar material (PCM)** (see Figure 10-8). Ironically, the relatively conspicuous centrioles seem to have little or no function in the process of mitosis, while the inconspicuous PCM appears to play a key role in governing the assembly of the microtubules of the spindle apparatus. This hypothesis is supported by studies on dividing plant cells that possess a mitotic apparatus that lacks centrioles entirely but contain material resembling the PCM found in animals. You might also note that the growing ends of all of the various microtubules depicted in Figure 10-8 are embedded in the PCM.

Assembly of the spindle apparatus takes place outside the nucleus at the same time the chromosomes are becoming compacted inside the nucleus. Toward the end of prophase, the nuclear envelope surrounding the compacted chromosomes becomes fragmented and disappears from view. At about the same time, the nucleolus disappears, and the chromosomes become attached to the ends of certain spindle fibers. By the end of prophase, the chromosomes have moved to the center of the cell.

METAPHASE: LINING UP THE CHROMOSOMES

During **metaphase** (*meta* = middle), the chromosomes of the dividing cell are aligned in a plane (called the *metaphase plate*) that typically lies at the cell's "equator"; that is, midway between the spindle poles (Figure 10-6). The chromosomes and spindle apparatus of the metaphase cell are interconnected, forming a structure depicted in Figure 10-8. One group of spindle fibers extends to a specialized attachment site on each chromosome, called the **kinetochore,** which is located in the narrowed centromere region of the chromosome (Figure 10-7).[1] During metaphase, the kine-

---

[1] When duplicated chromosomes were first observed in the light microscope in the late 1800s, the two chromatids were seen to be attached at a point where the chromosome was pinched inward, forming a "waist." This indented region was called the *centromere.* Later, when chromosomes were observed in the electron microscope, the point where the chromatids were attached to one another revealed a specialized structure called the *kinetochore.* These two terms are often confused; the kinetochore is a physical structure embedded within the body of the chromosome, whereas the centromere is that region of the chromosome marked by the indentation. The kinetochores are located within the centromere.

tochores of all the chromosomes are aligned within the metaphase plate, and the "arms" of the chromatids extend out in various directions (Figure 10-6). The alignment of kinetochores and their attachment to the spindle fibers ensure precise separation of sister chromatids during the next stage, anaphase.

ANAPHASE: SEPARATING THE CHROMATIDS

Anaphase (*ana* = backward) begins with the sudden, synchronous separation of the kinetochores present on sister chromatids. Separated chromatids (which, after separation, are called chromosomes) now move away from each other, toward opposite spindle poles (Figure 10-6). Since each daughter cell receives one copy of every chromosome, the two daughters are genetically identical to each other as well as to the mother cell from which they arose.

TELOPHASE: PRODUCING TWO DAUGHTER NUCLEI

Telophase (*telo* = end) begins when the chromosomes reach their respective spindle poles (Figure 10-6). The events of telophase are virtually the reverse of those that occur during prophase: The chromosomes uncoil, the spindle apparatus is dismantled, nucleoli reappear, and nuclear envelopes form around each chromosome cluster. Telophase ends when cytokinesis is completed and two separate daughter cells are produced.

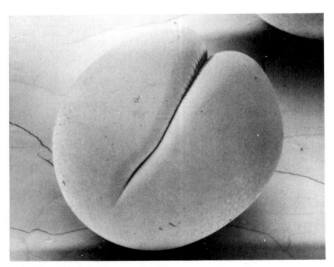

FIGURE 10-9

One cell splits into two. During cytokinesis, a frog egg divides into two cells as microfilaments contract and draw in the plasma membrane. Eventually, the microfilaments pull the edges of the plasma membrane together, pinching the cell's cytoplasm in two.

Mitosis is an intricate process that follows similar steps among diverse eukaryotic cells. These steps include packaging the chromosomal material to facilitate precise chromosome separation; assembling the machinery necessary for chromosome separation; and moving the duplicated chromosomes to opposite ends of the cell. (See CTQ #4.)

CYTOKINESIS: DIVIDING THE CELL'S CYTOPLASM AND ORGANELLES

Although mitosis is very similar in plant and animal cells, the way in which the cytoplasm is divided by cytokinesis is very different. If you watch a sea urchin or frog egg undergoing its first cell division, for example, you will see the first hint of cytokinesis as an indentation of the cell surface during late anaphase. Depending on the type of cell, the indentation may occur at one end (as in the frog egg depicted in Figure 10-9) or at points completely encircling the cell. As time passes, the indentation deepens and becomes a *cleavage furrow* that moves through the cytoplasm, pinching the cell in two. The plane of the furrow always lies in the same plane previously occupied by the metaphase chromosomes, thereby ensuring that the separated chromosomes will be partitioned in different cells.

Cytokinesis in animal cells is the result of contractions of a ring of microfilaments (page 104), which assembles just beneath the plasma membrane. As the ring contracts, it pulls the overlying membrane inward, constricting the cell, much like a purse string closes the diameter of a purse's opening. The ring of microfilaments disassembles after constriction is completed, and the subunits (composed of the protein actin) are put to use in the formation of microfilaments of the cytoskeleton.

Formation of a contractile ring would not be possible in plant cells because the rigid plant cell walls cannot be pulled inward. In a plant cell, cytokinesis begins with the formation of a **cell plate** between the daughter nuclei (Figure 10-10). The cell plate is formed by secretion vesicles containing polysaccharides produced by the nearby Golgi complex (see Chapter 5). As these vesicles accumulate in the center of the dividing cell, their membranes fuse together. The released polysaccharide forms the primary cell wall between the new cells, and the membranes of the fusing vesicles become incorporated into the plasma membranes of the adjoining daughter cells.

The exact separation of duplicated chromosomes during mitosis is complicated because it demands great precision. Dividing the cell's cytoplasm and organelles into daughter cells requires less precision, however, because equal amounts of these components are not necessary. How a dividing cell partitions its cytoplasm and organelles depends on whether or not the cell is surrounded by a cell wall. (See CTQ #5.)

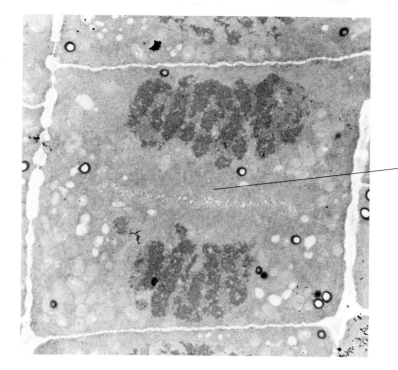

Formation of a
cell plate

FIGURE 10-10

Cytokinesis in an onion root cell. Secretion vesicles from nearby Golgi complexes become aligned midway between newly formed daughter nuclei. The membrane of the secretion vesicles will fuse to form the plasma membranes of the two daughter cells, and the contents of the vesicles will provide the material that forms the cell plate that will separate these cells. Additional materials for constructing a cell wall for each daughter cell will be delivered later by other secretion vesicles.

REEXAMINING THE THEMES

Relationship between Form and Function

The dramatic differences in the structure of the chromosomes and the cytoskeleton between interphase and M phase of the cell cycle reflect their different functions in these two stages. During interphase, the chromosomes are in an extended state, which allows them to send messages that direct cellular activities. During mitosis, these same chromosomes are compacted into a state where they could not possibly function as cell directors but can readily be moved around and separated into daughter cells. Similarly, the cytoskeleton of the interphase cell maintains cell shape and movement of cytoplasmic organelles and vesicles. During M phase, cytoskeletal elements are disassembled, and the same subunits are reused to construct a spindle apparatus, required for chromosome movement, and a contractile ring, required for cytokinesis in dividing animal cells.

Biological Order, Regulation, and Homeostasis

The cellular makeup of the complex tissues of multicellular plants and animals is maintained by strict controls over the rates of cell division. Many substances are involved in regulating cell cycles, including hormones and growth factors that bind to the outer surface of the cell, as well as substances (such as MPF) produced within the cell itself. Some substances stimulate cell division, while others inhibit the process. Mutations in any one of a number of different

genes can abolish growth control and lead to the formation of cancer.

Unity within Diversity

Every organism consists of one or more cells capable of undergoing division. Despite the great diversity of organisms and the types of cells they contain, there are only three basic types of cell division: prokaryotic fission, mitosis, and meiosis. These three processes provide the underlying basis for the reproduction and development of all organisms.

Evolution and Adaptation

The existence of similar processes occurring by similar mechanisms and utilizing similar molecules provides some of the strongest evidence for the evolutionary relationship of all organisms. Mitosis, for example, follows a similar sequence of steps among diverse eukaryotes, utilizing similar types of cellular organelles and regulatory mechanisms. MPF, which is present in organisms as diverse as yeast and humans, appears to be a universal trigger of mitosis. Many of the oncogenes—genes thought to control cell growth and division—are also found in diverse organisms. It appears that the mechanisms that regulate cell division in eukaryotes appeared at an early stage of evolution and have been retained ever since.

SYNOPSIS

Cells arise from other living cells by cell division. Cell division enables organisms to grow (by increasing the number of cells), to reproduce, and to repair or replace damaged or worn tissues.

There are three basic types of cell division: prokaryotic fission, mitosis, and meiosis. In prokaryotic cells, duplicate copies of the DNA attach to different sites on the plasma membrane; they are subsequently separated from each other by growth of the membrane and formation of a partitioning membrane and cell wall.

Eukaryotic cells divide their nuclear contents by the complex processes of either mitosis or meiosis. Mitosis produces genetically identical cells as part of the process of growth, repair, and asexual reproduction. Mitosis occurs in both diploid and haploid cells. Meiosis produces haploid daughter cells containing one-half the number of chromosomes as the diploid mother cell. The daughter cells from meiosis eventually develop into sex cells (gametes) for sexual reproduction.

The cell cycle is the sequence of stages through which a cell progresses from one division to the next. Most of the cell cycle is spent in interphase, a period divided into G_1, S, and G_2 stages, with S representing the period of DNA replication. The remainder of the cell cycle (M phase) includes mitosis and cytokinesis.

Mitosis is divided into several stages, based on the activities relating to the preparation and separation of the chromosomes. During prophase, the chromosomes undergo a coiling process in which the DNA-protein fibers become highly compacted and the cell assembles a mitotic spindle apparatus consisting of microtubules. During metaphase, the chromosomes are aligned at the center of the cell, attached to spindle fibers. Each chromosome consists of identical chromatids attached at their joined kinetochores. During anaphase, the chromatids split apart and move to opposite poles as separate chromosomes. During telophase, the cell returns to the interphase state.

Cytokinesis in animal cells occurs by the contraction of a ring of microfilaments, which splits the mother cell in two. In plant cells, the mother cell is divided by formation of an intervening cell plate resulting from the fusion of Golgi-derived vesicles.

Key Terms

cell division (p. 196)
mother cell (p. 196)
daughter cell (p. 196)
prokaryotic fission (p. 197)
cytokinesis (p. 197)
chromatid (p. 198)
centromere (p. 198)
homologue (p. 199)
diploid (2N) (p. 199)
haploid (1N) (p. 199)

mitosis (p. 199)
meiosis (p. 199)
gamete (p. 199)
cell cycle (p. 199)
M phase (p. 199)
interphase (p. 201)
oncogene (p. 202)
mitosis promoting factor (MPF) (p. 202)
prophase (p. 203)
mitotic chromosome (p. 204)

spindle apparatus (p. 204)
spindle fiber (p. 204)
centriole (p. 205)
pericentriolar material (PCM) (p. 205)
metaphase (p. 205)
kinetochore (p. 205)
anaphase (p. 206)
telophase (p. 206)
cell plate (p. 206)

Review Questions

1. Complete the figure by filling in the terms in their correct locations.

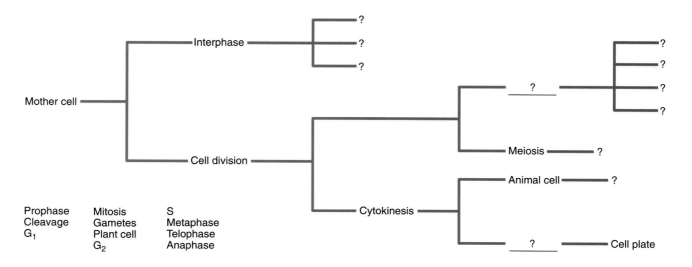

Prophase   Mitosis   S
Cleavage   Gametes   Metaphase
G_1 Plant cell   Telophase
    G_2 Anaphase

2. How does the process of chromosome coiling facilitate cell division? During what stage does it occur? During what stage does DNA replication occur?

3. Using pieces of colored yarn or crayons, set up the mitosis template (below) on a large sheet of paper and then physically run through the phases of mitosis.

4. How would you describe the genetic relatedness of sister chromatids? Of two chromosomes that were split apart during anaphase and moved to different daughter cells? Of the two members of a pair of homologous chromosomes?

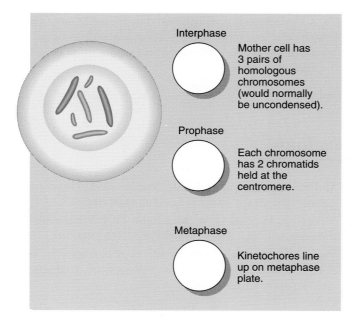

Interphase

Mother cell has 3 pairs of homologous chromosomes (would normally be uncondensed).

Prophase

Each chromosome has 2 chromatids held at the centromere.

Metaphase

Kinetochores line up on metaphase plate.

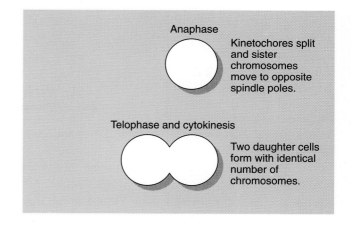

Anaphase

Kinetochores split and sister chromosomes move to opposite spindle poles.

Telophase and cytokinesis

Two daughter cells form with identical number of chromosomes.

Critical Thinking Questions

1. What do you expect the result would be if a zygote contained mutant MPF genes that produced a deficient MPF protein? Would you expect a liver cell or skin cell that had mutant MPF genes to be susceptible to developing into a cancer? Why or why not?

2. Complete the table below comparing mitosis and meiosis.

	Mitosis	Meioisis
chromosome number of mother cells		
number of cell divisions		
number of daughter cells produced		
chromosome number of daughter cells		
genetic composition of daughter cells (compared to mother cells)		

3. The graph opposite shows the results of measuring the DNA content of a suspension of cells by labeling the DNA with a fluorescent dye and then measuring the fluorescence of each cell as it passes through a sensitive detector. Identify the stage (or stages) of the cell cycle indicated by the letters A, B, and C.

4. Taxol is a drug used in the treatment of certain cancers. The drug acts by stabilizing microtubules, that is, pre-

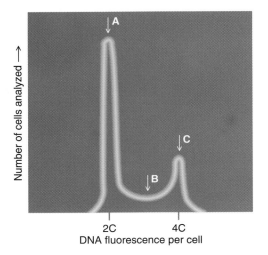

venting their disassembly. What effect would you expect taxol to have on cell division? Why? Why would taxol be useful in cancer treatment?

5. Explain how the manner in which cytokinesis occurs in animal cells and plant cells demonstrates the principle that structure and function are complementary.

6. The length of the cell cycle varies greatly from one cell type to another and for the same cell under different conditions. What do such differences tell you about the life span of the cell? In which parts of the human body would you expect to find cells with the short life cycles?

7. Cell division is usually blocked by the addition of cytochalasin and other drugs that cause the disassembly of actin filaments. Why would you expect these drugs to have this effect?

Additional Readings

Brooks, R. et al., eds. 1989. The cell cycle. *J. Cell Sci. Suppl.* 12. (Advanced)

Glover, D. M. et al., 1993. The centrosome. *Sci. Amer.* June:62–68. (Intermediate)

Hyams, J. S., and B. R. Brinkley, eds. 1989. *Mitosis: Molecules and mechanisms.* New York: Academic. (Advanced)

Marx, J. 1989. The cell cycle coming under control. *Science* 245:252–255. (Intermediate)

Marx, J. 1991. The cell cycle: Spinning farther ahead. *Science* 252:1490–1492. (Intermediate)

McIntosh, J. R., and K. L. McDonald. 1989. The mitotic spindle. *Sci. Amer.* Oct:48–56. (Intermediate)

Murray, A. W., and M. W. Kirschner. 1991. What controls the cell cycle. *Sci. Amer.* Mar:56–63. (Intermediate)

Pardee, A. B., et al. 1989. Frontiers in biology: The cell cycle. *Science* 246:603–640 (a collection of six research review papers). (Advanced)

Cell Division: Meiosis

STEPS
TO
DISCOVERY
Counting Human Chromosomes

MEIOSIS AND SEXUAL REPRODUCTION: AN OVERVIEW

The Importance of Meiosis

THE STAGES OF MEIOSIS

Meiosis I

Meiosis II

MITOSIS VERSUS MEIOSIS REVISITED

THE HUMAN PERSPECTIVE

Dangers That Lurk in Meiosis

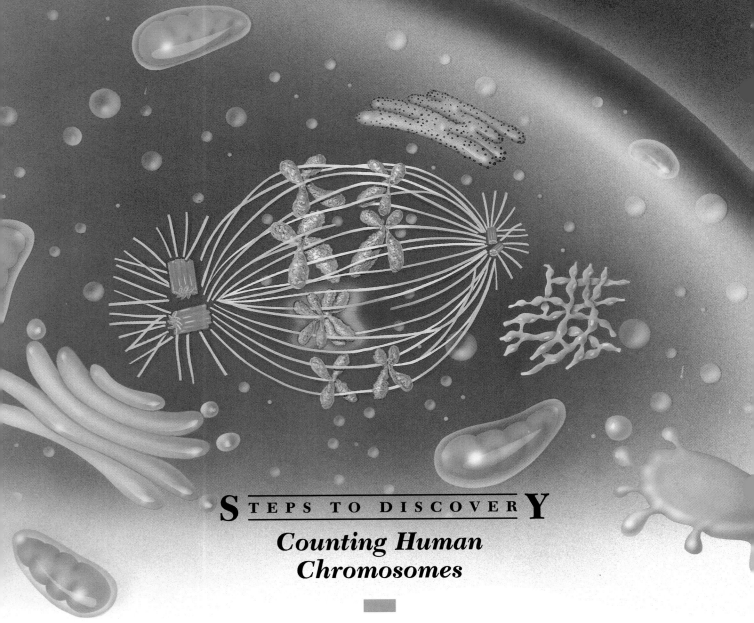

.   .   .   *"Thus man has 48 chromosomes."* This quote was taken from a biology book published in 1952. You might recall from the last chapter, however, that the diploid number of chromosomes in humans was stated to be *46*, not 48. But before you condemn the authors of this book for their inaccuracy, consider that, up until 1955, biologists accepted the "fact" that human cells contained 48 chromosomes.

Why did it take so long to discover the right number? Prior to the 1950s, chromosome numbers were determined primarily by examining sections of tissue in which cells were occasionally "caught" in the process of cell division. Trying to count several dozen chromosomes crammed together within a microscopic nucleus is like trying to count the

number of rubber bands present in a cellophane package without opening the package. It is actually quite remarkable that biologists got as close as 48! Before 1922, when Theophilus Painter, a geneticist at the University of Texas, arrived at this number, previous guesses had ranged from 8 to more than 50; 48 chromosomes then became the accepted value for more than 30 years until a remarkably simple discovery changed everything.

In 1951, times were tough in academia, and many new Ph.D.'s couldn't find teaching positions. Having just completed his doctorate on insect chromosomes at the University of Texas, T. C. Hsu fell into this distressing category. Reluctantly, Hsu accepted a postdoctoral research position

When a pair of homologous chromosomes fail to separate during meiosis, the resulting daughter cells will have an abnormal

in Galveston working on mammalian chromosomes, which were notoriously difficult to study. After several months of frustration, he was examining a new batch of cells with a microscope when, in Hsu's words "I could not believe my eyes when I saw some beautifully scattered chromosomes in these cells. I did not tell anyone, took a walk around the building, went to the coffee shop, and then returned to the lab. The beautiful chromosomes were still there. I knew they were real."

He tried to repeat the work, but his preparations regained their "normal miserable appearance." For many months after that, every attempt to discover what he had done "wrong" to get that beautiful preparation failed. Finally, Hsu tried pretreating the cells with a hypotonic (more dilute) salt solution. The cells expanded like balloons and exploded onto the slide, releasing the chromosomes and spreading them out so each was separated from its neighbors (review Figure 10-3). The earlier preparation of cells must have been accidentally washed with a dilute saline solution to which someone had failed to add enough salt. Since then, Hsu's hypotonic technique for treating cells has become a standard part of preparing chromosomes for microscopic examination. Ironically, Hsu did not use his technique to reexamine the question of the human diploid number; Painter had been one of his mentors at the University of Texas, and Hsu accepted the 48-chromosome count.

Within 3 years, however, Albert Levan and Jo Hin Tijo, working in the United States and Sweden, used the new hypotonic pretreatment technique on human cells treated with a drug called colchicine. Colchicine disassembles the mitotic spindle apparatus so that cells "freeze" in their metaphase configuration, providing many more mitotic cells to observe. Levan and Tijo carefully counted the chromosomes of these mitosis-arrested cells and found only 46. They repeated their observations and cautiously concluded that, at least in lung cells, the human diploid number was 46. The number was soon confirmed by other investigators on other human cell types and has been the accepted value ever since.

Meanwhile, Jerome Lejeune, a French clinician, had been studying children with Down syndrome for many years. Such children (discussed in the Human Perspective box) are characterized by a short stocky stature, distinctive folds of the eyelids (which gave rise to the earlier name "mongolism"), and mental retardation. After hearing a lecture by Jo Hin Tijo, who described how chromosomes could be counted using newly developed techniques, Lejeune decided to examine the chromosomes from a few of his Down syndrome patients. These children showed such a wide range of abnormalities that some alteration in the chromosomes was very likely responsible. Lejeune had never been involved in this type of research, however, and, in fact, did not even possess a microscope. He finally located one that had been discarded by the bacteriology lab. The microscope had been used so much that it would not remain in focus. In order to use it, Lejeune inserted a piece of tinfoil from a candy wrapper to hold the focusing knob in place. In 1959, Lejeune and two colleagues who had helped in cell preparation published a two-page paper indicating that the cells of nine different patients with Down syndrome all had 47 chromosomes, rather than 46.

Lejeune's paper opened the door to a new field of medical genetics. It was soon followed by a number of other reports in which patients with various types of disorders were shown to have an abnormal number of chromosomes. Examination of cells from fetuses that had spontaneously aborted revealed that many had three or more full sets of chromosomes in their cells. Thus, an abnormal fetal chromosome number was discovered to be a common cause of miscarriage. These insights into the effects of chromosome abnormalities revealed how important it was that the cells of developing embryos contain precisely the correct number of chromosomes. It soon became obvious that the formation of gametes with an abnormal number of chromosomes could be traced to a defect occurring during meiosis — the subject of this chapter.

number of chromosomes.

*E*ach organism is a living showcase, possessing thousands of inherited features. When you glance in a mirror, you can see dozens of inherited characteristics—the length and thickness of your eyelashes, the width of your smile, the depth of a dimple. The genetic blueprint containing the plans for your construction is contained in only 46 chromosomes, which are carried in virtually every cell of your body. Within these 46 chromosomes, there are actually two similar sets of genetic instructions: One haploid set of 23 chromosomes is derived from your father's sperm, and the other haploid set of 23 chromosomes originated from your mother's egg. At conception, the two sets of homologous chromosomes unite at fertilization to form a diploid cell, the **zygote.** The zygote and all of its progeny cells have dutifully duplicated those 46 chromosomes each time they divided, eventually growing into a remarkable cell collection—you.

But your parents are also collections of cells containing 46 chromosomes. How did they manufacture gametes (sex cells—sperm or eggs) with only 23 chromosomes? They could not have done so by mitosis, since mitosis produces cells with exactly the same number of chromosomes as the mother cell (see Figure 10-4). Rather, they were able to do so through **meiosis,** a type of nuclear division that divides the number of chromosomes in half, thereby forming haploid gametes for sexual reproduction.

▼ ▼ ▼

MEIOSIS AND SEXUAL REPRODUCTION: AN OVERVIEW

⬙ Reproduction may be the most important characteristic that distinguishes organisms from nonliving objects. The sequence of events that occur in the life of an organism from the time it is formed by reproduction until it produces offspring itself makes up the **life cycle** of that organism. The life cycles of nearly all eukaryotic organisms include sexual reproduction, a process that is just as important to the survival of trees and worms as it is to humans. Meiosis is a critical step in all forms of sexual reproduction because it generates haploid nuclei in cells that eventually form gametes. Gametes are produced differently in animals and plants, although meiosis is a key step in their production in both groups (Figure 11-1). In animals, meiosis produces haploid daughter cells that are directly transformed into sperm or eggs. In plants, meiosis produces haploid daughter cells called *spores*, which then divide by *mitosis* to grow into a multicellular haploid plant that produces the

gametes—either male pollen cells or female eggs. (Sexual reproduction in plants and animals is discussed further in Chapters 20 and 31, respectively.) Haploid gametes, in turn, come together at fertilization to generate a diploid zygote (a fertilized egg). Thus, meiosis and fertilization are critical events in the life cycle of sexual organisms, since both processes change the number of chromosome sets.

Sexually reproducing organisms are made up of two types of cells: germ cells and somatic cells. **Germ cells** are those cells that will undergo meiosis to form the gametes used in sexual reproduction. (In this case, the word "germ" does not refer to a disease-causing organism.) Germ cells are found in the *gonads* of animals—the female ovaries and the male testes. In most plants, germ cells are located in the female ovaries and the male anthers. All remaining cells that make up the body of an animal or plant—that is, all cells except the germ cells—are called **somatic cells.** Somatic cells divide only by mitosis, never by meiosis.

THE IMPORTANCE OF MEIOSIS

In 1883, Edouard van Beneden, a Belgian biologist, noted that the body cells of the roundworm *Ascaris* contained four large chromosomes, but the male and female gametes had but two chromosomes apiece. In 1887, August Weismann, a German biologist, proposed the existence of some type of "reduction division" that occurred in both the ovaries and the testes, allowing the chromosome number to be divided in half during the process of gamete formation. This reduction division was soon shown to constitute the first part of meiosis.

Consider for a moment what would happen if meiosis did not occur and the gametes contained the same number of chromosomes as did the somatic cells. Every time two gametes came together at fertilization, the chromosome number of the offspring would be twice that of their parents. For humans, the first generation of offspring would have 92 chromosomes (46 from the sperm plus 46 from the egg); the second generation would have 184; the third, 368; and so forth. Cells would soon become overloaded with surplus chromosomes and die after just a few generations. Thus, for organisms that reproduce sexually, the formation of reproductive cells containing only *one* set of chromosomes (rather than two sets, as in somatic cells) is crucial for ensuring a stable chromosome number for a species from generation to generation. The role of meiosis in maintaining a constant number of chromosomes in the cells of parents and offspring is another example of a mechanism that maintains biological order.

▥▶ The importance of meiosis goes beyond maintaining the chromosome number, however. As we will see later in the chapter, meiosis also builds *genetic variability* in a species. That is, meiosis increases the variety in characteristics among individuals that make up a species' population. A glance at a crowd of people instantly reveals the enormous variability that can exist within the human species. Different individuals may be better suited for different habitats.

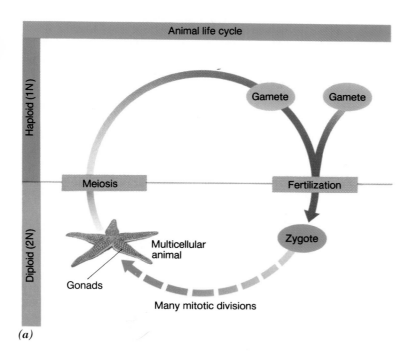

(a)

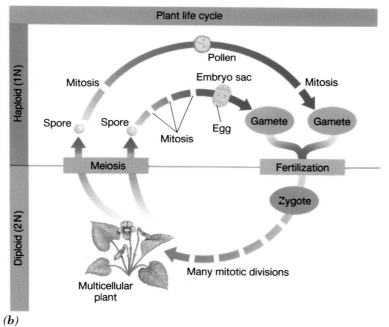

(b)

FIGURE 11-1

Life cycles of sexually reproducing plants and animals. Sexual reproduction requires both meiosis and fertilization. Meiosis reduces the number of chromosomes by half to form haploid nuclei that become incorporated into gametes. Gametes then fuse during fertilization to form a zygote (sometimes called a "fertilized egg"). *(a) Animal life cycle.* In animals, the haploid cells produced by meiosis directly form gametes. After fertilization, numerous mitotic divisions of the zygote produce a multicellular adult with gonads containing germinal tissues that divide by meiosis. *(b) Plant life cycle.* In plants, the haploid cells produced by meiosis form spores. The spores divide by mitosis to form a multicellular, haploid structure (in flowering plants, a pollen grain and embryo sac) that houses the sperm and egg gametes. Following fertilization, many mitotic divisions of the zygote produce a multicellular diploid plant with germinal tissues that divide by meiosis (as in animals), completing the cycle (Discussed in Chapter 20).

For example, shorter members of a population might be best suited for living in a forest whose trees have very low branches, while darker-skinned individuals may be better suited for living in bright, sunny climates, and so forth. A greater range of characteristics in a population improves the chances that some individuals will survive environmental changes, thereby increasing the likelihood that the species will be perpetuated.

Meiosis is not only important for the perpetuation of a species; it is also necessary for the formation of new species. Together with the appearance of new genetic characteristics by mutation, meiosis is one of the underlying mecha-

nisms that ensure genetic change in a population over time (discussed further in Chapters 33 and 34). If meiosis had never appeared, life on earth may well have been limited to bacteria and simple, unicellular eukaryotes.

Genetic variability increases during meiosis because chromosomes and genes are reshuffled to form new combinations before they are distributed to haploid daughter cells. Consequently, each gamete has a unique genetic composition. When gametes combine during fertilization, the organism produced is different from all others of the species. This explains why you *resemble* members of your family, yet you do not look *exactly* like any of them. The one exception to genetic variability among humans is identical twins; since identical twins develop from the same zygote, they have the same genes.

Life on earth has persisted over billions of years because species have the capacity to change. Change is facilitated by sexual reproduction, which requires meiosis as a step in the path toward formation of gametes. Meiosis provides genetic diversity that facilitates biological change while maintaining chromosomal numbers from one generation to the next. (See CTQ #2.)

THE STAGES OF MEIOSIS

There are many similarities between meiosis and mitosis (Chapter 10). Even the names of the stages are the same (prophase, metaphase, anaphase, and telophase). There are also some very important differences, however. This is not unexpected, considering that mitosis maintains chromosome number, whereas meiosis divides the number in half.

Like mitosis, meiosis is a continuous process—one that progresses through similar steps in all eukaryotic organisms. Biologists divide meiosis into stages based primarily on differences in chromosomal activity. Like mitosis, meiosis is preceded by an interphase period in which the chromosomes are duplicated. Unlike mitosis, in which the chromosomes are divided between two daughter nuclei, meiosis includes *two* sequential divisions—*meiosis I* and *meiosis II*—so that the chromosomes are distributed among four nuclei, rather than two. The entire sequence of meiotic stages is shown in Figure 11-2.

MEIOSIS I

Meiosis I is called the **reduction division** of meiosis because the nuclei of the two daughter cells contain half the number of chromosomes as does the nucleus of the mother cell. This reduction in the number of chromosomes is achieved by separating the members of each pair of homologous chromosomes into different nuclei. As a result, the daughter cells from meiosis I are haploid; that is, they contain only one complete set of chromosomes instead of two sets as in the diploid mother cell. In order to ensure that each of the daughter cells has only one member of each pair of homologues, an elaborate process of chromosome pairing occurs that has no counterpart in mitosis. During the period in which the homologous chromosomes are paired, some very important chromosome choreography takes place that increases genetic variability.

Prophase I

Unlike the prophase of mitosis, which usually lasts only a few minutes, the prophase I stage of meiosis can take hours, days, and sometimes even years. In humans, for example, all of the potential eggs that can ever be formed during the life of a woman have already entered prophase I by the time of her birth. Most of these germ cells will remain "stuck" in this stage of meiosis for decades, which is not without possible consequences.

Prophase I begins when the diffuse threads of the interphase chromosomes begin coiling. When the chromosomes have reached a stage of partial compaction, the members of each homologous pair recognize each other and become aligned together along their entire length. This process of chromosomal alignment is called **synapsis.** Since each chromosome is made up of two chromatids, a synapsed pair of homologous chromosomes is called a **tetrad,** which is a unit of four chromatids (Figure 11-3).

Early observers of meiosis noticed that the homologous chromosomes of the tetrads actually become wrapped around each other while they are synapsed (Figure 11-3). In 1909, F. A. Janssens proposed that this interaction might result in the breakage and exchange of pieces between different chromatids of the tetrad. Janssens' hypothesis proved to be basically correct; he had predicted the process now called **crossing over.**

Crossing Over and Genetic Recombination: An Increase in Genetic Variability Recall from page 199 that homologous chromosomes have identical sequences of genes along their length, although the corresponding genes on the two chromosomes may code for different characteristics of a particular trait. Crossing over between homologous chromatids can produce gene combinations different from those originally found in either chromatid. Let's see how crossing over can produce new gene combinations in homologous chromosomes with genes for plant height at one location and genes for flower color at another. These chromosomes are depicted in Figure 11-4.

In this example, one chromosome (on top) directs the development of a tall individual with yellow flowers, whereas its homologue produces a short plant with white flowers. Crossing over can *reshuffle* these genes (Figure 11-4)—a process known as **genetic recombination.** In our example of plant chromosomes, crossing over and recombination team the gene for a tall plant with the gene for white flowers on one chromatid and the gene for a short plant with the gene for yellow flowers on the homologous chromatid.

Genetic recombination greatly increases the genetic variability of the offspring by producing gametes with novel combinations of traits. Notice that the recombined genes are found on only two of the chromatids; the original gene combinations remain on the two chromatids that did not participate in crossing over. Genetic recombination during meiosis is one of the biological processes responsible for generating the diversity of life on earth.

The Molecular Basis of Genetic Recombination
Crossing over rearranges chromosomal material, producing chromatids that contain sections of both paternal and maternal origin (Figure 11-4). Diagrams and photographs of

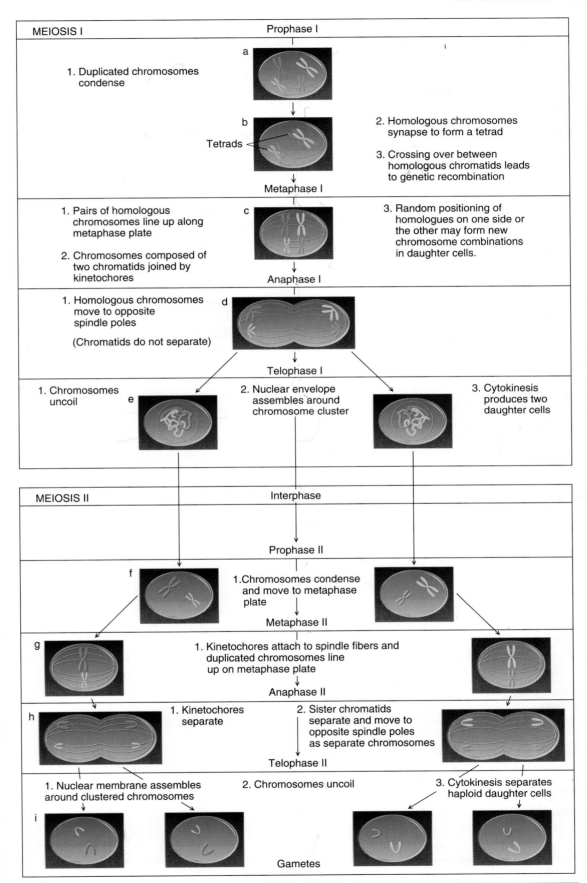

MEIOSIS I — Prophase I

1. Duplicated chromosomes condense

2. Homologous chromosomes synapse to form a tetrad

3. Crossing over between homologous chromatids leads to genetic recombination

Tetrads

Metaphase I

1. Pairs of homologous chromosomes line up along metaphase plate

2. Chromosomes composed of two chromatids joined by kinetochores

3. Random positioning of homologues on one side or the other may form new chromosome combinations in daughter cells.

Anaphase I

1. Homologous chromosomes move to opposite spindle poles

(Chromatids do not separate)

Telophase I

1. Chromosomes uncoil

2. Nuclear envelope assembles around chromosome cluster

3. Cytokinesis produces two daughter cells

MEIOSIS II — Interphase

Prophase II

1. Chromosomes condense and move to metaphase plate

Metaphase II

1. Kinetochores attach to spindle fibers and duplicated chromosomes line up on metaphase plate

Anaphase II

1. Kinetochores separate

2. Sister chromatids separate and move to opposite spindle poles as separate chromosomes

Telophase II

1. Nuclear membrane assembles around clustered chromosomes

2. Chromosomes uncoil

3. Cytokinesis separates haploid daughter cells

Gametes

F I G U R E 11-2
The stages of meiosis.

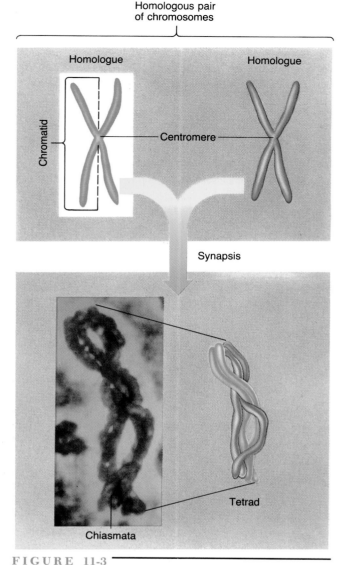

Homologue   Homologue

Chromatid

Centromere

Synapsis

Tetrad

Chiasmata

FIGURE 11-3

Homologous chromosomes forming a tetrad after synapsis. The points where the two homologous chromosomes are in contact are called *chiasmata*. These are sites where crossing over is believed to have occurred at an earlier stage.

this type give the erroneous impression that large sections of two chromatids simply break off and are physically exchanged with each other. Consider for a moment what this would entail. Because it is compacted, each chromatid of a dividing cell is so thick that it must contain thousands of DNA fibers lying side by side, as depicted in Figure 10-7. Are all of these fibers broken simultaneously? What about the genes contained in these DNA fibers? Recall what happens when just one alteration occurs in a gene that codes for a globin protein—the result can be sickle cell anemia (page 81). The breakage of an entire compacted chromatid would split *hundreds* of genes. Could all of these genes be reunited precisely with their missing segment when the two chromatids fused during crossing over? The answer is undoubtedly no.

Let us go back to an earlier stage of prophase I and reexamine the process of synapsis more closely. As the homologous chromosomes come together during synapsis, they can be seen in the electron microscope to be physically attached to a ladderlike structure called the **synaptonemal complex.** Each lateral element of the "ladder" is associated with one of the chromosomes of the tetrad, so that the homologues are not actually pressed against each other. Instead, they are separated by a space of about 100 nanometers (Figure 11-5*a*). The space in the middle of the ladder is not empty but contains loops of DNA extending from each of the chromosomes (Figure 11-5*b*). This suggests that the synaptonemal complex acts as a physical scaffold to which the chromosomes adhere, and that genetic recombination results from exchanges between the DNA loops extending into the center of the ladderlike structure. Viewed in this way, crossing over does not require the breakage and reunion of thick, compact meiotic chromatids but of individual DNA molecules.

Metaphase I and Anaphase I

Near the end of prophase I, the nucleolus and nuclear envelope become dispersed; spindle fibers become con-

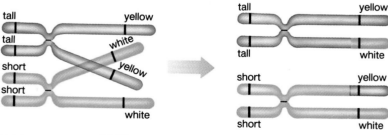

tall   yellow
tall
white
short   yellow
short
white

tall   yellow

tall   white

short   yellow

short   white

FIGURE 11-4

Genetic recombination. Meiotic chromosomes showing the two duplicates (sister chromatids) joined to one another. The sites (*loci*) on the chromosomes for the genes that govern two plant traits—height and flower color—are indicated. After crossing over, each chromosome contains one chromatid with the original combination of genetic characteristics and one chromatid with a mixture of the maternal and paternal characteristics.

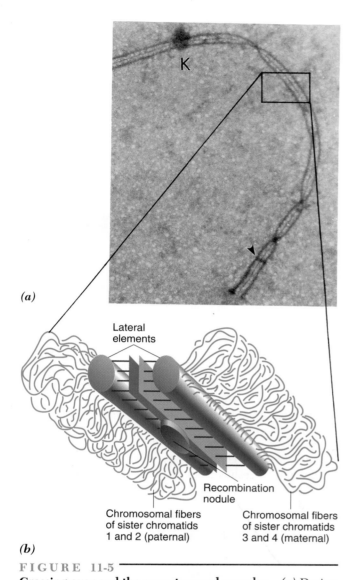

(a)

(b)

FIGURE 11-5

Crossing over and the synaptonemal complex. *(a)* During prophase I of meiosis, homologous chromosomes become paired with one another and may appear as parallel fibers joined by the kinetochore (K). *(b)* Closer examination shows that the homologous chromosomes are tightly associated with a ladderlike structure, the synaptonemal complex. The dense granules (*recombination nodules*) seen between the "rungs of the ladder" are thought to contain the enzymatic machinery required for genetic recombination.

nected to the kinetochores of the chromosomes; and the tetrads are moved to the metaphase plate (page 205). In metaphase I, the tetrads become aligned in such a way that the two homologous chromosomes of each tetrad are situated on opposite sides of the plate (Figure 11-2*c*). Anaphase I begins when the spindle fibers attached to each chromosome shorten, pulling homologous chromosomes of each tetrad to opposite spindle poles (Figure 11-2*d*). Unlike in anaphase of mitosis, the kinetochores do not split during

anaphase I of meiosis. Each homologous chromosome is therefore still made up of two chromatids joined by their fused kinetochores. As a result of anaphase I, each daughter cell now has half the number of chromosome sets of its mother cell.

Independent Assortment of Homologues: An Additional Increase in Genetic Variability When homologous chromosomes become aligned at the metaphase plate, each homologue can line up on one side of the plate or the other, depending entirely on chance. Hence, all the chromosomes originally derived from one parent (**maternal chromosomes** are those derived from the mother and **paternal chromosomes** are those derived from the father) very rarely line up on the same side of the metaphase plate. Since each pair of homologous chromosomes lines up independently of other pairs, the chromosomes of each parent are sorted independently during meiosis I (Figure 11-6). This **independent assortment** of homologues reshuffles homologous chromosomes to form new chromosome combinations in the daughter cells.

Let's look at the numbers. If the diploid number of chromosomes of a species happened to be four (two pairs of homologous chromosomes), four (2^2) different types of daughter cells could be produced (Figure 11-6*a*). If the diploid number is six, there can be eight (2^3) different daughter cells (Figure 11-6*b*). In each case, there are 2^n number of genetically different daughter cells, where *n* is the haploid number of chromosomes. In a human, there are more than 8 million (2^{23}) possible chromosome combinations, with 23 chromosomes from your father and 23 chromosomes from your mother. Thus, excluding crossing over, the possibility that two parents would produce two identical offspring in two different pregnancies would be about one in 70 million ($2^{23} \times 2^{23}$). Add to this the gene exchanges that occur during crossing over, and it is easy to see how meiosis vastly increases the genetic variability of a species. This is the evolutionary advantage of sexual reproduction.

Telophase I and Cytokinesis

Telophase I of meiosis (Figure 11-2*e*) produces less dramatic changes than does telophase of mitosis. Although in many cases the chromosomes undergo some uncoiling, they do not reach the extremely extended state of the interphase nucleus. Similarly, the nuclear envelope may or may not reform during telophase I. Telophase I is followed by cytokinesis, which splits the cell into two daughter cells. Because DNA does not replicate during the interphase between cytokinesis and prophase II, the period separating meiosis I and II (called *interkinesis*) is usually very brief.

MEIOSIS II

Since sister chromatids become separated into different nuclei during **meiosis II,** the events in meiosis II are similar to those that occur during mitosis. During prophase II, the chromosomes shorten and thicken as they journey toward

Distribution of two pairs of homologous chromosomes by independent assortment during meiosis

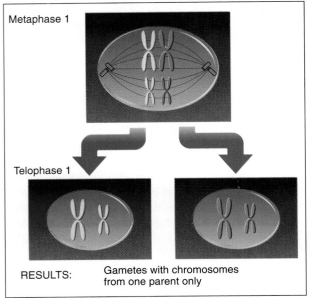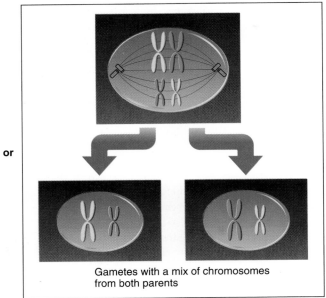

Distribution of three pairs of homologous chromosomes by independent assortment

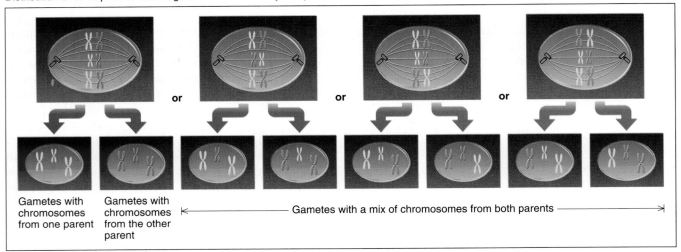

(b)

FIGURE 11-6

Independent assortment of homologous chromosomes during metaphase I of meiosis.
Paired homologous chromosomes line up randomly along the metaphase plate. Because each pair of
homologous chromosomes lines up independently of other pairs, new chromosome combinations are
often produced in the daughter nuclei, leading to increased genetic variability in a species. (To follow
independent assortment, the homologues originally derived from one parent are colored blue, and
those from the other parent are orange.) *(a)* In a cell with two pairs of homologous chromosomes,
there are four possible combinations of chromosomes and, thus, four genetically different daughter
nuclei that can form. Either both chromosomes from the same parent go into the same daughter cell,
or the chromosomes are mixed, and each daughter cell receives one chromosome from each parent.
(b) In a cell with three pairs of chromosomes, eight (2^3) genetically different daughter nuclei can
form, not including the added genetic variability introduced by crossing over.

◁ THE HUMAN PERSPECTIVE ▷
Dangers That Lurk in Meiosis

On occasion, a mistake occurs during meiosis. Homologous chromosomes may fail to separate from each other during meiosis I, or sister chromatids may fail to come apart during meiosis II. When either of these situations occurs, gametes are formed that contain an abnormal number of chromosomes—either an extra chromosome or a missing chromosome. What happens if one of these gametes happens to fuse with a normal gamete and forms a zygote with an abnormal number of chromosomes? In most cases, the zygote develops into an abnormal embryo that dies in the womb, causing a miscarriage. In a few cases, however, the zygote develops into an infant whose cells have an abnormal chromosome number. These infants usually develop characteristic deformities.

The most common congenital disorder resulting from a meiotic mistake is Down syndrome (Figure 1), a condition in which individuals have three copies of chromosome number 21 in their cells, rather than two. Chromosome 21 is the smallest chromosome (see Figure 10-3*b*), containing an estimated 1,500 genes. Consequently, you might think that having a bit of extra genetic information would not be harmful. Yet, as we mentioned earlier in the chapter, Down syndrome is characterized by serious problems. These include mental impairment, alteration in some body features, frequent circulatory problems, increased susceptibility to infectious diseases, a greatly increased risk of developing leukemia, and the early onset of Alzheimer's disease. All of these medical

FIGURE 1

Chris Burke, star of the television program *Life Goes On*, was born with Down syndrome.

problems are thought to result from the abnormal activity of genes located on chromosome 21. Seventy years ago, children with Down syndrome had an average life expectancy of about 10 years; most of them were shut away in mental institutions, where they withered and died. Today, most of these children are being raised at home and are encouraged to develop to their full potential; many attend "regular" schools and grow up to become working adults.

The likelihood of having a child with Down syndrome rises dramatically with the age of the mother, from 0.05 percent for mothers 19 years of age to nearly 2 percent for women over the age of 45. There are several possible explanations for this age-related increase. Some geneticists believe that the increased risk results primarily from aging of the germ cell as it remains for longer periods in the ovary. Another possibility is that younger females are more likely to abort a fetus bearing a chromosomal abnormality, whereas older females are more likely to carry an abnormal fetus to full term. Regardless of the reasons, older pregnant women are encouraged to undergo tests that examine the fetal cells for chromosomal abnormalities (Chapter 17). The effects of other chromosome abnormalities are described in Chapter 13.

---

the cell's metaphase plate (Figure 11-2*f*). In metaphase II, the chromosomes line up on the metaphase plate, and the kinetochores attach to the spindle fibers (Figure 11-2*g*). Anaphase II begins when the kinetochores split and the newly independent chromosomes are pulled toward opposite spindle poles (Figure 11-2*h*). During telophase II, a nuclear envelope is assembled around each of the four clusters of chromosomes, generating four nuclei. The two cells then divide by cytokinesis to produce a total of four haploid daughter cells (Figure 11-2*i*). Each daughter cell contains one-half of the chromosomes of the original diploid cell (although each chromosome may be heavily altered due to crossing over) and, therefore, has *one* complete set of hereditary instructions. The separation of chromosomes during meiosis must occur with great precision; the consequences of a failure in the process are considered in The Human Perspective: Dangers That Lurk in Meiosis.

Like mitosis, meiosis requires precision in order to separate duplicated chromosomes exactly. Because meiosis produces daughter cells that contain half the number of chromosomes as does the dividing cell, two sequences of cell division are required. (See CTQ #3.)

MITOSIS VERSUS MEIOSIS REVISITED

The major differences in the processes of mitosis and meiosis were summarized in Figure 10-4. In short, mitosis produces two daughter nuclei that are genetically identical to each other and to the parent cell, whereas meiosis produces four daughter nuclei that are genetically different from one another, each containing half the number of chromosomes of the parent cell. We have also discussed the different roles of mitosis and meiosis in the life cycles of plants and animals. Mitosis generates the nuclei of the body's cells, while meiosis generates the nuclei that ultimately become incorporated into the gametes.

Now let's go beyond these general comparisons and see how the differences between mitosis and meiosis can make a difference in the life of a specific organism. Suppose you were to visit a local pond, remove an eyedropper full of pond water, and put it in a vial. When you get back to your biology lab, you find you have captured a single ciliated protist that you identify as a member of the genus *Paramecium.* You place the contents of the vial into a container of purified water and add some algae or yeast to provide the organism with food. Over the next few weeks, your "domesticated" ciliate seems to thrive, dividing repeatedly and producing large numbers of genetically identical offspring by asexual reproduction (mitosis and cytokinesis). Periodically, you remove a certain volume of your culture and transfer it to fresh surroundings with fresh algae or yeast so that the culture remains unpolluted.

You might think that you could keep a culture of *Paramecium* alive indefinitely this way, but soon the organisms would begin to divide more slowly and to appear abnormal under the microscope; eventually, they would die off. This was the finding of Tracy Sonneborn of Indiana University, who studied the longevity of cultures of ciliates that reproduced solely by mitosis. A similar phenomenon has been observed when *cells* of humans and various animals are grown in culture. The cells divide mitotically for about 50 generations and then gradually lose their ability to divide, becoming unhealthy-looking, and eventually dying.

In contrast, meiosis seems to have the power to rejuvenate cells. While human cells growing in culture cannot undergo meiosis, the ciliates swimming in a pond can. Suppose that, instead of beginning with a single individual ciliate, you were able to capture a few dozen of these protists.

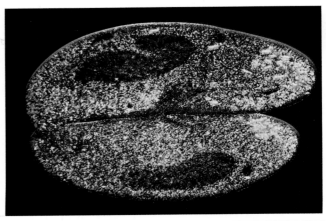

FIGURE 11-7

These protists are engaged in conjugation. A haploid nucleus from each individual passes into the other cell and fuses with the haploid nucleus of the recipient cell, forming a diploid nucleus with a new genetic constitution.

Although ciliated protists do not produce gametes that fuse to form a zygote, they do carry on a type of sexual reproduction called *conjugation* (Figure 11-7). During conjugation, two individuals of different strains of a species come together; the nucleus of each individual undergoes meiosis; and one haploid nucleus from each member of the conjugated pair travels across a bridge into the other member cell. There, it fuses with a haploid nucleus in the recipient, forming a diploid nucleus. The two ciliates then separate from each other, and each swims away.

▐▶ Although this type of sexual reproduction does not produce additional offspring, it does generate individuals with a new genetic composition. Some ciliates that have undergone conjugation are found to die rather rapidly; presumably, these individuals have become genetically unfit for their environment. Others, however, appear to do much better than the original stocks, suggesting that these individuals possess a genetic constitution that makes them better adapted to their surroundings. This example illustrates the advantage of meiosis. It generates genetically diverse individuals required for the perpetuation of a species.

Because of its capacity to generate offspring with a unique genetic composition, meiosis is important for the survival of a species. (See CTQ #5.)

REEXAMINING THE THEMES

Relationship between Form and Function

Genetic recombination requires that the DNA molecules of different chromatids be broken and precisely reunited. This process is facilitated by the synaptonemal complex, a scaffoldlike structure that allows the DNA molecules of homologous chromosomes to be brought into close enough proximity for crossing over to occur. At the same time, the synaptonemal complex prevents the chromatids from becoming intertwined, which would result in genetic chaos.

Biological Order, Regulation, and Homeostasis

Maintaining biological order requires that the cells of a species maintain a constant number of chromosomes. The importance of chromosome number is dramatically illustrated by Down syndrome, a human disorder resulting from the presence of just one extra chromosome. Meiosis is *the* process by which the number of chromosomes in sexually reproducing organisms is cut in half, ensuring that the number of chromosomes will remain constant from one generation to the next.

Unity within Diversity

Sexual reproduction occurs throughout the four eukaryotic kingdoms, and, in every case, the formation of haploid gametes requires meiosis. In the more than 1 million different kinds of sexually reproducing organisms that exist, the process of meiosis is remarkably similar. At the same time, meiosis is largely responsible for generating the genetic variability that, through natural selection, has led to the remarkable diversity of life on earth.

Evolution and Adaptation

There could be no sexual reproduction without meiosis. If reproduction occurred solely by mitosis, offspring would be virtually identical to their parents. Biological evolution by natural selection depends on the genetic variability produced by independent assortment and crossing over. It is not unreasonable to conclude that, without meiosis, life on this planet may well have been limited to bacteria and a smattering of primitive, single-celled protists.

SYNOPSIS

Unlike mitosis, which maintains chromosome number, meiosis divides the number of chromosomes in half by separating the homologues of a diploid mother cell into four haploid daughter nuclei. The haploid products ultimately become incorporated into gametes for sexual reproduction. The fusion of two haploid gametes restores the diploid number of chromosomes in the zygote, thus maintaining a stable number of chromosomes from generation to generation.

Meiosis includes two sequential nuclear divisions —meiosis I and meiosis II—without an intervening period of replication. The first division is called the reduction division because it separates homologous chromosomes into two different daughter nuclei. The second division separates attached chromatids, producing a total of four haploid nuclei.

Genetic variability in the daughter cells (when compared to the original mother cell) is ensured by two events: independent assortment and genetic recombination. Genetic recombination results from crossing over, which occurs during prophase I of the first meiotic division. In crossing over, homologous portions of maternal and paternal chromatids are exchanged with each other, thus producing chromosomes that contain both maternal and paternal regions. Since both homologues in a pair are separated (assorted) independently of each other during anaphase I, daughter cells receive mixtures of both maternal and paternal chromosomes.

Genetic recombination and independent assortment during meiosis increase the genetic variability of a species. Genetic variability enhances the chances that some members of the species will survive inevitable environmental changes.

Key Terms

zygote (p. 214)
meiosis (p. 214)
life cycle (p. 214)
germ cell (p. 214)
somatic cell (p. 214)
meiosis I (p. 216)

reduction division (p. 216)
synapsis (p. 216)
tetrad (p. 216)
crossing over (p. 216)
genetic recombination (p. 216)
synaptonemal complex (p. 218)

maternal chromosome (p. 219)
paternal chromosome (p. 219)
independent assortment (p. 219)
meiosis II (p. 219)

Review Questions

1. Match the activity with the phase of meiosis in which it occurs.

   a. synapsis
   b. crossing over
   c. kinetochores split
   d. independent assortment
   e. homologous chromosomes separate
   f. cytokinesis

   1. prophase I
   2. metaphase I
   3. anaphase I
   4. telophase I
   5. prophase II
   6. anaphase II
   7. telophase II

2. How do crossing over and independent assortment increase the genetic variability of a species?

3. Why is meiosis I (and not meiosis II) referred to as the reduction division?

4. Suppose that one human sperm contains x amount of DNA. How much DNA would a cell just entering meiosis contain? A cell entering meiosis II? A cell just completing meiosis II? Which of these three cells would have a haploid number of chromosomes? A diploid number of chromosomes?

Critical Thinking Questions

1. Why are disorders, such as Down syndrome, that arise from abnormal chromosome numbers, characterized by a number of seemingly unrelated abnormalities?

2. A gardener's favorite plant had white flowers and long seed pods. To add some variety to her garden, she transplants some plants of the same type, but with pink flowers and short seed pods from her neighbor's garden. To her surprise, in a few generations, she grows plants with white flowers and short seed pods and plants with pink flowers and long seed pods, as well as the original combinations. What are two ways in which these new combinations could have arisen?

3. Set up the meiosis template in the diagram below on a large sheet of paper. Then use pieces of colored yarn or pipe cleaners to simulate chromosomes and make a model of the phases of meiosis. (*See template on opposite page*)

4. Would you expect two genes on the same chromosome, such as yellow flowers and short stems, always to be exchanged during crossing over? How might they remain together *in spite of* crossing over?

5. Suppose paternal chromosomes always lined up on the same side of the metaphase plate of cells in meiosis I. How would this affect genetic variability of offspring? Would they all be identical? Why or why not?

Additional Readings

Chandley, A. C. 1988. Meiosis in man. *Trends in Gen.* 4:79–83. (Intermediate)

Hsu, T. C. 1979. *Human and mammalian cytogenetics.* New York: Springer-Verlag. (Intermediate)

John, B. 1990. *Meiosis.* New York: Cambridge University Press. (Advanced)

Moens, P. B. 1987. *Meiosis.* Orlando: Academic. (Advanced)

Patterson, D. 1987. The causes of Down syndrome. *Sci. Amer.* Feb:52–60. (Intermediate-Advanced)

White, M. J. D. 1973. *The chromosomes.* Halsted. (Advanced)

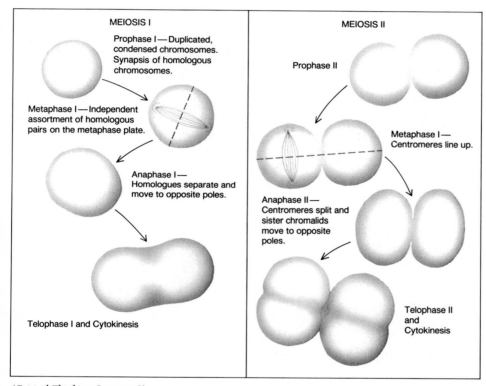

(*Critical Thinking Question 3*)

PART

◄ 3 ►

The Genetic Basis of Life

DNA stores the information that directs the development and maintenance of every organism. Such biological blueprints are fundamental to life itself. Changes in the DNA blueprint create new traits that propel evolution. Most of these random changes have no effect, are lethal, or create bizarre characteristics. Fruit flies, such as the one shown here, played an important role in exploring the genetic basis of life.

PART
· 3 ·

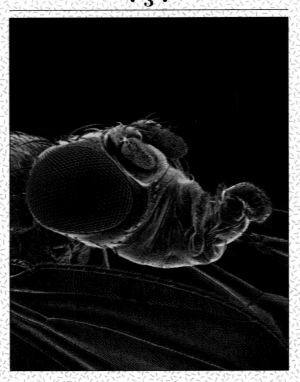

On the Trail of Heredity

**STEPS
TO
DISCOVERY**
The Genetic Basis of Schizophrenia

**GREGOR MENDEL: THE FATHER OF
MODERN GENETICS**

Getting Started

The First Generation of Hybrids

**Interpreting the Results and Formulating a
Hypothesis**

Predicting Patterns of Inheritance

**Crossing Plants That Differ by More
Than One Trait: Mendel's Law
of Independent Assortment**

MENDEL AMENDED

Incomplete Dominance and Codominance

Multiple Alleles

Combined Effects of Genes at Different Loci

Pleiotropy

Polygenic Inheritance

Environmental Influences

**MUTATION: A CHANGE IN THE
GENETIC STATUS QUO**

The Genetic Basis of Schizophrenia

The human mind remains one of the most intriguing frontiers in the biological sciences. Let us consider a question pondered by social and natural scientists alike: To what degree does a person's genetic makeup determine whether he or she develops a serious psychological disorder, such as schizophrenia or manic depression? In a sense, this question is part of the larger issue of how much of our personality is a result of our genes—nature—as opposed to influences from our environment—nurture.

Researchers unraveling the mysteries of schizophrenia—a disorder characterized by depression, delusions, hallucinations, and confusion—typify the scientists confronting the "nature-versus-nurture" debate. Early in this century,

investigators tried to determine if schizophrenia ran in families, as would be expected if it were an inherited condition. The results were unequivocal: The likelihood that a person born to a schizophrenic parent will be diagnosed with the disease was about one in ten, compared to about 1 in 100 in the general population. If both parents were schizophrenic, nearly half of the offspring were likely to share the same fate.

Had these results been obtained in rats or mice, rather than in humans, only one conclusion would have been drawn: Schizophrenia is genetically determined. But many psychiatrists of the period argued vehemently that the data could just as well be interpreted to mean that schizophrenic

Mounting scientific evidence points to a genetic (chromosomal) basis for schizophrenia. The maze symbolizes being trapped in

parents created an *environment* that fostered the development of the disorder in their children. The debate has important implications. On one side, the argument that schizophrenia has its roots in environmental influences opens the door to changing parental behavior to lessen the risk. On the other side, if the disease is genetically determined, there is the possibility that it results from a simple biochemical imbalance and can be treated with more effective drugs. How could these two interpretations be distinguished?

In the 1960s, Leonard Heston of the University of Oregon and Seymour Kety and his colleagues at Harvard University simultaneously hit on a new approach to the question—the study of adoptees. Adoption provides a "natural experiment" in which the two variables, heredity and environment, can be separated. The natural parents provide the genes, while the adoptive parents (who had no knowledge of the adopted child's parental schizophrenia) provide the environment. If the disease has a major genetic component, adoptees born to a schizophrenic mother should exhibit a high incidence of the illness even in homes free of schizophrenic influences. If genes are not a factor, the incidence of schizophrenia in the adoptees should be no higher than that in the general population.

The results of the researcher's studies were clear-cut: Adopted children whose natural mother was schizophrenic, but who were raised by a nonschizophrenic mother, had as high an incidence of schizophrenia as if they had been reared in the home of their natural parents. Conversely, those adoptees who were born to a nonschizophrenic mother but were raised in a home with a schizophrenic parent had no increased risk of the disease, although they often complained of the "crazy home they were raised in."

These results supported the hypothesis that defective genes cause schizophrenia. But there are more data to consider. If one member of a pair of identical twins is a diagnosed schizophrenic, there is only a 50 percent likelihood that the other twin will also suffer from the "full-blown" disorder (although as many as 85 percent show some schizoid behavior, which is characterized by less severe schizophrenic tendencies). Identical twins have identical genes; if schizophrenia is *strictly* determined by a person's genetic inheritance, both twins should have the identical disorder. There must be more to the story than which genes are acquired from the biological parents. Most mental-health experts believe that our genes provide a strong *predisposition* or vulnerability to the development of psychotic disorders but that environmental factors also affect whether or not the disease develops and its severity. In other words, nature and nurture work together.

Current investigations among biologists are focused on the specific genes that predispose a person to schizophrenia. In 1988, two papers were published in an issue of the journal *Nature*. One report described the use of newly available gene-locating techniques (Chapter 17) to discover a single gene on chromosome number 5, which is associated with schizophrenia in a number of Icelandic and British families. The second paper reported that this particular gene was *not* associated with schizophrenia in members of a large Swedish family suffering from the disease. Is one of these reports mistaken? Probably not. It is more likely the case that defects in several distinct genes can bring on the same symptoms. In other words, schizophrenia may result from different biochemical defects that generate a common disorder. The identification of these genes and the determination of the biochemical processes they control are two of the key areas of interest in clinical molecular biology.

a world of distorted reality, one of the symptoms of schizophrenia.

*I*n 1865, an obscure monk stood before the members of the Natural History Society of Brünn, Austria, and described the results of his 8-year study of **heredity**—the passage of traits from one generation to the next. The group had the rare privilege of being the first to be presented with the key that would unlock the door to one of life's great puzzles. Yet this esteemed audience responded not with questions or other signs of interest but with polite applause. They simply did not understand the significance of what they had just witnessed, which was nothing less than the birth of modern genetics. Unfortunately, the monk who presented his findings that day in 1865 would not receive any recognition for his scientific achievements during his lifetime. Gregor Mendel's work emerged before its time, preceding the discovery of chromosomes, genes, diploidy, and meiosis, all of which would have provided a physical basis for understanding his principles. His experiments were published in the Brünn society's journal in 1865, where they generated no interest whatsoever until 1900. In that year, three different European botanists *independently* reached the same conclusions, and all three rediscovered Mendel's paper, which had been sitting on the shelves of numerous libraries throughout Europe for 35 years.

▼ ▼ ▼

GREGOR MENDEL: THE FATHER OF MODERN GENETICS

Mendel (Figure 12-1) grew up on a small farm, where he learned methods for breeding plants. After twice failing an exam for accreditation as a high-school science teacher, Mendel was sent by his abbot from the monastery where he resided to the University of Vienna for 2 years to study natural science and mathematics. This early training in plant breeding and mathematics was a rare combination that was to prove of remarkable value. When he returned to the monastery, Mendel taught mathematics and, in his spare time, worked in his laboratory—a small garden plot on the grounds of the monastery.

Mendel's contemporaries supported the "blending" view of inheritance, according to which parents produce "hereditary fluids" that mix together to form progeny possessing a mixture of the characteristics of their parents. (The term "blood relative" is a holdover from this misconception.) However, since blending would always produce offspring that were intermediate forms, halfway between each parent, it failed to explain why children are sometimes

FIGURE 12-1
Gregor Mendel. The father of modern genetics.

"chips off the old block," closely resembling one parent but not the other, or why characteristics may skip a generation and reappear in grandchildren.

▐▶ This general ignorance about inheritance had consequences for other areas of biology. For example, one of the weaknesses in Darwin's theory of natural selection (see Chapter 1) stemmed from his inability to describe how favorable variations could be passed on to subsequent generations. Mendel's demolishment of the notion of genetic blending overcame this weakness by revealing how advantageous traits could be preserved in populations over subsequent generations.

GETTING STARTED

We don't know exactly what motivated Mendel to begin his studies, but he evidently had a clear experimental plan in mind: His goal was to mate, or *cross*, pea plants having different inheritable characteristics and to determine the pattern by which these characteristics were transmitted to the offspring. The reasons for Mendel's choice of pea plants are described in Figure 12-2.

Mendel understood the need for controls in properly evaluating the results of his experimental procedures (see Chapter 2). He also understood the value of quantifying (counting) his results so that he could determine whether his findings were significant or just the products of random chance. Other investigators had approached the problem of inheritance by examining the multitude of traits in whole organisms. Such an unmanageable number of variables prevented any possibility of sorting out the principles governing inheritance. Consequently, Mendel chose to focus

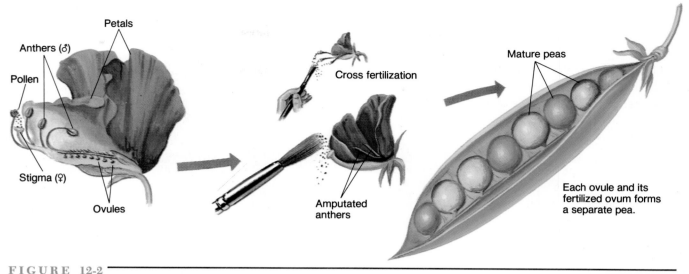

FIGURE 12-2

Reproduction in Mendel's pea plants. Mendel's choice of pea plants was strategic. He had two opportunities: (1) He could self-fertilize his plants by allowing pollen to fall from the *anthers* (the male reproductive organs) onto the *stigma* (the sticky area to which pollen adheres), and (2) he could prevent self-fertilization by amputating the pollen-producing anthers. These pollenless flowers were then cross-fertilized by brushing on pollen from another plant (pollination and fertilization are discussed in Chapter 20). In this way, Mendel was able to control which plants mated with each other. With accidental matings eliminated, Mendel was able to correlate parental traits with those of the progeny. Each pea seen in the pod is the product of a separate fertilization event (which is why both green and yellow peas can share the same pod). Seeds develop only following fertilization, so results are not complicated by "fatherless" offspring. Seeds are readily collected, grown into mature plants, and the inherited characteristics are determined. All of these properties contributed to Mendel's scientific success.

on seven clearly definable traits, each of which occurred in one of two characteristic forms (Table 12-1; the dominant or recessive nature of a characteristic is discussed in the next section).

First, Mendel had an important control experiment to perform: He had to be certain that each type of plant he was studying would *breed true*—that is, consistently produce the same characteristic—when the plants were self-ferti-

TABLE 12-1

SEVEN TRAITS OF MENDEL'S PEA PLANTS

Trait	Dominant Allele	Recessive Allele
Height	Tall	Dwarf
Seed color	Yellow	Green
Seed shape	Round	Angular (wrinkled)
Flower color	Purple	White
Flower position	Along stem	At stem tips
Pod color	Green	Yellow
Pod shape	Inflated	Constricted

lized. It was essential, for example, to determine that a plant from the original stocks that produced round seeds could not possibly transmit a wrinkled-seed characteristic.

THE FIRST GENERATION OF HYBRIDS

Once Mendel was satisfied that each of the original stocks bred true when self-fertilized, he was ready to cross-fertilize plants possessing different characteristics (as discussed in the legend of Figure 12-2) and to examine the offspring, called *hybrids* because they are the product of the differences between the parents. In the beginning, he limited his observations to a single trait. These single-trait experiments are called **monohybrid crosses.** He repeated this experiment for each of the seven traits. To simplify our discussion, we will focus on the results of just one set of crosses, that of seed shape.

When plants that produced wrinkled seeds were cross-fertilized with plants that produced round seeds, Mendel found that all the plants of the next generation had round seeds (Figure 12-3). Mendel called the original parental generation the **P generation,** and he called the first generation of offspring the **F$_1$ generation** (F$_1$ = first filial). Mendel referred to the characteristic that appeared in the

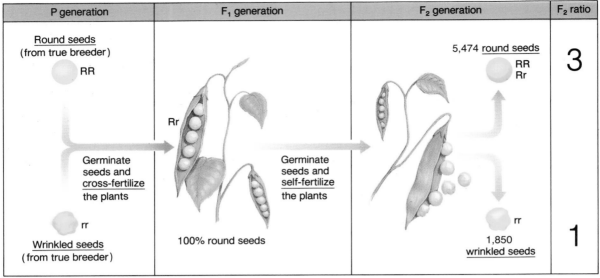

P generation	F₁ generation	F₂ generation	F₂ ratio

Round seeds
(from true breeder)
RR

Rr

Germinate seeds and cross-fertilize the plants

Germinate seeds and self-fertilize the plants

rr
Wrinkled seeds
(from true breeder)

100% round seeds

5,474 round seeds
RR
Rr

rr
1,850
wrinkled seeds

3

1

R = allele for round seeds
r = allele for wrinkled seeds

FIGURE 12-3

Mendel conducted monohybrid crosses to track the course of a single trait (in this case seed shape) through two generations.   In this experiment, all plants in the F₁ generation had round seeds, while the F₂ generation included plants with both round and wrinkled seeds in a 3 : 1 ratio. He found a similar pattern of inheritance for each of the seven traits. From this pattern, Mendel concluded that there are paired genetic factors, one for each alternate characteristic of the trait (shown here as *R* and *r*), which segregate during gamete formation. The factor in one gamete pairs up again with another factor from another gamete during fertilization. (As discussed in the following section of the text, genetic characteristics, such as round and wrinkled seeds, are determined by alternate versions of a gene, called *alleles*. In this case, the alleles are *R* and *r*.)

hybrids (in this case, round seeds) as the **dominant** characteristic, and the characteristic that had seemingly been lost (in this case, wrinkled seeds) as the **recessive** characteristic.

The following year, Mendel allowed the F₁ hybrids to self-fertilize. He then observed the nature of the subsequent offspring, which he called the **F₂ generation.** To his surprise, he found that the wrinkled characteristic that had previously disappeared in the F₁ generation had returned in a number of F₂ offspring! In fact, there were approximately three times as many F₂ plants with round seeds as with wrinkled ones (Figure 12-3). This same approximate 3 : 1 ratio was found for each of the seven traits he examined. Mendel concluded that hereditary patterns were not determined randomly but followed an orderly, predictable pattern. He was one of the first scientists to demonstrate that biological processes occur with a high degree of order; otherwise, the transmission of inherited characteristics would not have produced such consistent ratios.

INTERPRETING THE RESULTS AND FORMULATING A HYPOTHESIS

Even though each of the F₁ hybrid plants *appeared* to be identical to one of its original parents, it must have been different in some way. How else could Mendel explain the fact that the original parents with round seeds, when self-fertilized, produced only plants with round seeds, while the F₁ plants produced offspring with both round and wrinkled seeds? Furthermore, how could the wrinkled-seed characteristic disappear in the F₁ generation and then reemerge in the F₂ offspring? If sexual reproduction resulted in the blending of hereditary fluids, as was commonly believed, how could a genetic characteristic that was blended away possibly reappear?

Mendel concluded that the potential to produce the wrinkled characteristic was never actually lost but only hidden. He proposed that hereditary traits were governed by particulate factors (or *units*) of inheritance (later termed *genes*) which remained intact from generation to generation. Each trait in a plant was determined by the presence of two independent factors, one derived from each parent. The two factors could be of an identical or a nonidentical nature. Later, the term **allele** was used to refer to alternate forms of the same gene.

For each of the seven traits, Mendel found one of the two alleles to be dominant over the other, recessive allele. In the case of seed shape, the allele for round seeds is dominant over the allele for wrinkled seeds. Dominant alleles are usually represented by capital letters, recessive

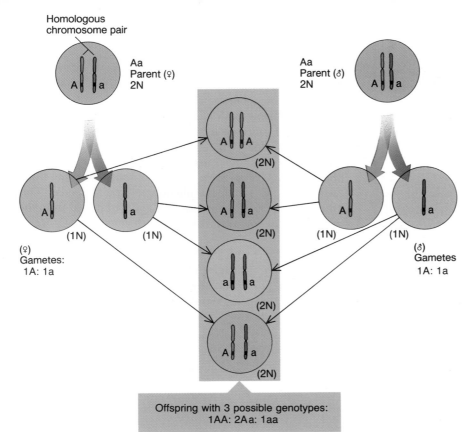

FIGURE 12-4

Meiosis explains Mendel's findings. Because they occur on separate homologous chromosomes, alleles (*A* and *a*) segregate from each other during meiosis to form gametes. Each gamete has an equal probability of receiving an *A* or an *a*. Fertilization randomly reestablishes the diploid state, so each allele has an equal chance of joining with either of the others. Each offspring will therefore acquire one of the indicated genotypes. It will be twice as likely to be an *Aa* genotype than either an *AA* or *aa*.

alleles by the same letter in lowercase. For example, the two alleles for seed shape are denoted by the symbols *R* (round) and *r* (wrinkled). Thus, the F$_1$ plants had round seeds (rather than some intermediate, blended shape) because the *R* allele was dominant over the *r* allele.

The original, true-breeding parents had two identical forms of the gene for each trait; the parental plants with round seeds were designated *RR*, and those with wrinkled seeds were labeled *rr*. When both alleles for a trait are the same, the individual is described as **homozygous** (*homo* = same) for that trait. In contrast to their parents, the F$_1$ plants had only one *R* and one *r* allele; consequently, they were designated *Rr*. When two different alleles for a trait are present, as in this case, the individual is described as **heterozygous** (*hetero* = different) for that trait.

In modern terminology we can say that the parental plants with round seeds and the F$_1$ hybrids had the same **phenotype**—physical appearance or measurable quality —but different **genotypes**—the underlying genetic makeup. Simply put, the genotype determines the phenotype. The genotype of the parent with round seeds was *RR*, and that of the F$_1$ was *Rr*. Mendel concluded that any plant exhibiting a wrinkled-seed phenotype must be homozygous for the recessive allele; that is, possessing a genotype of *rr*.

Mendel's Law of Segregation

Mendel concluded that each reproductive cell (gamete) could carry only one allele for each trait. Somehow, during formation of the gametes, each allele separated (*segregated*) in an orderly manner from its partner, the two always

going to different reproductive cells. Predictably, half the gametes produced by each F$_1$ plant carried the *R* allele, and the other half carried the *r* allele. This phenomenon is now referred to as Mendel's **law of segregation.** Validation of Mendel's "first law" came years after his death, with the discovery of homologous chromosomes and their separation during meiosis. The first meiotic division (page 216) is the physical event that brings about the ordered segregation of alleles during gamete formation and generates a predictable pattern of inheritance (Figure 12-4).

PREDICTING PATTERNS OF INHERITANCE

As we mentioned earlier, Mendel obtained an F$_2$ generation by self-fertilizing heterozygous F$_1$ plants. During meiosis in a heterozygote, each gamete has an equal chance of receiving either of the alleles for a particular trait. Then, during self-fertilization (or fertilization with a gamete from another heterozygous individual), each allele has an equal chance of being teamed up with either the same type or the alternative type of allele. Thus, if an *Rr* F$_1$ plant self-fertilizes, the random combination of *R*- and *r*-bearing gametes creates three possible genotypes in the F$_2$ generation: homozygous dominant (*RR*), heterozygous (*Rr*), and homozygous recessive (*rr*).

One way to visualize the results of such crosses is to use the **Punnett square** method, named after the British poultry geneticist, R. C. Punnet, who first used the method in the early 1900s. Punnett squares predict the possible genotypes and phenotypes (and their expected ratios) when there is an equal chance of acquiring either of the two alleles. A Punnett square of each of Mendel's monohybrid crosses predicts a 3 : 1 ratio of dominant to recessive phenotypes, just as Mendel had observed (Figure 12-5).

The ratios of genotypes and phenotypes predicted by Punnett squares are a matter of *probability;* they are validated by experiments only if the number of offspring is large enough to eliminate chance variations. The same is true for any random combination of events. For example, if you flipped a coin three times, and it happened to land on "heads" all three times, the results could be attributed to chance variation. In other words, three flips are not enough to conclude that the probability of heads is 100 percent. But if you flipped the coin 100 times, the likelihood of its landing on heads every time would be very, very small. Instead, the frequency of obtaining heads would likely be close to 50 percent. In genetics, the larger the number of offspring examined, the closer the results should approach the predicted ratio.

Punnett squares are convenient for determining the genotypes of the offspring when only one or two traits are involved, but they soon become unwieldy when genetic crosses involve many traits. Although Punnett squares are valuable because they offer a visual portrait of gamete combinations, it is easier to reach the same predictions by simple multiplication. Keep in mind that *the probability of two independent events occurring together is the product of the probability of each occurring alone.* In the case of garden peas, the chance of a gamete acquiring the recessive allele for wrinkled seeds from a heterozygous parent is one in two, or 1/2. In a cross between heterozygotes the probability of a zygote receiving both alleles for wrinkled seeds is one in four, or 1/4, since each gamete has a probability of 1/2 for having this allele (1/2 × 1/2 = 1/4). Thus, on the average, one of every four offspring of two heterozygous parents will be homozygous recessive (wrinkled seeds); the other three will show the dominant phenotype (round seeds). This brings us back to the 3 : 1 ratio Mendel obtained from the thousands of genetic crosses he performed.

Hereditary Patterns in Humans

Since the principles of dominance and segregation of alleles apply to all sexually reproducing diploid organisms, not just garden peas, Mendel's findings can help you explain hereditary patterns in humans. A dimple in the chin (Figure 12-6*a*), a "widow's peak" hairline (Figure 12-6*b*), achondroplastic dwarfness, webbed fingers, and the absence of fingerprints are all examples of traits determined primarily by dominant alleles (though other factors may also play a role). A person with just *one* of these dominant alleles will develop the corresponding characteristic, regardless of the second allele. A child of parents both of whom are heterozygous for one of these traits has a 3 in 4 probability (75 percent chance) of exhibiting the same dominant phenotype as that found in the parents. This does not mean that individuals with a dimple in their chin are three times more numerous than persons without them, however. In fact, for

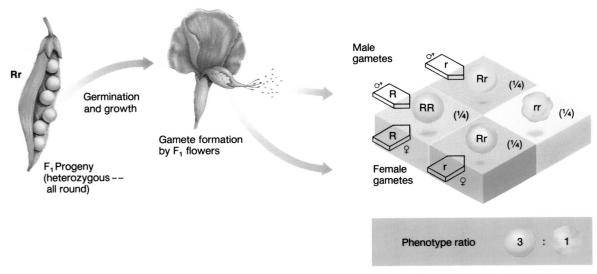

FIGURE 12-5

The Punnett square method of determining probability ratios. Following self-fertilization of a heterozygous pea plant, each F₁ gamete will contain either an *R* (round) or *r* (wrinkled) allele. In the Punnett square representation, all possible male gametes are listed along the top, and all female gametes along the side. The possible combinations of alleles following fertilization are shown in the boxes, each box representing the genotype created by the union of these two alleles. The genotype ratio is 1 : 2 : 1 *(RR : Rr : rr)*, meaning that there are twice as many *Rr* boxes as either *RR* or *rr* boxes. The 3 : 1 phenotype ratio explains why recessive (wrinkled) characteristics reappeared in one-fourth of Mendel's F₂ progeny after disappearing from the F₁ generation.

(a)   (b)

FIGURE 12-6

Dominant alleles. The dimple in Cary Grant's chin *(a)* and Marilyn Monroe's widow's peak hairline *(b)* result from dominant alleles.

some traits, the recessive phenotype is far more common than is the dominant phenotype. After all, most of us *do* have fingerprints, even though it is a recessive condition. In this case, the dominant allele is so rare that nearly everyone is homozygous recessive, and therefore has a complete set of fingerprints.

The Test Cross

As we saw earlier, outward appearance of a dominant phenotype cannot reveal whether an individual is homozygous or heterozygous for that trait. A pea plant with purple flowers, for example, may be homozygous or heterozygous for the trait of flower color. Mendel concluded that if his concept of dominant and recessive factors were correct, then it should be possible to determine genotype by crossing an individual in question with a mate that shows the recessive trait. This strategy is known as a **test cross,** and it reveals the presence of a "hidden" recessive allele if the individual is heterozygous for that trait, as explained in Figure 12-7.

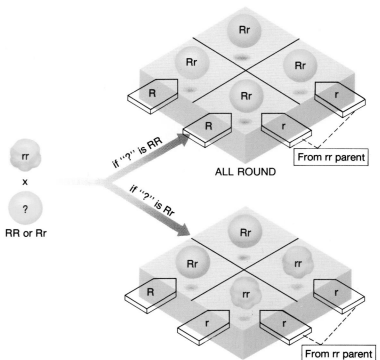

FIGURE 12-7

The test cross allows us to determine the genotype (*RR* or *Rr*) of an organism that shows the dominant characteristic. When crossed with an individual that shows the recessive characteristic (genotype of *rr*), a homozygous dominant (*RR*) individual will only yield progeny with the dominant phenotype, since all the offspring will have a dominant allele. In contrast, if the test organism is a heterozygous individual (*Rr*), about half the progeny will possess the recessive phenotype.

CROSSING PLANTS THAT DIFFER BY MORE THAN ONE TRAIT: MENDEL'S LAW OF INDEPENDENT ASSORTMENT

Once Mendel understood how a single trait was inherited, he was ready to examine the transmission of two traits simultaneously. To do this, he performed a series of **dihybrid crosses**—crosses between plants that differed from one another in two distinct traits, such as seed shape and seed color. Mendel crossed true-breeding plants having round, yellow seeds (genotype of *RRYY*) with plants having wrinkled, green seeds (*rryy*). As expected, the seeds of the F_1 offspring showed only the dominant characteristics (Figure 12-8); that is, seeds of a round, yellow phenotype (*RrYy* genotype). When he allowed these F_1 hybrids to self-fertilize, however, four different types of seeds were found among the next generation. Most of the seeds (approxi-

mately 9 out of 16) were round and yellow, while approximately 1 out of 16 were wrinkled and green. In addition, two types of seeds had a combination of traits not seen in either parent—round, green seeds and wrinkled, yellow seeds (both represented by approximately 3/16 of the population). Thus, the ratio of these four phenotypes was approximately 9:3:3:1.

In order to account for these results, Mendel made a final conclusion: The segregation of the pair of alleles for one trait had no effect on the segregation of the pair of alleles for another trait; that is, alleles for different traits segregate *independently* of one another. Just because an F_1 individual inherited one *R* and one *Y* allele from one gamete did not mean that those two alleles must remain together when that F_1 plant formed its own gametes. Rather, either of the alleles for seed shape could team up with either of the alleles for seed color (Figure 12-8). This would produce

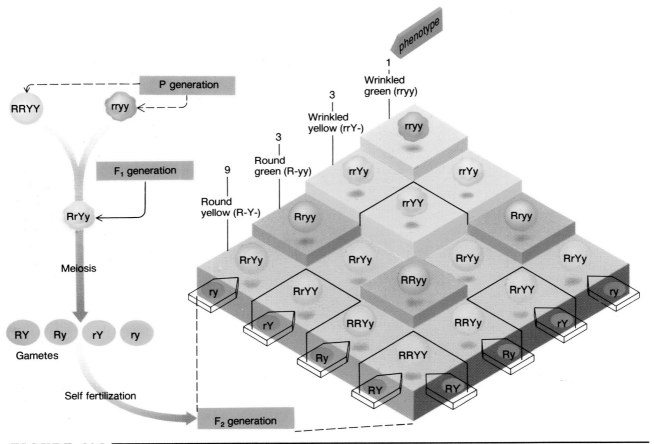

FIGURE 12-8

Mendel's dihybrid crosses. When plants that breed true for round, yellow (*RRYY*) seeds are crossed with those producing wrinkled, green (*rryy*) seeds, the F_1 progeny are heterozygous (*RrYy*) for both seed shape and color. Independent assortment allows either *R* or *r* to combine with either *Y* or *y*, generating gametes with four different possible combinations of alleles (*RY, Ry, rY, ry*). Self-fertilization of an F_1 plant produces an F_2 generation with the nine genotypes and four phenotypes shown in the Punnett square. When Mendel performed this experiment with hundreds of peas, he obtained ratios very close to the 9:3:3:1 phenotype pattern predicted here. Mendel also determined the genotypes of the F_2 progeny by allowing them to self-fertilize and then examining their offspring. The F_2 generation consisted of all nine genotypes in a ratio that approximated the predicted 1:2:1:2:4:2:1:2:1.

gametes with four possible combinations (*RY, Ry, rY, ry*), each in equal proportion. This conclusion is referred to as Mendel's **law of independent assortment** and, as we saw in the previous chapter, derives directly from the fact that homologous chromosomes become aligned independently in the metaphase plate during meiosis I.

Because the two traits are independent of each other, the probability of acquiring any phenotypic combination of both can be calculated by multiplying the probability of inheriting each phenotype alone:

Probability of Acquiring Each Characteristic

Round seeds = 3/4   Yellow seeds = 3/4
Wrinkled seeds = 1/4   Green seeds = 1/4

Probability of Combination of Characteristics

Round, yellow seeds = 3/4 × 3/4 = 9/16
Round, green seeds = 3/4 × 1/4 = 3/16
Wrinkled, yellow seeds = 1/4 × 3/4 = 3/16
Wrinkled, green seeds = 1/4 × 1/4 = 1/16

Through the use of carefully designed experiments, meticulous gathering of data, and insightful interpretation, Mendel discovered the "rules" by which genetic characteristics are transmitted. He correctly formulated why the patterns of transmission were observed and how they could be applied to predict the genetic outcome of particular matings. (See CTQ #3.)

MENDEL AMENDED

Even though Mendel's "laws" provided the foundation on which modern genetics was built, several exceptions to Mendel's principles were soon discovered as the science of genetics matured. For example, not all traits have two alternative forms of appearance or display the simple dominant–recessive relationship exhibited by the seven traits examined by Mendel. Furthermore, just as Mendelian patterns of inheritance are found in all types of sexually reproducing eukaryotic organisms, the same types of non-Mendelian patterns also appear in diverse organisms. The reason for the nearly universal application of these principles is simple: Regardless of the organism, genetic information is encoded in the same types of material and is expressed by the same types of molecular processes. This will become more apparent in the following section.

INCOMPLETE DOMINANCE AND CODOMINANCE

At first glance, some traits seem to follow the notion of "blending" rather than Mendelian inheritance. For example, when red-flowered snapdragons are crossed with white-flowered snapdragons, all of the progeny are pink, as though the traits had blended. But the next generation of plants clearly demonstrates that blending has *not* occurred.

Self-pollination of pink snapdragons produces an interesting mix of offspring—50 percent pink, 25 percent red, and 25 percent white. If blending had occurred, red and white varieties could never have been recovered from pink parents. How does this happen?

The answer is evident if we first introduce the concept of "gene products." Genes are not abstract factors that pass from parents to offspring; rather, they are portions of chromosomes that dictate the formation of specific proteins. This relationship between genes and proteins was first glimpsed in 1908 by a British physician, Archibald Garrod, who found that individuals suffering from the genetic disease *alcaptonuria* were missing a key enzyme essential for metabolizing certain amino acids. The defect, which does not result in serious health problems, is detected because the urine of the affected individual turns black upon exposure to air. This curious symptom results from the oxidation of a compound that is absent in the urine of normal individuals. In one of biology's most farsighted proposals, Garrod suggested that genes carried information directing the manufacture of specific enzymes.

Although Garrod's proposal proved to be true, it wasn't until biologists turned to less conspicuous organisms—including fruit flies and bacteria—that the universality of Garrod's conclusions came to be accepted. We can illustrate Garrod's principle using one of the traits studied by Mendel. The gene product for the purple flower allele in pea plants is an enzyme that catalyzes the formation of a purple pigment. The allele for white flowers is a variant of this gene which directs the formation of an inactive enzyme. This explains why purple is dominant over white in garden pea flowers: As long as there is one purple flower allele in the cells, enough pigment will be produced to turn the flower purple. The white color appears only in the homozygous recessive configuration because of the flower's inability to produce the purple pigment. Many genes control their traits in such a fashion; the recessive allele is simply unable to direct formation of an active gene product.

The case of snapdragons is a bit different, however. The red allele directs the formation of an active enzyme that produces red pigment, while the white allele cannot generate an active enzyme. Therefore, when two copies of the red allele are present, enough enzyme is manufactured to produce a true red phenotype. When only one red allele is present, however, only enough enzyme is manufactured to produce a paler (pink) color. This phenomenon is known as **incomplete** (or **partial**) **dominance.** Another example of this phenomenon is found in people who are heterozygous for the gene that causes familial hypercholesterolemia (page 146). Heterozygotes for this gene have a markedly elevated blood-cholesterol level, but one not nearly as high as that of homozygotes.

Heterozygous individuals are also distinguishable in cases of **codominance,** but, in these cases, the phenotype is not an intermediate (diluted) form. Rather, codominance means that both alleles are expressed simultaneously and are unmodified in the heterozygous individual. A classic example of codominance is seen in the ABO blood group antigens—proteins that are found on the surface of red blood cells and are used to determine human blood type (page 94). The codominant alleles for A and B, written as I^A and I^B, direct the blood cell to manufacture the A antigen and the B antigen, respectively, giving the individual type AB blood. How, then, do people acquire type O blood? The answer lies in multiple alleles.

MULTIPLE ALLELES

The human ABO blood group system can be used to illustrate another departure from classical Mendelian genetics —the existence of more than two possible alleles for a single trait. Although a single diploid individual can possess only two alleles, the existence of additional alleles increases the number of possible genotypes in the population. There are three different alleles in the ABO system, for example: I^A, I^B, and I^O. The I^O allele is a variant that, like the allele for white flowers in pea plants, directs the formation of a defective protein. A homozygous person with two I^O alleles would lack both A and B antigens. These individuals are blood type O. An allele such as I^O is recessive to both of the alternative alleles, which is sometimes why it is written as "*i*." Therefore, an $I^A I^O$ heterozygous pair would produce the same phenotype as would the homozygous $I^A I^A$—that is, a person with type A blood. (These atypical genetic symbols were coined many years ago by a pioneering Austrian physician—Karl Landsteiner—and remain in use today.)

COMBINED EFFECTS OF GENES AT DIFFERENT LOCI

Not only did Mendel's seven pairs of alleles segregate independently, they also *functioned* independently of one another. In many cases, however, the expression of a pair of alleles is *influenced* by the genotype at other *loci* (plural of

locus), or sites, on the chromosomes. This relationship is illustrated by human pigmentation. Human beings have separate genes for hair color, eye color, and skin color. Let's consider a person who has a genotype that dictates brown hair, brown eyes, and a relatively dark complexion. All of these phenotypes require the production of a dark pigment (known as *melanin*) which, in turn, requires the activity of a certain enzyme encoded by a gene at the *A* locus. If a person carries two alleles *(aa)* that produce an *inactive* enzyme, their hair, eyes, and skin will be devoid of color, regardless of the genotypes at the hair, eye, and skin color loci. Individuals who acquire this pair of recessive alleles *(aa)* are called *albinos*. Albinism (Figure 12-9) is an example of **epistasis,** a process whereby a pair of recessive alleles at one locus masks the effect of genes at other loci.

PLEIOTROPY

The product of one gene can have far-reaching secondary effects on many diverse characteristics, a condition called **pleiotropy.** The inherited disease *cystic fibrosis* is one example of this phenomenon. The diverse (and serious) symptoms of this disease result from the production of an abnormal membrane protein that causes thickened mucus, which impairs the functioning of the digestive, sweat, and respiratory mucous glands. This, in turn, produces widespread disorder within the body. For example, the digestive enzymes are poorly secreted, causing improper digestion and absorption of food; the sweat contains abnormally high levels of salt; and the mucus that lines the airways is too thick to be propelled out of the lower respiratory tract by its ciliated lining. This viscous mucus, laden with trapped microbes, settles into the lungs. As a result, pneumonia and other lung infections are the most frequent causes of death in those afflicted with cystic fibrosis (Chapter 17).

POLYGENIC INHERITANCE

If human height were determined by simple dominance at a single locus, people would come in two sizes—tall and short (dwarf)—much like the pea plants studied by Mendel. Height in peas is described as **discontinuous** because the phenotypes fall into two distinct categories. But many traits, such as human height and skin coloration, show **continuous variation** within the population. Such traits are determined by a number of different genes present at different loci, rather than by a single gene. This phenomenon is known as **polygenic inheritance** (Figure 12-10). The overall expression of the polygenic trait of skin color represents the sum of the influences of many contributing genes. Two people who differ in just one pair of alleles show only slight color differences, for example, as compared to two people who differ in several of the polygenic determinants.

Many of the traits currently being studied by geneticists are polygenic. These include animal and plant traits selected by breeders for improving livestock and crops, as

FIGURE 12-9

Albinism occurs throughout the vertebrates, including these crocodilians.

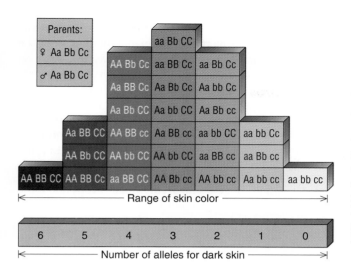

FIGURE 12-10
Polygenic inheritance and skin color. People don't come in two contrasting colors: dark and light. Most of us lie somewhere between these extremes. There are, at least, three genes—*A, B, C*—located at three different loci, that determine the amount of pigment in human skin. At each locus, there are two possible alleles of the gene, one for maximum pigment (the capital letter), and one for no pigment (the lowercase letter). This figure shows the 27 genotypes and 7 phenotypes that could be produced by two people who are heterozygous for all three genes (*AaBbCc*). The 27 genotypes are indicated within each of the boxes; the 7 different phenotypes are indicated by the 7 columns indicating different shades of darkness. The darkness of skin is determined by the total number of alleles for dark pigment (capital letters) in the genotype. The darkest phenotype would have all 6 genes represented by the dominant allele, while the lightest would have all six genes represented by the recessive allele.

well as such elusive human characteristics as susceptibility to cardiovascular disease and athletic prowess. Polygenic inheritance poses particular challenges to geneticists because the outcome of an experimental cross is the sum of many genotypic variables.

ENVIRONMENTAL INFLUENCES

Nearly all phenotypes are subject to modification by the environment. A genetically tall person, for example, will not achieve his or her maximum height potential if deprived of adequate nutrition. Furthermore, two organisms with the same genotype may have different phenotypes in different environments. For example, a light-skinned person living in a sunny, warm climate is likely to be considerably darker skinned than would a genetically similar individual living in a cold, overcast climate.

Environmental alteration of phenotype can affect single-gene traits as well as polygenic traits. An excellent example is the two-color pattern of the Siamese cat (Figure 12-11). This type of cat has only one gene for hair color, and the genotype (c^sc^s) at this locus is the same in all of the cat's cells, including those that produce light fur and those that make dark fur. The *siamese* allele (c^s) directs the synthesis of a temperature-sensitive variant of an enzyme that manufactures the dark pigment. This enzyme is operative only on the cooler areas of the body (feet, snout, tip of the tail, and ears). The rest of the cat's body is warm enough to inactivate the enzyme, causing the fur to remain light.

Each of the traits studied by Mendel was determined by two alternate alleles, one of which was dominant over the other. Following the rediscovery of Mendel's work, geneticists learned that not all traits followed such simple patterns of inheritance. The basis of these non-Mendelian patterns can be understood when interpreted in the light of more recent research on the nature of gene products. (See CTQ #4.)

MUTATION: A CHANGE IN THE GENETIC STATUS QUO

▶ It is fitting to end this chapter with the subject of mutation because, in a sense, we have been dealing with it all along. **Mutations** are rare but permanent changes in a gene that may alter a gene's product. Thus far, we have been describing the inheritance of alleles—alternate forms of a gene. Alleles arise by mutations in an existing gene. The original allele is termed the *wild type*, to distinguish it from the newer *mutant* form. If there were no mutations, there would be no new alleles, *and there would be no biological evolution.* This may seem overly dramatic, so let us explore it a bit further.

FIGURE 12-11
A Siamese cat provides a visual portrayal of the influence of environment on phenotype.

The segregation of alleles and their independent assortment during meiosis increase genotype diversity by promoting new combinations of genes. But the shuffling of existing genes alone does not explain the presence of such a vast diversity of life. If all organisms descended from a common ancestor, with its relatively small complement of genes, where did all the genes present in today's millions of species come from? The answer is mutation.

Most mutant alleles are detrimental; that is, they are more likely to disrupt a well-ordered, smoothly functioning organism than to increase the organism's fitness. For example, a mutation might change a gene so that it produces an inactive enzyme needed for a critical life function. Occasionally, however, one of these stable genetic changes creates an advantageous characteristic that increases the fitness of the offspring. In this way, mutation provides the raw material for evolution and the diversification of life on earth.

One of the requirements for genes is stability; genes must remain basically the same from generation to generation or the fitness of organisms would rapidly deteriorate. At the same time, there must be some capacity for genes to change; otherwise, there would be no potential for evolution. Alterations in genes do occur, albeit rarely, and these changes (mutations) represent the raw material of evolution. (See CTQ #7.)

REEXAMINING THE THEMES

Biological Order, Regulation, and Homeostasis

Mendel discovered that the transmission of genetic factors followed a predictable pattern, indicating that the processes responsible for the formation of gametes, including the segregation of alleles, must occur in a highly ordered manner. This orderly pattern can be traced to the process of meiosis and the precision with which homologous chromosomes are separated during the first meiotic division. Mendel's discovery of independent assortment can also be connected with the first meiotic division, when each pair of homologous chromosomes becomes aligned at the metaphase plate in a manner that is independent of other pairs of homologues.

Unity within Diversity

All eukaryotic, sexually reproducing organisms follow the same "rules" for transmitting inherited traits. Although Mendel chose to work with peas, he could have come to the same conclusions had he studied fruit flies or mice or had he scrutinized a family's medical records on the transmission of certain genetic diseases, such as cystic fibrosis. Although the mechanism by which genes are transmitted is universal, the genes themselves are highly diverse from one organism to the next. It is this genetic difference among species that forms the very basis of biological diversity

Evolution and Adaptation

Mendel's findings provided a critical link in our knowledge of the mechanism of evolution. A key tenet in the theory of evolution is that favorable genetic variations increase the likelihood that an individual will survive to reproductive age and that its offspring will exhibit these same favorable characteristics. Mendel's demonstration that units of inheritance pass from parents to offspring without being blended revealed the means by which advantageous traits could be preserved in a species over many generations. The subsequent discovery of genetic change by mutation revealed how new genes appeared in a population, thus providing the raw material for evolution.

SYNOPSIS

Gregor Mendel discovered the pattern by which inherited traits are transmitted from parents to offspring. Mendel discovered that inherited traits were controlled by pairs of factors (genes). The two factors for a given trait in an individual could be identical (homozygous) or different (heterozygous). In heterozygotes, one of the gene variants (alleles) may be dominant over the other, recessive allele. Because of dominance, the appearance (phenotype) of the heterozygote (genotype of *Aa*) is identical to that of the homozygote with two dominant alleles

(*AA*). An individual must possess two recessive alleles (*aa*) to exhibit the recessive phenotype. Although a particular recessive characteristic may disappear in heterozygotes for a generation or more, it can eventually reappear because the allele for that characteristic remains intact as it is passed from parents to offspring. Monohybrid crosses between heterozygous parents produce three times as many off-spring with the dominant characteristic as with the recessive characteristic.

Mendel's law of segregation explained the results of monohybrid crosses. Paired alleles segregate from each other during formation of the gametes. Each gamete receives only one of the two alleles, and a gamete has an equal chance of receiving either allele.

Mendel's law of independent assortment explained the results of dihybrid crosses. Dihybrid crosses between heterozygotes produce plants with four phenotypes, two of which were present among the parental plants. The ratios of the four phenotypes are approximately 9 : 3 : 3 : 1, which is exactly what would be expected if the alleles of the two genes are combining in a random manner, as predicted by simple probability equations (or by the Punnett square method). These findings led Mendel to conclude that the segregation of the pair of alleles for one trait had no effect on the segregation of the pair of alleles for another trait.

Not all inherited traits are transmitted according to Mendelian patterns. (1) Some alleles show incomplete dominance, whereby the pair of alleles "dilute" each other's effect in heterozygous individuals. (2) Both codominant alleles are fully expressed in heterozygous individuals. (3) Multiple (more than two) alleles for a particular trait can combine to form more than three genotypes and two phenotypes. (4) Alleles present at one locus can effect the expression of genes at a different locus. (5) A single gene may be pleiotropic, meaning that it can have many effects. (6) One trait may be due to multiple genes at several loci; this is known as polygenic inheritance. (7) Tracking the fate of alleles from generation to generation can be complicated by the effect of environment on phenotype. (8) New alleles can emerge by mutations—changes in the genetic message. Such changes have produced a rich diversity of living organisms.

Key Terms

heredity (p. 232)
monohybrid cross (p. 233)
P generation (p. 233)
F_1 generation (p. 233)
dominant (p. 234)
recessive (p. 234)
F_2 generation (p. 234)
allele (p. 234)
homozygous (p. 235)

heterozygous (p. 235)
phenotype (p. 235)
genotype (p. 235)
law of segregation (p. 235)
Punnet square (p. 235)
test cross (p. 237)
dihybrid cross (p. 238)
law of independent assortment (p. 239)

incomplete (partial) dominance (p. 239)
codominance (p. 239)
epistasis (p. 240)
pleiotropy (p. 240)
discontinuous variation (p. 240)
continuous variation (p. 240)
polygenic inheritance (p. 240)
mutation (p. 241)

Review Questions

1. Why can't the law of independent assortment be discussed without an initial understanding of the law of segregation?

2. How do Mendel's findings explain how a recessive genetic disorder, such as sickle cell anemia, may skip over one, or even two, generations?

3. How does each of the following increase genetic diversity: (a) meiosis, (b) fertilization, (c) mutation?

4. If a person has very light hair and skin but dark eyes, what might you conclude about the genotype of his or her melanism locus (*A*)?

Critical Thinking Questions

1. Do any of the non-Mendelian patterns of inheritance discussed in the chapter provide a useful hypothesis for understanding the fact that one identical twin may not display the same magnitude of disease as his or her twin? Explain your answer.

2. Why was it necessary for Mendel to begin his work by self-fertilizing his original stocks of pea plants? How would Mendel's conclusions have differed had he found that one of his parental plants produced offspring with two different characteristics? What might you conclude about the genotype of this particular plant? How might this genotype have arisen?

3. In peas, yellow seed color is dominant over green. The chart below shows the results of crossing pea plants of known phenotypes. Using the letters Y for yellow and y for green, identify the genotypes of each parent plant.

	Offspring	
Parents	Yellow Seeds	Green Seeds
yellow × green	104	100
yellow × yellow	156	54
green × green	0	112
yellow × green	92	0
yellow × yellow	89	0

4. Do the non-Mendelian patterns of inheritance, such as incomplete and codominance, multiple alleles, pleiotropy, and polygeny negate Mendel's Laws of Inheritance? How did Mendel's laws contribute to biologists' understanding of non-Mendelian patterns? Which of the non-Mendelian patterns would have been most helpful to Mendel and others had Mendel found it in his pea plants?

5. Design a mating experiment that would demonstrate that incomplete dominance is not evidence of genetic blending. Explain how your proposed experiment proves that genes remain intact, even in organisms with the intermediate phenotype.

6. Suppose Mendel had made a trihybrid cross between a pea plant that was tall with purple flowers and green seeds (genotype *TTPpyy*) with a pea plant that was tall with white flowers and green seeds (genotype *Ttppyy*). What fraction of the progeny would have the genotype *TT*? What fraction would have white flowers? What fraction would have the phenotype of dwarf with purple flowers and green seeds? What would be the ratio of genotypes if the parental plant with purple flowers was test-crossed? (Try to answer these questions without using a Punnett square.)

7. In his theory of evolution by natural selection, Darwin assumed that individuals in a population vary, but he could not explain the origin of those variations because the principles of genetics had not yet been discovered. If Darwin were to appear to you today, how would you explain the origin of variations and the relationship of mutations to evolution?

Additional Readings

Brennan, J. R. 1985. *Patterns of human heredity: An introduction to human genetics.* Englewood Cliffs, NJ: Prentice-Hall. (Intermediate)

Carlson, E. A. 1973. *The gene: A critical history,* 2d ed. Philadelphia: Saunders. (Intermediate-Advanced)

Gardner, E. J., M. J. Simmons, and D. P. Snustad. 1991. *Principles of genetics,* 8th ed. New York: Wiley. (Intermediate-Advanced).

Mendel, G. 1866. *Experiments in plant hybridization* (English translation). Cambridge: Harvard University Press (1951). (Intermediate)

Suzuki, D. T., et al., 1989. *Introduction to genetic analysis,* 4th ed. New York: Freeman. (Intermediate-Advanced)

Wender, P. H., and D. F. Klein. 1981. *Mind, mood, and medicine.* New York: Farrar, Straus & Giroux. (Introductory)

Young, P. 1988. *Schizophrenia.* New York: Chelsea House. (Introductory)

Genes and Chromosomes

STEPS
TO
DISCOVERY
The Relationship Between Genes and Chromosomes

THE CONCEPT OF LINKAGE GROUPS

LESSONS FROM FRUIT FLIES

Genetic Markers along the Chromosome

Crossing Over: Gene Exchange

Mapping the Chromosome

The Use of Mutagenic Agents

Giant Chromosomes

SEX AND INHERITANCE

Gender Determination

Sex Linkage

Balancing Gene Dosage

ABERRANT CHROMOSOMES

POLYPLOIDY

BIOLINE
The Fish That Changes Sex

THE HUMAN PERSPECTIVE
Chromosome Aberrations and Cancer

S TEPS TO DISCOVERY

The Relationship between Genes and Chromosomes

By the 1880s, a number of prominent European biologists were closely following the activities of cells and using rapidly improving light microscopes to observe newly discernable cell structures. None of these scientists was aware of Mendel's work, which had demonstrated the particulate nature of hereditary factors. Yet, they realized that whatever it was that governed inherited characteristics would have to be passed from cell to cell and from generation to generation. This, by itself, was a crucial realization; all the genetic information needed to build and maintain a human being, for example, had to fit within the boundaries of a single cell.

Observations of dividing cells revealed that the cytoplasmic contents were merely distributed among the two daughter cells, and often not very evenly. In contrast, the cell's nuclear material was always evenly divided. The contents of the nucleus become organized into thick, darkly stained strands, inspiring biologists in 1888 to give them the name "chromosomes," meaning "colored bodies."

During cell division, each chromosome appeared out

The genes for five of the traits studied by Mendel—seed shape, seed color, flower color, pod color, and pod shape—are located

of the formless nuclear mass, split down the middle into two separate entities (each of which passed into separate cells), and then disappeared again. Careful observations of the sizes and shapes of the mitotic chromosomes suggested that a chromosome that disappeared after one division seemed to be the same structure that reappeared before the next. Even though the chromosome could not be seen in the nucleus of the nondividing cell, it was presumed to be present in an "invisible" state. Just as cells arise only from preexisting cells, it appeared that chromosomes arose only from preexisting chromosomes.

Attention then turned to the gametes and the material present in these cells which might be responsible for endowing the offspring with its genetic inheritance. Even though the contribution from the male was just a tiny cell (a sperm), it was as genetically important as the much larger egg, contributed by the female. What was it that these two very different cells had in common? The most apparent feature was the nucleus and its chromosomes.

During the 1880s, two German biologists, Theodor Boveri and August Weismann, turned their attention to the chromosomes. Boveri discovered that sea urchin eggs possessing abnormal numbers of chromosomes (a result of his experimental treatments), developed into abnormal embryos. Boveri concluded that the orderly process of normal development is "dependent upon a particular combination of chromosomes; and this can only mean that the individual chromosomes must possess different qualities." Weismann focused on the germ cells and concluded that the number of chromosomes in these cells must be divided precisely in half prior to the formation of gametes. If not, the chromosome number would double every time an egg fused with a sperm at fertilization. Each generation would have twice the number of chromosomes as the previous generation, which obviously did not occur.

The discovery and confirmation of Mendel's work in 1900 had an important influence on the work of the cell biologists of the time. Whatever their physical nature, the carriers of the hereditary units would have to behave in a manner consistent with Mendelian principles. In 1903, Walter Sutton, a graduate student at Columbia University, published a paper that pointed directly to the chromosomes as the physical carriers of Mendel's genetic factors. Sutton followed the chromosomes during the formation of sperm cells in the grasshopper. He found that for each chromosome in one of the diploid body cells of a grasshopper, there was another chromosome in the cell that matched it perfectly in size and shape. In other words, the nucleus of diploid cells contained homologous pairs of chromosomes, a feature that correlated perfectly with the pairs of inheritable factors uncovered by Mendel. When he examined cells at the beginning of meiosis, Sutton found that the members of each pair of homologous chromosomes were joined together until they separated at the first meiotic division. This was the physical basis for Mendel's proposal that genes exist in pairs that segregate from each other upon formation of the gametes. Sutton concluded that genes are situated on chromosomes.

The reduction division observed by Sutton explained several of Mendel's other findings: Gametes could contain only one version (allele) of each gene; an equal number of gametes with each allele would be formed; and two gametes that united at fertilization would produce an individual with two alleles for each trait. But many questions remained unanswered. For instance, how were the genes organized within the chromosomes, and could the location of specific genes be determined?

at specific sites on particular chromosomes.

*T*hey have been called biological computers and living strings of pearls. Such comparisons diminish the importance of their two-pronged biological role, however, which is absolutely essential to life. In the first of these roles, they retain all of the information for the genetic traits of a species that are passed from generation to generation; in the second, they control the molecular activity in cells so that biological order prevails. They reside deep inside a cell's interior, dictating which inheritable features to manufacture and, in so doing, instructing an organism to develop into itself. They are the same cellular structures that the nineteenth-century European microscopists first saw "dancing" around within the nucleus just before a cell divided. They are the bearers of the genes. They are the chromosomes.

▼ ▼ ▼

THE CONCEPT OF LINKAGE GROUPS

Although Mendel provided convincing evidence that inherited traits were governed by discrete factors, his studies were totally unconcerned with the physical nature of these factors or their location within the organism. Mendel was able to carry out his entire research project without ever observing anything under a microscope. But genes are physical structures residing within the chromosomes of living cells. Mendel had shown that each trait was governed by two copies of a gene. These copies were located at the same locus on each homologous chromosome (Figure 13-1). If the individual were heterozygous for a particular trait, the homologous chromosomes would contain different alleles at the corresponding sites. If homozygous, the homologous chromosomes would contain identical alleles.

As clearly as Sutton saw the relationship between chromosomal behavior and Mendel's results on pea plants, he saw one glaring problem. Mendel had examined the inheritance of seven traits and found that the manner in which each of these characteristics was inherited was independent of the inheritance of the others. This formed the basis of his law of independent assortment. But if genes are packaged together on chromosomes, like pearls on a string, then each package of genes should be passed from each parent to its offspring, just as intact chromosomes are passed. In other words, genes on the same chromosome should act as if they are *linked* to one another; that is, they should be part of the same **linkage group.**

Let's reconsider the 9:3:3:1 ratio from the F_1 dihybrid cross ($RrYy \times RrYy$) discussed on page 238. Suppose the genes for seed color and seed shape lay near each other on the same chromosome; the two dominant alleles should stay together during gamete formation, as should the two recessive alleles. This should reduce the number of possible gametes from four (RY, Ry, rY, and ry) to two (RY and ry). Such crosses should yield progeny showing two possible phenotypes instead of four.

How is it that all of Mendel's seven traits demonstrated independent assortment? Were they all on different linkage groups; that is, on different chromosomes? As it turned out, the garden pea has seven different pairs of homologous chromosomes, and each of the traits on which Mendel re-

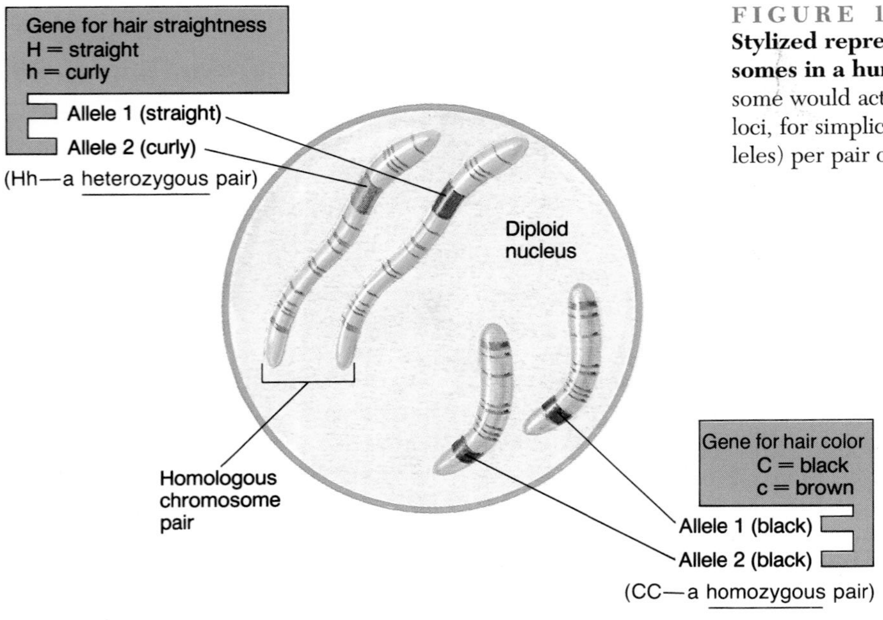

FIGURE 13-1

Stylized representation of alleles on homologous chromosomes in a human diploid nucleus. Although each chromosome would actually contain thousands of genes at different loci, for simplicity, we have identified only one gene (two alleles) per pair of chromosomes in this illustration.

Gene for hair straightness
H = straight
h = curly

Allele 1 (straight)
Allele 2 (curly)
(Hh—a heterozygous pair)

Diploid nucleus

Homologous chromosome pair

Gene for hair color
C = black
c = brown

Allele 1 (black)
Allele 2 (black)
(CC—a homozygous pair)

ported either occurred on a different chromosome or was so far apart on the same chromosome that it acted independently. Whether Mendel owed his findings to good fortune or simply to a lack of interest in any traits that did not fit his predictions is unclear.

Sutton's prophecy of linkage groups was soon transformed from speculation to fact. Within a couple of years, two traits (flower color and pollen shape) in sweet peas were shown to be linked. Other evidence would soon follow.

Because genes are physically connected to one another, like pearls on a strand, the various alleles on a single chromosome tend to remain linked together as they pass from cell to cell and from parents to offspring. (See CTQ #1.)

LESSONS FROM FRUIT FLIES

Genetic investigations on domestic plants or animals can be enormously time-consuming. Experimental matings yield results only when offspring appear, often a year or more after the parents mate. Analysis of subsequent generations requires additional years of study. Geneticists needed an organism that was easy to maintain, produced multiple generations every year, and was governed by the same "rules" of genetic inheritance exhibited by more complex organisms. They found just such an organism swarming around rotting fruit. Fruit flies (of the species *Drosophila melanogaster*) could easily be raised by the thousands in small

laboratory bottles, completing their entire reproductive cycle in 10 days. In 1909, the fruit fly seemed like the perfect organism to Thomas Hunt Morgan of Columbia University, as he began the research that initiated the "golden age of genetics," a period which has yet to end.

GENETIC MARKERS ALONG THE CHROMOSOME

There was one major disadvantage in beginning work with this insect: There was only one type of fly available, the *wild type* (Figure 13-2a). Inheritance in fruit flies, pea plants, or any other organism can be charted only if there is detectable evidence of genetic differences. Whereas Mendel had simply purchased his different genetic strains of peas from seed merchants, Morgan had to generate his own strains. Morgan expected that variants, or *mutants,* from the wild type might appear if he bred sufficient numbers of flies. Several months and thousands of flies later, he found his first mutant, a fruit fly with white eyes instead of the normal red ones.

By 1915, Morgan and his students had found 85 different mutants. As expected, some of these mutant alleles assorted independently; others did not (such as the linked characteristics illustrated on page 250). Taken together, the mutant alleles belonged to four different linkage groups, one of which contained very few genes (only two had been discovered by 1915). This discovery correlated perfectly with the finding of four different chromosomes in the cells of *Drosophila,* one of which was very small (Figure 13-2). There was little remaining doubt that genes resided on chromosomes.

(a)

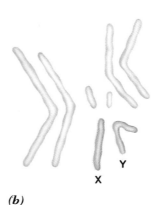

(b)

FIGURE 13-2

***Drosophila melanogaster,* the geneticist's ally.** *(a)* These petite fruit flies reproduce rapidly in bottles, dining on yeast that grows on a medium placed in the bottle. Each female produces hundreds of offspring, each of which can reproduce before they are 2 weeks old. Results from 1 year of *Drosophila* experimentation would require 30 years to obtain using an organism that reproduced only once a year. Note the red eyes and the other normal (wild-type) characteristics. *(b)* Fruit flies have four pairs of chromosomes, one of which is very small. The two dissimilar homologues are the sex chromosomes that determine gender.

CROSSING OVER: GENE EXCHANGE

Morgan and his students soon discovered that alleles in a linkage group occasionally segregated independently, as though on different chromosomes, ending up in separate gametes. The genes for body color and wing length, for example, reside on the same chromosome (Figure 13-3). Therefore, crossing a heterozygous gray-bodied, long-winged fly (*BbWw*) with a black-bodied, short-winged mate (*bbww*) should yield only two phenotypes, the same as the parents (Figure 13-4). Yet, contrary to these predictions, in about one-sixth of the progeny the two alleles in a pair appeared to have traded places with each other on homologous chromosomes. These offspring showed neither of the predicted phenotypes. Instead, they inherited a combination that would be possible only if the alleles were no longer connected, for example, a gray-bodied and short-winged fly (*Bbww*). In 1911, Morgan offered an explanation for this case of incomplete linkage—**crossing over.** Crossing over is the exchange of genetic segments between homologous chromosomes during meiosis and is described in Figure 13-5.

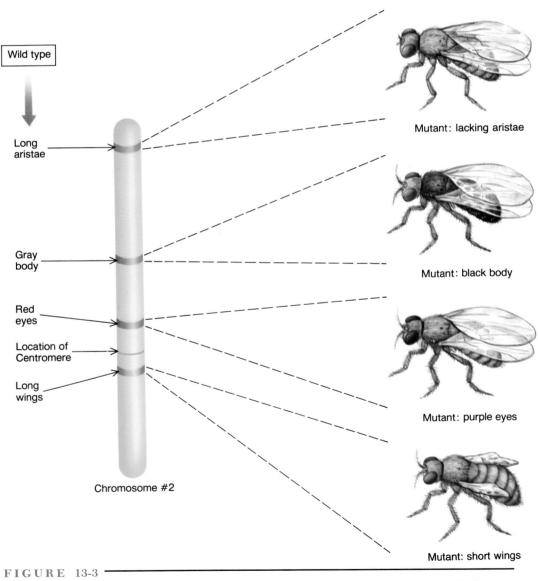

Wild type

Long aristae

Gray body

Red eyes

Location of Centromere

Long wings

Chromosome #2

Mutant: lacking aristae

Mutant: black body

Mutant: purple eyes

Mutant: short wings

FIGURE 13-3 —————

The mutation that produced each mutant phenotype pictured here occurs at a particular location on one of the chromosomes. Each wild-type characteristic is printed next to its location on the chromosome; the corresponding mutant characteristic is illustrated on the right.

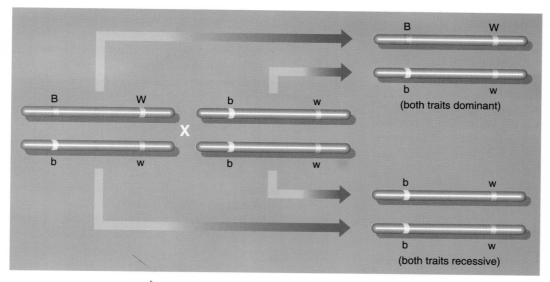

BbWw x bbww ━━━━━━━━▶ BbWw or bbww (*no offspring with combination of recessive and dominant traits*)

FIGURE 13-4 ━━━━━━━━━━

The pattern of inheritance if linkage were 100 percent complete. If two alleles on the same chromosome always remained together, as predicted by Mendel's law of independent assortment, then all of the offspring in the case depicted here would resemble one or the other parent. They would be either gray-bodied and long-winged (*BbWw*) *or* black-bodied and short-winged (*bbww*).

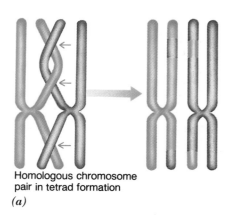

Homologous chromosome pair in tetrad formation
(a)

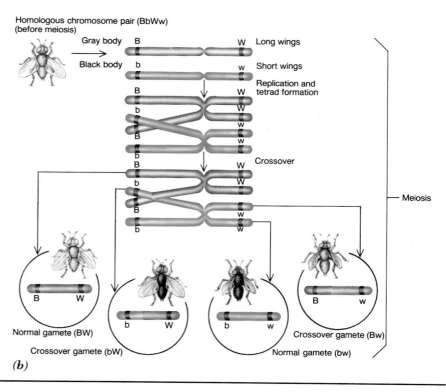

(b)

FIGURE 13-5 ━━━━━━━━━━

Crossing over provides the mechanism for reshuffling alleles between maternal and paternal chromosomes. *(a)* Tetrad formation during synapsis, showing three possible crossover intersections (*chiasmata*, indicated by red arrows). *(b)* Simplified representation of a single crossover in a *Drosophila* heterozygote (*BbWw*) at chromosome number 2, and the resulting gametes. If one of the crossover gametes participates in fertilization, the offspring will have a combination of alleles that was not present in either parent.

MAPPING THE CHROMOSOME

Studies on fruit flies also revealed that specific genes are arranged along a particular chromosome in a predictable linear sequence, much like towns along a highway. All individuals of a given species have chromosomes with the same sequence of genes. Since scientists cannot "drive" along a chromosome, recording the locations of genes and the distances that separate them, they had to discover an indirect way of determining gene location. Crossing over supplied this indirect approach.

We noted earlier that genes on a chromosome are like pearls along a strand. Consider what would happen if you were to pick up a linear strand of pearls, close your eyes, and randomly cut the strand once with a scissors. It is more likely that two pearls located close to each other will remain *together* on one of the cut pieces, than would two pearls located much farther apart. The same principle applies to crossing over, which involves *random* physical breakage of the chromosome. The more room between two sites on the chromosome for breakage to occur, the more likely a break will occur between those two sites. Since alleles of two genes at opposite ends of a chromosome have more room between them than do alleles of two genes that lie close to each other, the former are more likely to be separated during crossing over than are the latter. In other words, the crossover frequency between two genes reflects their relative distance from each other—the greater the distance, the greater the frequency. This information is used to construct genetic maps that show relative distances between genes on a chromosome. In *Drosophila*, for example, alleles of the linked genes for body color and wing length (Figure 13-3) become observably detached from each other (Figure 13-5) in about 180 of every 1,000 offspring (180/1,000 = 0.18 crossover frequency). To geneticists, 1 percent crossing is equated to one *map unit*. The alleles w and b are thus 18 map units apart ($180/1,000 \times 100 = 18\% = 18$ map units).

Mapping not only reveals relative distances between different genes but also delineates the sequence of those genes along a chromosome. We have already indicated that the alleles for body color (b) and wing length (w) show a crossover frequency of 18 percent. Let's consider a third "genetic marker"—pr, the allele for purple eyes—that shares the same chromosome as b and w (Figure 13-3). Mating experiments show that the crossover frequency between w and pr is 12 percent and the crossover frequency between b and pr is 6 percent. Only one sequence of these three genes along the chromosome is consistent with these data. That sequence is shown in the following genetic map:

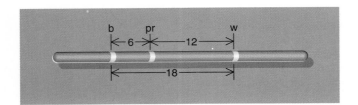

Using this technique, hundreds of gene loci have been mapped on this one fruit fly chromosome alone. The precise positioning of specific genes along specific chromosomes of a species represents another example of the high degree of order that exists in biological systems. Consider what would happen if the linear arrangement of genes differed from one individual to the next. As you know by now, diploid organisms contain homologous chromosomes derived from two different parents. If these homologous chromosomes had a different arrangement of genes, then crossing over between homologs would create chromosomes that were lacking certain genes. Offspring inheriting these chromosomes would be genetically deficient.

Crossover frequencies have been used to map genes on a diverse variety of organisms, from yeast to mice. As we will see in Chapter 17, genes on human chromosomes have also been extensively mapped—using very different techniques. Each month or so, researchers announce the discovery of the chromosomal location of a gene responsible for a different inherited disease, such as cystic fibrosis or muscular dystrophy. With each announcement comes the hope that the underlying cause of the disease will be determined and a new, more effective treatment forthcoming.

THE USE OF MUTAGENIC AGENTS

During the early period of genetics, the search for mutants was a slow and tedious procedure that depended on the *spontaneous* appearance of altered genes. In 1927, Herman Muller, a former undergraduate in Morgan's laboratory, discovered that flies subjected to sublethal doses of X-rays exhibited 100 times the spontaneous mutation frequency as did their counterparts who were not exposed to X-rays. This finding had several important consequences. On the practical side, the use of *mutagenic* (mutation-causing) agents, such as X-rays, provided a great increase in the number of mutants available for research using a variety of laboratory organisms, ranging from bacteria and yeast to mice. This discovery also pointed out the hazard of the increasing use of radiation in the industrial and medical fields, since most mutations are harmful. The work earned Muller the Nobel Prize in 1949.

GIANT CHROMOSOMES

In 1933, Theophilus Painter of the University of Texas rediscovered a phenomenon that had been observed many years earlier: giant chromosomes in certain insect cells (Figure 13-6). For example, cells from the larval salivary gland of *Drosophila* contain chromosomes that are about 100 times thicker than are the chromosomes found in most other *Drosophila* cells. During larval development, these salivary gland cells cease dividing, yet the cells keep growing in size. DNA replication continues, providing the additional genetic material needed to support the high level of secretory activity of these enormous cells. The duplicated DNA strands remain attached to each other in perfect side-by-

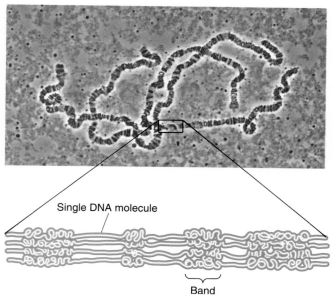

FIGURE 13-6

Giant chromosomes of the fruit fly—genes on display.
These giant polytene chromosomes from the salivary gland of
larval fruit flies show thousands of distinct, darkly staining
bands. Many of the bands have been identified as the loci of
particular genes. The inset shows how polytene chromosomes
consist of a number of individual DNA strands. The bands on
the chromosomes correspond to those sites where the DNA is
more tightly compacted.

side alignment (see inset of Figure 13-6), producing giant
chromosomes with as many as 1,054 more copies of the
DNA than are found in normal chromosomes. These unusual *polytene chromosomes,* as they are called, are rich in
visual detail, revealing about 6,000 dark bands when stained
and examined microscopically.

Painter realized that these giant chromosomes might
provide the opportunity to obtain a visual portrait of the
chromosomes of a species that had been mapped by breeding experiments over the past 20 years. Could a correlation
be made between the genetic maps obtained by crossing-
over analyses and the chromosomal banding patterns?
Painter and others found that it could. Many of the bands
correlated with specific genes, providing visual confirmation of the entire mapping procedure.

**Over the past 90 years, one organism, the fruit fly, has re-
vealed more about the chromosomal organization of genes
and their role in directing biological activities than has any
other organism. The primary value of fruit flies to biologists
stems from the great variety of mutants that have been isolated. Each new genetic mutant tells us something about the
activity of a normal gene product whose existence would
otherwise have gone undetected. (See CTQ #3.)**

SEX AND INHERITANCE

Genetically, humans are a lot like flies and, therefore,
most of the genetic breakthroughs revealed by *Drosophila*
apply to humans as well as to most other organisms. We
even resemble fruit flies in the fundamentals of sex; an
animal's gender is determined by specific chromosomes.

GENDER DETERMINATION

The chromosomes of males and females of the same animal
species typically look identical, with the exception of one
chromosomal pair. This pair is called the **sex chromosomes** to distinguish it from the other pairs of
chromosomes—the **autosomes**—which are identical in
both genders. In both fruit flies and mammals, the female's
cells harbor a pair of identical sex chromosomes, called **X
chromosomes,** whereas the male's cells possess one X
chromosome and a smaller **Y chromosome** (illustrated for
Drosophila in Figure 13-2 and in humans in Figure 10-3).

During gamete formation in humans and other mammals, the X and Y chromosomes synapse and then separate
at the first meiotic division. Accordingly, half the sperm
produced by males will contain the Y chromosome, and the
other half will possess the X chromosome. Since all gametes
produced by an XX female will contain an X chromosome,
the sex of each offspring depends on whether the egg is
fertilized by an X-bearing sperm (forming an XX *female*
offspring) or a Y-bearing sperm (forming an XY *male* offspring). Gender in humans is thus determined by the male
(Figure 13-7). King Henry VIII of England, who disposed
of many wives because they failed to bear him a son, should
have pointed his lethal finger squarely at himself. It was *his*
royal gametes that determined the gender of his progeny!

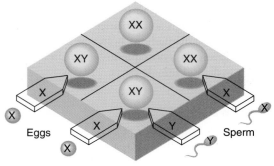

FIGURE 13-7

Sex determination in *Drosophila*. As in humans, in *Drosophila,* males have an X and a Y chromosome, while females
have two X chromosomes. Males produce two types of gametes
in equal numbers, while females produce only one. As depicted
in this Punnett square, the sex of each offspring is determined
by the chromosome composition of the fertilizing sperm.

◁ B I O L I N E ▷
The Fish That Changes Sex

In vertebrates, gender is generally a biologically inflexible commitment: An individual develops into either a male or a female as dictated by the sex chromosomes acquired from one's parents. Yet, even among vertebrates, there are organisms that can reverse their sexual commitment. The Australian cleaner fish (Figure 1), a small animal that sets up "cleaning stations" to which larger fishes come for parasite removal, can change its gender in response to environmental demands. Most male cleaner fish travel alone rather than with a school. Except for a single male, schools of cleaner fish are comprised entirely of females. Although it might seem logical to conclude that maleness engenders solo travel, it is actually the other way around: Being alone fosters maleness. A cleaner fish that develops away from a school *becomes* a male, whereas the same fish developing in a school would have become a female.

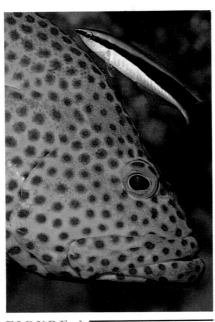

FIGURE 1

The small Australian wrasse (cleaner fish) is seen on a much larger grouper.

But what of the one male in the school—the one with the harem? He may have developed as a solo fish and then found a school in need of his spermatogenic services. But there is another way a school may acquire a male. If the male in a school dies (or is removed experimentally), one of the females, the one at the top of a behavioral hierarchy that exists in each school, becomes uncharacteristically aggressive and takes over the behavioral role of the missing male. She begins to develop male gonads, and within a few weeks, the female becomes a reproductively competent male, indistinguishable from other males. Furthermore, the sex change is reversible. If a fully developed male enters the school during the sexual transition, the almost-male fish developmentally backpedals, once again assuming the biological and behavioral role of a female.

Not all organisms follow the mammalian pattern of sex determination. In some animals, most notably birds, the opposite pattern is found: The female's cells have an X and a Y chromosome, while the male's cells have two Xs. An exception to this rule of a strict relation between sex and chromosomes is discussed in the Bioline: The Fish That Changes Sex. Although some plants possess sex chromosomes and gender distinctions between individuals, most have only autosomes; consequently, each individual produces both male and female parts.

SEX LINKAGE

For fruit flies and humans alike, there are hundreds of genes on the X chromosome that have no counterpart on the smaller Y chromosome. Most of these genes have nothing to do with determining gender, but their effect on phenotype usually *depends on* gender. For example, in females, a recessive allele on one X chromosome will be masked (and not expressed) if a dominant counterpart resides on the other X chromosome. In males, it only takes one recessive allele on the single X chromosome to determine the individual's phenotype since there is no corresponding allele on the Y chromosome. Inherited characteristics determined by genes that reside on the X chromosome are called **X-linked characteristics.**

So far, some 200 human X-linked characteristics have been described, many of which produce disorders that are found almost exclusively in men. These include a type of heart-valve defect (*mitral stenosis*), a particular form of mental retardation, several optical and hearing impairments, muscular dystrophy, and red-green colorblindness (Figure 13-8).

One X-linked recessive disorder has altered the course of history. The disease is **hemophilia,** or "bleeder's disease," a genetic disorder characterized by the inability to produce a clotting factor needed to halt blood flow quickly following an injury. Nearly all hemophiliacs are males. Although females can inherit two recessive alleles for hemophilia, this occurrence is extremely rare. In general, women who have acquired the rare defective allele are heterozygous **carriers** for the disease. The phenotype of a carrier

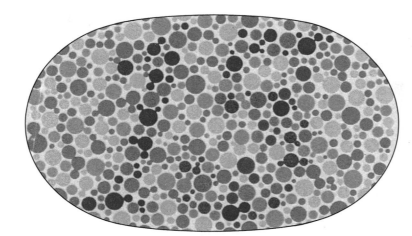

FIGURE 13-8
Distinctively male. There are hundreds of X-linked recessive characteristics that appear in males but rarely in females, who would have to have the recessive allele on both X chromosomes. Red-green colorblindness is such a trait. Males with this defect in vision cannot see the number "15" in the above pattern.

discloses no suggestion of the presence of the recessive allele because the normal allele on the other X chromosome directs formation of enough blood-clotting factor to assure a normal phenotype. Since the Y chromosome has no allele for producing the clotting factor, a boy who inherits the defective allele from his heterozygous mother will develop hemophilia.

The mutant hemophilia gene in the royal family of England can be traced to Queen Victoria, who transmitted the recessive allele to four of her children—a son (who eventually died of hemophilia) and three daughters (who were heterozygous carriers). The disease plagued several royal families when male descendants married Victoria's daughters; their descendants were, in turn, afflicted, in some cases killing the heir to the throne. The son of Nicholas, the last Russian czar, was a hemophiliac, having acquired the disease from his mother, Alexandra, a granddaughter of Queen Victoria (Figure 13-9). Desperate to save their son, Nicholas and Alexandra put their faith in the infamous monk Rasputin, who convinced them he was able to cure the young Alexis. It has been argued that Rasputin's influence on the monarchy contributed to the overthrow of the czar during the Russian Revolution. In case you are wondering about members of the current royal family of England, they are descended from Edward VII, one of Victoria's sons who did *not* inherit her "tainted" X chromosome.

The path of inheritance of specific traits is best revealed through a diagram called a **pedigree,** such as that depicted in Figure 13-10. Pedigrees reveal that X-linked recessive characteristics tend to skip a generation. When a male with an X-linked recessive characteristic has children, none of them shows evidence of the trait (except for the unlikely circumstance where his mate also carries the defective allele) because all male progeny receive a normal allele from the mother. All daughters will be carriers. If a carrier and a normal mate conceive offspring, the chance that a male will inherit the sex-linked characteristic is 50 percent, versus 0 percent for female offspring; half of the daughters will be carriers.

Although nearly all known sex-linked disorders are associated with a gene located on the X chromosome, a small number of genes are carried on the Y chromosome and determine **Y-linked characteristics.** In mammals, male reproductive anatomy is Y-linked, the Y chromosome carrying genetic information that instructs the embryonic gonads to develop into testes. The testes then produce the hormone testosterone which, in turn, directs the embryo to develop other male reproductive structures. In the absence of a Y chromosome, a female will develop (Chapter 17).

FIGURE 13-9
The last Russian royal family. Alexis, great-grandson of Queen Victoria, is seen in front of his mother in this family photo taken in 1914.

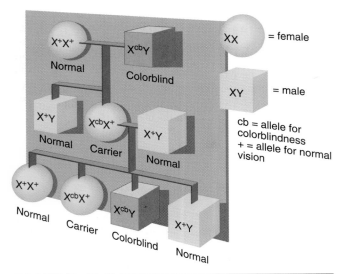

FIGURE 13-10

The "generation gap." Pedigrees show the tendency of X-linked traits—in this case, colorblindness—to skip a generation, producing a phenotypic gap: a generation of normal-visioned men and normal-visioned, carrier women, but no individuals showing the abnormal characteristic. Hemophilia and other X-linked traits follow this basic inheritance.

BALANCING GENE DOSAGE

The amount of gene product manufactured by a cell is influenced by the "gene dosage"—the number of copies of the allele that directs its synthesis. Increasing gene dosage usually intensifies the expression of the trait. It would seem, then, that a mammalian female who is homozygous for a trait on the X chromosome would express that trait with twice the intensity as would a male who has only one copy of the gene. Yet, this does not happen; most X-linked traits are expressed more or less equally in both sexes. This puzzling observation was explained by Mary Lyon, a British geneticist, who suggested that one of the X chromosomes in each cell of females was *inactivated* during embryonic development. The inactivated chromosome forms a dense mass in the interphase nuclei of female cells (Figure 13-11*a*), which is absent in males. Thus, like the cells in males, each cell in a female has only one *functional* copy of each X-linked gene.

This phenomenon may appear to contradict what you have just learned about sex-linked traits. If the cells of females have only one active X chromosome, why don't females show X-linked diseases such as hemophilia? They would if the same X chromosome were inactivated in all of the female's cells. But this isn't the case. Inactivation occurs randomly in the cells of early embryos. In about half the cells, one X remains active, while in the other cells, the homologous X prevails, creating two genetically distinct types of cells scattered randomly throughout the embryo.

(a)

(b)

FIGURE 13-11

Chromosome in retirement. *(a)* The inactivated X chromosome in the nucleus of a woman's cells appears as a darkly staining structure, named a *Barr body* after its discoverer. *(b)* Random inactivation of either X chromosome in different cells during early embryonic development creates a mosaic of tissue patches that descend from each cell after inactivation. These patches are visually evident in calico cats, which are heterozygous for fur color—the allele for black coat color residing on one X chromosome, and that for yellow on the other X. This explains why male calico cats are virtually nonexistent since all cells in the male will have the same color allele.

The millions of cells that descend from each of these early embryonic cells form a patch of tissue that may be genetically and phenotypically different from adjacent tissue patches in which the other X chromosome directs the action. As a result, every female is a *mosaic* of genetically different tissue patches, some of which may have an active dominant allele, and others that may have an active recessive allele for a particular gene. X-chromosome inactivation appears to occur in all mammals and is reflected in the patchwork coloration of the fur of some mammals, including the calico cat (Figure 13-11*b*). Pigmentation genes in humans are not located on the X chromosome, hence the absence of "calico women." Mosaicism due to X inactivation can be demonstrated in women, however. For example, if a narrow beam of red or green light is shone into the eyes of a woman who is a heterozygous carrier for red-green colorblindness, patches of retinal cells with defective color vision can be found interspersed among patches with normal vision.

Sex and chromosomes are closely related. Most animal species have distinct male and female individuals whose gender and sexual characteristics are determined by the chromosomes they inherit. Many nonsexual genetic traits are also carried on the sex chromosomes, which produces striking differences between males and females in the frequency of many genetic characteristics. (See CTQ #5.)

ABERRANT CHROMOSOMES

In addition to mutations that alter the information content of a single gene, chromosomes may be subjected to more extensive alterations that typically occur during cell division. Pieces of a chromosome may be lost or exchanged between nonhomologous chromosomes, or extra segments may appear (Table 13-1). Since these **chromosomal**

TABLE 13-1
MODIFICATIONS IN CHROMOSOMAL STRUCTURE

Type of Alteration	Example of How Change May Occur	Some Possible Effects ● Favorable ● Harmful
Deletion		Rarely favorable; perhaps elimination of detrimental genes
		Loss of critical genes is lethal; disrupts chromosome separation during meiosis
Duplication		Provides raw material for evolution of new proteins as part of a family of related proteins
		May interfere with chromosome separation; may disrupt gene function if duplication occurs within a gene
Inversion		Increases genetic diversity by changing gene positions
		Reduced fertility; loss of control of gene expression
Translocation		Enormous genetic changes may generate rapid evolutionary advances
		May activate genes that cause cancer; reduce fertility; may result in gain or loss of part or whole chromosome

aberrations follow chromosomal breakage, their incidence is increased by exposure to agents that damage DNA, such as a viral infection, X-rays, or exposure to chemicals that can break the DNA backbone.

The consequence of a chromosomal aberration depends on the circumstances involved. If the aberration occurs during a mitotic cell division, the effects on the organism are generally minimal since only a few cells of the body will usually be affected. On rare occasions, however, a daughter cell that inherits the aberration may be transformed into a malignant cell, which can grow into a cancerous tumor.

▮▶ Many chromosome aberrations occur during meiosis, particularly as a result of abnormal crossing over, and can be transmitted to the next generation. When an aberrant chromosome is inherited through a gamete, all cells of the offspring will have the aberrant chromosome, and the individual generally does not survive embryonic development. However, like single-gene mutations, chromosomal aberrations may occasionally confer advantageous traits. These structural modifications in chromosomes produce large-scale genetic changes that may propel evolution forward in giant steps. There are several types of chromosomal aberrations (Table 13-1), including:

- *Deletions.* A **deletion** occurs when a portion of a chromosome is lost. Forfeiting a portion of a chromosome usually results in a loss of critical genes, producing severe consequences. The first correlation between a human disorder and a chromosomal deletion was made in 1963 by Jerome Lejeune, the same French geneticist who had earlier discovered the chromosomal basis of Down syndrome (Chapter 11). Lejeune discovered that a baby born with a variety of facial malformations and abnormal development of the larynx (voice box) was missing a portion of chromosome number 5. A defect in the larynx caused the infant's cry to resemble the sound of a suffering cat. Consequently, the scientists named the disorder *cri du chat syndrome*, meaning cry-of-the-cat syndrome.

- *Duplications.* A **duplication** occurs when a portion of a chromosome is repeated. Duplications have played a very important role in biological evolution. Many proteins are present as families consisting of a variety of closely related molecules. Consider the adult hemoglobin molecule illustrated in Figure 4-15*d*, which consists of two alpha-globin chains and two beta-globin chains. Examination of the amino acid sequence of alpha- and beta-globin chains reveals marked similarities, indicating that they both evolved from a single globin polypeptide present in an ancient ancestor. The human globin family contains several other members that form the hemoglobin molecules of embryos and infants. The amino acid sequences of these globins are also closely related to each other and to the alpha and beta forms.

The first step in the evolution of a protein family is thought to be the duplication of the gene. Over time, the two copies of the gene evolve in different directions, generating polypeptides with different, but related, amino acid sequences. In some instances, the polypeptides encoded by a given family of genes have evolved divergent functions, even though their amino acid sequences still show their ancestral relationship. Growth hormone and prolactin (a hormone that stimulates milk production), for example, are pituitary hormones that evoke completely different responses in different target cells, yet their amino acid sequences indicate that they have evolved from a common ancestral gene.

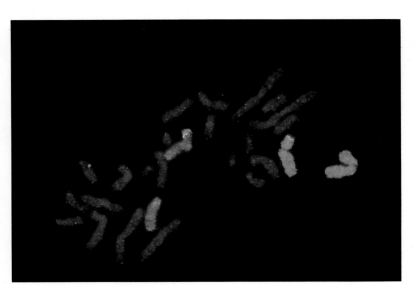

FIGURE 13-12

Translocation. Using special fluorescent stains, investigators can demonstrate when a piece of one chromosome breaks away and becomes attached to another chromosome. In this case, chromosome number 12 (green) and chromosome number 7 (red) have exchanged pieces.

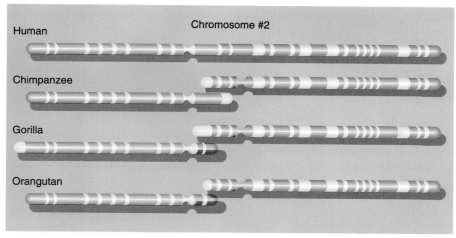

FIGURE 13-13

Translocation and evolution. Translocation may have played an important role in the evolution of humans from their apelike ancestors. If the only two ape chromosomes that have no counterpart in humans are "fused," they match human chromosome number 2, band for band.

- *Inversions.* Sometimes a chromosome is broken in two places, and the segment between the breaks becomes resealed into the chromosome in reverse order. This aberration is called an **inversion** and can interfere with the way that neighboring genes are expressed. It is possible, for example, for an inversion to disrupt the orderly expression of an oncogene (page 202), leading to the formation of a malignant tumor.

- *Translocations.* When all or a piece of one chromosome becomes attached to a nonhomologous chromosome, the aberration is called a **translocation** (Figure 13-12). Certain translocations increase the incidence of cancer in humans. The best-studied example is the *Philadelphia chromosome,* which is found in individuals with certain forms of leukemia, a malignancy of the white blood cells. The Philadelphia chromosome, which is named for the city in which it was discovered in 1960, is a shortened version of human chromosome number 22. For years, it was believed that the missing segment represented a simple deletion, but with improved techniques for observing chromosomes, the missing genetic piece was found translocated to another chromosome (number 9). This change in the positions of genes causes the cell to become malignant. Chromosome number 9 contains a gene that codes for a protein kinase (page 129) which plays a role in cell proliferation. As a result of translocation, one small end of this protein is replaced by about 600 extra amino acids encoded by a gene carried on the translocated piece of chromosome number 22. This new, greatly enlarged protein apparently retains the catalytic activity of the original version but is no longer subject to the cell's normal regulatory mechanisms. In other words, it cannot be turned off. As a result, the cell is no longer subject to normal regulatory mechanisms that prevent

one type of cell from overgrowing the body and it continues to proliferate. The result is a cancerous transformation. Several other types of cancer are also associated with chromosomal aberrations (see The Human Perspective: Chromosome Aberrations and Cancer).

Translocations also play an important role in evolution, generating large-scale changes that may be pivotal in the branching of separate evolutionary lines from a common ancestor. Such a genetic incident probably happened during our own recent evolutionary history. A comparison of the 23 pairs of chromosomes in human cells to the 24 pairs of chromosomes in the cells of chimpanzees, gorillas, and orangutans reveals a striking similarity (Figure 13-13). But how did humans "lose" the twenty-fourth pair of chromosomes? A close examination of human chromosome number 2 reveals that we didn't really lose it at all. If the two ape chromosomes that have no counterpart in humans were to fuse together to form a single chromosome, they would form a perfect match, band for band, with human chromosome number 2 (Figure 13-13). At some point during the evolution of humans, an entire chromosome was translocated to another, creating a single fused chromosome, reducing the haploid number from 24 to 23.

Chromosomes are the repository of an organism's genetic endowment, and different parts of a chromosome contains different information. The maintenance of a well-ordered, smoothly functioning organism requires virtually all of its inherited information. It is not surprising, therefore, that alterations in a cell or organism's chromosomal composition almost invariably lead to a disruption of normal activities. (See CTQ #6.)

◁ THE HUMAN PERSPECTIVE ▷
Chromosome Aberrations and Cancer

Malignant cells can often be identified by the abnormal structure of their chromosomes. In some cases, as with the Philadelphia chromosome, alterations in chromosome structure are thought to play a role in the original transformation of a normal cell to a cancerous state. Careful analysis of the banding patterns of chromosomes (see Figure 10-3) from human malignant cells indicates that the cancer-causing aberrations are not random but tend to occur in approximately 100 specific bands. It is presumed that these bands contain genes that play a special role in the formation of malignant tumors.

We have already discussed one group of cancer-causing genes, called *oncogenes.* When *activated* ("turned on"), oncogenes cause cells to proliferate out of control, leading to tumors (Chapter 10). More recently, we have discovered a different group of genes, called **tumor-suppressor genes** (or anti-oncogenes), whose normal role is to *block* the formation of cancerous cells. Both copies of a tumor-suppressor gene in a cell must be knocked out in order to eliminate their protective function, allowing the transformation of the cell to the malignant state.

Several examples of this phenomenon have been documented. For instance, a rare childhood cancer of the eye, called *retinoblastoma,* occurs in a small number of families. Examination of normal cells from children suffering from retinoblastoma indicated that one of the thirteenth pair of homologous chromosomes was missing a small piece from the interior portion of the chromosome. The missing piece normally contains the *retinoblastoma gene* which, when altered or absent, predisposes the child to the eye cancer. The cancer itself develops only in those individuals where a retinal cell accidently loses the second copy of the gene on the homologous chromosome, leaving the individual with a cell completely lacking the tumor-suppressive capacity. Recent evidence suggests that the retinoblastoma gene directs formation of a protein that inhibits cell division; the protein loses this inhibitory function when it is phosphorylated. Tumor-suppressor genes have been implicated in a variety of common malignancies and are of widespread importance.

The discovery of tumor-suppressor genes has important clinical implications. Rather than using toxic chemotherapeutic agents to kill malignant cells, it might be possible to treat cells with the protein encoded by the tumor-suppressor gene, a protein that might stop tumor growth, as it normally does in the body. Identification of the products and functions of tumor-suppressor genes is one of the foremost goals of cancer biologists.

POLYPLOIDY

We saw in Chapter 11 how the presence of an extra chromosome due to a failure of chromosome separation during meiosis can cause abnormal development, leading to Down syndrome. On occasion, an entire set of chromosomes fails to separate during meiosis, resulting in a zygote with an increased number of chromosomal sets—a condition called **polyploidy.** In most higher animal species, polyploidy is lethal. In humans, for example, a high percentage of miscarriages are due to a polyploid fetus. Plants are much more tolerant of polyploidy. In fact, half of all known plant species are polyploid or contain some polyploid tissue. It is important that polyploid plants have an even number of chromosomal sets (4N, 6N, etc.) so that every chromosome has a chance to pair with a homologue during meiosis. As a result, gametes receive equal shares of chromosomes, and the plant retains fertility. Plants that have odd-numbered sets (such as triploid—3N—plants) have meiotic problems and tend to be sterile. Such species may still survive by asexual reproduction, as discussed below.

▐▶ Polyploidy in plants may confer advantages over their diploid counterparts, especially when the extra set of chromosomes is the result of hybridization of two closely related plant species. The polyploid hybrid is often hardier than either of the diploid species since it combines the adaptations of both species (for example, disease resistance of one and drought resistance of the other). In addition to enhanced hardiness, polyploid plants have several other adaptive advantages over their diploid relatives, such as increased yields and production of fruit with special qualities. Many of our crop plants, including wheat, strawberries, and potatoes, are polyploid. The development of new polyploid strains of wheat, rice, and corn has helped raise the worldwide level of food production. Some seedless fruits, including bananas, are produced by triploid plants. Although these plants are usually sterile—hence, their inability to make seeds—they can be propagated vegetatively by cuttings or grafts. Polyploids of pears, apples, and grapes produce giant fruit that show commercial promise.

Some organisms may possess more than a diploid number (two sets) of chromosomes. In some cases, these extra sets disrupt the normal chromosomal balance in a cell and prove disastrous. In many plants, however, extra sets of chromosomes provide an opportunity for enhanced genetic variation and genetic vigor. (See CTQ #8.)

REEXAMINING THE THEMES

Relationship between Form and Function

Chromosomes are the physical carriers of the genes. The structure of chromosomes correlates with their genetic function and helps explain the basis of the principles of inheritance discovered by Mendel and others. Chromosomes provide a discrete location for each gene and a mechanism for transmitting copies of genes to offspring. The existence of chromosomes as pairs of homologs explains why each body cell has two copies of each allele. The separation of homologous chromosomes during meiosis accounts for the presence of only one of a pair of alleles in a gamete.

Biological Order, Regulation, and Homeostasis

The chromosomes of a higher plant or animal may contain tens of thousands of genes, yet each particular gene is predictably located at a particular site (locus) on that chromosome in virtually every member of that species. This ordered arrangement allows the positions of genes to be mapped by crossing over and other techniques. If different individuals of a species had chromosomes with different arrangements of genes, then crossing over between homologous chromosomes would result in the loss of genes in the gametes and death of the ensuing offspring. Similarly, the loss of parts of a chromosome, or rearrangements within or between chromosomes, disrupts the orderly expression of genes and can lead to biological disorders, including abnormal development and cancer.

Unity within Diversity

The chromosomes of all eukaryotic organisms are constructed in a similar manner and consist of linear arrays of genes comprising linkage groups. Crossing over occurs during meiosis in all sexually reproducing diploid organisms; therefore, crossover frequencies can be used to map the relative distances between genes in a wide variety of organisms. While chromosomes have the same internal construction in all eukaryotes, their number, size, and shape vary greatly among diverse organisms. Fruit flies, for example, have a haploid number of 4 chromosomes, whereas humans have 23. Both fruit flies and humans have one pair of sex chromosomes, and in both organisms, X-linked recessive characteristics are passed from mothers to male offspring.

Evolution and Adaptation

The duplication of genes within a chromosome is one of the most important cellular processes underlying the process of evolution. Once a gene is duplicated, the two copies of the gene tend to evolve in different directions, often generating distinct proteins with related amino acid sequences. In some cases, the two proteins retain similar functions, as occurs among the various polypeptides that can form a hemoglobin molecule. In other cases, the two proteins evolve different functions, as occurs with the pituitary hormones prolactin and growth hormone.

Because evolution depends on changes in genes, and genes are carried on chromosomes, changes in chromosomes can provide a direct visual display of the evolutionary process. Although mammalian chromosomes don't exist in a giant polytenic form, they can be stained to reveal banding patterns that can be used to identify chromosomes and to provide an indication of genetic relatedness, as between apes and humans, whose chromosomes are nearly identical.

SYNOPSIS

Genes reside on chromosomes in fixed linear orders. The physically connected set of alleles on the same chromosome constitutes a linkage group—a group of alleles that do not assort independently.

Linked genes don't always stay linked. Crossing over during meiosis shuffles alleles between homologous chromosomes, creating new combinations that are not found in the parents. Crossing over increases genetic diversity beyond that resulting from independent assortment.

The positions of genes on a chromosome can be mapped by analysis of crossing over data. The farther apart two genes are from each other on a chromosome, the more likely they will be separated by crossing over between the two. Genetic maps of chromosomal loci can be constructed by determining the frequency with which alleles become "unlinked" during meiosis.

Gender is determined by sex chromosomes in most animals. In humans (and fruit flies), females possess X chromosomes (XX), whereas males possess an X and a Y chromosome (XY). X-linked recessive characteristics, such as hemophilia and muscular dystrophy, are normally expressed when males inherit a recessive allele on the X chromosome from their mother. In female mammals, one of the X chromosomes is inactivated in each cell. Inactivation of one of the X chromosomes occurs randomly during embryonic development so the cells of adult females are a mosaic of X-linked alleles.

Chromosomes are subject to alterations that change their genetic content. Chromosomal aberrations include the loss of a segment (a deletion), the acquisition of extra genes (a duplication), a reversal of the order of genes in part of a chromosome (an inversion), and the transfer of a chromosome (or a part of a chromosome) to a nonhomologous chromosome (a translocation). Each of these aberrations can have serious effects on the cell and even the whole organism.

Key Terms

linkage group (p. 248)
crossing over (p. 250)
sex chromosome (p. 253)
autosome (p. 253)
X chromosome (p. 253)
Y chromosome (p. 253)

X-linked characteristic (p. 254)
carrier (p. 254)
pedigree (p. 255)
Y-linked characteristic (p. 255)
chromosomal aberration (p. 258)
deletion (p. 258)

duplication (p. 258)
inversion (p. 259)
translocation (p. 259)
tumor-suppressor gene (p. 260)
polyploidy (p. 260)

Review Questions

1. The traits of one well-studied organism fall into seven distinct linkage groups. How many chromosomes are in the *somatic* cells of this organism?

2. Draw the processes of synapsis, crossing over, and segregation, and indicate the severity of the consequences if different individuals of a species were to have a different order of genes on their chromosomes.

3. Why do X-linked traits tend to skip a generation? Under what circumstances would this *not* happen? (For instance, how could a colorblind man have a child that is also colorblind?)

4. What is the likelihood that Nicholas would acquire hemophilia, knowing that his grandmother, Queen Victoria, was a carrier of the mutant allele?

5. If you were offered a calico kitten, why would you not have to ask whether it is a male or female?

Critical Thinking Questions

1. Sutton was able to provide visual evidence for Mendel's law of segregation. Why would it have been impossible for him visually to confirm or refute Mendel's law of independent assortment?

2. Draw a simple map showing the gene order and relative distances among the genes X, Y, and Z, using the data below:

Crossover Frequency	*Between These Genes*
36%	X–Z
10%	Y–Z
26%	X–Y

3. Make a list of reasons why the fruit fly has been an ideal organism to use in genetic studies.

4. Alleles on opposite ends of a chromosome are so likely to be separated by crossing over between them that they segregate independently. How would we ever know that these two genes belong to the same linkage group? (Hint: If A is linked to B, and B is linked to C, then A is linked to C.)

5. Below is a map of the human X-chromosome, showing the location of known genes. Why are the traits associated with these genes said to be "sex-linked" even though most of them have nothing to do with determining the sex of individuals? Why do such traits often skip a generation? What circumstances must occur for the traits not to skip a generation?

6. Genes are often compared to beads on a string; each bead is independent of the other beads. How do results of studies on translocations of chromosomes demonstrate that genes are not independent of other genes?

7. Why do you suppose the Philadelphia chromosome causes only one type of cancer—leukemia—rather than other types as well, such as colon cancer or skin cancer?

8. In a diagram, show how a plant with a 3N chromosome number could arise? 4N number? Why are polyploids more likely to survive mitosis than meiosis? Which would you expect to have more chance of reproducing, the 3N or the 4N organism? Why? Can you think of at least one reason why polyploids are scarcer among animals than among plants?

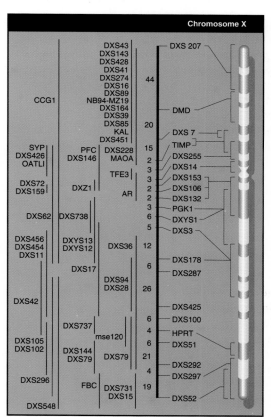

Source: *National Institutes of Health Human Genome; New York Times*, Tuesday, October 6, 1992.

Additional Readings

Edey, M. A., and D. C. Johanson. 1989. *Blueprints: Solving the mystery of evolution.* Boston: Little Brown (Introductory)

Marx, J. 1991. How the retinoblastoma gene may inhibit cell growth. *Science* 252:1492. (Intermediate)

Moore, J. A. 1972. *Heredity and development.* 2d ed. New York: Oxford. (Intermediate)

Weinberg, R. A. 1988. Finding the anti-oncogene. *Sci. Amer.* Sept:44–51. (Intermediate)

Wills, C. 1989. *The wisdom of the genes.* New York: Basic Books. (Intermediate)

The Molecular Basis of Genetics

STEPS TO DISCOVERY
The Chemical Nature of the Gene

CONFIRMATION OF DNA AS THE GENETIC MATERIAL

THE STRUCTURE OF DNA

The Watson-Crick Proposal

DNA: LIFE'S MOLECULAR SUPERVISOR

Function 1: Storage of Genetic Information

The Organization of DNA in Chromosomes in Prokaryotes and Eukaryotes

Function 2: Passage of Genetic Information to Descendants Through DNA Replication

Function 3: Genetic Expression from DNA to RNA to Protein

THE MOLECULAR BASIS OF GENE MUTATIONS

THE HUMAN PERSPECTIVE
The Dark Side of the Sun

The Chemical Nature of the Gene

*B*y the end of the 1930s, biologists were well versed in the alignment of genes along chromosomes and the ways these genes were reshuffled during meiosis and transmitted from one generation to the next. By 1940, the major question confronting geneticists was very simple: What is the identity of the genetic material?

Ironically, the genetic substance had been hiding in "plain sight" for over 70 years. DNA was discovered in 1869 by the Swiss physician Friedrich Miescher, who first purified DNA (which he called "nuclein") from pus cells obtained from discarded surgical bandages and later from salmon sperm. He chose these unlikely cells because they had large nuclei surrounded by very little cytoplasm.

In 1923, the German biochemist Robert Feulgen developed a procedure that specifically stained DNA in tissues. Observing stained cells under the microscope, Feul-

gen demonstrated the presence of DNA in mitotic chromosomes. Since protein had also been shown to be present in chromosomes, a new question emerged: Was it the DNA or the protein that carried the hereditary instructions? The answer seemed obvious. During the first half of the century, the complexity, versatility, and importance of proteins had been determined. In contrast, DNA was thought to be composed of a monotonous repeat of only four nucleotide subunits (page 270). A molecule of such simple construction could hardly be considered a candidate for the master molecule that directs life itself!

A few years later, Fred Griffith, a meticulous, soft-spoken bacteriologist who worked at the British Ministry of Health, was conducting research on techniques that would distinguish different strains of pneumococcus, the bacterium that causes pneumonia. Griffith was studying two ge-

A DNA timeline ties the findings of studies on pea plants, fruit flies, and bacteria that led to the discovery of DNA as the

netically stable strains of pneumococcus. One strain produced capsules that surrounded the cell, enabling the bacteria to evade the defense system of both a mouse and a human, thereby causing pneumonia. The other strain produced no capsule but was identical to its counterpart in every other way. These nonencapsulated bacteria were quickly destroyed by the host's defenses before they could cause disease. Griffith grew a batch of the encapsulated strain and heated the cells until they died. As expected, injections of these heat-killed bacteria into mice were harmless, as were injections of the live, nonencapsulated strain.

In 1928, Griffith injected both these preparations into the same mouse; remarkably, the mouse contracted pneumonia and died. Furthermore, the dead mouse contained live, encapsulated bacteria, even though none had been injected into it. Since there was no possibility that the heat-killed bacteria had been brought back to life, Griffith concluded that the presence of the dead encapsulated cells had *transformed* the nonencapsulated cells into an encapsulated strain. The transformed bacteria and their progeny continued to produce capsules; thus, the change was permanent and *inheritable*. In 1928, however, the scientific community had yet to recognize that bacteria possessed genes, and the genetic significance of Griffith's results went unappreciated.

Meanwhile, a physician named Oswald Avery decided to give up his clinical practice and shift his attentions to bacteriologic research. In 1928, Avery was 51 years old and an expert on the immunological properties of pneumococcal capsules. At first, Avery was skeptical of Griffith's results and had a young scientist in his laboratory attempt to confirm them. Not only did Avery's colleague confirm the findings, but he discovered that the transformation of nonencapsulated cells to the encapsulated state did not require the use of a host animal (such as a mouse). Rather, transformation could be accomplished in a culture dish simply by adding a soluble extract of the encapsulated cells to the medium in which the nonencapsulated cells were growing. For the next decade, Avery was preoccupied in purifying the substance responsible for transformation and determining its chemical nature.

Extensive chemical analysis by Avery and his colleagues Colin MacLeod and Maclyn McCarty eventually identified DNA as the transforming substance. In addition, an enzyme (*DNase*) that selectively destroys DNA was the only enzyme the scientists found that was capable of abolishing the substance's transforming activity; protein-degrading enzymes had no effect. In 1944, Avery's group published a paper on their findings in the *Journal of Experimental Medicine*. The paper was written with scrupulous caution and made no dramatic statements that genes were made of DNA rather than protein. Avery was determined not to repeat the embarassing mistake of Richard Willstater (Chapter 6), who had claimed that enzymes

could not be made of protein simply because he could not detect proteins in his most active preparations.

The paper drew remarkably little attention. Maclyn McCarty, one of the three coauthors, describes an incident in 1949 when he was asked to speak at Johns Hopkins University along with Leslie Gay, who had been testing the effects of the new drug Dramamine for the treatment of seasickness. The large hall was packed with people and "after a short period of questions and discussion following [Gay's] paper, the president of the Society got up to introduce me as the second speaker. Very little that he said could be heard because of the noise created by people streaming out of the hall. When the exodus was complete, after I had given the first few minutes of my talk, I counted approximately 35 hardy souls who remained in the audience because they wanted to hear about pneumococcal transformation or because they felt they had to remain out of courtesy." But Avery's awareness of the potential of his discovery was revealed in a letter he wrote in 1943 to his brother Roy, also a bacteriologist:

> If we are right, & of course that's not yet proven, then it means that nucleic acids are not merely structurally important but functionally active substances in determining the biochemical activities and specific characteristics of cells—& that by means of a known chemical substance it is possible to induce predictable and hereditary changes in cells. This is something that has long been the dream of geneticists. . . . Sounds like a virus —may be a gene. But with mechanisms I am not now concerned—one step at a time. . . . Of course the problem bristles with implications. . . . It touches genetics, enzyme chemistry, cell metabolism & carbohydrate synthesis—etc. But today it takes a lot of well documented evidence to convince anyone that the sodium salt of deoxyribose nucleic acid, protein free, could possibly be endowed with such biologically active & specific properties & that evidence we are now trying to get. It's lots of fun to blow bubbles,—but it's wiser to prick them yourself before someone else tries to.

Why were Avery's findings so broadly overlooked? And what finally convinced the world that DNA was the genetic material?

genetic material.

*I*f ever there was a fleeting period of discovery that changed the face of biology, it was the period between the early 1940s and the mid-1960s. Virtually all that you will read about in this chapter was discovered in that scant period of about 25 years—a period of biological revolution. Before this revolution, we knew that genes were carried on chromosomes, and we knew the rules by which these genes were transmitted from generation to generation. But we knew little else. By the time the dust had settled, we understood how it all worked.

The revolution began with Avery's discovery of DNA as the substance responsible for transforming one type of pneumococcus into another. Many articles and passages in books have dealt with the reasons why Avery's findings were not met with greater acclaim. Part of the reason may be due to the subdued manner in which the paper was written and the fact that Avery was a bacteriologist, not a geneticist. Some biologists were persuaded that Avery's preparation was contaminated with miniscule amounts of protein and that the contaminant, not the DNA, was the active transforming agent. Others questioned whether studies on bacteria had any relevance to the field of genetics, and they viewed the phenomenon of transformation as a bacterial peculiarity. In other words, Avery's findings were ahead of their time.

⚠ During the years following the publication of Avery's paper, the climate in genetics changed in a very important way. The existence of the bacterial chromosome was recognized, and a number of prominent geneticists turned their attention to these prokaryotes. These scientists believed that knowledge gained from the study of the simplest cellular organisms would shed light on the mechanisms that operate in the most complex plants and animals. In addition to being much simpler than fruit flies, pea plants, or mice, bacteria are haploid, so dominant traits cannot hide the presence of recessive alleles. Bacteria also have the advantage of being unicellular and capable of rapidly increasing in number; when grown under rich nutrient conditions, a culture of bacteria can double every 20 minutes. Millions of genetically identical cells can be grown in small containers in a matter of hours.

With bacteria as their research tool, geneticists began to hunt for clues at the most basic level of life—the molecular level. As the quest for answers drew scientists into the molecular domain, a new field of science was created: **molecular biology.** By 1950, the emerging field of molecular biology still had a central question to answer: What was the chemical nature of the gene?

▼ ▼ ▼

CONFIRMATION OF DNA AS THE GENETIC MATERIAL

Seven years after the publication of Avery's paper on bacterial transformation, Alfred Hershey and Martha Chase of the Cold Spring Harbor Laboratories in New York turned their attention to an even simpler system—**bacteriophage**—viruses that infect bacterial cells. By 1950, researchers recognized that every virus had a genetic program. The genetic material was injected into the host cell, where it directed the formation of new virus particles inside the infected cell. Within a matter of minutes, the infected cell broke open, releasing new bacteriophage particles, which infected neighboring host cells.

It was clear that the genetic material directing the formation of viral progeny had to be either DNA or protein because these were the only two molecules the virus contained. Hershey and Chase reasoned that the virus's genetic material must possess two properties: First, if the material were to direct the development of new bacteriophage during infection, it must pass into the infected cell. Second, it must be passed on to the next generation of bacteriophage. The molecular biologists prepared two batches of bacteriophage to use for infection. One batch contained radioactively labeled DNA (^{32}P-DNA); the other batch contained radioactively labeled protein (^{35}S-protein). Since DNA lacks sulfur (S) atoms, and protein usually lacks phosphorus (P) atoms, these two radioisotopes provided specific labels for the two types of macromolecules.

The course of infection of a bacterial cell by a bacteriophage containing labeled DNA and protein is illustrated in Figure 14-1. Of the two labeled molecules, only the ^{32}P-DNA entered the infected cell and was passed on to the next generation of viruses. By 1952, the scientific community finally accepted DNA as the genetic material responsible for storing an individual's hereditary message.

The first important step in the newly emerging field of molecular biology was to convince the scientific community that DNA was the genetic material. This was ultimately accomplished using bacteriophage, the simplest system known to contain a genetic program. (See CTQ #2.)

THE STRUCTURE OF DNA

Before Hershey and Chase published their findings, DNA was far from a preoccupation of biologists. In fact, only one group of biologists—located at King's College in England—was working full-time in 1951 trying to determine the structure of DNA. One member of this group, Rosalind Franklin, was busy firing X-rays through DNA crystals, trying to learn how the atoms of the DNA molecule were arranged. This is the same technique (X-ray crystallogra-

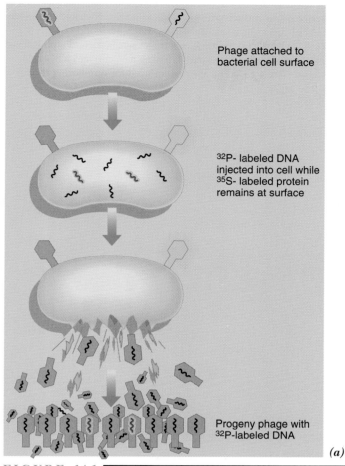

Phage attached to bacterial cell surface

^{32}P- labeled DNA injected into cell while ^{35}S- labeled protein remains at surface

Progeny phage with ^{32}P-labeled DNA

(a)

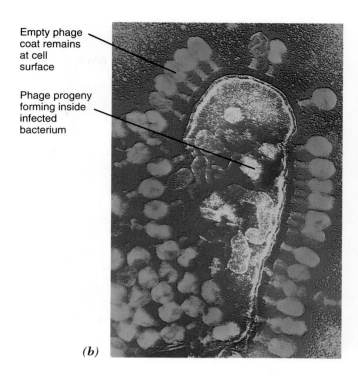

Empty phage coat remains at cell surface

Phage progeny forming inside infected bacterium

(b)

FIGURE 14-1

The Hershey-Chase experiment. *(a)* Bacterial cells were infected with bacteriophage that contained radioactively labeled protein (blue) or radioactively labeled DNA (red). Hershey and Chase found that none of the labeled protein entered the infected cells (it remained outside the bacteria in the viral coats). In contrast, labeled DNA entered the cells and was recovered in the viral progeny. *(b)* Micrograph showing a bacterial cell infected with bacteriophage. New viruses are being formed within the cell, and empty viral coats are seen attached to the outer cell surface.

phy) that played such an important role in determining the shape of proteins (page 67). Although Franklin didn't interpret her pictures correctly, a couple of other scientists eventually did.

By 1953, the stage was set for the central event in the DNA story, a crowning achievement by two young scientists at Cambridge University in England whose names were about to become household words—James Watson and Francis Crick. Revealed in the seemingly shapeless smudges of Franklin's X-ray photographs was the very essence of the DNA molecule—a helix. You will recall from Chapter 4 that DNA consists of nucleotides, each of which contains a five-carbon sugar called **deoxyribose,** a phosphate group, and a nitrogenous (nitrogen-containing) base (Figure 14-2). All nucleotides in DNA contain the same sugar and phosphate, but they differ in their nitrogenous base. Two of the bases, adenine (A) and guanine (G), are

purines, while the other two bases, cytosine (C) and thymine (T), are *pyrimidines* (Figure 14-3).[1] Using molecular-shaped cutouts of these nucleotides, Watson and Crick constructed a helical model of the DNA molecule which, 40 years later, remains the centerpiece of molecular genetics.

Nucleotides are monomers that become linked together to form a long strand of nucleic acid polymer. Each nucleotide has a polarized structure (Figure 14-2). One edge, where the phosphate is located, is called the *5′ end* (pronounced 5-prime end), while the other edge is called the *3′ end*. We saw in Chapter 4 how a strand of nucleic acid is formed by linkages between the sugar of one nucleotide

---

[1] It might help to remember the acronym "PurAG," signifying that A and G are purines. Purines have a shorter name than pyrimidines, but a more complex molecular structure (two rings, rather than one).

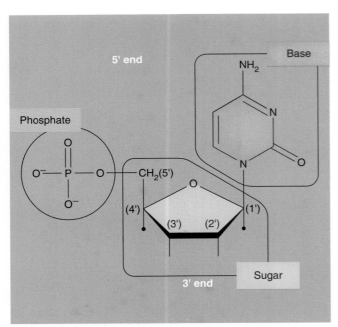

FIGURE 14-2

The structure of a nucleotide. Nucleic acids (DNA and RNA) are composed of repeating nucleotide units. Each nucleotide in DNA consists of the sugar deoxyribose, a nitrogenous base, and a phosphate group. Each nucleotide is polarized, having a 5′ end and a 3′ end. These numbers are based on the system used for numbering the carbons of the sugar.

FIGURE 14-3

Nitrogenous bases in DNA. There are four different nucleotides in DNA, depending on the nitrogenous base linked to the sugar. Adenine and guanine are purines; thymine and cytosine are pyrimidines.

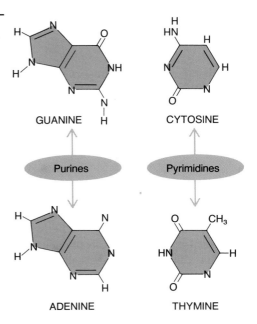

and the phosphate group of the adjoining nucleotide. Since each of the stacked nucleotides in a strand faces the same direction, the entire strand has a direction. In this regard, a DNA strand is like a line of people waiting to get into a theatre: One end—identifiable by a face—is the front of the line, while the other end—identifiable by the back of a head—is the rear of the line. For a strand of nucleic acid, one end is the 3′ end, the other is the 5′ end. These distinctions are useful in understanding the mechanisms for encoding and utilizing genetic information.

THE WATSON-CRICK PROPOSAL

The Watson-Crick model of DNA structure first proposed in 1953 is shown in Figure 14-4. A few of its main elements are listed below.

- A DNA molecule is composed of two chains of nucleotides that coil around each other to form a double helix.

- The two chains of a helix run in opposite directions, like two lines of people standing side by side, but facing opposite directions.

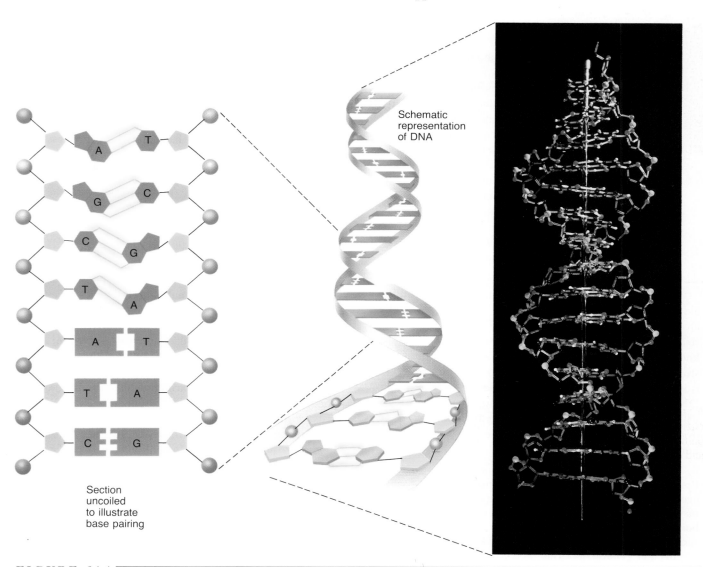

Schematic representation of DNA

Section uncoiled to illustrate base pairing

FIGURE 14-4

The double helix. Base pairing creates a double-stranded molecule that twists into a helix, resembling a spiral staircase. The nitrogenous bases of the nucleotides interact with one another by forming hydrogen bonds. The nucleotide sequence in the two strands is complementary. Guanine (G) binds to cytosine (C) by three hydrogen bonds, whereas adenine (A) binds to thymine (T) by two hydrogen bonds. Consequently, if you were given the sequence in one strand, you could predict the nucleotide order in the other strand. The three base pairs at the lower end of the left side of the figure depict the way nucleotides are illustrated in many of the following figures. The model on the right shows the spatial relationship between the two strands of the DNA molecule.

- The backbone (—sugar—phophate—sugar—phosphate—) of each chain is located on the outside of the molecule, whereas the bases project toward the center, giving the molecule the appearance of a spiral staircase.

- The two chains are held together by hydrogen bonds that form between the nitrogenous bases. Since individual hydrogen bonds are quite weak, the two strands of the double helix can separate during certain essential biological activities.

- An adenine on one chain always pairs with a thymine on the other chain, while a guanine on one chain always pairs with a cytosine on the other chain. Watson and Crick came to this conclusion about **base pairing** based on the way the molecular cutouts of the nucleotides fit with one another and on earlier findings in the late 1940s by Erwin Chargaff of Columbia University. Chargaff had studied the composition of the four different nucleotides in DNA molecules isolated from a variety of organisms. He found that in any given sample of DNA, the percentage of nucleotides that contained adenine always equaled the percentage that contained thymine (A = T). Similarly, the percentage of nucleotides containing guanine always equaled the percentage containing cytosine (G = C). Watson and Crick were the first to grasp the significance of Chargaff's "rules" and, in so doing, laid the cornerstone for their model of DNA structure.

If someone were to tell you the sequence of nucleotides along one strand of a DNA molecule, you could immediately tell him or her the sequence of bases in the other strand. Whenever the person mentioned a G, you would know that a C was situated across the way in the other strand. This relationship is known as **complementarity.** G is said to be complementary to C; one entire strand in a DNA molecule is complementary to the other stand. As we will see later on, the concept of complementarity is of overriding importance in nearly all the activities in which nucleic acids are involved.

> **Form and function go hand in hand. Once DNA was discovered to be a helical molecule consisting of two complementary strands of nucleotides held together by hydrogen bonds, biologists were finally poised to understand how such a molecule could participate in all of the activities required of genetic material. (See CTQ #3.)**

DNA: LIFE'S MOLECULAR SUPERVISOR

From the time biologists first considered DNA as the genetic material, DNA was expected to fulfill three primary functions:

1. **Storage of genetic information.** DNA is the molecular "blueprint," a stored record of precise instructions that determine all the inheritable characteristics an organism can exhibit.

2. **Self-duplication and inheritance.** Since DNA contains the entire blueprint for an organism, it has to contain the information for its own *replication* (duplication). DNA replication provides the means by which genetic instructions can be transmitted from one cell to its daughter cells or from one organism to its offspring.

3. **Expression of the genetic message.** For many years, scientists had suspected that individual genes carried the information for specific proteins. Some mechanism had to exist by which the information stored in a gene was actually put to use to direct the synthesis of a specific polypeptide.

The elucidation of the structure of DNA by Watson and Crick provided an instantaneous stimulus for determining how such a structure facilitated these three essential functions. We will consider the structure of DNA as it relates in turn to each of them.

FUNCTION 1: STORAGE OF GENETIC INFORMATION

The term "information" is difficult to define and is subject to many different interpretations. In biology, information usually refers to a kind of "instruction manual" situated within the DNA that makes up the genes. *The information content of the DNA is encoded in the precise order (linear sequence) of its nucleotides.* The nucleotides of a strand of nucleic acid are like chemical "letters" of an alphabet; they encode genetic information, just as the sequence of printed letters on this page encode the factual information you are now learning. Unlike the 26 letters in the English language, however, the genetic "alphabet" consists of only four letters (G, C, A, and T, corresponding to the four types of nucleotides in DNA). But four "letters" are all that are needed to "write" an unlimited variety of genetic messages. For example, a portion of the double helix only ten nucleotides long can be arranged in more than a million different sequences. Imagine the number of possible sequences you could form with human DNA, which contains about 3 billion base pairs!

With the publication of the Watson-Crick model, the gene was no longer seen as just a vague portion of a chromosome; it had become a specific stretch of nucleotides within a highly elongated DNA molecule. Organisms are highly three-dimensional—from their entire bodies, to the smallest subcellular organelles, to the proteins that bestow them with life. Yet, all of this three-dimensional structure is encoded in a one-dimensional array of nucleotide building blocks. The linear sequence of nucleotides dictates a linear sequence of amino acids in a polypeptide chain, which folds into a complex three-dimensional protein. Virtually all else in biology follows from this organization.

THE ORGANIZATION OF DNA IN CHROMOSOMES IN PROKARYOTES AND EUKARYOTES

The bacterial chromosome consists of a single, circular molecule of DNA, typically about 1 mm long, packaged into a cell that is only about 1 μm in length. One millimeter of DNA is sufficient to encode approximately 3,000 genes, of which about 1,000 have been localized in the chromosome of the common intestinal bacterium *Escherichia coli* (*E. coli*). Although the packaging of the bacterial chromosome into such a small cell is a remarkable feat (as evidenced by the photograph in Figure 14-5), it pales by comparison to that in the packaging of a eukaryotic cell. In every nucleated cell in your body, about 2 meters (6 feet) of DNA are packed into the nucleus, a sphere vastly smaller than the dot on the letter *i*. Moreover, a cell's DNA is not simply stuffed into the nucleus but is present in an orderly arrangement that allows the molecule to direct protein synthesis, to duplicate without tangling, and to coil periodically into compacted chromosomes in preparation for cell division.

The differences in structure between eukaryotic and prokaryotic chromosomes are summarized in Figure 14-5. Eukaryotic chromosomes contain a rich supply of proteins, including a group of small basic proteins called **histones.** Histones are described as basic because they contain large amounts of the basic amino acids arginine and lysine (see Figure 4-14). The positive charges on the R groups of these amino acids form ionic bonds with the negative charges of DNA's phosphate groups, which facilitates the packaging process. Two loops of DNA are always wrapped around a central cluster of eight histone molecules, forming a unit called a **nucleosome** (Figure 14-6). Not only does the DNA wrap around the histone "spools," it strings the nucleosomes together like beads on a necklace (Figure 14-6).

The relationship between histones and DNA is highly ordered and constant throughout all eukaryotic kingdoms. It appears that the nucleosome has worked well as a DNA-organizing unit for more than a billion years. Once again, the unity of life supplies evidence for the evolutionary kinship of all organisms as descendants of the same early ancestor.

Winding the DNA around nucleosomes shortens its length by approximately one-sixth. The nucleosomal filament is shortened even further by its coiling into thicker fibers, analogous to a coiled telephone cord, which, in turn, are bent into "looped domains" (Figure 14-6). Taken together, these structural properties of the chromosome allow the chromosomal material to fit into the tiny nucleus of a nondividing cell. The preparation of duplicated chromosomes for separation during mitosis requires an additional series of compaction steps, as illustrated in Figure 14-6. The highly coiled DNA of the interphase chromosome condenses around a "scaffold" (Figure 14-7) composed of structural proteins to form a mitotic chromosome. This final packaging step allows a dividing cell to parcel out identical allotments of the genetic material.

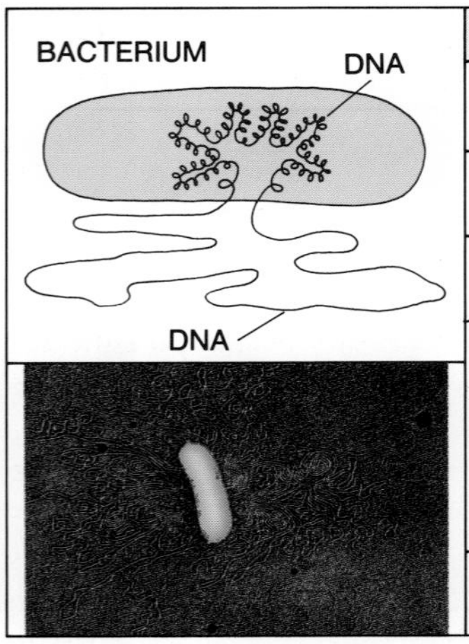

	CHARACTERISTIC	PROKARYOTIC	EUKARYOTIC
	Configuration of DNA	Circular	Linear (open-ended)
	Length (average)	1000 μm	1.8 meters (in humans)
	Number of chromosomes per cell	1	At least 2, up to 1,262
	Associated proteins	Enzymes and regulatory proteins transiently associated	Histones "permanently" associated. Also transiently associated enzymes and regulatory proteins
	DNA housed in nucleus	No	Yes

FIGURE 14-5

The prokaryotic chromosome is a model of both complexity and simplicity. It is a single circular DNA molecule having a diameter about 1,000 times that of the cell in which it is tightly packed. The adjacent table compares the chromosomes of prokaryotic and eukaryotic cells. Whereas the DNA of prokaryotes is only transiently associated with various proteins, the DNA of eukaryotic cells occurs in permanent association with structural proteins called *histones* (see Figure 14-6). The photo is of *Escherichia coli* broken open and spilling out its chromosome.

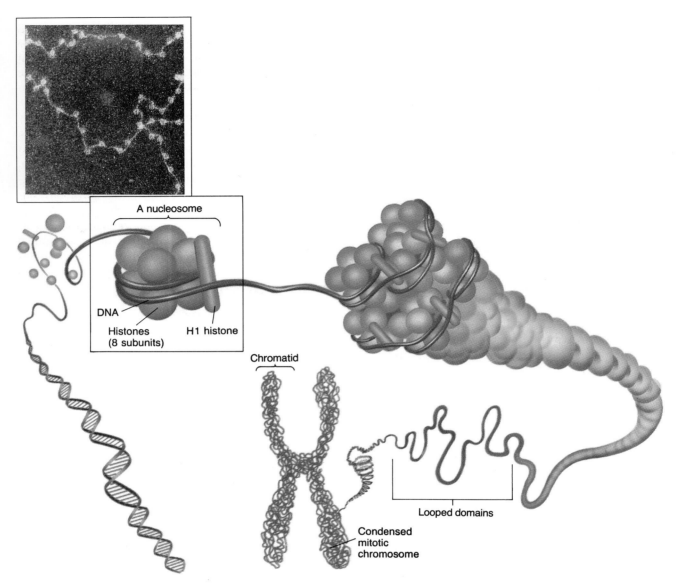

FIGURE 14-6

Packaging DNA into nucleosomes. Current model depicting packaging of DNA into condensed chromatin. The nucleosome (boxed area) is the fundamental packing unit. Each nucleosome consists of eight histone molecules encircled by approximately two turns of DNA. A different type of histone (called H1) locks the nucleosome complex together so that the DNA cannot unwind from the histone core. The nucleosomes are then coiled into thicker fibers that bend into "looped domains." These thickened strands can coil even further, forming the arms of a condensed mitotic chromosome. Nucleosomes are visible in this electron micrograph of uncondensed chromatin. The nucleosomes are the "beads" on the DNA "string."

each "daughter" molecule. As a result of replication, two new DNA molecules are formed, each containing precisely the same genetic message as that stored in the original molecule. This is the mechanism by which genetic instructions can be passed on from generation to generation.

The Mechanism of Replication

As in the case of other metabolic processes, much of the pioneering work on replication was carried out on bacterial cells. The first important steps in unraveling the mysteries of replication were the purification and analysis of the enzyme **DNA polymerase** by Arthur Kornberg and his colleagues at Washington University in 1957. DNA polymerase is the enzyme that moves along each template strand of the open helix, reading the nucleotide in the template and

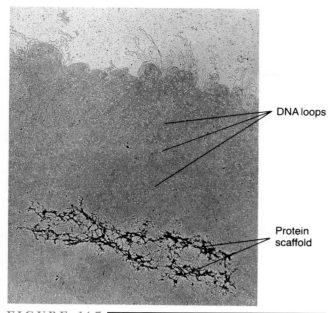

FIGURE 14-7 ———————

The protein scaffold of a mitotic chromosome. When mitotic chromosomes are treated so as to remove the histones, the freed DNA is seen to form giant loops that are attached at their base to the protein scaffold (which can be seen to retain the shape of the mitotic chromosome).

DNA loops

Protein scaffold

FUNCTION 2: PASSAGE OF GENETIC INFORMATION TO DESCENDANTS THROUGH DNA REPLICATION

The ability to reproduce is one of the most fundamental properties of all living systems. This process of duplication can be observed at several levels: Organisms duplicate by asexual or sexual reproduction; cells duplicate by cellular division; and the genetic material duplicates by **replication.**

The initial publication of Watson and Crick's proposal on the structure of DNA was accompanied by a proposed explanation of how such a molecule might replicate. The scientists suggested that during replication the hydrogen bonds holding the two strands of the DNA helix were sequentially broken, causing the gradual separation of the strands, much like the separation of two halves of a zipper. Each of the separated strands, with its exposed nitrogenous bases, would then serve as a **template** (a mold or physical pattern), directing the order in which complementary nucleotides become assembled to form the complementary strand. When complete, the process would have generated two identical molecules of double-stranded DNA, each containing one strand from the original DNA molecule and one newly synthesized strand (Figure 14-8). This form of DNA synthesis is called **semiconservative replication** because half the original DNA molecule is conserved in

FIGURE 14-8 ———————

Semiconservative DNA replication. During replication, the double helix unwinds, and each of the strands serves as a template for the assembly of a new complementary strand. Following replication, each new DNA molecule consists of one strand from the original duplex and one newly constructed strand. This arrangement, which was first predicted by Watson and Crick in their first publication of the structure of DNA, is called *semiconservative replication.*

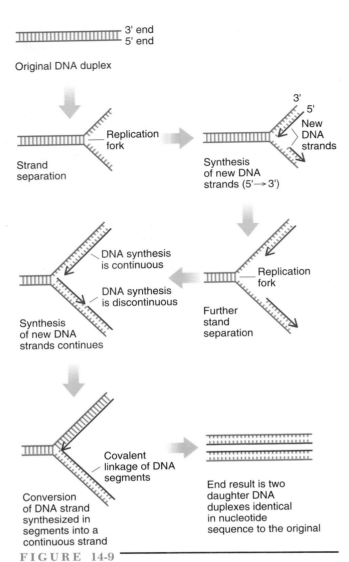

3' end
5' end

Original DNA duplex

Replication fork

Strand separation

3'
5'
New DNA strands

Synthesis of new DNA strands (5'→3')

DNA synthesis is continuous

DNA synthesis is discontinuous

Synthesis of new DNA strands continues

Replication fork

Further stand separation

Covalent linkage of DNA segments

Conversion of DNA strand synthesized in segments into a continuous strand

End result is two daughter DNA duplexes identical in nucleotide sequence to the original

FIGURE 14-9

The mechanism of replication. DNA consists of two strands that run in opposite directions. One of the strands runs in the 3' to 5' direction, and the other runs in the 5' to 3' direction. DNA polymerase molecules are only capable of moving along a template in one direction, toward the 5' end of the template. Consequently, polymerase molecules (and their associated proteins) move in opposite directions along the two strands. As a result, the two newly assembled strands also grow in opposite directions, one growing toward the replication fork, and the other growing away from it. One strand is assembled in continuous fashion, the other in segments that must be joined together by an enzyme.

covalently joining the complementary nucleotide onto the end of the new strand, which thereby grows in length.

It was initially thought that DNA polymerase molecules would move along both template strands toward the *replication fork,* the site where the helix was unzipping. If this were the case, both of the new strands would simply grow continuously in length by addition of nucleotides at the end near the fork. It was soon shown, however, that polymerase molecules move in opposite directions along the two template strands—toward the replication fork on one side, and away from the fork on the other side (Figure 14-9). As a result, the two new strands must be assembled in different ways. The new strand that grows toward the replication fork is constructed by continuous assembly, while the new strand that grows away from the replication fork is assembled by discontinuous assembly—in pieces that are subsequently linked together.

The process of replication is actually much more complex than that shown in Figure 14-9, and a number of proteins besides DNA polymerase are involved. For example, proteins are needed to unwind the helix, to keep the strands separated, and to join segments of DNA together into a continuous strand. Many of these proteins are clustered together to form a giant enzyme complex that moves along the template strand, much as a locomotive moves along a railroad track.

In bacteria, replication begins at one site, the **origin,** and progresses outward in both directions, as indicated in Figure 14-10*a*. At 37°C, the replication machinery of *E. coli* moves along the DNA at an astounding rate of 850 nucleotides per second, reading the template and incorporating a complementary nucleotide at each step along the way. In contrast, the chromosomes of eukaryotic cells contain much more DNA than do those of bacteria, so it would take days to replicate the DNA of a large chromosome if there were only one origin of replication. Instead, replication of the DNA of a plant or animal chromosome begins at many sites simultaneously, proceeding outward from each site in both directions (Figure 14-10*b*).

The Accuracy of Replication

The maintenance of biological order and stability from one generation to the next requires that the process of replication occur with a minimal number of mistakes. Imagine that you are a DNA polymerase molecule moving along a template, selecting complementary nucleotides out of a bag simply by their shape. How many mistakes do you think you would make? What do you think the consequences of picking a mismatched base pair would be? Remarkably, the DNA polymerase of bacteria makes a mistake only about once in every billion nucleotides it incorporates, which is less than once every 100 cycles of replication. One of the reasons for this extraordinary accuracy is that DNA polymerase is one of a handful of proteins that is actually two enzymes in one; it has one active site for polymerization, and another active site for "proofreading." If the first active site happens to incorporate a noncomplementary

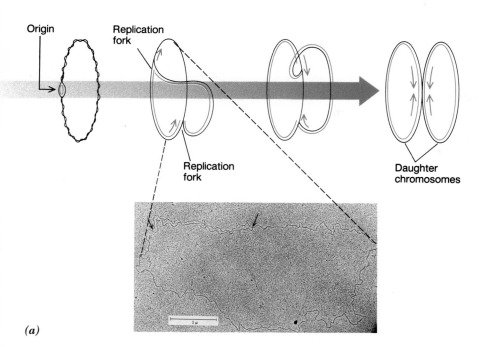

(a)

FIGURE 14-10 ————

Comparison of bacterial and eukaryotic chromosomal replication. In both types of cells, replication occurs at "forks" that travel away from each other in opposite directions. Replication is therefore bidirectional. **(a)** The circular chromosome of bacteria has only two replication forks. Chromosomal duplication is completed when the forks meet each other halfway around the circle. The chromosome in the photo has completed about one-sixth of the process. **(b)** In eukaryotes, replication begins at many points along the chromosome, each site forming two replication forks that travel away from each other until they meet another fork. Five distinct replication sites are apparent in the photo of a single DNA molecule from a mammalian cell. The dark lines are areas with newly replicated DNA.

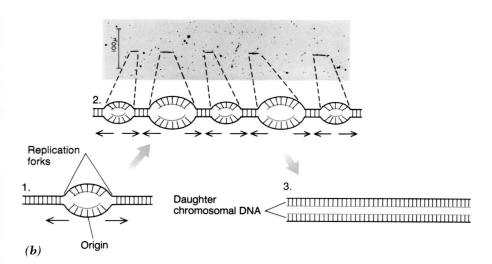

(b)

FUNCTION 3: GENETIC EXPRESSION FROM DNA TO RNA TO PROTEIN

monomer, the mistake is immediately recognized by the second active site, which removes the incorrect nucleotide. If the mistake happens to slip by the "proofreading" mechanism, it may result in a permanent change, or genetic mutation, in the information content of the DNA. The consequences of a mutation are discussed later in the chapter.

FUNCTION 3: GENETIC EXPRESSION FROM DNA TO RNA TO PROTEIN

An organism is the manifestation of its particular constellation of proteins, each of which is made up of one or more polypeptide chains. A single gene encodes the information for a single polypeptide chain. But just how does the linear order of nucleotides in the DNA lead to the assembly of a linear order of amino acids in a polypeptide? During the

late 1950s and early 1960s, it was discovered that the relationship between DNA and protein involves an intermediate—RNA. In other words, genetic information in a cell flows from DNA to RNA to protein.[2] This flow of encoded information is depicted in Figure 14-11. The cell first *transcribes* (copies) a gene's encoded instructions into a molecule of RNA, which is then sent to the "construction site" (ribosomes). These instructions direct the activities at the site, telling the "workers" (various proteins and RNA molecules) which polypeptide to build. The workers must be able to *translate* the instructions in the RNA into the exact gene product ordered for construction.

—————

[2] This flow is reversed under certain circumstances—from RNA to DNA — as occurs in a cell infected by HIV, the virus that causes AIDS (Chapter 30).

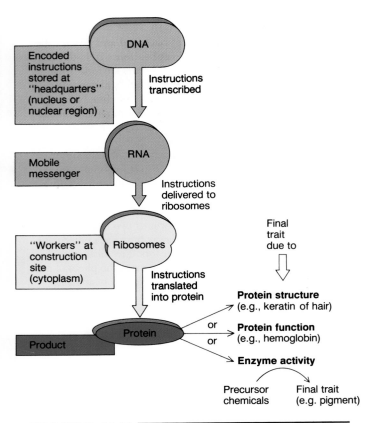

FIGURE 14-11

A summary of the flow of genetic information from DNA to RNA to protein.

Two questions of particular importance will occupy most of the remainder of this chapter. First, how is the genetic information stored in DNA transcribed into RNA? Second, how is the information contained in the RNA translated into the corresponding polypeptide? In order to understand the answers to these questions, we must reconsider the structure of RNA (originally described on page 82). RNA differs structurally from DNA in three ways.

- The nucleotides in RNA contain *ribose,* a sugar that has one more oxygen atom than does DNA's sugar (*deoxyribose*). This is an important difference because it allows enzymes to distinguish between the two types of nucleotides.

- As in DNA, RNA has four distinct nitrogenous bases, but one of them—*uracil*—is unique to RNA. Uracil replaces thymine as the base that is complementary to adenine.

- RNA is typically a single-stranded molecule, so its bases are generally exposed and available for interaction with other molecules.

Transcribing the Message

The information stored in DNA is carried to the ribosomes by an RNA molecule aptly called **messenger RNA (mRNA).** RNA molecules are assembled by **transcription,** a process that, in some ways, resembles the synthesis of a new DNA strand during replication. During transcription (Figure 14-12), the double helix temporarily separates, and a complementary strand of mRNA assembles along one of the single DNA strands that acts as a template. **RNA polymerase,** the enzyme that directs the process, distinguishes between deoxyribose- and ribose-containing nucleotides, polymerizing only the latter type into the growing chain. The enzyme also distinguishes between the two strands of the DNA molecule, selecting the **sense strand;** that is, the strand with the appropriate sequence for protein synthesis.

In this way, genetic information stored in a gene is transcribed into an mRNA molecule that contains the DNA's message by virtue of its complementary sequence of nucleotides. This mobile "messenger" leaves the DNA template carrying the information to the ribosomes, where it directs the synthesis of a specific polypeptide. Messenger RNA is only one of three major types of RNA synthesized by cells. As we will see shortly, the other two types—transfer RNA and ribosomal RNA—are also involved in protein synthesis, but their function is very different from mRNA.

The Genetic Code

In a sense, the genetic basis of life is a matter of "reading, copying, and following instructions" on a molecular scale. Just as a reader of a sentence deciphers an encoded message by translating a linear string of letters into a meaningful thought, cells use a language—a *genetic code*—which they translate into genetic characteristics. As you translate the line of characters in this sentence, you do so by recognizing groups of letters (words) that have specific meanings. In constructing proteins, the linear array of nucleotides in mRNA is also translated in groups, called **codons,** which could be considered molecular "words." Each codon is three nucleotides long and specifies the insertion of one—and only one—of the 20 different amino acids at a specific point in the polypeptide being built. With a four-letter genetic alphabet (A, G, C, and U), 64 triplet combinations (4 × 4 × 4) are possible; 64 combinations are more than enough to assign at least one unique codon to each of the 20 different amino acids.

By 1961, the general properties of the genetic code were known, but one great task remained: breaking the code itself. What is the "language" that cells use to instruct ribosomes to insert the correct amino acids in the proper sequence in a growing polypeptide chain? At the time, most experts believed it would take 5 to 10 years to decipher the entire code. But the codebreakers received a terrific boost from the development of techniques used to synthesize "artificial mRNAs" having known nucleotide sequences. The first of these artificial mRNAs, which was synthesized and tested by Marshal Nirenberg and Heinrich Matthaei of

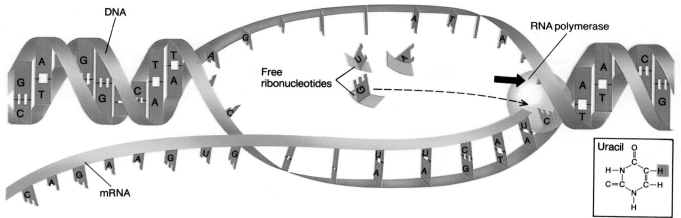

FIGURE 14-12

Transcription: dispatching the molecular messenger. Only one strand of DNA (called the "sense strand") encodes information in the message sent to the ribosomes. During transcription, the RNA polymerase binds to the DNA, the double helix is temporarily separated, the RNA polymerase recognizes the sense strand as the proper template, and the enzyme then assembles a complementary strand of RNA which grows toward its 3' end. (The uracil in RNA—see inset—that replaces thymine in DNA differs by the single chemical group shaded for emphasis.)

the National Institutes of Health, contained nucleotides with only one type of nitrogenous base—uracil. When these synthetic RNA strands (called "poly-U") were added to a test tube containing a bacterial extract with all 20 amino acids plus the materials necessary for protein synthesis (ribosomes and various soluble factors), the system followed the artificial messenger's instructions and manufactured a polypeptide. This polypeptide turned out to be polyphenylalanine, a polymer of the amino acid phenylalanine. Nirenberg had thus shown that the codon UUU specifies "phenylalanine."

◬ Over the next 4 years, synthetic mRNAs were used to test all 64 possible codons to determine which amino acids would be incorporated. The result was the universal decoder chart, or "genetic code," shown in Figure 14-13. (Instructions for reading the chart are provided in the accompanying figure legend.) The chart is "universal" because regardless of the type of cell—whether a bacterium, yeast, mushroom, redwood tree, or human—the same codons specify the same amino acids.[3] The codon CAC, for example, will always specify the insertion of histidine at the corresponding point in the polypeptide being assembled. The universality of the genetic code is powerful evidence that all organisms on earth have evolved from a common

ancestor at some early point in the history of life on this planet.

Translating the Message

Let us return to a natural mRNA molecule containing a spectrum of different codons. The presence of a particular codon triplet in mRNA orders the insertion of the corresponding amino acid in the growing polypeptide chain. The next codon in the mRNA would specify the insertion of the next amino acid in the polypeptide being synthesized. In this way, the entire message is read—codon by codon—until the polypeptide chain is completely assembled according to the instructions originally encoded in the DNA.

The mRNA-directed assembly of a polypeptide is called **translation** and is the most complex synthetic activity occurring in a cell. Translation is the process of protein synthesis; it requires mRNA, amino acids, numerous enzymes, ribosomes, and chemical energy in the form of ATP (and GTP). It also requires another type of RNA that decodes mRNA's encoded message (written in codons) and translates it into the language of proteins (amino acids). This "molecular decoder" is called **transfer RNA (tRNA).**

▨ Transfer RNAs are small molecules (as RNAs generally go), and each one is able to fold in a characteristic manner (Figure 14-13, right panel, Figure 14-14). The function of tRNAs is closely correlated with their three-dimensional shape. Once it has folded, one end of each tRNA molecule contains a unique sequence of three nucleotides which can form base pairs with one of the mRNA codons. The nucleotide triplet of the tRNA that recognizes and attaches to an mRNA codon is called the **anticodon.** On the side opposite

---

[3] Minor variations have been discovered in a few microorganisms. In *Paramecium,* for example, UAG = glutamine rather than "stop." Mitochondria, which have their own protein-synthesizing machinery, also have some minor differences in codon recognition. It is believed that these alterations evolved from the standard genetic code.

◁ THE HUMAN PERSPECTIVE ▷
The Dark Side of the Sun

We owe our lives to the rays of the sun. The energy in these rays are captured by photosynthetic protists and plants and used to manufacture complex organic molecules upon which all heterotrophic organisms—you included—depend. The sun is also important in less tangible ways: We thrive on warm, sunny days; we spend our vacations on sunlit beaches; and we become depressed when the weather turns dark and "gloomy." But the sun also emits a constant stream of ultraviolet rays that ages and mutates the cells of our skin. Ultraviolet radiation damages DNA by causing adjacent thymine bases to become covalently bonded to one another, forming a "dimer."

UV damaged DNA
(contains T-T dimer)

— C-C-T-A-T-T-A-G-C-A —

— G-G-A-T-A-A-T-C-G-T —

To repair the damage, the region containing the dimer must be cut out of the damaged strand, and the original nucleotides replaced.

The hazardous effects of the sun are most dramatically illustrated by the rare recessive genetic disorder, *xeroderma pigmentosum* (*XP*). Victims of XP possess a deficient repair system that is unable to remove segments of DNA damaged by ultraviolet light. Because of their genes, children with this disease are forced to live in a continually dark environment. Even brief exposure to the ultraviolet rays in sunlight increases the danger of severe skin damage and promotes the formation of skin tumors and other fatal cancers. Sleeping during the day behind blackened windows, playing indoors under conditions of low interior light, going outside only at night when there is no possibility of exposure to ultraviolet light—these children can die from

enjoying what most of us take for granted, a day in the sun.

Before you conclude that because you don't have xeroderma pigmentosum you don't have anything to worry about from exposure to the sun, consider the following statistics. Over 600,000 people develop one of three forms of skin cancer every year in the United States; most of these cases are attributed to overexposure to the sun's ultraviolet rays. Fortunately, the two most common forms of skin cancer—basal cell carcinoma and squamous cell carcinoma—rarely spread to other parts of the body and can usually be excised in a doctor's office. Both of these types of cancer arise from cells that form the bulk of the epidermis, the outer layer of the skin (Chapter 26).

Malignant melanoma, the third type of skin cancer, is a potential killer. Unlike the other types of skin cancer, melanomas

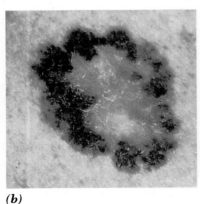
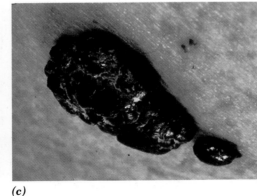

(a)   *(b)*   *(c)*

FIGURE 1
Stages in the growth of a melanoma. These malignant skin lesions are often characterized by rapid growth, an irregular boundary, variegation in color, and a tendency to become crusty and bleed.

the anticodon on the tRNA is the site that carries a specific amino acid. Amino acids are covalently linked to tRNAs by a special set of enzymes (referred to as "charging enzymes" in Figure 14-18).

Once the amino acid is linked to the tRNA, base pairing between the tRNA's anticodon and the mRNA's codon brings the corresponding amino acid into position for in-

corporation into the polypeptide being synthesized. For example, the codon UCU in mRNA binds only with the anticodon AGA of the tRNA carrying the amino acid serine. Therefore, UCU in mRNA instructs the cell to insert serine at that point in the newly forming polypeptide. In this way, tRNA bridges the language gap between the nucleotide codon and the amino acid.

develop from pigment cells in the skin. They may arise in an existing, nonmalignant mole, or they may appear where no previous mole is present. Every year, nearly 30,000 Americans are diagnosed with melanoma, and the number is climbing at an alarming rate due to the increasing amount of time people have been spending in the sun over the past few decades. Many scientists predict that the rate of melanoma will climb even more rapidly in the future if the UV-absorbing ozone layer of the atmosphere continues to deteriorate.

If a melanoma is detected and removed at an early stage, when it is small and has not yet penetrated into the deeper layers of the skin, the prognosis is very good. Unlike most tumors, melanomas appear on the surface of the skin where they can be observed, so there is no reason for such cancers to go undetected. Everyone should know the warning signs of a melanoma, which are depicted in Figure 1.

Light-skinned individuals are particularly susceptible to developing melanoma, as are those with close relatives who have had the disease; those who received a severe, blistering sunburn as a child; or those

who live in sunnier regions. For example, the highest incidence of melanoma is found in Australia, which has a sunny, tropical climate and is populated largely by light-skinned individuals of northern-European descent. Similarly, in the United States, the incidence of melanoma in Arizona is over twice that found in Michigan. The best way to avoid developing melanoma is to avoid sunbathing (Figure 2) and to wear sunblock creams when you plan to spend time in the sun. Most important: Avoid serious sunburns and be sure your children do the same.

FIGURE 2 ⎯⎯⎯

Four codons warrant special mention. Three (UGA, UAG, and UAA) have no corresponding amino acid (Figure 14-13). These triplets, called **stop codons** (or *nonsense codons*), spell "stop!" when mRNA is being translated into protein. They are used to terminate synthesis at the completion of a polypeptide. Another codon, AUG (specifying the amino acid methionine), is the "start" codon. AUG always appears at the beginning of the coding portion of an mRNA and initiates the assembly of amino acids into the polypeptide.

Translation occurs at ribosomes—complex particles composed of several dozen different proteins and a number of different RNA molecules. The RNAs of the ribosome, called **ribosomal RNAs (rRNAs),** constitute the third major type of RNA made by cells. Whereas mRNAs carry encoded information, and tRNAs serve to decode this information, rRNAs are primarily structural molecules functioning as a scaffold to which the various proteins of the ribosome can attach themselves. The proteins that make up the ribosome have varied functions: Some play a structural role in holding the particle together; others bind to the mRNA or tRNAs; still others act as enzymes involved directly in protein synthesis.

The ribosome is a nonspecific component of the translation machinery—sort of a "workbench"—in that any ribosome can serve as a translation site for any mRNA. This is why bacterial cells can be used as "pharmaceutical factories" to churn out proteins encoded by human mRNAs (Chapter 16). A functioning ribosome consists of a large and a small subunit (Figure 14-15). The ribosome assembles from its subunits at the start of polypeptide synthesis and then disassembles when synthesis has been completed.

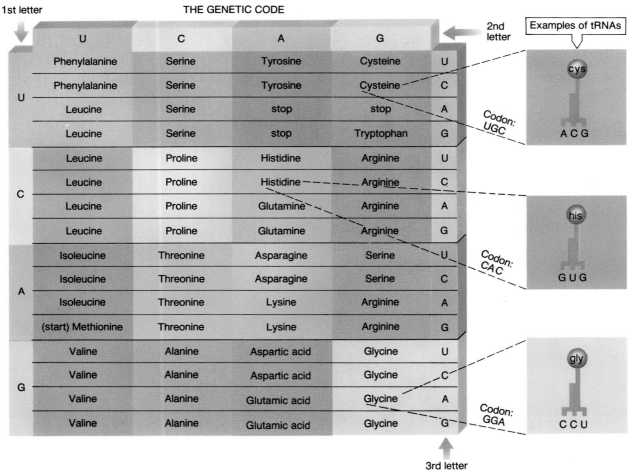

FIGURE 14-13

The genetic code. The genetic code is a universal biological language. The correlation between codon and amino acid indicated in this decoder chart is the same in virtually all organisms. To use the chart to translate the codon UGC, for example, find the first letter (U) in the indicated row on the left. Follow that row to the right until you reach the second letter (G) indicated at the top, then find the amino acid that matches the third letter (C) in the row on the right. UGC specifies the insertion of cysteine. Each amino acid (except two) has several codons that order its insertion. These are genetic "synonyms," backup systems that reduce the danger of lethal mutations disrupting the cell. A change in a single nucleotide in the codon's third position, for example, would change which amino acid would be incorporated into the polypeptide, unless the new codon were synonymous with the original (which happens frequently). In that case, the same amino acid would be incorporated, and the mutation would have no phenotypic effect. As discussed in the following section, decoding in the cell is carried out by tRNAs, a few of which are illustrated in the right side of the figure.

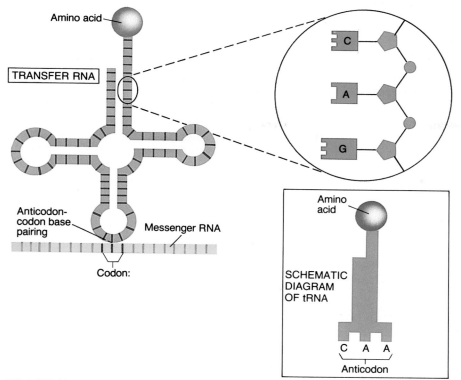

FIGURE 14-14

Molecular decoder. Transfer RNA molecules translate one "language" (a three-nucleotide codon "word") into another (a specific amino acid) by using anticodons. Each anticodon in a tRNA recognizes a single codon in mRNA. Similarly, each tRNA carries a specific amino acid. In this way, a particular tRNA associates the correct amino acid with its corresponding codon in mRNA. The "cloverleaf" structure illustrates how the anticodon base pairs with its complementary codon. The attachment of a specific amino acid to its corresponding tRNA is catalyzed by a set of amino acid "charging" enzymes.

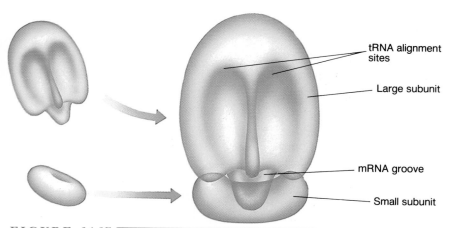

FIGURE 14-15

Assembly of a functional ribosome. Two subunits, one large and one small, fit together to form a groove for mRNA and create sites for accepting two tRNAs at a time. A component of the large subunit catalyzes the formation of a peptide bond that connects the two aligned amino acids. Thus, ribosomes are more than mere "workbenches" on which proteins are synthesized, they are more like "workers" that help assemble the protein.

An assembled ribosome has a groove through which the mRNA molecule can travel. The ribosome also has sites that position two tRNA molecules so that their amino acids end up adjacent to each other on the large subunit. The large subunit contains a factor that covalently links the adjacent amino acids. As the ribosome moves along the mRNA strand, amino acids are incorporated into the growing polypeptide chain in the order specified by the mRNA. The incorporation of each amino acid is accompanied by the hydrolysis of GTP (a high-energy compound similar to ATP), which provides the energy driving the assembly process. A simplified version of the events of translation is depicted in Figure 14-16. (Many of the enzymes and accessory molecules have been omitted for clarity.) The three stages of translation are initiation, chain elongation, and chain termination.

Step 1: Initiation The process of translation begins when the small subunit of a ribosome binds to the mRNA near its 5′ end. This binding always occurs at the initiation codon, AUG. The binding of the small subunit is followed by the attachment of the first tRNA, the *initiator tRNA,* whose anticodon is UAC. The initiator tRNA always brings a methionine as the first amino acid of the assembling polypeptide. The large subunit soon joins the complex, and the assembly of the polypeptide begins. The binding of the ribosome to the AUG codon fixes the "reading frame," assuring that translation begins with the correct nucleotide.

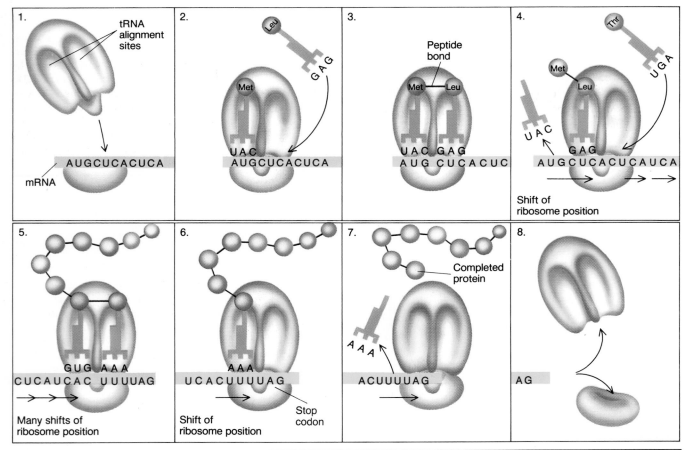

FIGURE 14-16

General steps in translation. Each of the three major steps in translation—initiation, elongation, and termination—is discussed in the text. (The mRNA in the figure is only a short portion of the entire molecule, which, for most proteins, would exceed 600 nucleotides in length.) The methionine inserted by the initiator tRNA as the first amino acid is usually clipped from the polypeptide by an enzyme.

If the message were initiated one or two nucleotides over, all the remaining triplets would be incorrectly read, and the wrong amino acids would be inserted, producing a totally useless polypeptide. The brackets below show the incorrect codons that would be produced if initiation began two nucleotides over from the proper site.

```
correct     →        met   leu   his   pro
                     ⌐⊓   ⌐⊓   ⌐⊓   ⌐⊓
mRNA        →   — A U G C U G C A U C C A —
                        ⌐⊓   ⌐⊓   ⌐⊓
incorrect   →            ala   ala   ser
```

Steps 2–5: Chain Elongation

With the initiator tRNA in place, a second free site remains on the ribosome, where another tRNA can align with the next codon in the mRNA (which is CUC, in this illustration). CUC pairs with the anticodon GAG of the tRNA carrying the amino acid leucine. The two amino acids become aligned next to each other, and the amino acid attached to the first tRNA is enzymatically transferred to the second amino acid, forming a covalent (peptide) bond. The first tRNA, which has lost its amino acid cargo, then departs, and the ribosome moves down the mRNA by three nucleotides, bringing the third codon into position. A tRNA molecule with a complementary anticodon binds to the third codon, orienting its amino acid next to the previous one, and a peptide bond now forms between them. The growing polypeptide chain is now three amino acids long. The second tRNA is then released, and the ribosome moves down the mRNA, bringing the fourth codon into position. This process continues, adding amino acids in the proper sequence until the entire polypeptide chain is synthesized according to the original genetic instructions.

Steps 6–8: Chain Termination

The completion of a polypeptide chain is signaled by the presence of a stop codon in the mRNA strand. Since these triplets do not specify amino acids, their presence produces a region on the mRNA to which no tRNA can bind. Therefore, the amino acid inserted just before the stop codon becomes the terminal member of the chain.

Protein synthesis occurs very rapidly, and its efficiency is increased by simultaneous translation of each mRNA by numerous ribosomes. As soon as one ribosome has translated the first few codons, the mRNA initiation site is again available for another ribosome to become attached. As a result, ribosomes are often found in chains held together by an mRNA strand. This complex is called a **polysome** (Figure 14-17). Since amino acids are being incorporated at each ribosome along the polysome, a single mRNA may be generating dozens of identical polypeptide chains simultaneously.

The ordered flow of genetic information from its stored form in DNA to its transcription into RNA and its expression as a specific protein is summarized in Figure 14-18.

Within 20 years after Avery's first indication of the importance of DNA, scientists had worked out the mechanism by which genetic information is stored, replicated, and used to run cellular activities—which is nothing less than the molecular basis of life itself. In so doing, these scientists have delivered a deeply satisfying message: The universe inside every living organism is within our cognitive reach. Life is explainable in physical and chemical terms, without the need to evoke mystical forces. (See CTQ #5.)

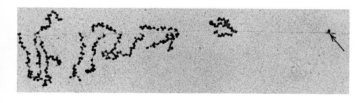

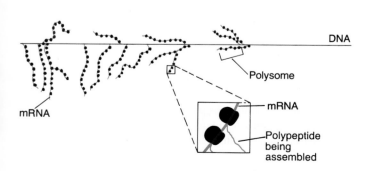

FIGURE 14-17

Gene expression—caught in the act. In bacteria, ribosomes attach to an mRNA molecule as it is being synthesized and begin translation (as shown in this photograph). In both prokaryotic and eukaryotic cells, a single mRNA is translated by a number of ribosomes that follow each other along the mRNA. The complex of ribosomes held together by an mRNA molecule is called a polysome. Colored lines emerging from the ribosomes depicted in the box indicate that each ribosome is synthesizing a polypeptide chain.

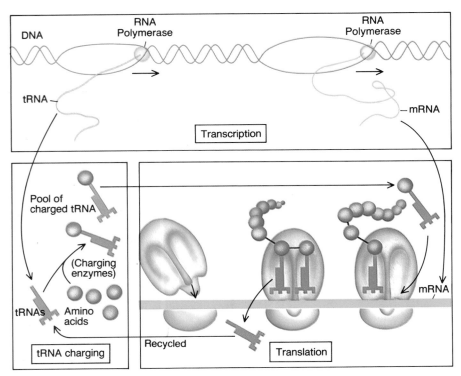

DNA
RNA Polymerase
RNA Polymerase
tRNA
mRNA
Transcription

Pool of charged tRNA
(Charging enzymes)
tRNAs   Amino acids
mRNA
Recycled
tRNA charging
Translation

FIGURE 14-18
Overview of protein synthesis. All the processes depicted here occur in a single cell. The steps are compartmentalized for clarity. The genetic information in DNA ultimately dictates the amino acid sequence in protein, explaining why cells are biologically "obedient" to their genes. A charged tRNA is a tRNA with an attached amino acid, and a charging enzyme is an enzyme that attaches a specific amino acid to the appropriate tRNA (one with the appropriate anticodon).

THE MOLECULAR BASIS OF GENE MUTATIONS

If genetic information is stored in the linear sequence of nucleotides along the strands of a DNA molecule, it follows that changes in that sequence will alter the DNA's information content. This is the basis of a **gene mutation.** As we saw in Chapter 4, sickle cell anemia is a genetic disease that results from the substitution of one amino acid (a valine) for another (a glutamic acid) in the oxygen-carrying protein hemoglobin. Looking back at Figure 14-13, notice the various codons for glutamic acid and valine. All codons for both amino acids begin with G, but the second nucleotide for the glutamic acid codons is A, while that for valine is U. Herein lies the basis for the gene mutation that causes sickle cell anemia. (Remember, codons represent letters in the mRNA.) A change from CTC to CAC in the DNA produces a change from GAG to GUG in the mRNA codon, which, in turn, produces a change from a glutamic acid to a valine in the corresponding polypeptide.

The substitution of one base for another in the DNA is called a **point mutation** because it only affects a single "point" in the gene. This type of mutation can occur during replication, when an incorrect nucleotide is incorporated into the growing DNA chain, or it may occur as the result of damage to the DNA of a nonreplicating cell. The likelihood of mutation is greatly increased if the DNA happens to come into contact with a **mutagen**—a chemical or physical agent that induces genetic changes.

A more serious mutation occurs to a gene when one or two nucleotides are added or deleted because such an alteration throws off the reading frame of the rest of the mRNA molecule, changing all the codons "downstream" from the point of change. As a result, the message is read in the wrong sequence, generating a useless protein that contains a string of incorrect amino acids beyond the shift point. This type of *frameshift mutation,* as it is called, is responsible for certain cases of human *thalessemia,* a type of anemia (a deficiency of red blood cells) resulting from the production of abnormal hemoglobin molecules.

Any agent capable of causing gene mutations is also a potential carcinogen (cancer-causing agent) since the alteration of certain types of genes (including oncogenes and tumor-suppressor genes) can lead to the transformation of a normal cell into a malignant cell. Because of this relationship between mutation and cancer, large numbers of substances can be screened for their carcinogenicity by determining whether or not they are capable of causing mutations. To carry out this test—called the *Ames test,* after Bruce Ames of the University of California, who developed the procedure—a culture of bacteria is simply exposed to the chemical in question, and the number of detectable mutant cells produced is counted. This number is then compared to the number of mutants that appear in a control culture treated the same way but without exposure to the chemical being tested. An increased mutation rate in the chemically treated bacteria flags the substance as a potential human carcinogen. This procedure has detected cancer-causing potential in many substances to which people are frequently exposed, including materials in hair dyes, cured meats, artificial food colors, and cigarette smoke. Ames has become a controversial figure in the current debate over the danger of carcinogenic environmental pollutants by determining that many of the "natural" foods we eat, such as sprouts and mushrooms, contain rather high levels of compounds capable of causing mutations.

Chemicals are not the only agents capable of causing mutations: DNA is also very susceptible to damage by radiation. In fact, one of the most common mutagenic agents is ultraviolet radiation—the subject of The Human Perspective: The Dark Side of the Sun.

Mutagens do not only arise in the external environment; they are also produced in large numbers within the body as the result of normal metabolic reactions. To appreciate the magnitude of the mutation "problem," consider the estimate that, on the average, several thousand bases are lost from the DNA of *every human cell every day!* How is the DNA of a cell able to maintain a nucleotide sequence necessary for life while absorbing such molecular punishment? The answer is that cells contain a diverse array of DNA **repair enzymes** that patrol the DNA, searching for alterations and distortions they can recognize and repair. In this regard, a cell might be likened to a car driving around at all times with a team of mechanics inside who monitor and repair problems as they develop, while the car carries out all of its normal activities. DNA repair systems provide another example of the mechanisms that maintain the order and complexity of the living state.

■▶ While genetic mutations are generally thought to be harmful, resulting in decreased fitness, some mutations will inevitably be beneficial, providing an individual with an increased chance to survive and produce offspring. As we discussed in Chapter 12, mutations are the raw material of evolution since they introduce new genetic information into a population. Without these spontaneous changes in the DNA, the level of genetic variability would be greatly limited, and evolution, as we know it, could not have occurred.

Life on earth is exposed to a variety of destructive agents that can act on DNA and alter its information content. If organisms lacked mechanisms to repair genetic alterations, life could not continue. Yet, if all of these alterations were repaired, there would be no mutation and, thus, no variation or biological evolution. (See CTQ #6.)

REEXAMINING THE THEMES

Relationship between Form and Function

⚄ Although relatively simple, the structure of DNA explains a great deal of biological function. Most importantly, it reveals how information—genetic instructions—can be stored in the linear sequence of nucleotides. Elucidation of the structure of DNA immediately suggested a likely mechanism of replication; the two strands would separate, and each would act as a template for the assembly of a complementary strand. The *translation* of the information in DNA into a polypeptide chain is more complex. The assembly of a polypeptide requires the intervention of a tRNA decoder—a molecule whose structure allows it to translate nucleotide languages into amino acid languages. Each tRNA carries an anticodon at one end, which interacts with an mRNA codon, and an amino acid at the other end, which is incorporated into the assembling polypeptide.

Biological Order, Regulation, and Homeostasis

🔄 Biological information is encoded in highly ordered sequences of nucleotides. The deletion of a single nucleotide in a DNA molecule can lead to the formation of a messenger RNA molecule that will direct the assembly of a meaningless string of incorrect amino acids. The ordered sequence of nucleotides in the DNA is maintained by a highly accurate mechanism of replication, which includes a molecular proofreading system, and a battery of repair enzymes that patrol the genetic material, discovering and repairing nucleotide damage. Maintaining the stability of genetic information can be thought of as a type of molecular homeostasis.

Unity within Diversity

🔺 Nowhere is the unity and shared ancestry of life better revealed than in the genetic code. Pick any three nucleotides to make up a codon, and they spell the same amino acid (or stop message) in a bacteriophage, a human, or a lilac bush. Introduce human mRNAs for insulin into a bacterial cell, and that bacterium begins producing human insulin indistinguishable from that made by your own pancreas. The diversity of life is also derived from the genetic code. Over periods of time, mutations cause changes in the sequence of nucleotides in a gene, thereby changing the codons of the gene and the amino acids incorporated into the corresponding polypeptide. The cumulative effect of gradual changes in polypeptides over evolutionary time has been the generation of life's diversity.

Evolution and Adaptation

▮▶ Evolutionary change from generation to generation depends on genetic variability. Much of this variability arises from reshuffling maternal and paternal genes during meiosis, but somewhere along the way *new* genetic infor-

mation must be introduced into the population. New genetic information arises from mutations in existing genes. Some of these mutations arise during replication; others occur as the result of unrepaired damage as the DNA is just "sitting" in a cell. Mutations that occur in an individual's germ cells can be considered the raw material on which natural selection operates; whereas harmful mutations produce offspring with a reduced fitness, beneficial mutations produce offspring with an increased fitness.

SYNOPSIS

Experiments in the 1940s and 1950s established conclusively that DNA is the genetic material. These experiments included the demonstration that DNA was capable of transforming bacteria from one genetic strain to another; that bacteriophages injected their DNA into a host cell during infection; and that the injected DNA was transmitted to the bacteriophage progeny.

DNA is a double helix. DNA is a helical molecule consisting of two chains of nucleotides running in opposite directions, with their backbones on the outside, and the nitrogenous bases facing inward like rungs on a ladder. Adenine-containing nucleotides on one strand always pair with thymine-containing nucleotides on the other strand, likewise for guanine- and cytosine-containing nucleotides. As a result, the two strands of a DNA molecule are complementary to one another. Genetic information is encoded in the specific linear sequence of nucleotides that make up the strands.

DNA replication is semiconservative. During replication, the double helix separates, and each strand serves as a template for the formation of a new, complementary strand. Nucleotide assembly is carried out by the enzyme DNA polymerase, which moves along the two strands in opposite directions. As a result, one of the strands is synthesized continuously, while the other is synthesized in segments that are covalently joined. Accuracy is maintained by a proofreading mechanism present within the polymerase.

Information flows in a cell from DNA to RNA to protein. Each gene consists of a linear sequence of nucleotides that determines the linear sequence of amino

acids in a polypeptide. This is accomplished in two major steps: transcription and translation.

During transcription, the information spelled out by the gene's nucleotide sequence is encoded in a molecule of messenger RNA (mRNA). The mRNA contains a series of codons. Each codon consists of three nucleotides. Of the 64 possible codons, 61 specify an amino acid, and the other 3 stop the process of protein synthesis.

During translation, the sequence of codons in the mRNA is used as the basis for the assembly of a chain of specific amino acids. Translating mRNA messages occurs on ribosomes and requires tRNAs, which serve as decoders. Each tRNA is folded into a cloverleaf structure with an anticodon at one end—which binds to a complementary codon in the mRNA—and a specific amino acid at the other end—which becomes incorporated into the growing polypeptide chain. Amino acids are added to their appropriate tRNAs by a set of enzymes. The sequential interaction of charged tRNAs with the mRNA results in the assembly of a chain of amino acids in the precise order dictated by the DNA.

Mutation is a change in the genetic message. Gene mutations may occur as a single nucleotide substitution, which leads to the insertion of an amino acid different from that originally encoded. In contrast, the addition of one or two nucleotides throws off the reading frame of the ribosome as it moves along the mRNA, leading to the incorporation of incorrect amino acids "downstream" from the point of mutation. Exposure to mutagens increases the rate of mutation.

Key Terms

molecular biology (p. 268)
bacteriophage (p. 268)
deoxyribose (p. 269)
base pairing (p. 272)
complementarity (p. 272)
histone (p. 273)
nucleosome (p. 273)
replication (p. 275)
template (p. 275)

semiconservative replication (p. 275)
DNA polymerase (p. 275)
origin (p. 276)
messenger RNA (mRNA) (p. 278)
transcription (p. 278)
RNA polymerase (p. 278)
sense strand (p. 278)
codon (p. 278)
translation (p. 279)

transfer RNA (tRNA) (p. 279)
anticodon (p. 279)
stop codon (p. 281)
ribosomal RNA (rRNA) (p. 281)
polysome (p. 285)
gene mutation (p. 286)
point mutation (p. 286)
mutagen (p. 286)
repair enzyme (p. 287)

Review Questions

1. How is complementary base pairing important to inheritance and to the expression of genetic messages?

2. Describe at least two similarities and two differences between the processes of DNA replication and transcription.

3. If a portion of DNA had the nucleotide sequence AGCAGGCAGC, would you be able to predict the next nucleotide in the chain? Why or why not?

4. Why are tRNA molecules, rather than mRNA molecules, considered the decoders?

5. Considering the importance of maintaining the correct reading frame during translation, what would be the effect of (1) a single nucleotide deletion from the DNA, or (2) the loss of three successive nucleotides?

Critical Thinking Questions

1. The use of DNase was an important part of Avery's studies, pointing to DNA as the transforming principle. DNase is purified from the pancreas, an organ that produces a variety of digestive enzymes. With this in mind, can you think of any other possible explanations for Avery's results? Is there any control that Avery might have run to eliminate this possibility?

2. If you were teaching biology to college students in 1942, would you tell them that the genetic material was protein or nucleic acid? What evidence would you have used to support your position? If you were still teaching in 1952, what would you tell them, and what evidence would you draw on?

3. How do you think the discovery of the molecular structure and function of DNA affects the argument between vitalists and mechanists? (See Critical Thinking Question, #2, Chapter 1.)

4. Suppose that protein, not DNA, were the genetic material. How would the results of the Hershey-Chase experiment have been different?

5. Using the genetic code chart in Figure 14-13, construct the polypeptide chain coded for in a strand of DNA with the following sequence of bases: TACGGATCGCCTACG. (Remember to include the mRNA sequence.)

6. If mutations play an important role in evolution, why are many scientists concerned about the mutagenic effects of x-rays, radiation from nuclear power plants, chemicals that cause mutations, etc.?

7. What would be the effect on the offspring if a DNA polymerase were absolutely foolproof in its proofreading activity? What would be the long-term effect on biological evolution?

8. Suppose the first synthetic polynucleotide that Nirenberg synthesized had been poly A. What type of polypeptide would have been assembled in his experiment? What if he had been able to synthesize the polynucleotide AUAUAUAU . . . by polymerizing the dinucleotide AU? What polypeptide would that polynucleotide encode?

Additional Readings

Alberts, B., et al. 1989. *Molecular Biology of the Cell.* 2d ed. New York: Garland. (Intermediate-Advanced)

Ames, B. N., et al. 1987. Ranking possible carcinogenic hazards. *Science* 236:271–280. (Intermediate-Advanced)

Dubos, R. J. 1976. *The professor, the institute, and DNA: Oswald T. Avery.* New York: Rockefeller University Press. (Intermediate)

Jaroff, L. 1990. Special report on skin cancer. *Time,* July 23: 68–70. (Introductory)

Jaroff, L. 1993. Happy birthday, double helix. *Time,* March 15:56–59. (Introductory)

Judson, H. F. 1979. *The eighth day of creation.* New York: Simon & Schuster. (Intermediate)

McCarty, M. 1985. *The transforming principle: Discovering that genes are made of DNA.* New York: Norton. (Intermediate)

Prescott, D. 1988. *Cells.* Boston: Jones and Bartlett. (Intermediate)

Radman, M., and R. Wagner, 1988. The high fidelity of DNA duplication. *Sci. Amer.* Feb:40–47. (Intermediate)

Watson, J. D. 1969. *The double helix.* New York: Dutton. (Introductory)

Watson, J. D., and F. H. C. Crick, 1953. Molecular structure of nucleic acids. A structure of deoxyribose nucleic acid. *Nature* 171:737–738. (The original paper describing the structure of DNA.)

Orchestrating Gene Expression

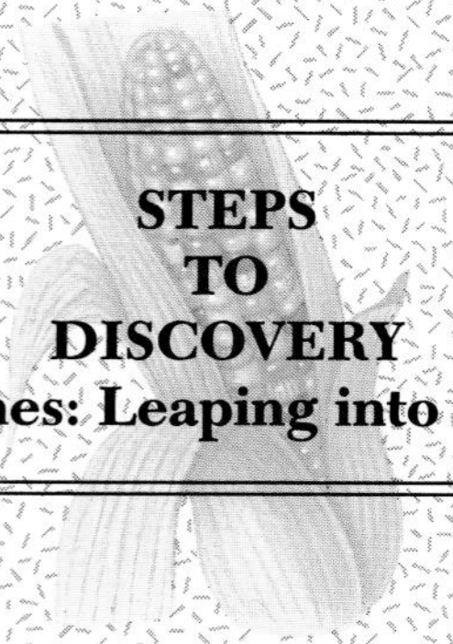

STEPS
TO
DISCOVERY
Jumping Genes: Leaping into the Spotlight

**WHY REGULATE GENE
EXPRESSION?**

The Basis of Gene Regulation

**GENE REGULATION IN
PROKARYOTES**

The Bacterial Operon

**GENE REGULATION IN
EUKARYOTES**

Cell Differentiation and Specialization

DNA that is Never Expressed

**LEVELS OF CONTROL OF
EUKARYOTIC GENE EXPRESSION**

Regulating Gene Expression at the
Transcriptional Level

Regulating Gene Expression at the RNA
Processing Level

Regulating Gene Expression at the
Translational Level

BIOLINE

RNA as an Evolutionary Relic

THE HUMAN PERSPECTIVE

Clones: Is There Cause For Fear?

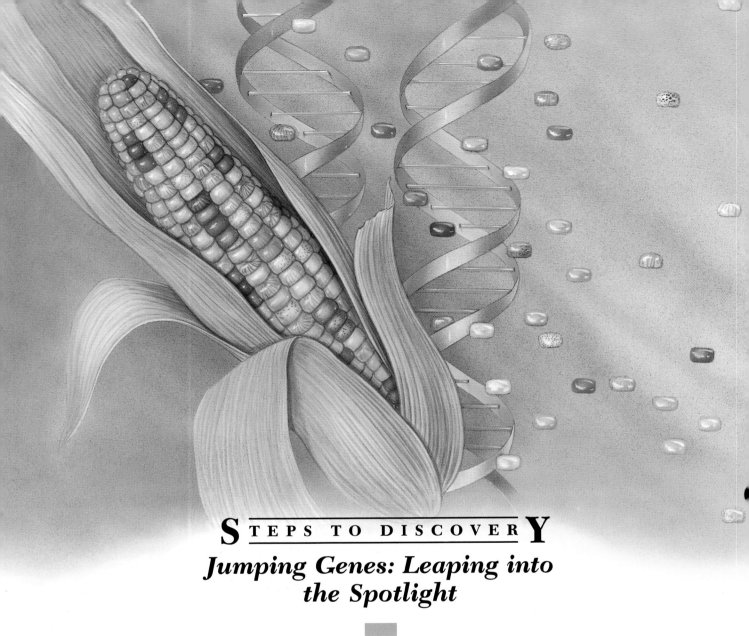

Jumping Genes: Leaping into the Spotlight

*D*uring the first quarter of this century, it was very difficult for women to gain admission to graduate schools in science at major U.S. universities. One exception was Cornell University, which had earned a reputation for its fair treatment of women; it was there that Barbara McClintock received her Ph.D. in botany in 1927. McClintock's research thesis was on the genetics of maize (*Zea mays*), the indigenous species from which agricultural corn was developed. McClintock was the first to identify and characterize the ten different chromosomes of the species and to localize a variety of genetic markers at specific sites on these chromosomes.

In 1931, McClintock, and a student, Harriet Creighton, published one of the classic papers in genetics, one that finally established direct visual "proof" that crossing over during meiosis actually involved the physical exchange of segments of homologous chromosomes. Until that time, it had been impossible to distinguish between maternal and paternal homologues under the microscope, even though the reshuffling of genetic characteristics in offspring (page 250) pointed *indirectly* to chromosomal exchange. McClintock and Creighton developed strains in which maternal and paternal homologues carried extra pieces that enabled them to be microscopically distinguished. In this way, they were able to show the actual exchange of chromosomal segments.

As difficult as it was for women to receive advanced degrees in science in the 1920s and 1930s, it was even more difficult for them to secure academic research positions. Harriet Creighton accepted a teaching position at Connecticut College for Women, a small college far from the genetics limelight. McClintock was able to continue her

Changes in the patterns of inheritance of colors and streaks on kernels of maize led Barbara McClintock to the conclusion

research by accepting a series of postdoctoral fellowships in the laboratories of colleagues. It wasn't until 1936, after 9 years of postdoctoral work had given her a worldwide reputation, that she was finally offered a faculty position at the University of Missouri—at a salary well below that of her male colleagues of lesser experience. When she wasn't promoted, McClintock left Missouri in 1941 and accepted a research position at the Cold Spring Harbor Laboratories in New York, where she remained throughout the rest of her career.

By 1945, McClintock had become the President of the Genetics Society of America and only the third female member of the prestigious National Academy of Sciences. About the same time, she began a new series of experiments aimed at determining the mechanism by which gene expression was regulated. The previous year, Avery had reported his findings that DNA was the substance responsible for transformation—a finding that, in the minds of many geneticists, marked the beginning of molecular biology. Within a few years, most of the bright young geneticists had turned from studying fruit flies and maize to examining bacteria and bacterial viruses. Just as the geneticists of an earlier period had failed to accept the work on bacteria, this new crop of "molecular biologists" were paying relatively little attention to more traditional genetic studies on plants and animals.

Genetic traits in maize are expressed primarily as changes in the patterns and markings in leaf and kernel coloration. McClintock carefully monitored such changes from generation to generation and discovered that certain mutations did not remain stable over time. After several years of careful study, she concluded that certain genes were moving from one place in a chromosome to an entirely different site, affecting gene expression. She called this movement of genetic elements *transposition* and suggested that these elements were involved in gene regulation.

Meanwhile, molecular biologists working with bacteria were finding no evidence of gene transposition. To them, genes appeared as stable elements situated in a linear array on the chromosome, an array that remained constant from one individual to another and from one generation to the next. McClintock's results were ignored. It didn't help any that her papers were written in a dense, difficult-to-understand style and concerned an organism whose genetics were very complex and totally unfamiliar to other prominent investigators. Molecular biologists were talking about DNA sequences, and McClintock was talking about patterns of colored streaks in the leaves and kernels of corn! Just as Gregor Mendel had spoken to his colleagues in "a different language" 100 years earlier, Barbara McClintock was having the same trouble presenting her foreign ideas to her peers. According to Evelyn Keller, McClintock's biographer, "Her talk at the Cold Spring Harbor Symposium that summer (1951, at a major symposium on genes and muta-

tions) was met with stony silence. With one or two exceptions, no one understood. Afterward, there was mumbling —even some snickering—and outright complaints. It was impossible to understand. What was this woman up to?"

Despite the fact that McClintock's hypothesis concerning transposition was an elegant model backed by years of rigorous experimental evidence, her ideas were all but forgotten until the late 1960s and early 1970s. At that time, genetic elements were discovered in bacteria which were capable of moving from one place on the chromosome to another. From a practical viewpoint, the most important of these so-called jumping genes was an antibiotic-resistance gene in the bacterium *Salmonella* (an agent responsible for food poisoning), which seemed to move around the chromosome. Because of its capability for movement, this drug-resistance gene is able to spread from one bacterium to another, producing strains of disease-causing bacteria that are difficult to treat with antibiotics.

Within a few years, the movement of genetic elements within and between chromosomes was also shown to occur in fruit flies. It was also demonstrated to be the mechanism by which antibody genes are assembled during the differentiation of cells of the immune system (Chapter 30) and was postulated to be one of the key mechanisms driving the formation of new genes during evolution. McClintock's work on transposition was finally recognized, and she received the Nobel Prize in 1983 at age 81.

Although mobile genetic elements do not appear to play a wide role in gene regulation, new evidence of their importance in other genetic functions continues to be uncovered. For example, Margaret Kidwell and her colleagues at the University of Arizona have recently found evidence that parasitic mites (tiny, ubiquitous arthropods) may carry jumping genes from one species of insect host to another. The movement of genes across species' barriers provides the opportunity for the input of new genes into a species population and the possibility of rapid evolutionary changes. Haig Kazazian and his colleagues at Johns Hopkins University have recently discovered two patients with hemophilia, whose disease is a result of a mobile genetic element that has "jumped" into the middle of one of the key genes involved in blood clotting. This is the first example of a human disease being caused by a mobile genetic element.

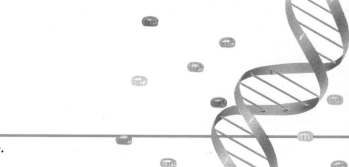

that genes were capable of changing their location on a chromosome.

*I*f it happened to you, it would be a disaster. But to a salamander, losing a portion of a leg is a relatively small sacrifice to make to escape becoming a hungry predator's dinner. The salamander soon regenerates the missing limb (Figure 15-1) by performing a feat of genetic "alchemy," turning one type of cell into another. For example, muscle cells remaining in the stump may lose their specialized cellular organelles and become transformed into unspecialized cells, which then form cartilage cells in the regenerating limb. A muscle cell could only become a cartilage cell if it retained the genes needed for cartilage formation—genes that were "silenced" while the cell was part of a muscle.

The ability of a muscle cell to become transformed into a cartilage cell is a small feat when compared to certain specialized plant cells that can give rise to an entire plant (see The Human Perspective: Clones: Is There Cause for Fear?). The conversion of one type of cell into another type requires a major shift in gene expression, but these are exceptional cases. Let's begin with a discussion of the need for regulating gene expression under more typical conditions.

▼ ▼ ▼

WHY REGULATE GENE EXPRESSION?

In spite of their differences in form and function, the various cells in a multicellular organism—including the nerve cells, liver cells, and cartilage cells—have a complete set of genes. The genetic information present in each of these cells can be compared to a book of blueprints prepared for the construction of a large, multipurpose building. During the construction process, all of the blueprints will probably be needed, but only a small subset of this information needs to be consulted during the work on a particular floor or room. Similarly, cells carry many more instructions (genes) than they will use at any given time.

Consequently, cells contain mechanisms that allow them to express their genetic information *selectively*, following only those instructions needed for the tasks at hand. The distinct characteristics of each cell depend on which genes are expressed and which remain "silent" (Figure 15-2).

Selective gene expression presents one of the most compelling puzzles in biology: How are different genes turned on and off in response to internal or external signals? This question goes beyond determining whether a cell will become a muscle or a cartilage cell during embryonic development or regeneration. Selective gene expression is also essential for the homeostatic function of an organism. Here are just a few of the phenomena that depend on controlled gene expression:

- *Response to changes in the chemical environment* Organisms conserve energy and materials by leaving most genes inactive until their gene products are needed. For example, certain digestive enzymes that function in your intestine are produced only when there is food in the digestive tract ready to be digested.

- *Response to changes in the physical environment* Fluctuations in temperature, moisture, pH, and photoperiod (day length) can trigger changes in gene expression (Figure 15-2b). For example, even though a particular plant has the genetic capacity to produce flowers, the genes remain unexpressed until the correct season.

- *Response to hormones* Many hormones trigger metabolic or structural changes in target cells which require the activation or suppression of gene activity. For example, the white of a chicken egg consists largely of the protein ovalbumin. In an immature hen, the gene that codes for this protein is not expressed. As the bird matures, however, its ovary produces the hormone progesterone, which acts on target cells of the reproductive tract, which produce and secrete large amounts of ovalbumin.

THE BASIS OF GENE REGULATION

The mechanism by which a cell awakens a "sleeping" gene is one of the most important, complex, and intensively studied processes in molecular biology. The earliest studies

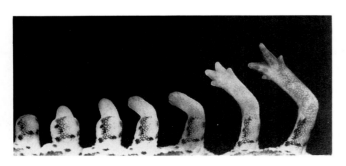

FIGURE 15-1

Limb regeneration. If a portion of the limb of a salamander is lost, the animal simply regenerates the missing part. During the regeneration process, cells that are differentiated as one cell type, such as a muscle cell, are capable of losing their differentiated properties and redifferentiating into a different cell type, such as a cartilage cell. This transformation would presumably require the deactivation of muscle-specific genes and the activation of cartilage-specific genes.

focused on bacterial cells and led to rapid rewards; by the end of the 1960s, biologists had a rather clear understanding of the mechanisms by which prokaryotic gene expression is controlled. Understanding selective gene expression in eukaryotic cells has proven to be a much more formidable task but, after more than 2 decades, biologists are finally gaining insight into the underlying mechanisms involved. Although fundamental differences exist in the ways pro-

karyotes and eukaryotes control gene expression, there is one unifying similarity: In both types of cells, specific genes are turned on or off as a result of direct physical interaction with **gene regulatory proteins**. In most cases, the regulatory sites in the DNA are located close to, but outside of, the gene itself.

We saw in Chapter 6 how the shapes of enzymes enable them to recognize a specific substrate. Similarly, gene reg-

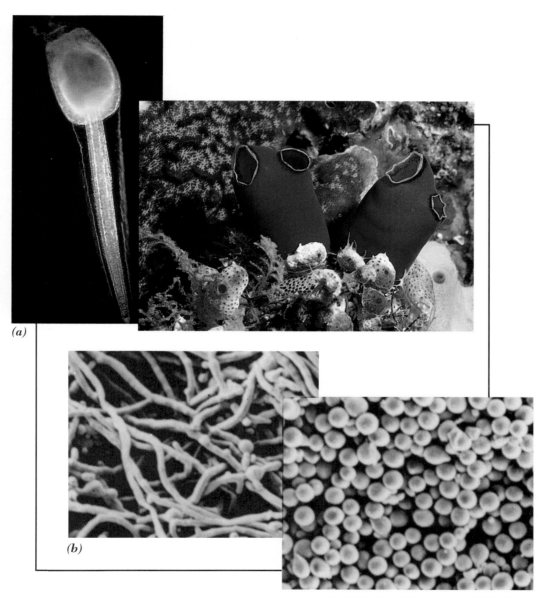

FIGURE 15-2

Two examples of gene orchestration. *(a) During normal development.* The life cycle of a sea squirt (a member of the same phylum containing humans) is characterized by two strikingly different body forms. The fertilized egg develops into a free-swimming larva that resembles the familiar tadpole. After swimming in the ocean for some time, the larva settles and metamorphoses into the adult body form shown on the right. The striking distinctions in form and function are due to the selection of different genes for expression as growth and development proceed. *(b) In response to environmental change.* Many fungi can "turn off" genes for growth in the mold form (fluffy and filamentous) and "switch on" a different block of genes, those for unicellular growth (the yeast phase). Moisture and temperature determine which block of genes is expressed.

◁ THE HUMAN PERSPECTIVE ▷
Clones: Is There Cause for Fear?

"The right Hitler for the right future. A Hitler for the 1980s, '90s, 2000s." These words weren't spoken by a Nazi official during World War II but by Gregory Peck in his role as Josef Mengele in the movie "The Boys From Brazil." In the film, Nazi hunter Ezra Lieberman (played by Laurence Olivier) stumbles across two 14-year-old boys who look *exactly* alike, one living in Germany and the other in the United States. The boys are not simply identical twins but two of 94 genetically identical *clones* created by Mengele in the jungles of Brazil. What is a clone? Is it possible to produce one? If so, what would be required?

Clones are asexually produced offspring that are genetically identical to their one-and-only parent. People have been cloning certain types of organisms for thousands of years; nature has been doing it for billions of years! Every time a "cutting" is taken from a plant for vegetative propagation, for example, an organism is being cloned. The resulting plant is genetically identical to the original from which it was cut, reproduced by mitotic cell divisions alone. Mitosis diligently preserves the organism's genetic makeup rather than scrambling it, as occurs during meiosis and fertilization. In raising oranges, apples, avocados, and dozens of other types of fruits, cutting techniques replaced the practice of growing plants from seeds long ago.

Scientists have also attempted to develop laboratory techniques for cloning organisms. Their primary goal has not been to produce identical organisms with particular genetic traits but to answer a basic biological question: Does a cell that has acquired specialized properties, such as a leaf cell of a tree or a skin cell of a mammal, still contain all of the genetic information necessary to generate an entire individual? In 1958, Frederick Steward and his colleagues at Cornell University isolated root cells from a mature plant and placed the

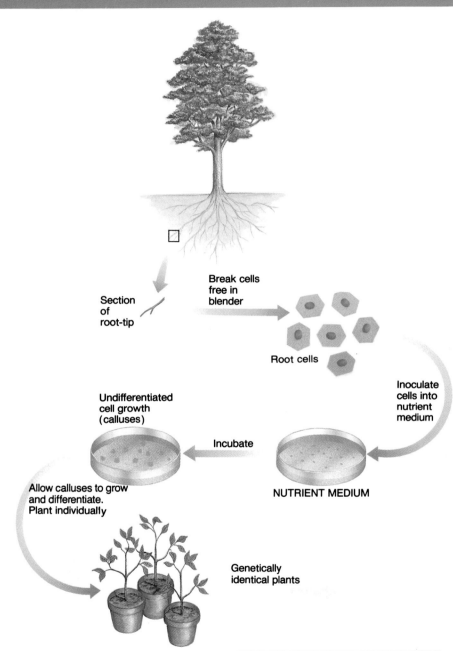

Break cells free in blender

Section of root-tip

Root cells

Inoculate cells into nutrient medium

Undifferentiated cell growth (calluses)

Incubate

NUTRIENT MEDIUM

Allow calluses to grow and differentiate. Plant individually

Genetically identical plants

FIGURE 1

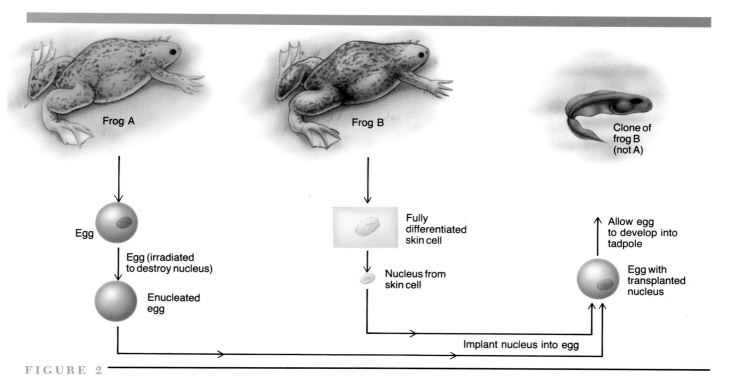

F I G U R E   2

cells in an appropriate growth medium, where they proliferated into a tumorlike mass called a *callus*. The calluses could then be grown into a fully developed plant, containing all the various cell types normally present (Figure 1). We can conclude that the original root cell placed into culture is **totipotent**—capable of giving rise to any of the organism's cell types or even of generating a whole individual. It is evident that plants can be cloned by this technique, but is the same true for animals?

To date, no one has been able to induce a differentiated animal cell to develop into a whole animal. Rather than trying to clone animals from isolated cells, scientists have focused on isolated nuclei in an attempt to demonstrate that the nucleus of a fully differentiated cell retains all of the genes originally present in the zygote. To carry out this experiment, a single nucleus is isolated from a cell to be tested and then transplanted into the cytoplasm of an egg cell that has no nucleus. (The egg's own nucleus is previously destroyed by irradiation with ultraviolet light.) The development of this egg then proceeds under the

direction of the genes from the transplanted nucleus.

The result of one such experiment involving the transplantation of a nucleus from an adult skin cell is shown in Figure 2. The egg that receives this transplanted nucleus does not develop into an adult frog but forms a well-developed tadpole that contains a variety of fully differentiated cells, in addition to skin cells. It is believed that these tadpoles do not develop into adults due to damage sustained by the nuclei following transplantation into the egg. Regardless, the nucleus of the skin cell must retain genes needed to produce many other types of cells.

To date, experiments with mammals have produced results similar to those with amphibians: Investigators have been unable to obtain an adult animal from an egg containing a nucleus transplanted from an *adult* cell. For the sake of argument, let's assume that the present failure in cloning mammals from transplanted nuclei is the result of technical problems that can be overcome. The question remains: Is there cause for fear? What are the prospects that

some future totalitarian regime may try to utilize these techniques to produce a "master race"? Obviously, this type of question is impossible to answer.

Members of the scientific community are currently banned from carrying out any type of nuclear-transplantation studies using human embryos but not other mammals. Many researchers believe that nuclear-transplantation techniques can be developed to improve livestock, cloning thousands of copies of an especially hardy individual or a prolific milk producer. Most informed citizens believe that preventing avenues of research that can provide answers to basic biological questions or increase agricultural output is not the answer. Rather, the scientific community must maintain constant vigilance and open discussion to ensure that such technologies are not put to an improper use. In this regard, the more informed the average citizen is about both the benefits and dangers of modern technology, the better he or she will be able to influence the way in which governments use or abuse this technology.

"Finger" of gene
regulatory protein

(a)

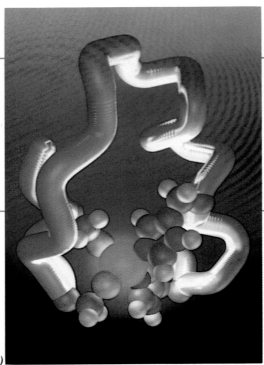

(b)

FIGURE 15-3

Interaction of a gene regulatory protein with the DNA double helix. *(a)* The shape of this gene regulatory protein enables it to recognize a specific nucleotide sequence in the DNA. Many of these proteins, such as the one depicted here, contain "fingers" that fit precisely and specifically into successive grooves in the DNA molecule. *(b)* Three-dimensional model of one of the protein's fingers. The greenish sphere is a zinc ion that helps hold the shape of the polypeptide.

ulatory proteins (Figure 15-3) have such a precisely determined shape that they are able to recognize one specific sequence of nucleotides in the DNA while ignoring millions of other sequences constructed out of the same four nucleotides. While different regulatory proteins are shaped to recognize different sequences, many of them have "fingers" that fit into the grooves of the DNA at the site of recognition. These fingers allow the protein to hold onto the DNA in a manner analogous to the way your hand would use its fingers to grip a bowling ball. The evolution of sequence-specific DNA-binding proteins was an important step in the pathway leading to complex cellular life forms.

While each living cell contains a complete set of genetic instructions, only part of that information is usefully expressed in a particular cell at a particular time. That part of the genetic blueprint expressed by a cell may change over time. (See CTQ #2.)

GENE REGULATION IN PROKARYOTES

Bacterial cells live in environments whose chemical composition may undergo drastic change. A particular nutrient molecule—sucrose, for example—may be available at one time but absent at another time. Selective gene expression enables these cells to use their limited space and resources efficiently. This can be illustrated by considering the challenges faced by a population of *E. coli*, a species of bacteria that normally lives in the human intestine. Let us consider two situations.

1. It's a hot day and you're not very hungry; you decide you'll have a milkshake for lunch. Milk and ice cream contain the disaccharide lactose (page 70), which your intestinal bacteria can use as an energy source. First, however, the bacterium must split the disaccharide into its two component sugars, glucose and fructose, a reaction catalyzed by the enzyme beta-galactosidase. *Before you had lunch*, a "check" on your bacterial cells would have indicated the absence of the enzyme beta-galactosidase. A cell growing in the absence of lactose would be wasting valuable resources and energy if it manufactured an enzyme, such as beta-galactosidase, which was not needed at the time. If you were to check on these same bacterial cells after you have consumed a milkshake, you would find that each contained thousands of beta-galactosidase molecules. The presence of lactose has **induced** the synthesis of this enzyme. (We will return to lactose's effect on gene expression shortly.)

2. After having such a light lunch, you are ready for a steak dinner. The bacterial cells in your intestine have gone all day with very little protein available in their envi-

ronment. Since protein in *your* diet provides the bacterial cells with their amino acids, your lack of protein intake requires that these cells synthesize the amino acids needed to assemble their proteins. Let's consider one amino acid, tryptophan, which is synthesized via a metabolic pathway that includes a number of enzymes. Once you have eaten your dinner, and proteins are entering your intestine, the bacterial cells no longer have to synthesize their own tryptophan since it is now available in their diet. Within a few minutes, the production of the enzymes of the tryptophan pathway would stop. In the presence of tryptophan, the genes that encode these enzymes are **repressed**.

The advantage of inducing or repressing synthesis of specific enzymes is apparent. But just how are bacterial cells able to do this?

THE BACTERIAL OPERON

To understand how something functions you must have knowledge of the structure of the materials that accomplish the task. In bacteria, the genes that contain the information for constructing the enzymes of a metabolic pathway are usually clustered together on the chromosome in a functional complex called an **operon**. All the genes of an operon are regulated as a functional unit. The nature of bacterial operons was first described in 1961 by Francois Jacob and

Jacques Monod of the Pasteur Institute in Paris, corecipients of the 1965 Nobel Prize in Medicine. A typical bacterial operon (Figure 15-4) consists of structural genes, a promoter region, an operator region, and a regulatory gene.

- *Structural genes* **Structural genes** code for the bacterial enzymes themselves. The structural genes of an operon usually lie adjacent to one another, and the RNA polymerase moves from one structural gene to the next, transcribing them into a single mRNA. This giant mRNA is then translated into distinct polypeptides, representing the various enzymes. Consequently, turning on one gene turns on all the enzyme-producing genes of an operon.

- *A promoter and an operator region on the DNA* The **promoter** is the site where the RNA polymerase binds to the DNA prior to beginning transcription. The **operator** is situated between the promoter and the first structural gene; the operator serves as the binding site for a regulatory protein, called the **repressor**.

- *A regulatory gene* The **regulatory gene** codes for the protein repressor.

The key to operon expression lies with the repressor. When the repressor binds to the operator (Figure 15-4), the promoter is shielded from the polymerase and transcription of the structural genes is switched off. Whether or not the repressor can bind to the operator depends on the shape of

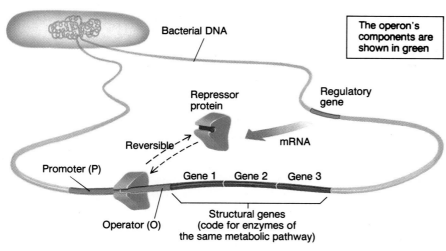

FIGURE 15-4

The bacterial operon: a model of the regulation of gene expression. Clustered structural genes (1, 2, and 3) lie "downstream" from a promoter (the mRNA polymerase attachment site) and an operator. A regulatory gene produces a repressor protein that determines whether or not the structural genes are expressed. When the repressor is attached to the operator, the polymerase cannot join with the promoter to initiate transcription of the operon's structural genes. As a result, no enzymes are produced. In contrast, when the repressor is not attached to the operator, the genes are transcribed. In this way, the repressor protein acts as the molecular "switch" that enables cells to turn genes on and off.

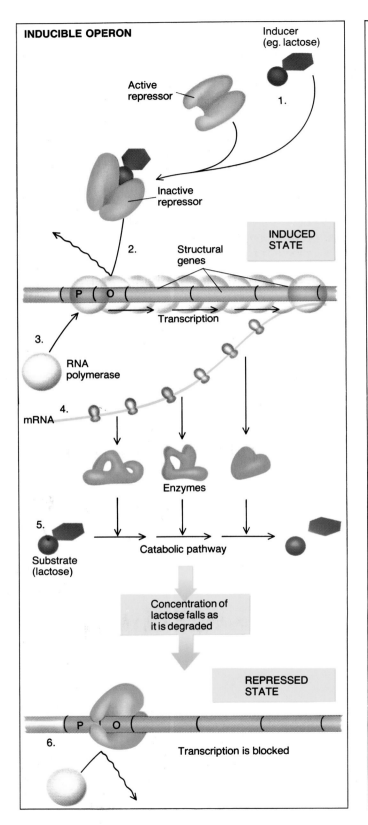

INDUCIBLE OPERON

Inducer (eg. lactose)

Active repressor

1.

Inactive repressor

2.

INDUCED STATE

Structural genes

P O

Transcription

3.

RNA polymerase

4.

mRNA

Enzymes

5.

Substrate (lactose)

Catabolic pathway

Concentration of lactose falls as it is degraded

REPRESSED STATE

P O

6.

Transcription is blocked

RNA polymerase

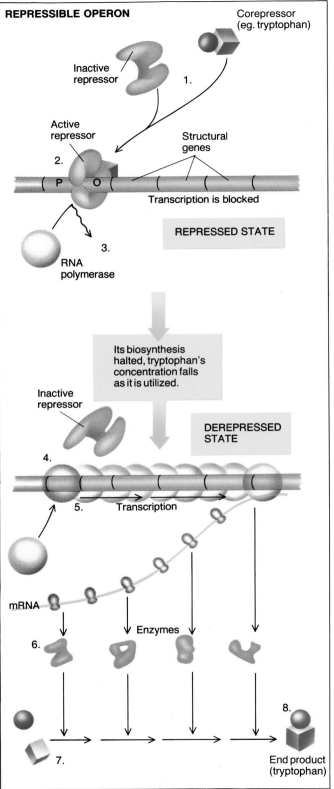

REPRESSIBLE OPERON

Corepressor (eg. tryptophan)

Inactive repressor

1.

Active repressor

2.

Structural genes

P O

Transcription is blocked

REPRESSED STATE

3.

RNA polymerase

Its biosynthesis halted, tryptophan's concentration falls as it is utilized.

Inactive repressor

DEREPRESSED STATE

4.

5.

Transcription

mRNA

6.

Enzymes

7.

8.

End product (tryptophan)

◄ FIGURE 15-5 ────────────────────────────

Gene regulation by operons. Inducible and repressible operons work on the same principle: If the repressor is able to bind to the operator, genes are turned off; if the repressor is inactivate and unable to bind to the operator, genes are expressed.

Inducible operons: (1) In high concentration, the inducer (in this case, the disaccharide lactose) binds with the repressor protein and (2) prevents its attachment to the operator (O). (3) Without the repressor in the way, RNA polymerase attaches to the promoter (P) and (4) transcribes the structural genes. Thus, when the lactose concentration is high, the operon is induced, and the needed sugar-digesting enzymes are manufactured. (5) Sugar is utilized by the enzymes encoded by the structural genes. If the sugar is not replenished, its concentration dwindles until the enzymes are no longer needed. (6) Now there is not enough lactose present to combine with the repressor, which then regains its ability to attach to the operator and prevent transcription. When the inducer concentration is low, the operon is repressed (turned off), preventing synthesis of unneeded enzymes.

Repressible operons: In a repressible operon, the repressor, by itself, is *unable* to bind to the operator, and the structural genes coding for the enzymes are active. In the case of the tryptophan operon, the cells are able to produce this essential amino acid when it is not available in the environment. (1) When ample tryptophan is available, tryptophan molecules act as a corepressor by binding with the *inactive* repressor and (2) changing its shape so it can attach to the operator, (3) preventing transcription of the structural genes. Thus, when tryptophan concentration is high, the operon is repressed, preventing overproduction of tryptophan. (4) When the tryptophan concentration is low, most of the repressor remains unbound by the corepressor and therefore fails to attach to the operator. Transcription proceeds, (5) genes are transcribed, (6) enzymes are synthesized, and (7) the needed end product (tryptophan) is manufactured. (8) As biosynthesis proceeds, tryptophan accumulates to high enough concentration to again repress the operon (step 1).

the repressor, which, in turn, depends on the presence or absence of a key compound in the metabolic pathway being regulated. It is the concentration of this metabolic substance (such as lactose or tryptophan) that determines if the operon is active or temporarily halted.

Induction in the lac Operon

The interplay among these various elements is illustrated by the *lac operon*—the cluster of genes that regulates production of the enzymes needed to metabolize lactose in bacterial cells. The lac operon is an example of an *inducible operon;* that is, one in which the presence of the key metabolic substance (in this case, the milk sugar lactose) induces transcription of the structural genes (Figure 15-5). The lac operon contains three adjacent structural genes that encode the enzymes responsible for the uptake and breakdown of lactose. In the absence of lactose, the lac repressor produced by the regulatory gene binds to the operator site, blocking the transcription of the structural genes. If lactose becomes available, the molecules enter the cell and bind to

the repressor, which changes the repressor's shape, making it unable to attach to the operator site. In this state, the mRNA is transcribed, the enzymes are synthesized, and the lactose molecules are consumed.

In an inducible operon, such as the lac operon, the repressor protein is active in the absence of the inducer (e.g., lactose). Conversely, in a *repressible operon* (such as the tryptophan, or trp, operon) the repressor becomes active only when bound to a corepressor (such as tryptophan). The mechanism by which both types of operons operate is described in Figure 15-5.

─────────────────────────────

In bacteria, proteins that act together in a common function, such as enzymes that make up a metabolic pathway, tend to be encoded by genes that are clustered on the chromosome and organized into a single functional unit. Transcription of all genes in the unit can be switched on or off simultaneously by the action of a single gene regulatory protein. (See CTQ #3.)

GENE REGULATION IN EUKARYOTES

▲ As molecular biologists learned more about the workings of bacterial operons, a large-scale search for similar mechanisms in eukaryotic cells was launched without success. Eventually, scientists concluded that operons did not exist in eukaryotic cells. It appeared that, although prokaryotes and eukaryotes employ the same fundamental mechanisms for storing, transmitting, and transcribing genetic information, they have evolved different strategies for controlling which genes they express.

CELL DIFFERENTIATION AND SPECIALIZATION

One bacterial cell in a species population looks like all the others. In contrast, multicellular eukaryotes contain a diverse array of cells possessing very different structures and functions. The human body contains several hundred recognizably different cells, each far more complex than a single bacterial cell, and each possessing a distinct set of proteins that allows it to carry out its specialized activities.

All of us began life as a single fertilized egg that divided by mitosis to produce trillions of cells containing the same genes. During this early embryonic development, the cells of the body become specialized for different activities, a process called **cell differentiation**. Those cells that become liver cells express a specific set of "liver genes," for example, while those that develop into nerve cells express a specific set of "nerve genes." Furthermore, even though a nerve cell may live for 80 years in a human brain, it will *never* express the genes for hemoglobin, the oxygen-carrying protein found only in red blood cells.

The fact that most eukaryotic DNA is transcriptionally silent does not mean that the genes cannot respond to change; they can. But eukaryotic cells can also respond to changes in nutrients, but many of the changes in the cells of multicellular organisms occur in response to hormones, growth factors, and other regulatory signals.

The Magnitude of the Problem

◯ The average bacterial cell contains enough DNA to encode approximately 3,000 polypeptides, of which about one-third are typically expressed at any particular time. Compare this to a human cell that contains enough DNA (3 billion base pairs) to encode several million different polypeptides, of which a few thousand may actually be expressed at one time. Because of the tremendous amount of DNA found in a eukaryotic cell and the vast number of different proteins being assembled, regulating eukaryotic

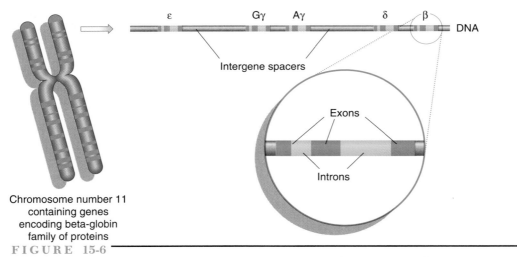

FIGURE 15-6

A map of human globin genes along a portion of the DNA of chromosome number 11, showing how much "space" exists both between and within the genes. Most of this DNA has no known function. The genes illustrated here code for a family of related proteins: the beta-globin family. Only the gene at the far right codes for a product found in adult hemoglobin molecules; the other four are synthesized during embryonic development and infancy. Each of the genes consists of three exons (the expressed sequences) and two introns (the intervening sequences). This arrangement is illustrated in the expanded view of the beta-globin gene. All five genes shown here have the same organization because they are presumed to have originated from duplication of a single ancestral globin gene that had this arrangement of exons and introns. Of the entire stretch of DNA shown in this drawing, only about 15 percent (corresponding to the red segments) actually codes for part of a polypeptide chain.

gene expression is an extraordinarily complex process. Consider the situation facing a cell developing into a red blood cell inside the marrow of a human leg bone. Hemoglobin accounts for more than 95 percent of the protein of a red blood cell, yet the genes that code for the hemoglobin polypeptides represent less than one-millionth of the total DNA. Not only does the cell have to find this genetic "needle" in the chromosomal "haystack," it has to regulate its expression to such a high degree that production of these few polypeptides becomes the dominant synthetic activity of the cell.

DNA THAT IS NEVER EXPRESSED

Virtually all of the DNA of a bacterial cell consists of either structural genes or regulatory genes. In contrast, the amount of DNA found in eukaryotic cells, particularly in higher plants and animals, is vastly greater than that which has any known function. While a human cell may have enough DNA to encode several million different average-sized polypeptides, less than 10 percent of this genetic information is thought to produce a gene product anywhere in the organism. While some of the remaining DNA has a regulatory function, most of it is "forever silent." Some biologists think that this DNA has a function; others think it is simply "genetic garbage" resulting from processes that randomly increase the amount of DNA each generation, such as gene duplication (page 258).

Split Genes: Exons Separated by Introns

Until 1977, biologists had assumed that genes that code for polypeptides consisted of an uninterrupted strip of genetic information transcribed into a molecule of mRNA of the same length as the gene. This was shown to be the case in bacteria, and it was a logical expectation for eukaryotic genes as well. All of the excess DNA in a eukaryotic chromosome—the DNA that is never expressed—was assumed to lie *between* the genes. Then, in 1977, a startling discovery was made that shattered this concept. It was shown that not all of the unexpressed DNA resided between genes after all; some was actually situated smack in the middle of the genes themselves, separating each gene into discontinuous parts (Figure 15-6). It soon became apparent that most genes contain a number of these *intervening sequences* that split the coding portion of the gene into separate sections. A few genes are even known to be split into 50 to 100 different pieces! Investigators named the intervening sequences **introns**. The segments of DNA that are transcribed and translated into portions of the amino acid chain are called **exons**, for "expressed sequences."

▐▶ At first, the existence of "split" genes seemed to make very little sense. What possible selective advantage could be gained by separating a gene into discontinuous pieces? The best explanation to date has been provided by Walter Gilbert, a Nobel laureate from Harvard University, who had earlier isolated the repressor of the lac operon. Gilbert proposed that introns facilitate evolution. The ability of

DNA segments to move around (as discussed in the chapter opener and on page 259) is an important component of Gilbert's hypothesis. To understand his hypothesis, let's look at the globin gene, which consists of three disconnected exons (Figure 15-6). According to Gilbert's hypothesis, at some time in the distant past, each of these exons was a separate gene that coded for an entire polypeptide chain. Over the course of evolution, pieces of chromosomes became rearranged so that in some distant ancestor, these different genes happened to end up next to each other and became part of a single gene—a beta-globin gene. Viewed in this way, eukaryotic genes have a modular construction made up of subunits (exons) that can be shuffled between different parts of the DNA.

If this hypothesis is correct, evolution need not occur only by the slow accumulation of mutations but can move ahead by "quantum leaps" as new proteins appear (literally) overnight from new combinations of exons. Over time, natural selection would lead to changes in these "composite" genes, improving the efficiency of the encoded polypeptide. The discovery of split genes has reminded biologists to expect the unexpected. Just because an observation doesn't fit the generally accepted scheme doesn't necessarily mean that the observation is erroneous; it may just signify that the generally accepted scheme needs revising.

Eukaryotes generally consist of numerous specialized cells, each of which expresses its own unique subset of genes; eukaryotic cells contain large amounts of DNA dispersed among numerous chromosomes; eukaryotic DNA contains large stretches that have no coding function; and eukaryotic genes contain intervening sequences. For all of these reasons, gene regulation in eukaryotes is much more complex than is gene regulation in prokaryotes. (See CTQ #4.)

LEVELS OF CONTROL OF EUKARYOTIC GENE EXPRESSION

↻ The assembly of a functional protein is a complex process that requires many different steps. Earlier in this chapter, we compared genetic information to an architect's blueprint. If the final product, whether an office building or a protein, is going to resemble the plan encoded in the blueprint, then every step of the construction process must be carefully regulated. The blueprints must be read properly; the sequence of construction must proceed in the proper order; the proper materials must be delivered to the assembly site at the right time; the activities of the construction workers must be coordinated; and so forth. Eukaryotic cells regulate their biochemical activities at three fundamental levels of gene supervision: the transcriptional level, the processing level, and the translational level.

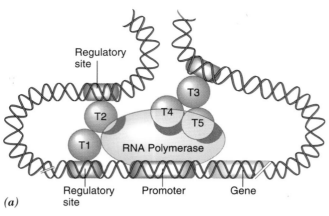

(a)

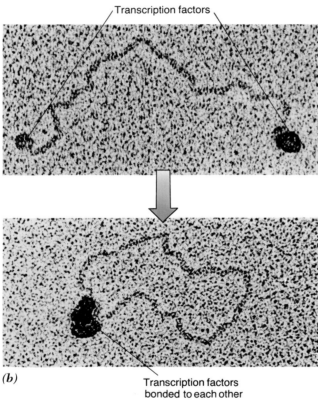

(b)

FIGURE 15-7

The interaction between gene regulatory proteins and DNA in eukaryotic cells. *(a)* A hypothetical example, whereby an interaction among several gene regulatory proteins (T1–T5) controls transcription. Several different proteins attach to different sites in the DNA, including the promoter (which binds the RNA polymerase prior to transcription). Proteins situated at widely spaced DNA sites can interact with one another as a result of the bending of the double helix into a hairpin loop. *(b)* This pair of electron micrographs illustrates how transcriptional factors located at different sites on a DNA molecule (top) can interact with each other, forming a loop in the DNA (bottom).

1. **Transcriptional-level control** mechanisms regulate gene expression by determining whether a particular gene will be transcribed to form an RNA transcript and, if so, how often.

2. **Processing-level control** mechanisms regulate gene expression by determining whether the RNA transcript is converted *(processed)* into a messenger RNA that can be translated into a polypeptide.

3. **Translational-level control** mechanisms regulate gene expression by determining whether or not a particular mRNA is actually translated.

We will now take a closer look at the activities involved at each level.

REGULATING GENE EXPRESSION AT THE TRANSCRIPTIONAL LEVEL

As we noted earlier, all cells of a multicellular plant or animal contain all the genes of that organism, but only a small fraction of that information is put to use in any given cell type. How does a developing red blood cell "know" to transcribe the genes for manufacturing hemoglobin, while brain cells and muscle cells "ignore" these genes? Just as in bacterial cells, control of gene transcription in eukaryotes depends on the presence or absence of specific gene regulatory proteins. Each type of cell has its own unique combination of regulatory proteins which allows that cell to transcribe a unique set of genes.

One of the challenges facing molecular biologists is the large number of sites in the DNA which bind regulatory proteins (Figure 15-7*a*). Transcription of a globin gene, for example, is controlled by at least five distinct regions of the DNA, located on both sides of the gene itself. Moreover, some of these regulatory sites on the DNA may be tens of thousands of base pairs away from the gene they regulate. How can gene expression be regulated from such a great distance? The formation of large loops in the DNA is thought to bring distant regions of the DNA into close contact with one another (Figure 15-7*b*). Some of the best-understood gene regulatory proteins mediate the response to steroid hormones, including testosterone, the male sex hormone.

Steroid Hormones and Gene Activation

Cells that respond to testosterone must contain a protein called the **testosterone receptor**. Normally, the receptor resides in the cytoplasm of specific target cells, including the cells of the male reproductive tract. When the testosterone level in the blood rises, hormone molecules diffuse into all cells, but only testosterone-sensitive cells contain the appropriate receptor, so only these cells are able to respond.

Just as lactose changes the shape of the lac repressor when it binds to the protein, a testosterone molecule

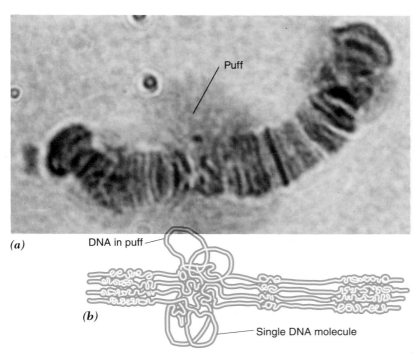

(a)

DNA in puff

(b)

Single DNA molecule

FIGURE 15-8

Puffing in giant polytene chromosomes. *(a)* Photograph of a polytene chromosome of a fly larva, taken with a light microscope. The image reveals the huge size of these structures. One large puff (and two smaller puffs) are evident, revealing areas of active transcription. *(b)* The DNA in the region of the puff is extended, providing sites accessible to RNA polymerase.

changes the shape of the testosterone receptor upon binding. No longer a "simple" receptor, the receptor–testosterone complex becomes a gene regulator (see Figure 25-14*b*) and moves from the cytoplasm into the nucleus, where it binds to specific sites on the DNA, activating the expression of nearby genes. For example, hair cells on the face of both men and women possess testosterone receptors. In general, only men produce high enough concentrations of testosterone to bind with the receptor and convert it into a gene activator, turning on the genes required for hair formation. Even though women possess competent hair follicles on their chins, these cells remain relatively inactive in the absence of testosterone, explaining why bearded women are so uncommon. Facial hair formation is stimulated in female athletes who inject themselves with testosteronelike steroids to enhance their muscle bulk.

The importance of the testosterone-receptor protein in gene regulation can be illustrated by a somewhat shocking example. On occasion, an adolescent female who fails to begin puberty, or a woman who has not been able to become pregnant, will receive some extraordinary news from her doctor—she is genetically a male! That is, she has one X and one Y sex chromosome rather than two X chromosomes characteristic of females. She is likely the victim of *androgen-insensitivity syndrome*. Although she appears externally as a mature woman, inside, she possesses a defective gene for the testosterone receptor. This person has a pair of internal testes and a normal concentration of testosterone in the blood, but none of the body's cells can respond to the male hormone, and differentiation proceeds in a female direction. These individuals are usually treated with estrogen to promote maturation of adult female characteristics, but they will never be able to have children.

Visualization of Gene Activation: Chromosomal Puffs

The old adage "seeing is believing" is just as true for molecular biologists as it is for anyone else. Most studies of gene expression employ techniques that provide indirect data in the form of spots on a gel or numbers on a piece of paper, indicating amounts of radioactivity in various samples. Fortunately, gene expression can be visualized directly through a few systems, one of which is the polytene chromosome. Recall that polytene chromosomes (page 253) in insect cells consist of hundreds of parallel strands of DNA organized so that individual genes are aligned across the entire width of the chromosome. When examined under the microscope, polytene chromosomes display a characteristic "puffing" at various points (Figure 15-8). These **chromosomal puffs** are sites where the DNA has unraveled and transcription is occurring. The presence of a puff provides direct visual evidence that the gene is being expressed. Like mammals, insects secrete steroid hormones that alter the patterns of gene expression. One of these hormones, *ecdysone*, triggers insect metamorphosis—from a pupa into an adult, for example (Figure 15-9). If salivary glands are removed from a larva and experimentally incubated in ecdysone, the puffing pattern along the polytene chromosomes undergoes dramatic changes within minutes. Within an hour or so, new proteins appear in the cytoplasm—

◁ B I O L I N E ▷
RNA as an Evolutionary Relic

▣▶ Which came first, the protein or the nucleic acid? Evolutionary biologists have been arguing this point for decades. The dilemma arose from the seemingly non-overlapping functions of these two types of macromolecules. Nucleic acids store information, whereas proteins catalyze reactions. With Cech's discovery of ribozymes, it became apparent that one type of molecule—RNA—could do both.

These findings have led to speculation that neither DNA nor protein existed during an early stage in the evolution of life. Instead, RNA molecules performed double duty; they served as genetic material, and they catalyzed necessary enzymatic reactions. Only at a later stage in evolution were these activities "turned over" to DNA and protein, leaving RNA to function primarily as a "go-between" in the flow of genetic information.

How feasible is this proposition? So far, only a few chemical reactions have been found to be catalyzed by RNA *in the cell*. Many biologists were shocked when it was demonstrated in 1992 that the "enzyme" that catalyzes the reaction in which amino acids are polymerized into polypeptides was actually an RNA molecule present within the large subunit of the ribosome (p. 282). It had long been assumed that this reaction was catalyzed by one of the ribosome's proteins. Several research groups are now exploring the catalytic *potential* of RNA. These investigators are modifying RNAs in various ways and searching for previously unknown catalytic properties. It has recently been shown that some of these modified RNAs can carry out RNA replication—the duplication of other RNA molecules. Since self-duplica-tion is a fundamental property of life, this finding provides support for the existence of an ancient "RNA world."

In addition to fortifying our understanding of evolution, the practical considerations of enzymatically active RNAs are particularly exciting. Competing patent applications have been filed in both the United States and Australia that capitalize on the ability of ribozymes to recognize specific sequences in other RNA molecules and to cut the RNA strand. These applicants hope to be able to synthesize ribozymes that can recognize and destroy certain mRNAs, such as those produced by the AIDS virus, while leaving the rest of the cell's RNA unharmed. The first ribozyme to be subjected to clinical trials will probably be aimed at the herpes virus responsible for human skin lesions.

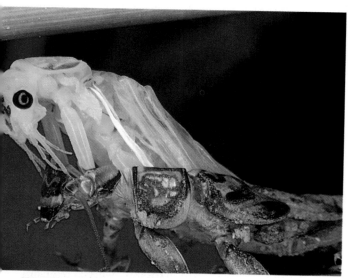

FIGURE 15-9

The metamorphosis of this stone fly was induced by the steroid hormone ecdysone.

products of the newly transcribed genes. These new proteins would normally participate in those activities that carry the insect beyond the larval stage.

REGULATING GENE EXPRESSION AT THE RNA PROCESSING LEVEL

The presence of introns within the boundaries of genes introduces serious obstacles in the path of gene expression. Consider what would happen if these intervening sequences were present in an mRNA molecule being translated. We saw in the last chapter that a ribosome moves along an mRNA from the initiation codon to the stop codon, producing a continuous polypeptide chain. The presence of intervening sequences in mRNA would cause the message to be translated into a polypeptide with disruptive stretches of amino acids within its midst. By 1977, when introns were discovered, a number of eukaryotic mRNAs had already been isolated and sequenced, and it was evident that they contained an unbroken "coding sequence." It seemed obvious, therefore, that even though a *gene* may contain noncoding introns, the mRNA formed from that gene does not. Subsequent investigations revealed that the entire split gene (including introns and exons) is transcribed into a giant RNA molecule, called a *primary transcript*, which is subsequently *processed* into the mature mRNA.

RNA processing removes those segments of the primary transcript that correspond to the introns (Figure 15-10). But investigations of the RNA processing reactions revealed further unexpected revelations.

In 1982, Thomas Cech of the University of Colorado was studying the processing of RNA in a ciliated protozoan. Cech discovered that RNA molecules possessed the ability to catalyze their own processing. Self-processing is not a simple reaction: The RNA has to cut itself into pieces; those pieces that are not needed must be eliminated; and the remaining pieces must be joined together to form the final mRNA product (Figure 15-10). At first, Cech was skeptical about his results, but he soon confirmed his findings. The concept that RNA molecules, as well as proteins, could possess enzyme activity emerged. The question was whether this was some peculiar property of a few RNAs in a strange protozoan or a universal process.

Further studies have revealed that RNA plays a key role in RNA processing in all eukaryotes, but not in the manner first discovered by Cech. Apparently, as a gene is being transcribed, the RNA transcript becomes associated with particles that contain both RNA and protein. These particles move along the primary transcript, cutting the RNA chain, removing the intron sequences, and joining (*splicing*) the exon sequences together (Figure 15-10). As anticipated by the work on the protozoan, the RNAs in the particles carry out the cutting and splicing reactions; the proteins play only a supportive role. Because of their enzymelike properties, these catalytic RNAs are referred to as **ribozymes**.

For his unexpected discovery (as well as further investigations), Cech was awarded the 1989 Nobel Prize in Chemistry. The discovery that RNA molecules can have catalytic activities has also led to a revision in our thinking about the origin of life (see Bioline: RNA as an Evolutionary Relic). The fact that proteins are not life's only catalysts illustrates once again that science is not a fixed, unchangeable body of knowledge. Rather, through the scientific method, our concepts are constantly revised as new information becomes available.

Alternative Processing

Cells may exert control over gene expression by processing RNA transcripts differently. For example, in one type of cell, a primary transcript may be processed into a cytoplasmic mRNA molecule, whereas in another type of cell, the transcript is simply degraded in the nucleus without ever producing a translatable mRNA. Cells can also process the same transcript in different ways; consequently, the same gene can have different forms of expression. An example of this *alternative processing* is provided in Figure 15-11.

REGULATING GENE EXPRESSION AT THE TRANSLATIONAL LEVEL

Just because a mature messenger RNA is able to escape from the nucleus and enter the cytoplasm does not guaran-

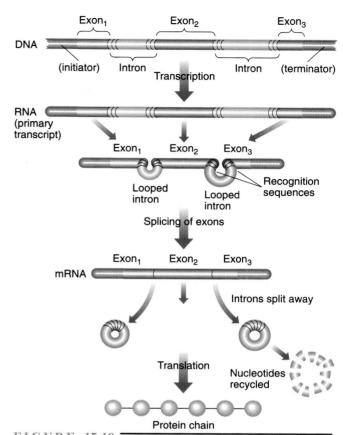

FIGURE 15-10

Editing a gene's message. As an RNA polymerase moves along a split gene, it generates a primary transcript whose length is equivalent to that of the gene itself. The final mRNA product is produced by RNA processing, whereby the sections corresponding to the introns are removed from the RNA transcript, and the remaining sections corresponding to the exons are spliced (joined) together. RNA processing must occur with absolute precision so that the nucleotides that code for necessary amino acids are not accidentally removed from the exons. Specific nucleotide sequences identify the beginning and the end of each intron to be removed. These signals pinpoint the location where the RNA is to be cut. As a result, the mRNA formed by the splicing process contains a continuous sequence of codons which specifies the sequence of amino acids for a polypeptide chain.

tee that it will be translated immediately. The cell regulates gene expression at the translational level as well. In some situations, mRNAs are temporarily "masked" by proteins, so they cannot be translated. Consequently, huge amounts of mRNA for certain proteins can accumulate in dormant cells, such as an unfertilized egg awaiting activation by a sperm. Once the dormant cell is activated, the mRNAs are "unmasked" by removal of the blocking proteins, and synthesis of the corresponding polypeptides takes place.

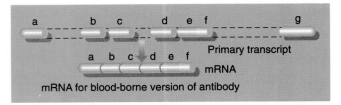

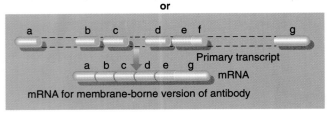

mRNA for membrane-borne version of antibody

FIGURE 15-11

Orchestrating gene expression by alternative processing.
A single region of the DNA can be processed in more than one way so as to generate different mRNA molecules. The case depicted here concerns antibodies: disease-fighting proteins that occur (1) as blood-borne proteins and (2) as part of the plasma membrane of certain white blood cells. These two forms of an antibody—the blood-borne and the membrane-borne forms—have different amino acids at one end of the polypeptide. The two forms of the protein are translated on different mRNAs that are derived from the same primary transcript as a result of alternate processing. The two different mRNAs have different terminal sections (as indicated by the differently colored portions of the message).

Cells also regulate gene expression by controlling the life span of mRNAs. For example, protecting specific mRNAs from enzymatic degradation increases the number of times the mRNA can be translated into a polypeptide. Prolactin—the hormone that triggers milk production in mammary glands—operates by this mechanism. The cells of the mammary glands of a nursing mother must produce tremendous quantities of milk proteins in a sustained manner over a period of months or even years. It would be wasteful for these cells continually to synthesize and destroy the mRNAs that carry the message for milk proteins. Prolactin stabilizes these mRNAs, thus augmenting their translation. Withdrawal of prolactin decreases the stability of these mRNAs, which quickly deteriorate, causing milk protein production to stop almost immediately.

Understanding the mechanisms of gene control in eukaryotes is a primary goal of biological investigators. Not only does an understanding of gene control contribute to our comprehension of life on one of its most fundamental levels, but it also helps us learn how and why these controls malfunction in ways that lead to human disease.

Before a polypeptide can be synthesized in a eukaryotic cell a gene must be transcribed; the RNA transcript must be cut and spliced; and the mRNA product must enter the cytoplasm and become associated with a ribosome to be translated. Each of these activities is complex and serves as a step where gene expression can be regulated. (See CTQ #5.)

REEXAMINING THE THEMES

The Relationship between Form and Function

Regulation of gene expression at the transcriptional level requires gene regulatory proteins (such as the lac repressor or the testosterone receptor), whose shape can precisely recognize one specific sequence of nucleotides in the DNA among the millions of possible sequences present. While different regulatory proteins are shaped to recognize different sequences, many of them have "fingers" that fit into the grooves of the DNA at the site of recognition.

Biological Order, Regulation, and Homeostasis

Cells contain much more genetic information than they can possibly use at any one time. Bacterial cells typically express only those genes that code for proteins that are useful under existing environmental conditions. Eukaryotic cells typically express only those genes that are appropriate for that particular cell type. A host of different regulatory mechanisms intervene at various points along the path leading from gene to polypeptide, ensuring that only the appropriate proteins are manufactured.

Acquiring and Using Energy

In addition to preventing the breakdown of biological order, gene regulation in prokaryotic cells helps avoid needless waste of energy (and chemical resources) by preventing the synthesis of enzymes and metabolic products that are not useful at that time. For example, it would be a waste of energy and resources for a bacterial cell to produce enzymes that metabolize lactose when there is no lactose in the medium.

Unity within Diversity

Gene regulatory proteins that bind to specific DNA sequences regulate gene expression at the transcriptional level in all cells. The binding of these proteins either activates or represses the transcription of a nearby gene. Prokaryotic and eukaryotic cells have fundamentally different mechanisms by which these regulatory proteins act, however. Prokaryotes possess operons in which a sequence of adjacent genes is controlled by a single operator site. Large, multicellular eukaryotes may contain hundreds of different cell types, each with its own unique set of gene regulatory proteins.

Evolution and Adaptation

■▶ The discovery that genes contain introns (noncoding, intervening sequences) has led to an important hypothesis concerning the mechanism by which new genes can arise during evolution. According to this hypothesis, each exon represents a distinct ancestral polypeptide. As the exons moved around from one place to another in the chromosomes, they came together in new combinations. Some of these combinations led to the formation of new genes whose "combined" polypeptides had new or improved functions, which gave the organism a selective advantage.

SYNOPSIS

All cells regulate which part of their genetic endowment is expressed. Every cell has far more genes than it uses at any particular time. Cells possess mechanisms that determine which genes are turned on and which are turned off at a given time. Consider, for example, a bacterial cell living in a medium whose composition suddenly changes, or a eukaryotic cell transforming from an undifferentiated stage into a liver cell, or a cell in your reproductive tract being exposed to higher levels of a hormone. In each of these cases, the cell responds by activating new genes and repressing the expression of others.

In prokaryotes, the fundamental regulatory mechanism is the operon. Operons are clusters of structural genes (and their control elements O, P, and R) that typically code for different enzymes in the same metabolic pathway. Since all of the structural genes are transcribed into a single mRNA, their expression can be regulated in a coordinated manner. The level of gene expression is controlled by a key metabolic compound, such as the inducer lactose, which attaches to a protein repressor and changes its shape. This event alters the ability of the repressor to bind to the operator site on the DNA, thereby allowing transcription.

Regulation of gene expression in eukaryotes is more complex than in prokaryotes. Eukaryotic cells have much more DNA than do prokaryotes, and much of this DNA is never expressed. Most of the nonexpressed DNA lies between genes, but some of it (in the form of introns) lies within the coding regions of the genes. Introns require their removal from the RNA transcript during processing, but they are thought to have increased the rate of evolution by facilitating the formation of new genes by exon shuffling.

Eukaryotic gene expression is regulated primarily at three levels. (1) Gene regulation at the transcriptional level determines whether or not a gene will be transcribed and, if so, how often. Gene regulatory proteins, such as the testosterone-receptor complex, are capable of binding to specific DNA sites and controlling the rate of transcription of nearby genes. (2) Genes are transcribed into large primary transcripts that must be cut and spliced to form the mature mRNA. Control of these activities constitutes regulation at the processing level. (3) Control of gene expression at the translational level centers on whether or not an mRNA is translated and how long the mRNA will survive.

Key Terms

gene regulatory protein (p. 295)
clone (p. 296)
totipotent (p. 297)
induce (p. 298)
repress (p. 299)
operon (p. 299)
structural gene (p. 299)

promoter (p. 299)
operator (p. 299)
repressor (p. 299)
regulatory gene (p. 299)
cell differentiation (p. 302)
intron (p. 303)
exon (p. 303)

transcriptional-level control (p. 304)
processing-level control (p. 304)
translational-level control (p. ?)
testosterone receptor (p. ?)
chromosomal puff (p. ?)
RNA processing (p. ?)
ribozyme (p. 307)

Review Questions

1. Describe the cascade of events responsible for the sudden changes in gene expression in a bacterial cell following the addition of the milk sugar lactose.

2. Which of the following statements are true? (1) The regulatory gene produces the promoter. (2) The inducer binds to the operator site. (3) The repressor binds to DNA in the presence of lactose. (4) The structural genes are transcribed sequentially into one long mRNA.

3. Cite an example of an RNA molecule forming hydrogen bonds between complementary nucleotides in the same molecule.

4. Describe the different levels at which gene expression is regulated to allow a beta-globin gene with the following structure to direct the formation of a protein that accounts for over 95 percent of the protein of the cell.

exon #1/intron/exon #2/intron/exon #3

5. How are the functions of a bacterial lac repressor and a human testosterone receptor similar? How are they different?

6. What is the primary difference between a primary transcript and the mRNA to which it gives rise?

Critical Thinking Questions

1. Suppose you were studying tobacco plants and found a gene that coded for a protein very similar to that of a bacterial cell that infects these plants. Can you suggest two alternate mechanisms by which these proteins could be so similar between these widely divergent organisms?

2. What are the adaptive advantages to organisms of selective gene expression that can change over time?

3. What is the advantage of clustering structural genes so that all the enzymes for a metabolic pathway are regulated together rather than independently?

4. In eukaryotes, only about 1 percent of the DNA ever codes for mRNA that is subsequently translated into proteins. In prokaryotes, the figure is above 90 percent. Explain how gene structure and gene expression in eukaryotes account for this difference. What are the advantages of the eukaryotic system? What are the disadvantages?

5. Hormones dramatically alter gene expression, as exemplified by the distinctions between men and women, which are fundamentally the result of sex hormones selecting different genes for activation. Create a model by which one hormone might temporarily activate a gene and another hormone might turn off that gene later in development. Your model can use any of the mechanisms discussed in this chapter.

Additional Readings

Beardsley, T. 1991. Smart genes. *Sci. Amer.* Aug:86–95. (Intermediate-Advanced)

Gibbons, A. 1991. Molecular scissors: RNA enzymes go commercial. *Science* 251: 521. (Intermediate)

Hoffman, M. 1991. Brave new (RNA) world. *Science* 254: 379. (Intermediate)

Keller, E. F. 1983. *A feeling for the organism: The life and work of Barbara McClintock.* New York: W. H. Freeman. (Introductory-Intermediate)

McKnight, S. L. 1991. Molecular zippers in gene regulation. *Sci. Amer.* April:54–64. (Intermediate)

Ptashne, M. 1989. How gene activators work. *Sci. Amer.* Jan:40–47. (Intermediate)

Rhodes, D. and Klug, A. 1993. Zinc fingers. *Sci. Amer.* Feb:56–65. (Intermediate)

Singer, M. and P. Berg. 1991. Genes and genomes. University Science Books. (Advanced)

DNA Technology: Developments and Applications

STEPS
TO
DISCOVERY
DNA Technology and Turtle Migration

GENETIC ENGINEERING

Genetically Engineered Cells and Human Proteins

Genetically Engineered Cells and Industrial Products

Recombinant Organisms at Work in the Field

Genetic Engineering of Domestic Plants

Genetic Engineering of Laboratory and Domestic Animals

Controversy over the Perceived Dangers of Genetic Engineering

DNA TECHNOLOGY I: THE FORMATION AND USE OF RECOMBINANT DNA MOLECULES

Tools for Assembling Recombinant DNA Molecules

Amplification of Recombinant DNAs by DNA Cloning

Expression of a Eukaryotic Gene in a Host Cell

DNA TECHNOLOGY II: TECHNIQUES THAT DO NOT REQUIRE RECOMBINANT DNA MOLECULES

The Separation of DNA Fragments by Length

Enzymatic Amplification of DNA

USE OF DNA TECHNOLOGY IN DETERMINING EVOLUTIONARY RELATIONSHIPS

BIOLINE

DNA Fingerprints and Criminal Law

THE HUMAN PERSPECTIVE

Animals That Develop Human Diseases

BIOETHICS

Patenting a Genetic Sequence

STEPS TO DISCOVERY
DNA Technology and Turtle Migration

A well-studied population of green turtles lives most of the year off the coast of Brazil. Every winter, the adults from this population begin a month-long journey that takes them from Brazil to Ascension Island, a remote site in the mid-Atlantic Ocean. There, the turtles mate; the females lay their eggs on the beaches; and the animals return to their feeding grounds 2,000 kilometers (1,240 miles) away. Why don't these turtles simply lay their eggs in Brazil? Why do they make such a long, and seemingly unnecessary, journey?

About 20 years ago, Archie Carr, of the University of Florida, offered an explanation. An authority on sea turtles, Carr based his hypothesis on the theory of continental drift, which states that the earth's continents were not always located in the same places as they are today. Rather, at one time—more than 100 million years ago—all of the continents were situated in relatively close proximity (see Chapter 35).

Carr further based his hypothesis on *natal homing*—the phenomenon whereby a green turtle always returns to the beach where it was hatched. According to Carr, a green turtle travels from Brazil to Ascension Island to breed because it was hatched on this island, as were its ancestors, and their ancestors, and so on, going back millions of years. In the distant past, however, the trip from the turtle's feeding grounds to its nesting grounds was very short because the continents were close together. As the continents gradually moved apart, natural selection favored animals that could

By using electrophoresis to compare the DNA patterns from different populations of sea turtles, biologists determined

make longer and longer journeys back to the place where they were hatched. What had once been a short commute eventually became a long sea odyssey.

A few years after Carr proposed his hypothesis, he was challenged by the evolutionary biologist Stephen Jay Gould. Gould argued that it was too much to expect that over a period of 100 million years conditions would remain suitable, year after year, for turtles to return to nesting grounds on Ascension Island. All it would take to break the link would be a series of "bad" years when conditions prevented a generation of turtles from making their long journey back home. If they could not reach Ascension Island, the turtles would have to start colonies at other sites, or perish as a population. Gould argued that the colonization of the Ascension Islands' breeding grounds was a relatively recent event.

You may be wondering what any of this has to do with DNA technology—the subject of this chapter. In fact, the techniques used to create and study DNA molecules can be used to probe other biological questions that are far removed from the subject of molecular biology, such as the mystery of turtle migration. As you will see in this chapter, the nucleotide sequence of a DNA molecule can be readily determined. This technique of DNA sequencing allows investigators to compare the genetic relatedness among various populations of animals on the basis of similarities in their nucleotide sequences, which brings us back to the green turtles.

Ascension Island is not the only site where green turtles go to breed; other populations nest on the beaches of Florida and Venezuela. If Carr's hypothesis is correct, and animals have returned to the same hatching site for hundreds of thousands of generations, then populations with different nesting grounds should have remained reproductively isolated from each other over a very long period of time. Reproductive isolation, in turn, leads to genetic divergence that should be reflected in differences in the nucleotide sequences in the DNA of members of different populations. How similar are the DNA sequences from animals that nest in these different parts of the Atlantic Ocean? A former graduate student of Carr's named Anne Meylan,

working with Brian Bowen and John Avise, two evolutionary biologists from the University of Georgia, recently investigated this question. The team indeed found distinct differences in the DNA from one population of turtles to another, indicating that they had been reproductively isolated from one another. But the differences were not nearly as great as one would expect if the isolation had continued for millions of years. Rather, the populations are estimated to have remained distinct for a few tens of thousands of years, at most.

The results suggest that the various turtle populations investigated are all descended from a population of ancestral turtles that once laid their eggs on a particular beach in the Atlantic Ocean. Over time, a number of descendants of these ancestral turtles failed to make it back to the beach where they were hatched. Instead, they made a "wrong turn" and ended up at a new nesting site, which is probably how the Ascension Island breeding ground became established. Regardless of how it came into being, the connection between Brazil and Ascension Island appears to be of relatively recent origin, which still leaves us wondering why these turtles travel so far to carry out their breeding activities. We can see from this example how biologists are able to use the recent advances in DNA technology as a tool for determining the evolutionary relationships among organisms—whether among populations of the same species or among different species that are only distantly related.

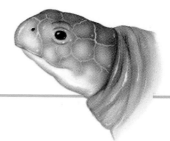

the approximate time the populations had separated.

Watchers of Wall Street were stunned. The company's stock began to sell the moment the opening bell sounded. Regarded by many as a novelty stock, it was the first company of its kind to be listed on the stock exchange, and its only products were unmarketable medicines years away from federal approval. Yet the stock caused a frenzy of buying and selling. By the end of its first day in 1976, it was worth $80 million!

The company was Genentech, the world's first stockholder-backed *biotechnology* organization, a company devoted to applying some of the recent lessons of molecular biology to manufacturing products of value in medicine, agriculture, and industry. Genentech was created as an offshoot of the scientific breakthroughs of a group of geneticists who have created new ways of putting organisms to work for people and the environment. In doing so, these geneticists have captured the attention of industrial leaders, members of the news media, and even representatives of the legal community. An important step in this "biotechnology revolution" came in 1980, when the U.S. Supreme Court ruled that "life" could be patented; that is, a company (or an individual scientist) can literally own the rights to the new life forms it "invents" in its laboratories. This issue is discussed further in Bioethics: Patenting a Genetic Sequence.

▼ ▼ ▼

GENETIC ENGINEERING

For thousands of years, people have been selectively breeding organisms to increase their value. They do so by mating individuals with desirable characteristics with each other to produce improved progeny. Such artificial breeding programs may even cross species lines, as illustrated by the emergence of a mule from the mating of a female horse and a male donkey. Although sterile, mules possess a useful combination of the genetic characteristics of both their parental species. Modern *genetic engineering* goes one step further, however; an organism's genotype can be modified according to human design by introducing genes that have never been present in the chromosomes of that particular species. This feat is accomplished using **recombinant DNA** molecules—molecules that contain DNA sequences derived from different biological sources that have been joined together in the laboratory. The techniques by which specific pieces of DNA are isolated, recombined, and intro-

duced into host cells are discussed in the second section of this chapter. Let us first describe some of the medical, agricultural, and industrial applications to which this technology is being put to use.

GENETICALLY ENGINEERED CELLS AND HUMAN PROTEINS

The earliest successes in genetic engineering were achieved by the creation of strains of bacteria that would act as microscopic "factories," churning out molecular products specified by a newly acquired gene. Today, genetically engineered cells are used to manufacture biological products ranging from drain cleaners to medicines. One biotechnologist has referred to these cells as "the largest nonunion workforce in the world."

Insulin was the first usable human protein to be produced in bacterial cells. The gene was first synthesized chemically, nucleotide by nucleotide, in 1978 and was subsequently introduced into bacterial cells, which produced the human insulin molecule and released it into the culture medium. Even though it is produced in bacterial "factories," the product of the human gene is identical to human insulin and is used to treat tens of thousands of diabetics, who suffer from a deficiency in insulin.

Prior to genetic engineering, human proteins with medical uses were either unavailable or were harvested in very low quantities from whole organs, blood, or "nonengineered" cells grown in the laboratory. The protein being sought was usually present at very low levels amidst a sea of thousands of different, unwanted proteins. Separating the desired protein from the unwanted proteins was very expensive, and adequate purity was impossible to achieve. Just consider the number of hemophiliacs who received the AIDS virus from contaminated blood products and you can see the danger inherent in treating patients with these substances.

Human proteins are now being produced safely and efficiently through the use of genetically engineered cells. Dozens of proteins are awaiting approval by regulatory agencies, while others have become available in doctors' offices and hospitals throughout the world. For example, the protein tissue plasminogen activator (TPA) is used widely in emergency rooms to dissolve blood clots that block coronary arteries, causing heart attacks. Patients treated quickly with TPA are much less likely to suffer permanent damage to heart tissue than are untreated patients. Similarly, hemophiliacs can now receive pure blood-clotting factors without the risk of infection from blood-borne viruses, and children who are unusually short for their age can receive recombinant human growth hormone to help them grow to a normal height. Other products currently produced by recombinant DNA technology are listed in Table 16-1. Virtually any protein manufactured by any organism is a candidate for production by genetically engineered microbes or cultured cells.

◁ B I O E T H I C S ▷
Patenting a Genetic Sequence
By ARTHUR CAPLAN
Division of the Center for Biomedical Ethics at the University of Minnesota

Scientists at the National Institutes of Health (NIH) in Washington, D.C., have been very busy recently. They have been rummaging through cells obtained from a collection of human brains. By extracting the DNA from these cells, they can locate the genes that permit a stew of chemicals to become a bit of brain. The rapid progress being made in mapping these genes has set some other brain cells working—those located in the heads of the lawyers, venture capitalists, Wall Street analysts, corporate executives, and government officials who are wondering who will own these gene maps.

In 1991, the NIH scientists planned to publish some of their initial discoveries in the prestigious journal *Science.* Just before they did, the attorneys responsible for patents at the NIH got wind of the scientists' publication plans. The lawyers realized that if the newly discovered genetic sequences appeared in print without a patent having been filed, then anyone could use them without having to pay fees or royalties. Consequently, the NIH attorneys postponed publication until they could file patent applications on the genetic sequences.

Many members of the scientific community were—and remain—outraged at the decision to seek patents on the information stored in brain cell genes. They feel that information about genetic language ought to be freely available to anyone who wishes to use it. Others feel that, while morally high-minded, it is doubtful whether this view can prevail. The free exchange of information may prove to be no competition when the stakes are in the tens of billions of dollars. Whoever controls knowledge about the genetic makeup of humans, plants, and animals will have a huge edge in getting products to the market. Many people feel that the lawyers at the NIH know this and that is why they are seeking patents even though their scientists have got only a partial list of the genetic sequences in brain cells. Furthermore, without practical plans for applying this new knowledge, the patent may not be granted so soon. Demonstrating utility is one of the key requirements for getting a patent.

The patenting of genetic information seems inevitable, however. If that is the case, then even though the NIH is probably a bit premature, it might be onto something important. Do you feel it would be best if no proprietary ownership were granted over the genetic code? Assuming the huge sums of money to be made from licensing the genetic code are cycled back to the NIH to fund future biomedical research, how do you feel about the leading U.S. scientific institution owning the code?

TABLE 16-1
SOME PRODUCTS OF RECOMBINANT DNA TECHNOLOGY

Product	Activity
Interferons	Fight viral infection; boost the immune system; possibly effective against melanoma (a form of skin cancer) and some forms of leukemia; may help relieve rheumatoid arthritis.
Interleukin 2	Activates the immune system and may help in treating immune-system disorders. Although it produces serious side effects, the drug is proving valuable in treating kidney cancer.
Tumor necrosis factor (TNF)	Attacks and kills cancer cells. Presently being used in the first experimental attempt to treat human cancer by introducing cells carrying foreign genes (page 648).
Erythropoietin	Stimulates red blood cell production; may be used to combat anemia.
Beta-endorphin	The body's "natural morphine"; used to treat pain.
Metabolic enzymes	Perform a multitude of services, from catalyzing chemical reactions in the pharmaceutical industry to replacing defective human enzymes.
Vaccines (e.g., hepatitis B)	Stimulate the body's immunity to protect against disease-causing viruses and bacteria.

GENETICALLY ENGINEERED CELLS AND INDUSTRIAL PRODUCTS

Fuel shortages may be relieved by genetically engineered microbes that inexpensively store solar energy in the chemical bonds of combustible organic compounds, providing a virtually inexhaustible fuel source. Genetic engineers are attempting to improve on the efficiency of natural systems that already accomplish this energy conversion (Figure 16-1a). In addition, genetic engineers have created "oil-producing" microorganisms that generate highly combustible organic compounds that are equivalent to petroleum in their potential energy. Another supplier of energy is cellulose—the earth's most abundant renewable resource. Each year, well over 3.5 billion tons of paper, cardboard, and other cellulose-based products are manufactured in the United States (15 tons for every person in the country). Nearly 900 million tons of this material ends up as refuse. Microbes can convert this huge resource to one of two valuable fuels: ethanol (the combustible alcohol found in "gasohol") or methane (natural gas, Figure 16-1b).

(a)

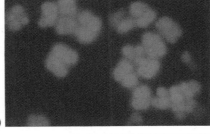

(b)

FIGURE 16-1

Two natural fuel producers. *(a)* Generally considered a nuisance that chokes waterways, this aquatic fern is being used to generate combustible fuel, methane. *(b)* Bacteria like these green fluorescent stained cells are especially prolific methane producers. Most biological fuel generators depend either on these kinds of cells or on different microbes that produce flammable ethanol.

RECOMBINANT ORGANISMS AT WORK IN THE FIELD

Some genetically engineered organisms must first make their way out of the lab and into the world before they can perform the task for which they were designed. One such organism has been constructed to reduce food loss when crops are subjected to freezing temperatures. Certain bacteria of the genus *Pseudomonas* that normally reside on plants produce a protein that acts as "seed crystals" around which water freezes; these bacteria are referred to as "ice-plus" cells (Figure 16-2). Freezing water expands and forms sharp crystals that crush cells and slash membranes. This type of cellular damage nourishes the bacteria residing on the plants, while possibly destroying the crop. The bacterial molecule that induces such water solidification is called the *ice-nucleating protein.*

Through recombinant DNA technology, a strain of bacteria has been developed that lacks the gene for producing the ice-nucleating protein. Called the "ice-minus" bacterium, this strain of *Pseudomonas* fails to induce ice formation, even when temperatures drop below − 4°C (25°F). When sprayed on plants, the engineered ice-minus bacteria compete with normal ice-forming bacteria, displacing them and reducing crop loss. In 1987, after a lengthy court battle over the risks involved in releasing genetically engineered organisms in the environment, ice-minus bacteria (under the trade name Frostban) were sprayed on a plot of strawberry plants in California, becoming the first genetically engineered microorganism to be legally released into the environment.

Genetically engineered microbes can also increase crop yields in other ways. We add fertilizer to plants to provide them with nitrogen-containing compounds for use in manufacturing protein. Some crops, such as soybeans and peas, are able to manufacture their own nitrogen-containing compounds from the nitrogen gas (N_2) in the air—a process known as *nitrogen fixation.* It is not the plant cells themselves that are able to "fix" N_2 but bacteria of the genus *Rhizobium* that live in nodules on the plant's roots. Plants that harbor these bacteria are called *legumes.* Genetically engineered strains of *Rhizobium* have increased soybean yields by 50 percent. Similarly, genetic engineers are trying to develop a strain of *Azotobacter,* a free-living, nitrogen-fixing bacterium, which would grow on the roots of nonleguminous plants, such as corn, freeing these plants from their need for applied nitrogenous fertilizer.

Genetic engineers are also releasing their microbial work force against pollution and toxic wastes, which pose a serious threat to the biosphere. Many of today's toxic pollutants are new synthetic compounds. Organisms capable of decomposing these chemicals have not yet had a chance to evolve. Consequently, the compounds accumulate to dangerous levels in the environment. Genetic engineers are trying to accelerate evolution in the laboratory by developing bacteria that can quickly degrade several types of poi-

FIGURE 16-2
The ice crystals seen on these strawberry leaves form around a bacterial *ice-nucleating protein.* This frost damage could have been prevented if the plants had been sprayed with the genetically engineered "ice minus bacterium."

sonous compounds; an oil spill, for example, would be one big meal for these bacteria (Figure 16-3).

GENETIC ENGINEERING OF DOMESTIC PLANTS

Biotechnologists are also attempting to modify the genetics of multicellular organisms. For example, genetic engineering can be used to improve plants by producing crop species that are more resistant to drought, disease, poor soil conditions, and chemical pesticides and herbicides. It has been known for many years that plants inoculated with a mild form of a virus may become resistant to more devastating strains. While the mechanism of plant "immunity" is unclear, the phenomenon has led to successful attempts by biotechnologists to induce viral resistance in certain crop plants. For example, Roger Beachy and his colleagues at Washington University introduced a gene into tobacco plants that codes for the outer coat protein of tobacco mosaic virus (TMV). Once incorporated into the chromosomes of the plants, the tobacco plants exhibited a marked resistance to TMV infection. The same technique has been used to protect tomato plants and potato plants from similar viruses.

▐▶ Genetic engineers are also developing plants that produce insect-killing toxins — "one bite and the pest dies." The best-studied protein toxins are produced by the bacterium *Bacillus thuringiensis* (BT). Different strains of the bacterium produce different varieties of the toxin, which are effective against different insect pests, including the tobacco hornworm and the gypsy moth. In 1986, a plant that had been genetically engineered with a BT toxin gene suc-

cessfully passed a field test, growing undamaged by pests that ravaged the "unprotected" plants close by. More recently, however, agricultural researchers are finding that insect pests are becoming resistant to the toxin produced by the genetically engineered plants, just as they would have become resistant if the substance had simply been sprayed over the field in the form of a pesticide. Researchers are hoping to stay one step ahead of the insects' evolutionary adaptations by engineering variations in the toxin's structure so that the insects do not have an opportunity to become immune.

FIGURE 16-3
Oil-eating bacteria. Workers are spraying a fertilizer solution on this oil-contaminated shore to stimulate the activity of oil-eating bacteria. This procedure greatly accelerates the rate of oil removal.

In another effort, genetic engineers have recently developed a new strategy to improve the quality of tomato plants. "Mushy" tomatoes result from the presence of an enzyme (polygalacturonase) that breaks down the plants' cell walls. In order to keep the tomatoes firm, farmers pick the fruits before they are ripe, redden them artificially, and then transport them to market. A tastier product could be marketed if the tomatoes were allowed to ripen on the vine and then kept from becoming mushy during transport. This is now possible through genetic engineering. Recall from Chapter 14 that only one strand of the DNA (the *sense strand*) is transcribed by an RNA polymerase; this strand contains the information for the encoded polypeptide. Consider what might happen if the other strand (the *antisense strand*) were also transcribed. The cell would contain RNA molecules with complementary sequences that have the potential to bind to one another, forming a double-stranded RNA molecule that could not be translated into the encoded protein. Recombinant DNA technologists

have been able to create genes in which the "wrong" strand is transcribed, including the gene that codes for the enzyme that causes mushy tomatoes. When this "reversed" gene is introduced into the cells of a tomato plant, the antisense RNA from the foreign gene interferes with translation of the sense RNA from the original gene, and production of the troublesome enzyme is reduced by 99 percent. The result is firm, "mush-resistant" tomatoes.

Taken together, these various approaches to improving the quality of crop plants provide an unparalled opportunity to increase the quantity of food that can be produced on a given amount of land. However, we should not lose sight of the fact that technological solutions often create unforseen problems that require new technological solutions. In the 1950s and 1960s, for example, plant geneticists created new strains of certain plants, particularly rice and wheat, which grew much more rapidly than did the original strains and led to much greater crop yields in many underdeveloped countries. Some of these countries became grain exporters,

(a)

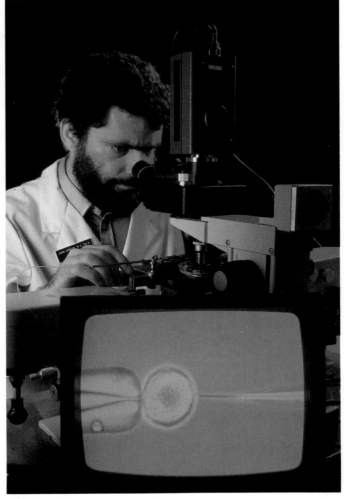

(b)

FIGURE 16-4

Transgenic mice. *(a)* A pair of littermates at 10 weeks of age. The larger mouse carries copies of rat growth hormone genes in all of its cells, which has caused this transgenic mouse to be much larger than its littermate, who lacks the rat gene. *(b)* The procedure by which a fertilized egg is injected with foreign DNA containing a gene that will become integrated into the egg's chromosomes. The video screen reveals the image seen by the researcher injecting the DNA. The egg is seen near the middle of the screen; the glass tube on the left holds the egg in place by suction, while the glass micropipette on the right contains the DNA to be injected.

rather than grain importers. This "Green Revolution," as it has been called, was not without problems. For example, the new strains relied heavily on the use of artificial fertilizers, which may be difficult to obtain and can contribute to serious pollution problems. Furthermore, increases in human population growth have virtually obliterated the gains of the Green Revolution, illustrating that no technological advance in agriculture will be successful until worldwide population control has been achieved.

GENETIC ENGINEERING OF LABORATORY AND DOMESTIC ANIMALS

In 1982, a litter of mice was born that included individuals unlike any mouse that had ever been born before; the chromosomes of these mice contained rat genes! Within a few weeks, those mice containing the rat genes were huge, compared to their littermates (Figure 16-4a). The increased size was not due to the fact that rats are larger than mice but to the fact that the foreign genes that had become integrated into the mouse chromosomes coded for the protein growth hormone (GH). When these rat GH genes were activated, the mice produced excess quantities of growth hormone, which led to their increased size and weight.

How did this rat gene get into the chromosomes of unborn mice? In order to accomplish this feat, Ralph Brinster of the University of Pennsylvania and Richard Palmiter of the University of Washington isolated fertilized mouse eggs from the reproductive tracts of female mice. They injected the nucleus of each egg with about 600 copies of the rat GH gene (Figure 16-4b). Some of these injected DNA fragments became integrated into the chromosomes of the injected eggs. These eggs were then implanted in the reproductive tracts of "surrogate mothers," where they developed normally to term. Because the foreign DNA was integrated into the chromosomes, it was passed on during mitosis to every cell of the fetus. When this first litter of genetically engineered mice reached maturity, the GH genes passed into the gametes and were transmitted to the next generation of offspring, which again exhibited exceptional growth. Animals that possess genes of a different species, such as these mice, are called **transgenic animals.**

These early experiments with transgenic mice have laid the groundwork for the development of genetically engineered livestock carrying genes that improve their food value. For example, pigs born with foreign growth hormone genes incorporated into their chromosomes (Figure 16-5) grow much leaner than do control animals lacking the genes. The meat of the former animals is leaner because the excess growth hormone stimulates the conversion of nutrients into protein rather than fat. Transgenic animals are also playing an important role in clinical laboratories (see The Human Perspective: Animals That Develop Human Diseases).

FIGURE 16-5
Transgenic "pork." A pair of young pigs who have developed from eggs that had been injected with foreign genes. Transgenic pigs containing foreign growth hormone genes produce leaner meat.

◁ THE HUMAN PERSPECTIVE ▷
Animals That Develop Human Diseases

(a)

(b)

FIGURE 1

(a) This woman suffers from Waardenburg's syndrome, a rare genetic disorder characterized by deafness, mismatched, widely spaced eyes, and a white forelock of hair. *(b)* The Splotch mouse shown on the left is an animal model of Waardenburg's syndrome. This mouse has a white forelock and tiny eyes compared with a normal mouse (right).

Animal models are laboratory animals that are susceptible to a particular human disease. Animal models provide one of the most important tools in evaluating the effectiveness of new drugs and therapies against human disorders. For example, a certain strain of rabbit has a defective gene for the LDL receptor (page 146) and, consequently, has very high blood-cholesterol levels. Another example is illustrated in Figure 1. Animal models are an important tool in the development of drugs for treating the corresponding condition in humans. But where do we turn when there are no animals with disorders similar to ours? Recent advances in molecular biology have opened the door to the possibility of "creating" animal models.

Sickle cell anemia is a disease from which no animal other than humans suffers—at least this was the case before 1990. In order to produce an animal model for sickle cell anemia, transgenic mice that carried the sickle cell version of the human globin gene were developed. Two mice were born that produced relatively high levels of the sickle cell type of hemoglobin in their red blood cells, and some of these cells exhibited the characteristic sickled shape. For unknown reasons, however, the mice showed no evidence of either anemia or of an elevated rate of red blood cell destruction, both characteristics of the human disease. These mice probably failed to produce enough of the human protein to trigger the disease's symptoms. Whatever the reason, the procedure is being refined, and animal models for the disease will likely be developed in the near future. Similar work is under way for other human genetic diseases, and it is hoped that a new era may be dawning in which experimental drugs can be more rapidly and safely tested on genetically engineered animal models.

Transgenic animals (and transgenic plants) have also been tested as "living factories" for the production of human proteins. As a result of pioneering work at the University of Edinburgh in Scotland, transgenic sheep have been developed which carry human genes for either Factor IX (a blood-clotting factor absent in some hemophiliacs) or alpha-antitrypsin (a protein used in the treatment of emphysema, a lung disease in which patients cannot inhale sufficient air). In both cases, the human gene is joined to a sheep gene that codes for a milk protein. DNA containing the fused genes is then injected into fertilized sheep eggs, which develop into sheep whose milk contains sufficient quantities of the human protein to be purified for clinical use. The next time you see farm animals grazing on the countryside, you may be looking at a living pharmaceutical plant!

CONTROVERSY OVER THE PERCEIVED DANGERS OF GENETIC ENGINEERING

In spite of its successes, releasing genetically engineered organisms into the environment has elicited considerable opposition from groups concerned that such practices might upset the balance of ecosystems or may create other unforseen catastrophies. For example, the release of ice-minus bacteria elicited a great deal of opposition. Some local farmers feared an accidental contamination of their fields with these engineered bacteria. Most of the opposition arose from a public fear of genetic engineering in general, a technology many suspected would lead to a world full of harmful genetic "creations." Even though the ice-minus bacterium showed no evidence of being harmful, injunctions, court battles, and even vandalism plagued the project. Finally, the project directors decided to isolate a naturally occurring ice-minus bacterium to use in place of the genetically engineered strain.

Use and release of genetically engineered organisms is subject to careful scrutiny by regulatory agencies and by the scientific community itself. Even though thousands of experiments have been carried out without any apparent danger, concern over this new technology is still prevalent among the general public. Perhaps the answer to realizing the benefits of biotechnology without promoting fear in the general community is to keep the general population better informed and to create well-publicized regulatory agencies to monitor the development and use of genetically engineered organisms. Regulatory controls should ensure public safety while also guarding against an atmosphere that retards progress in biotechnology.

In addition to working with plants and animals, genetic engineers are studying ways to cure inherited human diseases by "gene therapy"—implanting an effective gene in place of a defective or missing one. We will return to this provocative new development in medicine in Chapter 17.

In the meantime, let's turn to some of the specific techniques that have made these practical applications of genetic engineering possible.

Scientists can now isolate specific genes, which can be incorporated into the chromosomes of host cells to create new, genetically engineered life forms. Using this technology, cells can be converted into living pharmaceutical plants; crop plants and domestic animals can be modified to provide additional food for a hungry human population; and it is hoped that one day, individuals suffering from debilitating genetic diseases may be cured. (See CTQ #2.)

DNA TECHNOLOGY I: THE FORMATION AND USE OF RECOMBINANT DNA MOLECULES

All of the biotechnological successes described thus far depended on the introduction of "new" genes into a recipient "host" cell. This feat is not as simple as it may sound and actually required years of research before it was successfully accomplished. First, a recombinant DNA molecule must be formed by inserting, or **splicing,** the gene in question into a larger molecule of DNA. This DNA then acts as a transport vehicle, or *vector,* carrying the desired gene into the host cell. Common vectors include viruses and bacterial plasmids. Viruses carry the foreign gene into a host cell during an infection; the viral DNA, together with the foreign gene, then becomes integrated into the host cell's chromosome. **Plasmids** (Figure 16-6) are small, circular DNA molecules found in bacteria, which are separate from the main bacterial chromosome. Plasmids often contain genes that make bacterial cells resistant to antibiotics such as ampicillin and tetracycline, which are commonly used to treat bacterial infections.

TOOLS FOR ASSEMBLING RECOMBINANT DNA MOLECULES

One of the first steps in constructing most recombinant DNA molecules is to fragment large DNA molecules into smaller pieces. This is accomplished by **restriction enzymes,** special DNA-cutting enzymes that are found in bacteria. Unlike eukaryotic DNA-digesting enzymes (such as the DNase produced by your pancreas), restriction enzymes are *sequence specific,* meaning that they recognize specific DNA sequences four to six nucleotides long and make their incision within that sequence (Figure 16-7). Since particular nucleotide sequences of this length occur quite frequently simply by chance, they appear in all types of DNA—viral, bacterial, plant, and animal—and are thus susceptible to fragmentation by these enzymes. Bacteria

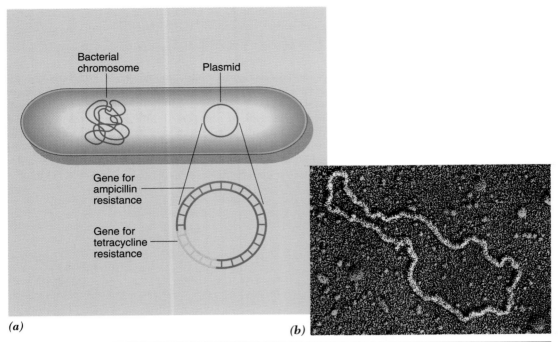

(a)   *(b)*

FIGURE 16-6

A bacterial plasmid. In addition to its chromosome, many bacteria contain small circular strands of DNA, called plasmids. The plasmid shown in part *(a)* contains genes that make the bacterial cells resistant to two antibiotics, tetracycline and ampicillin (a relative of penicillin). Consequently, a bacterium with this plasmid can live in the presence of these antibiotics. This particular plasmid is commonly used in recombinant DNA experiments. The photograph *(b)* is an electron micrograph of a bacterial plasmid.

FIGURE 16-7

Restriction enzymes are the molecular biologist's DNA-cutting scissors. In the case depicted here, a restriction enzyme called *Eco* R1 cuts both strands of a DNA double helix at a particular DNA sequence (in this example, after every "TTAA" sequence). This restriction enzyme makes staggered cuts that generate DNA fragments possessing "sticky ends."

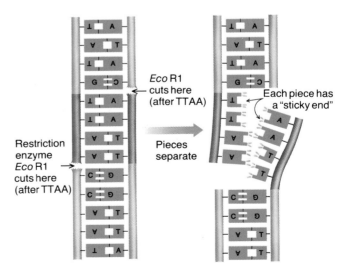

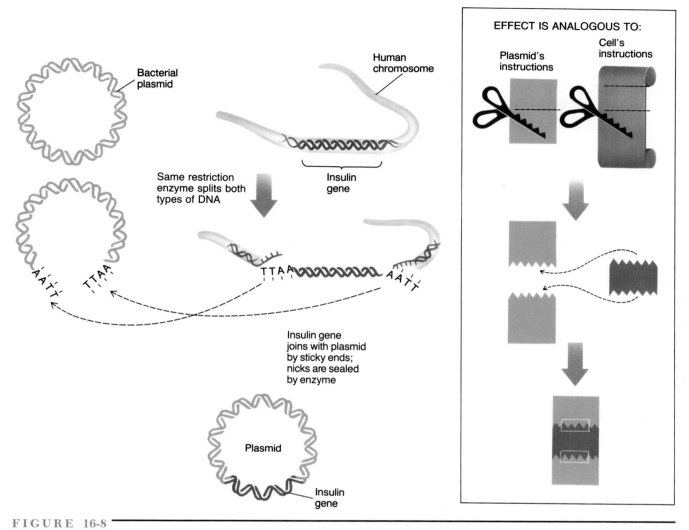

FIGURE 16-8

Formation of recombinant DNA molecules. In this example, both plasmid DNA and human DNA containing the insulin gene are treated with the same restriction enzyme so that the DNA from the two sources will have complementary "sticky ends." As a result, the two DNA molecules become joined to each other and are then covalently sealed by DNA ligase, forming a recombinant DNA molecule.

use their restriction enzymes to destroy ("restrict") the DNA of invading viruses, while genetic engineers have used the enzymes as molecular "scissors" to "cut and paste" DNA molecules obtained from different sources.

As illustrated in Figure 16-7, many restriction enzymes generate a "staggered cut" in the DNA because the two strands of the double helix are nicked at different sites along their length. Staggered cuts leave short, single-stranded tails that act as "sticky ends" that can bind with a complementary single-stranded tail on another DNA molecule to restore a double-stranded structure.

Since a particular restriction enzyme always cuts a DNA molecule at the same site in a sequence, it creates the same sticky ends regardless of the organism that donated the DNA. For example, the sticky ends of a genetic segment

cut from a human chromosome readily adhere to the complementary single-stranded tails of a bacterial plasmid that has been cut with the same enzyme (Figure 16-8). When these two DNA preparations are mixed together, the complementary sticky ends join the isolated human genetic segment to the plasmid DNA, much as an extra paragraph may be taped into a sheet of instructions. As long as both papers are cut with scissors that make the same pattern, the cuts fit, and the pieces can be joined together. The "tape" that permanently bonds the joined DNA fragments together is **DNA ligase,** an enzyme that seals the gaps in the DNA backbone by forming a covalent bond (page 55). The result is a recombinant DNA molecule—in this case, a bacterial plasmid containing a gene that codes for a human protein. The first recombinant DNA molecule—a frog gene joined to a bacterial plasmid—was formed in this manner in 1973 by Stanley Cohen of Stanford University and Herbert Boyer of the University of California, marking the birth of modern genetic engineering.

AMPLIFICATION OF RECOMBINANT DNA BY DNA CLONING

Before the recombinant DNA molecule formed by the procedure discussed above can be put to much use, it is necessary to generate a relatively large quantity of the gene(s) needed. This DNA amplification process can be accomplished by a procedure called **DNA cloning** (Figure 16-9). The first step in this procedure is to get the recombinant DNA molecule into a host bacterial cell. This is surprisingly easy since bacteria readily take up DNA from their medium.

Once inside a bacterium, the plasmid DNA molecule replicates in each cell before the cell divides in two; therefore the number of recombinant DNA molecules increases, as does the number of bacterial cells. A geneticist can begin with a single recombinant plasmid inside a single bacterial cell and, after a period of time, millions of copies of the plasmid will be formed. Once the desired amount of DNA replication has been reached, the amplified recombinant DNA can be purified (Figure 16-10) and used in other procedures.

EXPRESSION OF A EUKARYOTIC GENE IN A HOST CELL

◭ In addition to serving as cellular "copying machines" for generating large quantities of recombinant DNA, host bacterial cells can transcribe and translate a human gene residing within a bacterial plasmid, producing large amounts of high-quality human protein. This process requires that the DNA encoding a human protein be placed in the proper position next to a bacterial promoter (otherwise, the bacterial RNA polymerase cannot attach to the DNA and transcribe it). The human DNA used in these processes must also be free of introns (page 303) since bacterial cells do not have the machinery required to remove the intervening sequences from the transcribed message.

Biotechnologists are not restricted to using bacterial cells as such living pharmaceutical "factories." Similar techniques have been developed in which mammalian cells serve as hosts for human genes. These cells generally prove to be better suited than do bacteria for manufacturing human gene products and, like bacteria, can be grown to high density in large vats, where they secrete human proteins (such as those listed in Table 16-1) into the medium.

Work involving recombinant DNA technology has once again revealed the basic unity of life. Enzymes isolated from bacterial cells are used to prepare fragments of bacterial, viral, or eukaryotic DNA and to join them into a single,

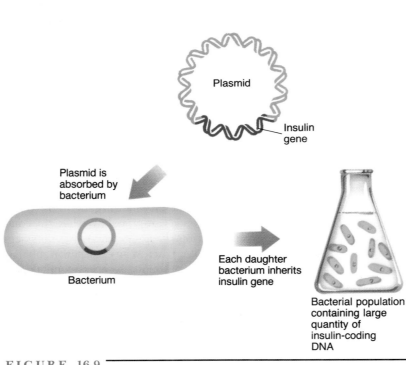

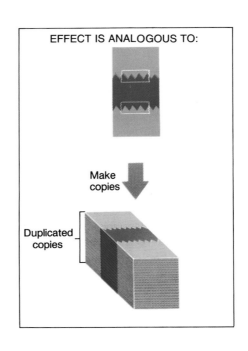

FIGURE 16-9

DNA cloning produces large quantities of a particular recombinant DNA molecule inside bacterial cells. The recombinant molecules are taken up by the bacterial cells and become amplified as the bacterial cells proliferate.

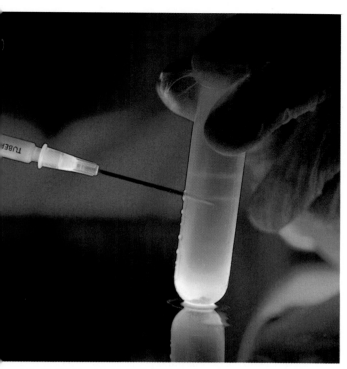

FIGURE 16-10

Purifying DNA. This centrifugation tube can be seen to contain two separate bands. One band represents plasmid DNA carrying a foreign DNA segment that has been cloned within the bacteria, while the other band contains chromosomal DNA from these same bacteria. The two types of DNA are separated during centrifugation. The researcher is attempting to remove the DNA from the tube with a needle and syringe. The DNA in the tube is made visible by using a fluorescent dye.

continuous DNA molecule. These recombinant DNA molecules can be introduced into either a bacterial or a eukaryotic cell, where they direct the formation of the same gene product as that which is in the cells of the species from which they were taken.

Recombinant DNA molecules form the cornerstone of the recent advances in genetic engineering. Formation and use of recombinant DNA have required the development of a new arsenal of molecular techniques in which specific DNA segments can be isolated, amplified, and inserted into suitable vectors. (See CTQ #3.)

DNA TECHNOLOGY II: TECHNIQUES THAT DO NOT REQUIRE RECOMBINANT DNA MOLECULES

The examples of biotechnology described in the beginning of this chapter require the formation of a recombinant

DNA molecule so that a foreign gene can be expressed in a host cell. Several other techniques that are also of great importance to the "biotechnology revolution" do not require the use of recombinant DNA. These techniques are also important for understanding human genetics, which will be discussed in the next chapter.

THE SEPARATION OF DNA FRAGMENTS BY LENGTH

As you will recall, the DNA molecules present in eukaryotic chromosomes are much larger than DNA molecules that make up the circular chromosome of a bacterial cell. When molecular biologists treat a sample of eukaryotic DNA with a restriction enzyme, thousands of different DNA pieces of varying lengths are generated. These pieces are called **restriction fragments.** It is important for most purposes to separate restriction fragments so that individual fragments can be identified. For example, suppose you have just treated a sample of human DNA with a restriction enzyme and you want to isolate any restriction fragments that contain part of the insulin gene. The easiest way to separate a large variety of restriction fragments is by a technique called **gel electrophoresis** (Figure 16-11), whereby DNA fragments move through a porous gel in response to an electric current. While the negative charge of the DNA fragment causes the DNA fragment to move, it is the length of the fragment that determines the actual speed of its movement. The smaller the DNA fragment, the faster it can "work its way" through the pores in the gel. Consequently, smaller fragments move farther down the gel than do larger fragments.

After electrophoresis, it is possible to determine where in the gel a particular fragment, such as one containing the insulin gene, is located. One DNA fragment is distinguished from others by virtue of differences in its nucleotide sequence. The search for specific DNA fragments in a giant "haystack" of different DNAs requires a **DNA probe** —a single-stranded DNA possessing a sequence complementary to the fragments being sought. The probe is like a guided missile, capable of searching out and binding to a particular target. In this case, the target in the gel is a single-stranded fragment of the insulin gene. Since the DNA probe contains radioactive atoms, an investigator can find the insulin gene by locating the radioactive emissions given off by the bound probe. The radioactive band(s) in the gel can be detected after electrophoresis by pressing the gel against a piece of X-ray film.

Determining DNA Nucleotide Sequences

Reported in 1965, the first nucleic acid to be sequenced was a strand of transfer RNA (tRNA) that was 77 nucleotides long and obtained from bacteria. It had taken approximately 7 years of painstaking work to sequence this nucleic acid, earning Robert Holley of Cornell University a Nobel Prize. Today, a fragment of DNA several times this length can be sequenced within an hour using a totally automated apparatus. Attempts to determine the complete sequence of two multicellular eukaryotes—the fruit fly *Drosophila* and a nematode worm—are now under way, and plans to launch a full-scale offensive to determine the entire nucleotide sequence of human DNA has been the subject of great plans and debate for several years. This undertaking is discussed in the Bioline box in Chapter 17.

ENZYMATIC AMPLIFICATION OF DNA

Within the past few years, a new technique has become available for amplifying a specific DNA fragment (such as the insulin gene), which does not require the use of a bacte-

rial cell. As we noted on page 131, the DNA-synthesizing enzyme (DNA polymerase) from hot springs bacteria is stable at temperatures of 90°C. Biotechnologists have taken advantage of this enzyme to develop a technique called **polymerase chain reaction (PCR),** whereby the temperature is raised far above that which would destroy the activity of other DNA polymerases. Using PCR, a single region of DNA, present in vanishingly small amounts, can be amplified cheaply, rapidly, and extensively.

To carry out PCR, a sample of DNA is mixed with the temperature-resistant DNA polymerase, along with short, synthetic DNA fragments (called *primers*) that are complementary to DNA sequences at either end of the region of the DNA to be amplified. The mixture is then heated above

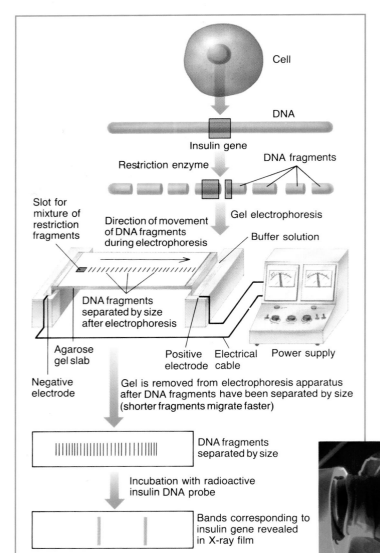

(a)

(b)

FIGURE 16-11

Separation of DNA restriction fragments by gel electrophoresis. *(a)* In the experiment outlined here, the researcher is attempting to determine the size of the restriction fragment(s) containing the insulin gene. To carry out gel electrophoresis, a thin slab of a gel (made of the polysaccharide agarose, obtained from seaweed) is prepared, and the sample containing the mixture of DNA fragments is placed in a small slot, or "well," located at one end of the gel. The slab (and the DNA sample) is then subjected to an electric current. Since DNA molecules have many negatively charged phosphate groups, they move through the porous gel toward the oppositely charged positive pole. But the gel itself acts like a sieve so that smaller fragments are able to make their way through the molecular obstacle course more easily than larger fragments. Consequently, the smaller the fragment, the faster it moves. After a period of time, the fragments become separated according to their length. In order to locate the DNA sequences that make up the insulin gene, the separated fragments are incubated in a liquid containing a radioactive insulin DNA probe that binds to the complementary DNA fragment(s). The radioactive band(s) can be located by laying the DNA-containing sheet onto X-ray film in a lightproof container. The radioactive particles expose the X-ray film, revealing the location of two distinct fragments that make up the insulin gene. *(b)* In *(a)*, specific bands of DNA were located following electrophoresis using a radioactive probe. In *(b)*, the DNA in the gel is revealed following electrophoresis by using a fluorescent dye that stains all of the DNA fragments.

75°C, which is hot enough to cause the DNA molecules in the sample to separate into their two component strands (Figure 16-12). The mixture is then cooled to allow the primers to bind to the two ends of the single-stranded target DNA. The polymerase then binds to the primers and selectively copies the intervening region, forming a new complementary strand. The temperature is raised once again, causing the newly formed and the original DNA strands to separate from each other. The sample is then cooled to allow the synthetic primers in the mixture to bind to the ends of the target DNA, which is now present at twice the original amount. This cycle is repeated over and over again, each time doubling the amount of the specific region of DNA that is flanked by the bound primers. Billions of copies of this one specific region can be generated in just a few hours.

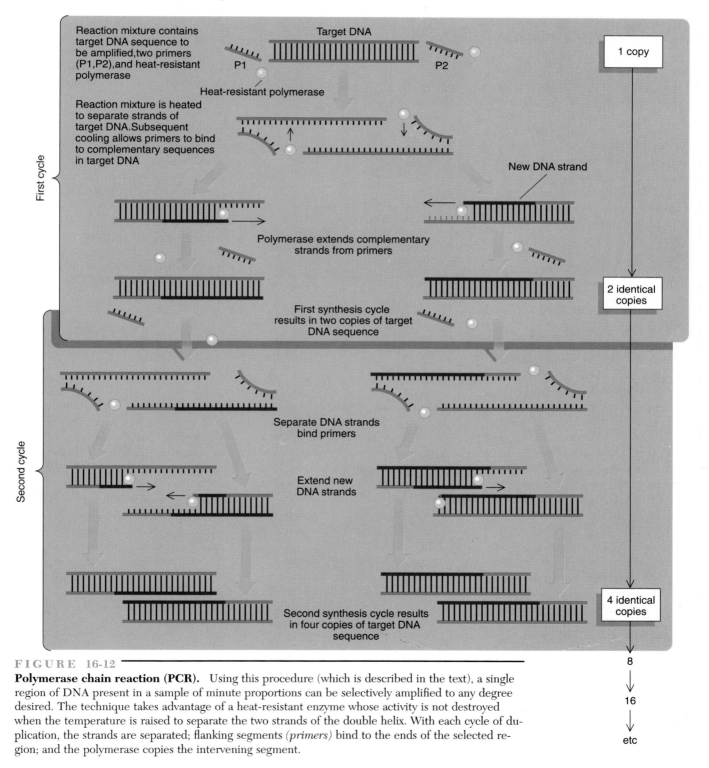

FIGURE 16-12

Polymerase chain reaction (PCR). Using this procedure (which is described in the text), a single region of DNA present in a sample of minute proportions can be selectively amplified to any degree desired. The technique takes advantage of a heat-resistant enzyme whose activity is not destroyed when the temperature is raised to separate the two strands of the double helix. With each cycle of duplication, the strands are separated; flanking segments (*primers*) bind to the ends of the selected region; and the polymerase copies the intervening segment.

◁ B I O L I N E ▷
DNA Fingerprints and Criminal Law

On February 5, 1987, a woman and her 2-year-old daughter were found stabbed to death in their apartment in the New York City borough of the Bronx. Following a tip, the police questioned a resident of a neighboring building. A small bloodstain was found on the suspect's watch, which was sent to a laboratory for DNA fingerprint analysis. The DNA from the white blood cells in the stain was amplified using the PCR technique and was digested with a restriction enzyme. The restriction fragments were then separated by electrophoresis, and a pattern of labeled fragments was identified with a radioactive probe. The banding pattern produced by the DNA from the suspect's watch was found to be a perfect match to the pattern produced by DNA taken from one of the victims. The results were provided to the opposing attorneys, and a pretrial hearing was called in 1989 to discuss the validity of the DNA evidence.

During the hearing, a number of expert witnesses for the prosecution explained the basis of the DNA analysis. According to these experts, no two individuals, with the exception of identical twins, have the same nucleotide sequence in their DNA. Moreover, differences in DNA sequence can be detected by comparing the lengths of the fragments produced by restriction-enzyme digestion of different DNA samples. The patterns produce a "DNA fingerprint" (Figure 1) that is as unique to an individual as is a set of conventional fingerprints lifted from a glass. In fact, DNA fingerprints had already been used in more than 200 criminal cases in the United States and had been hailed as the most important development in forensic medicine (the application of medical facts

FIGURE 1
Alec Jeffreys of the University of Leicester, England, examining a DNA fingerprint. Jeffreys was primarily responsible for developing the DNA fingerprint technique and was the scientist who confirmed the death of Josef Mengele.

to legal problems) in decades. The widespread use of DNA fingerprinting evidence in court had been based on its general acceptability in the scientific community. According to a report from the company performing the DNA analysis, the likelihood that the same banding patterns could be obtained by chance from two *different* individuals in the community was only one in 100 million.

What made this case (known as the Castro case, after the defendant) memorable and distinct from its predecessors was that the defense also called on expert witnesses to scrutinize the data and to present

their opinions. While these experts confirmed the capability of DNA fingerprinting to identify an individual out of a huge population, they found serious technical flaws in the analysis of the DNA samples used by the prosecution. In an unprecedented occurrence, the experts who had earlier testified *for the prosecution* agreed that the DNA analysis in this case was unreliable and should not be used as evidence! The problem was not with the technique itself but in the way it had been carried out in this particular case. Consequently, the judge threw out the evidence.

In the wake of the Castro case, the use of DNA fingerprinting to decide guilt or innocence has been seriously questioned. Several panels and agencies are working to formulate guidelines for the licensing of forensic DNA laboratories and the certification of their employees. In 1992, a panel of the National Academy of Sciences released a report endorsing the general reliability of the technique but called for the institution of strict standards *to be set by scientists*.

Meanwhile, another issue regarding DNA fingerprinting has been raised and hotly debated. Two geneticists, Richard Lewontin of Harvard University and Daniel Hartl of Washington University, coauthored a paper published in December 1991, suggesting that scientists do not have enough data on genetic variation within different racial or ethnic groups to calculate the odds that two individuals—a suspect and a perpetrator of the crime—are one and the same on the basis of an identical DNA fingerprint. The matter remains an issue of great concern in both the scientific and legal communities and has yet to be resolved.

In addition to its use in the amplification of specific DNA fragments, PCR is able to generate large amounts of DNA from minute starting samples. For example, PCR has been used in criminal investigations to generate quantities of DNA from a spot of dried blood left on a crime suspect's clothing or even from the DNA present in part of a single hair follicle left at the scene of a crime (see Bioline: DNA Fingerprinting and Criminal Law). In 1992, PCR was used to amplify DNA extracted from the exhumed bones of a man who had drowned several years earlier while swimming off the coast of Brazil. Some people claimed that the man was Dr. Josef Mengele, the infamous Nazi doctor who had carried out sadistic experiments on inmates at the Auschwitz concentration camp. Many skeptics doubted the identification, however. To confirm the man's identity, DNA from the bones was isolated, amplified by PCR, and compared to a DNA sample donated by Mengele's son, using DNA fingerprinting. The results were conclusive: The drowning victim was indeed Josef Mengele, closing the chapter on one of the most intensive international manhunts in history.

DNA fragments generated by restriction enzymes have applications beyond genetic engineering. Separation and analysis of DNA restriction fragments form the basis for DNA fingerprinting, constructing detailed genetic maps, and determining evolutionary similarity among species. (See CTQ #4.)

USE OF DNA TECHNOLOGY IN DETERMINING EVOLUTIONARY RELATIONSHIPS

▕▶ In the beginning of this chapter, we saw how DNA sequencing techniques were used to shed light on an ecological question concerning turtle migration. The technique has also been used by evolutionary biologists. Mutations lead to changes in the nucleotide sequence of DNA—changes that serve as the raw material for biological evolution. Since mutations in DNA occur continuously over time, the longer the amount of time that passes since two organisms diverged from a common ancestor, the greater the differences in their DNA sequences. The analogy might be made to two people who happen to meet on a bike trip through Europe and then set out on their bikes independently to see different parts of the continent. The longer the time that passes since the departure from their point of common origin, the greater the distance the cyclists are likely to be separated. Similarly, the more time that passes after two organisms have diverged from a common ancestor, the greater the differences in their DNA sequences.

An evolutionary question of interest to humans is: Who is our closest living relative? According to sequencing studies of a large region of DNA near the beta-globin gene (see Figure 15-6) carried out by Michael Miyamoto and his colleagues at Wayne State University, the answer is chimpanzees. While this answer may not be surprising, Miyamoto's group made another, unexpected finding. It has generally been assumed that chimpanzees and gorillas are more closely related to each other than either is to the human species. However, Miyamoto found that there were fewer differences in the DNA nucleotide sequences between humans and chimpanzees (a 1.6 percent nucleotide sequence difference) than between chimpanzees and gorillas (a 2.1 percent difference). According to these data, chimpanzees are more closely related to humans than they are to gorillas.

Until recently, biologists had to rely solely on phenotypic differences among organisms to assess evolutionary relatedness. With advances in DNA technology, evolutionists can now focus directly on differences in genotypes, the source of biological variation. (See CTQ #5.)

REEXAMINING THE THEMES

Unity within Diversity

⬥ The entire technology used in the formation and use of recombinant DNAs is based on the universality of the genetic code. DNAs from such diverse sources as humans, viruses, and bacteria can be spliced together to form a single, continuous DNA molecule. This recombinant DNA can then be introduced into a host cell (usually either bacterial or mammalian). Once in the host cell, the human DNA will direct the formation of a human protein that is indistinguishable from the same protein produced by a human cell.

Evolution and Adaptation

▕▶ Techniques are now available to alter the genetic composition of any organism, creating, in effect, a new type of life form. The advent of this new age of biotechnology reveals the power of science to alter genetic inheritance and even the course of evolution itself. At the same time, techniques described in this chapter are being used to study the natural course of evolution. Comparisons of DNA sequences between diverse organisms provide a measure of the evolutionary relatedness of the organisms. This experimental approach rests on the assumption that organisms that have diverged more recently from a common ancestor will have more similar DNA sequences than will organisms that diverged from one another at a more distant point in the past.

SYNOPSIS

Genetic engineering and the "biotechnology revolution" has depended on the formation of recombinant DNA. Recombinant DNA molecules contain portions derived from different sources and joined together in the laboratory. Recombinant DNAs have been used to manufacture human proteins with clinical applications, such as insulin, growth hormone, and plasminogen activator. In such cases, a gene encoding the desired product is spliced into a bacterial plasmid, and the recombinant molecule is introduced into a bacterial or a mammalian host cell, which transcribes and translates the foreign gene, releasing the gene product into the medium.

Genetic engineers have developed organisms that carry recombinant DNA. The recombinant DNA encodes a product that alters the organism's characteristics. For example, genetic engineers have developed plants that contain a bacterial gene that codes for an insect-killing protein, microbes that can produce enzymes to digest oil spills, crops and livestock with improved nutrient value, and animals that are capable of manufacturing human proteins of clinical value.

The biotechnology revolution has been built on the development of new techniques in DNA technology. The isolation of particular genes of interest in medicine and agriculture is achieved by using restriction enzymes — bacterial DNA-digesting enzymes that recognize a particular stretch of nucleotides. Once the DNA is enzymatically fragmented, and the gene of interest is isolated, the gene is spliced into a suitable vector, such as a bacterial plasmid, and the recombinant molecule is amplified either by DNA cloning or by PCR. When amplified by DNA cloning, the plasmid is introduced into bacterial cells, where it is replicated as the bacterial cells proliferate. When amplified by PCR, a particular gene is experimentally and selectively targeted for replication by a heat-resistant DNA polymerase.

DNA fragments can be separated from one another, and their nucleotide sequences determined. Created by treatment of a DNA sample with a restriction enzyme, DNA fragments are separated by gel electrophoresis, whereby the DNAs move through a sievelike gel in response to an electric current. The pattern of fragments generated by electrophoresis forms the basis for comparing DNA from different individuals of a species. For example, distinguishing the pattern of restriction fragments from different humans forms the basis for the forensic technique of DNA fingerprinting.

Comparing nucleotide sequences of DNAs obtained from individuals of different species provides a measure of the evolutionary relatedness of the species. Based on this technique, chimpanzees appear more closely related to humans than to gorillas.

Key Terms

recombinant DNA (p. 314)
transgenic animals (p. 319)
splicing (p. 321)
plasmid (p. 321)

restriction enzyme (p. 321)
DNA ligase (p. 323)
DNA cloning (p. 324)
restriction fragment (p. 325)

gel electrophoresis (p. 325)
DNA probe (p. 325)
polymerase chain reaction (PCR) (p. 326)

Review Questions

1. Distinguish between the following: bacterial plasmid and chromosome; conventionally produced insulin and insulin produced by genetically engineered *E. coli;* restriction enzyme and human pancreatic DNase; transgenic animal and a "normal" animal; a sense strand and an antisense strand of DNA; amplification by cloning and amplification by PCR.

2. Why is the production of "sticky ends" important in recombinant DNA technology?

3. Describe three ways crop plants might be improved by genetic engineering.

4. How are investigators able to use PCR to amplify one particular gene, while all the other parts of the DNA remain at their original amount?

5. Why do larger fragments of DNA migrate more slowly than do smaller fragments during gel electrophoresis?

Critical Thinking Questions

1. Would you expect that the ability of a sea turtle to return to the beach where it was hatched is somehow built into the animal's genes? That is, is natal homing an instinctive behavior? How might you test whether or not such behavior is learned from other turtles? Can you think of any clues a turtle could use to navigate across the ocean to a particular island where it was hatched?

2. Given the many benefits of genetic engineering illustrated in this chapter, why are some people opposed to it? List at least four arguments against genetic engineering. For each, develop a counterargument. After considering these pros and cons, what is your position on genetic engineering, and why?

3. Compare the mechanism for bacterial transformation discovered by Frederick Griffith (page 266) and the production of human insulin inside an "engineered" bacterial cell. What are the similarities and differences?

4. Until recently, bacteria that cause tuberculosis (TB) were susceptible to a variety of antibiotics. Now, some of these bacteria have developed resistance to many drugs. Most resistant strains of bacteria have genes that code for enzymes that destroy antibiotics, but the TB-resistant bacteria are the result of a deletion mutation. Can you suggest a mechanism by which a deletion could lead to drug resistance?

5. The graph below shows the percentage of nucleotide substitutions between the DNA and three proteins (fibrin, hemoglobin, and insulin) of the cow, sheep, and pig. Which pair of animals is most closely related?

Which of the four types of molecules shows the greatest rate of change? the least? What can you conclude from these data about the rate of change of different genes?

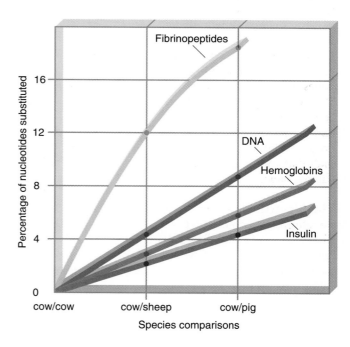

6. Do you harbor any fears about recombinant DNA technology? If so, what steps would you want taken to allay those fears?

Additional Readings

Barinaga, M. 1992. Knockout mice offer first animal model for CF. *Science* 257:1046–1047. (Intermediate)

Barton, J. H. 1991. Patenting life. *Sci. Amer.* May:84–91. (Intermediate)

Gibbons, A. 1992. Moths take the field against biopesticide. *Science* 254:646. (Introductory)

Kessler, D.A., et al. 1992. The safety of foods developed by biotechnology. *Science* 256:1747–1832. (Intermediate)

Kolberg, R. 1992. Animal models point the way to clinical trials. *Science* 256:772–773. (Intermediate)

Lewin, R. 1989. New look at turtle migration mystery. *Science* 243:1009. (Intermediate)

Lewontin, R. C., and D. L. Hartl. 1991. Population genetics in forensic DNA typing. *Science* 254:1745–1750. (Advanced)

McElfresh, K.C., et. al. 1993. DNA-based identity testing in forensic science. *Bioscience* March:149–157. (Intermediate)

Moffat, A. S. 1991. Making sense of antisense. *Science* 253:511. (Intermediate)

Moffat, A. S. 1992. Plant biotechnology explored in Indianapolis. *Science* 254:25. (Intermediate)

Moffat, A. S. 1992. Transgenic animals may be down on the pharm. *Science* 254:35–36. (Intermediate)

Neufeld, P. J., and N. Coleman. 1990. When science takes the witness stand. *Sci. Amer.* May:46–53. (Intermediate)

Roberts, L. 1991. Fight erupts over DNA fingerprinting. *Science* 254:1721–1723. (Intermediate)

Suzuki, D., and P. Knudson. 1990. *Genethics: The clash between the new genetics and human values.* Cambridge, MA: Harvard Univ. Press. (Intermediate)

CHAPTER
◄ 17 ►

Human Genetics: Past, Present, and Future

STEPS TO DISCOVERY
Developing a Treatment for an Inherited Disorder

THE CONSEQUENCES OF AN ABNORMAL NUMBER OF CHROMOSOMES

An Abnormal Number of Sex Chromosomes

DISORDERS THAT RESULT FROM A DEFECT IN A SINGLE GENE

New Techniques for Mapping Defective Genes

Cystic Fibrosis: A Lesson in Evolution

PATTERNS OF TRANSMISSION OF GENETIC DISORDERS

Disorders that Result from Autosomal Recessive Alleles

Disorders that Result from Autosomal Dominant Alleles

Disorders that Result from X-Linked Alleles

Complex Diseases

SCREENING HUMANS FOR GENETIC DEFECTS

Screening Fetal Cells For Genetic Disorders

Ethical Considerations

BIOLINE
Mapping the Human Genome

THE HUMAN PERSPECTIVE
Correcting Genetic Disorders by Gene Therapy

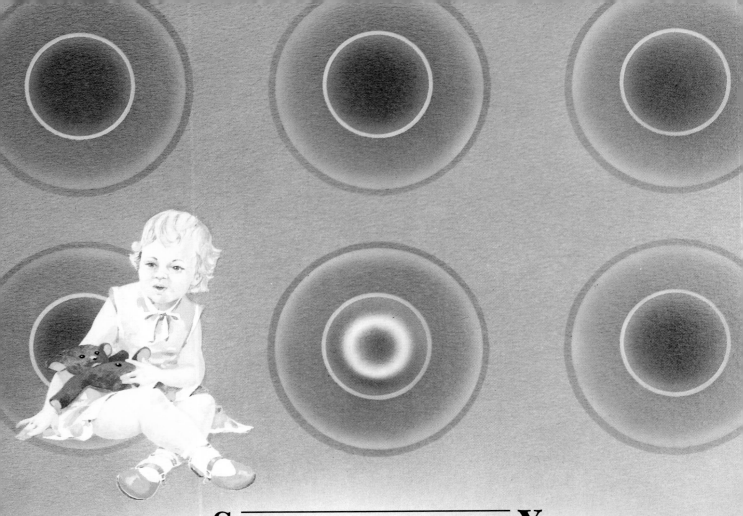

Developing a Treatment for an Inherited Disorder

*I*n 1934, Asbjorn Folling, a Norwegian physician, reported on a study of two mentally retarded infants. The infants' mother had complained that her children emitted a continual musty odor. Chemical tests of the infants' diapers revealed the presence of high levels of phenyl ketones, toxic compounds formed from the metabolic breakdown of an essential amino acid, phenylalanine. Because of the presence of these compounds, the childrens' disease was called *phenylketonuria (PKU)*. Subsequent research indicated that the condition was caused by the inherited deficiency of

phenylalanine hydroxylase, an enzyme that normally converts phenylalanine to another amino acid, tyrosine. As a result of this deficiency, the blood of these infants contained high levels of phenylalanine, which caused mental retardation.

In 1953, working at a British children's hospital, Evelyn Hickmans and co-workers Horst Bickel and John Gerrard, hypothesized that if the crippling effects of PKU were due to the presence of high levels of phenylalanine in the blood, then it might be possible to treat PKU children by restricting their dietary intake of phenylalanine. The researchers tested their hypothesis on a 2-year-old PKU victim who was "unable to stand, walk, or talk . . . and spent her time groaning, crying, and banging her head." The child was put on a diet containing only enough phenylalanine to support the synthesis of vital proteins. Over the next few months, the little girl improved dramatically; she learned to stand and climb on chairs, and she stopped crying and banging her head. Other studies soon followed, confirming the benefits of a low-phenylalanine diet for infants born with PKU. For treatment to be effective, however, early diagnosis is essential so that the infant can be placed on the diet before permanent damage to the nervous system occurs.

In 1961, Robert Guthrie of the University of Buffalo published a one-page "letter" outlining a procedure by which newborn infants could easily be screened for PKU. Guthrie's procedure took advantage of the fact that the blood of newborn infants with PKU contains about 30 times the level of phenylalanine as does that of normal infants. Using the Guthrie test, a drop of blood from a newborn infant is dried on a small piece of filter paper, which is then added to a well in a culture dish containing a low concentration of bacteria that require phenylalanine to grow. When the filter paper contains blood from an infant with PKU, the high phenylalanine content promotes the growth of the bacteria, producing a visible "halo" around the well. In contrast, blood from a normal infant does not promote the growth of the bacteria, so no halo will be observed.

Since the development of the Guthrie test, the vast majority of infants born in the United States and other western countries are automatically screened for PKU. Those infants who test positively for the inherited condition (about 1 in 18,000 newborns) can be placed on the prescribed diet. Once these children reach a certain age, their nervous systems are no longer susceptible to damage by high levels of phenylalanine in their blood, and they can begin eating a normal diet.

Although early diagnosis and treatment had virtually eliminated this rare form of mental retardation, PKU has returned to the spotlight in recent years. Successful screening and dietary treatment for PKU have allowed children with the disorder to develop into normal adults, who are having children of their own. Even though a high level of phenylalanine in the blood has little effect on the mother, it can produce terrible damage to the developing fetus. Damage to the fetus is best prevented by the mother's return to her strict, low-phenylalanine diet *before* she becomes pregnant to ensure that the developing baby will have a safe environment throughout the entire gestation period. This potentially serious problem illustrates the importance of genetic counseling for people at risk of having children with genetic diseases.

Dietary treatment of PKU prevents development of the effects of the disease without altering the root of the genetic problem itself. It can be said that a diet low in phenylalanine changes a person's phenotype without altering his or her genotype. In this chapter, we will examine new approaches in the treatment of genetic disorders where the genotype, rather than the phenotype, is the target for modification.

bacteria around blood samples, producing a halo.

*H*uman genetics is stocked with tales of tragedy and triumph. One strange tragedy comes in the form of a bizarre disease called Lesch-Nyhan syndrome. Victims of this syndrome, who are almost all boys, begin to mutilate themselves during their second year of life by biting off their lips and fingertips. Although the behavior was once attributed to lunacy, or "demonic possession," scientific examination of pedigrees of victims and controlled analysis revealed that the underlying defect was not a dysfunctional family life or some childhood trauma but something completely biological—a deficiency in a single enzyme, one needed for normal purine metabolism. Without this enzyme (hypoxanthine-guanine phosphoribosyltransferase, or HGPRT) children exhibit a buildup of uric acid in their blood and urine. In fact, the urine of these children often contained "orange sand," the precipitated uric acid crystals. The disorder was inherited from the children's mothers, carriers who showed no signs of the disorder.

Sadly, the triumph in this tragedy has yet to be realized, although through genetic counseling potential parents can discover their probability of having a baby with the syndrome. If a woman is pregnant, some of the methods described in this chapter can determine whether the incubating fetus is a victim of Lesch-Nyhan syndrome. Someday we may score a real triumph over this genetic disease by developing the ability to replace the faulty gene with a fully functional copy. Such *gene therapy* is the frontier that holds the most promise for genetic researchers, for physicians, and most importantly, for people who must cope with the consequences of acquiring one of the many diseases caused by defective human genes. Altogether, over 3,500 distinct genetic disorders have been described, 15 of which are described in Table 17-1.

▼ ▼ ▼

THE CONSEQUENCES OF AN ABNORMAL NUMBER OF CHROMOSOMES

↻ We saw in Chapter 11 how homologous chromosomes separate during meiosis I and sister chromatids separate during meiosis II. On a rare occasion, homologous chromosomes fail to come apart during meiosis I or sister chromatids fail to detach during meiosis II. If either of these types of meiotic **nondisjunction** occurs, gametes with either an extra chromosome or a missing chromosome are formed (Figure 17-1). Normal development depends on the proper number of chromosomes. If one of these abnormal gametes

fuses with a normal gamete, the resulting zygote usually develops into an abnormal embryo that dies in the womb and is miscarried. In a few cases, however, the zygote develops into a viable offspring, whose cells have an abnormal chromosome number. The most common disorder resulting from meiotic nondisjunction is Down syndrome, which, as discussed in Chapter 11, occurs in individuals having three homologues of chromosome number 21 in their cells, rather than two, which is why the disease is also called *trisomy 21*.

AN ABNORMAL NUMBER OF SEX CHROMOSOMES

Occasionally, a baby is born with an abnormal number of sex chromosomes due to meiotic nondisjunction. A zygote with three X chromosomes (XXX) develops into a relatively "normal" female, although she will likely have subaverage intelligence and experience menstrual irregularities. A zy-

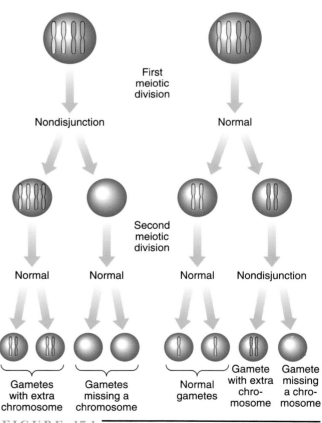

First meiotic division

Nondisjunction   Normal

Second meiotic division

Normal   Normal   Normal   Nondisjunction

Gametes with extra chromosome | Gametes missing a chromosome | Normal gametes | Gamete with extra chromosome | Gamete missing a chromosome

FIGURE 17-1

Meiotic nondisjunction occurs when chromosomes fail to separate from each other during meiosis. If the failure to separate occurs during the first meiotic division, all of the gametes will have an abnormal number of chromosomes. If nondisjunction occurs during the second meiotic division, only two of the four gametes will be affected.

gote with only one X chromosome and no second sex chromosome (X0) develops into a female with *Turner syndrome*, in which genital development is arrested in the juvenile state; the ovaries fail to develop; and body structure is slightly abnormal.

Males may also develop with abnormal numbers of sex chromosomes. Since a Y chromosome is male determining, persons with at least one Y chromosome develop as males. A male with an extra X chromosome (XXY) will develop *Klinefelter syndrome*, which is characterized by mental retardation, underdevelopment of genitalia, and the presence of feminine physical characteristics (such as breast enlargement). Alternatively, a zygote with an extra Y (forming an XYY male) develops into a man who appears normal in every way except that he will probably be taller than average. Considerable controversy has developed surrounding claims that XYY males tend to exhibit more aggressive, antisocial, and criminal behavior than do XY males, but this correlation has never been proven.

Human cells typically possess two copies of each gene, one on each homologue, except for those genes on the sex chromosomes for which a cell typically has only one functional copy. The well-being of many human cells depends on this "gene dosage"; when it is altered by the absence of a chromosome or the presence of an extra chromosome, the normal balance within the cells is disturbed. (See CTQ #2.)

TABLE 17-1
GENETIC DISORDERS

Genetic Disorder	Cause	Nature of Illness	Incidence	Inheritance
Down syndrome	Extra chromosome number 21	Mental retardation, body alterations	1 in 800	Sporadic
Klinefelter syndrome	Male with extra X chromosome	Abnormal sexual differentiation	1 in 2,000	Sporadic
Cystic fibrosis	Abnormal chloride transport	Complications of thickened mucus	1 in 2,500 Caucasions	Autosomal recessive
Huntington's disease	Unknown	Progressive neurological degeneration	1 in 2,500	Autosomal dominant
Duchenne muscular dystrophy	Deficient muscle protein dystrophin	Progressive muscle degeneration	1 in 7,000 males	X-linked
Sickle-cell anemia	Abnormal beta-globin	Weakness, pain, impaired circulation	1 in 625; mostly African descent	Autosomal recessive
Hemophilia	Deficiency in one of a number of clotting factors	Uncontrolled bleeding	1 in 10,000 males	X-linked
Phenylketonuria	Deficiency in enzyme phenylalanine hydroxylase	Mental retardation	1 in 18,000	Autosomal recessive
Tay-Sachs syndrome	Deficiency in enzyme acetylhexosaminidase	Deposition of fatty materials in brain, infant death	1 in 3,000 Ashkenazic Jews	Autosomal recessive
Lesch-Nyhan syndrome	Deficiency in enzyme HGPRT	Mental retardation, self-mutilation	1 in 100,000 males	X-linked
Galactosemia	Deficiency in enzyme galactose transferase	Mental retardation, digestive problems	1 in 60,000	Autosomal recessive
Xeroderma pigmentosum	Deficiency in a UV repair enzyme	Sensitivity to sunlight, cancer	1 in 250,000	Autosomal recessive
Severe combined immunodeficiency (SCID)	Deficiency in enzyme adenosine deaminase or others	Absence of immune defenses	Extremely rare	Autosomal recessive
Ehlers-Danlos syndrome	Deficiency in collagen-hydroxylating enzyme	Abnormal connective tissues, joint problems	1 in 100,000	Autosomal recessive
Familial hypercholesterolemia	Deficiency in LDL receptors	Cardiovascular disease	1 in 100,000	Autosomal recessive

DISORDERS THAT RESULT FROM A DEFECT IN A SINGLE GENE

Every gene encodes a product, either a polypeptide or an RNA (such as a transfer RNA), each of which has some function. Some gene products play a relatively minor role in the life of an organism; others are critical. When genes become altered so that they produce inactive products, the consequences are often serious. One of the first steps involved in studying the properties of a defective gene is determining the chromosome on which the gene is located and the relative position of the gene within that chromosome. This process of gene mapping has undergone a revolution in recent years, made possible by the DNA technology discussed in Chapter 16.

NEW TECHNIQUES FOR MAPPING DEFECTIVE GENES

Geneticists map genes along a chromosome by determining the frequency of crossing over (page 252); the closer two genes reside on a chromosome, the less likely their alleles will become separated from one another during crossing over. Using this mapping procedure, the different forms (alleles) of each gene act as **genetic markers,** serving as signposts that help reveal the relative locations of other genetic sites along the chromosome (see Bioline: Mapping the Human Genome). Genes that are known to exist in several allelic forms, such as those that determine blood type, are called **polymorphic genes,** and they make the best markers.

In 1978, a group of geneticists gathered to describe their research on the inheritance of diseases within large human families. The scientists were disturbed that there were so few known polymorphic genes that could be used as genetic markers to locate disease-causing genes within human chromosomes. During a presentation, two members of the audience—David Botstein of MIT and Ronald Davis of Stanford University—began talking together and announced that they knew a way to provide genetic markers located throughout the chromosomes. The markers Botstein and Davis had in mind were not genes but DNA fragments.

As you will recall, treatment of DNA with a restriction enzyme produces a collection of different-sized fragments (page 325). The restriction fragments generated from the DNA *of a given individual* form a precisely defined set of fragments. When restriction fragments from different people (even siblings) are compared, the patterns of fragments are similar, but not identical, due to differences in nucleotide sequences from one person to the next. Let's look more closely at these genetic differences among humans.

Recent analysis of human DNA molecules has revealed the presence of short, repeating sequences that lie adjacent to each other in clusters within the vast stretches between genes. One such cluster of these repeating sequences is outlined by the colored boxes in Figure 17-2a. In this illustration, the two DNA molecules represent comparable portions of two homologous chromosomes from a man with a wife and two daughters (Figure 17-2b). Note the different number of the repeating sequences (indicated by the boxes) on the two DNA molecules. Differences of this type are very common and, in fact, account for much of the genetic variation that occurs in the human population. While these differences play no apparent role in biological function, they have proven invaluable as markers for locating human genes.

Differences in the number of repeating sequences in a cluster produce restriction fragments of different length after cleavage by a restriction enzyme. In the example illustrated in Figure 17-2a, corresponding segments of DNA located on two homologous chromosomes are cleaved into different-sized fragments—1,400 versus 2,400 base pairs in length—because of the differences in the number of times a small sequence happens to be repeated. Such differences are called **restriction fragment length polymorphisms (RFLPs),** or simply "riflips." RFLPs produce distinctive differences in banding patterns following electrophoresis (Figure 17-2c–d) which can be used to identify individuals or to track the inheritance of particular genes.

The first scientist to search for RFLPs was Raymond White, then a recent arrival at the University of Massachusetts and a colleague of Botstein's. By 1979, White and co-worker Arlene Wyman had identified the first RFLP (Figure 17-2d). White and Wyman examined the DNA from a large Mormon family chosen because the genealogical relationships among the members of the family were well established. When DNA from 56 different people in the study group was subjected to digestion by the same restriction enzyme, a labeled probe identified fragments of 30 different sizes among the population. It was like finding 30 different alleles for a gene among 56 related individuals. In order to illustrate how this type of variability can be put to use in gene mapping, we will describe the use of RFLPs in locating the gene responsible for cystic fibrosis.

Locating the Gene Responsible for Cystic Fibrosis

Cystic fibrosis is the most common debilitating inherited disease among Caucasians (about 1 per 2,500 newborns), but it is almost nonexistent among members of other races. Among a variety of symptoms, victims of cystic fibrosis (CF) produce a thickened, sticky mucus that is very hard to propel out of the airways leading from the lungs. As a result, these individuals typically suffer from chronic lung diseases, including potentially fatal infections. Children with cystic fibrosis once faced near-certain death before the age of 5, but recent therapies to help clear congested airways (Figure 17-3) and antibiotics to fight infections have allowed these patients to live into their adult years.

In the summer of 1989, a press conference was held to announce the isolation of the gene responsible for cystic

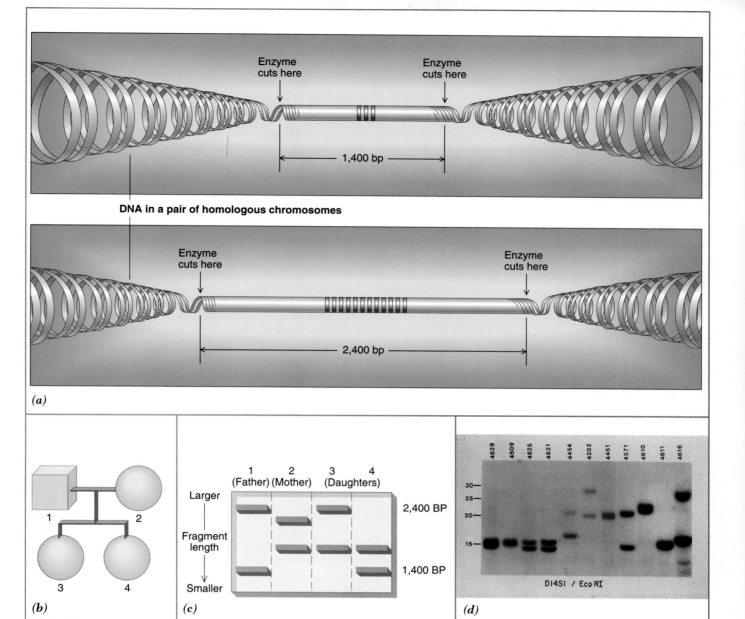

FIGURE 17-2

Restriction Fragment Length Polymorphisms (RFLPs). (*a*) Homologous segments of DNA from a pair of homologous chromosomes from one individual. The arrows represent sites in the DNA that are attacked by a particular restriction enzyme. The colored boxes denote short, repeating sequences in the DNA. In this example, the DNA on one chromosome has only three repeats of this sequence, whereas the DNA on the homologous chromosome has 12. When these two DNA molecules are treated with the restriction enzyme, one DNA molecule yields a 2,400 base pair fragment, while the other DNA molecule yields a fragment of only 1,400 base pairs. This is an example of a restriction fragment length polymorphism (RFLP). (*b*) A short pedigree of four family members: a father, mother, and their two daughters. The father's DNA is shown in part (*a*). (*c*) The pattern of DNA fragments from the four family members. Each lane (corresponding to a different person) has two bands —one from each homologous chromosome. The two bands in lane 1 identify the DNA fragments of 1,400 and 2,400 base pairs generated by enzyme treatment of the father's DNA from part (*a*). The DNA fragments of the mother are different from those of the father, reflecting differences in the nucleotide sequence of her DNA. Each of the daughters has inherited one chromosome from her mother and one from her father, generating the patterns in lanes 3 and 4. (*d*) The first demonstration by Wyman and White that RFLPs could be used to detect genetic differences in a human population. The number at the top of each lane stands for a particular individual. Note the marked differences among these banding patterns. Those lanes that have only one dense band are from individuals whose two homologous fragments either (1) are similar enough in size so that the bands merge into a single thicker band or (2) happen to be homozygous for this particular RFLP, so that both homologues produce the same-sized fragments.

◁ B I O L I N E ▷
Mapping the Human Genome

In 1986, a group of scientists from around the world gathered at the Cold Spring Harbor Laboratories in Long Island, New York, to discuss the possibility that all of the genetic information stored in human chromosomes might be decoded as the result of a giant, international scientific collaboration. Within a few years, governmental agencies in the United States, Europe, and Japan had decided to proceed with the effort, which has become known as HGP, the Human Genome Project. The term "genome" refers to the information stored in all the DNA of a single set of chromosomes. The human genome contains approximately 3 billion base pairs. If each base pair in the DNA were equivalent to a single letter on this page, the information contained in the human genome would produce a book approximately 1 million pages long!

The goal for the Human Genome Project is to sequence the DNA of all 24 human chromosomes (22 autosomes, plus the X and Y sex chromosomes) by the year 2005. Centers committed to achieving this goal have been established throughout the United States and other countries. During the first few years of the project, researchers will prepare highly detailed genetic maps of each human chromosome, with markers located every 100,000 or so bases. They will then isolate overlapping fragments of DNA that can be arranged in an order that spans the entire length of each chromosome. By the time these efforts are completed, DNA sequencing techniques should be substantially improved so that the isolated DNA fragments can eventually be sequenced in less time for less money than is currently possible. Advances in biocomputer technology will

also be required to enable researchers to sift through the massive amounts of data in order to come up with meaningful conclusions about the functions of gene products.

The overall cost for the project is estimated at approximately $3 billion — $1 for each base pair to be determined. Why spend so much money and effort simply to determine a list of four nucleotide letters running a million pages in length? The goal of the project is not simply to compile a sequence of letters but to learn what the sequence of letters "stands for." It is hoped that the Human Genome Project will

- Reveal the amino acid sequence of each polypeptide encoded in our genes. Using reverse genetics (page 342), biologists can then learn about the probable function of these polypeptides. This information could shed light on virtually every aspect of human biology, from the biochemical basis of memory to the gene products responsible for a cell's loss of growth control and consequent malignancy.

- reveal detailed information about the sequences involved in controlling gene expression.

- identify those genes that cause or contribute to human diseases — including complex diseases such as cancer, hypertension, Alzheimer's, and mental illness — and determine the changes in sequence that trigger the disease.

- reveal insights into evolution at the molecular level by comparing nucleotide and amino acid sequences in humans to those of more distant relatives. Studies of this type will also reveal which gene products are unique to humans, helping to explain how we are biologically different from — and similar to — other organisms.

- allow investigators to receive samples of DNA containing any particular gene, thereby providing an unlimited opportunity for continuing research in human molecular biology.

- provide biotechnologists with the opportunity to manufacture previously unknown human proteins that have potential medical value.

The Human Genome Project is not without its critics. Some scientists believe the project is too costly, at least at our current level of DNA sequencing expertise, and it will siphon money from other research endeavours. In addition, most of the work will be tedious and noncreative — not the kind of project attractive to most scientific minds. Sydney Brenner, one of England's pioneering molecular biologists, facetiously suggested that the work would be suitable punishment for convicts of a penal colony. Other critics argue that knowledge of human DNA sequences opens the door to further invasion of our privacy. It is feasible, for example, that insurers or employers could demand information about the genetic makeup of potential employees or insurees to determine their predisposition toward particular illnesses. Still others fear that knowledge about the nucleotide sequences responsible for certain genetic characteristics could lead unscrupulous governments to attempt to alter the human genome in desirable directions. Since it is inevitable that the human genome will be sequenced, it is up to all of us to remain vigilant and ensure that this genetic information is used for the betterment of the human species and not misused by those who feel they can profit from it.

F I G U R E 17-3

Coping with cystic fibrosis requires dislodging as much of the thickened mucus from the airways as possible. The standard therapy for the disease is to pound the person's back. In 1992, a new experimental treatment was introduced in which the patient inhales a mist containing the DNA-digesting enzyme, DNase, which is manufactured using recombinant DNA technology. The enzyme degrades the DNA that contributes to the viscosity of the mucus. The DNA present in the mucus is derived from disintegrating inflammatory cells that move into the respiratory tract.

fibrosis. For the first time, victims of cystic fibrosis saw a glimmer of light at the end of a long, dark tunnel; there was hope that a cure for their crippling disease might be developed. To understand how the search for RFLPs was a critical factor in the isolation of the CF gene, consider the following analogy: Imagine that you live in a neighborhood where the outsides of the houses are similar and there are no addresses to help you locate your house. Each block has a single commercial building at one corner, each housing a different business. One block has a laundry, another a grocery store, and so forth. You are fortunate to have a small coffeehouse on your corner, where you can have a cup of coffee and read your biology text. Every day when you come home, you drive along the main street until you see the coffeehouse then turn right and proceed to the eighth house, which is where you live.

The similar-looking houses are analogous to the genes that lie along the chromosome. Just as the houses are virtually identical from one street to the next, human genes, for most purposes, are identical from one person to the next. In contrast, the commercial properties, which are analogous to

RFLPs, are highly polymorphic—they vary from block to block. Once you have located a commercial property that is a given distance from the particular house you live in, you can always find your house (or CF gene) using the commercial property (or linked RFLP) as a guidepost. In other words, RFLPs serve as genetic markers for locating genes whose positions would otherwise be undetectable.

The goal of the researchers searching for the CF gene was to find a RFLP that was very close to the CF gene itself. How does a gene mapper know when he or she has identified such a RFLP? Remember that two genes located very close to each other are very unlikely to become separated from each other during crossing over. Consequently, if a particular RFLP and the CF gene are very close to each other, then all of the individuals in a family that are actually stricken with the disease will probably have the same version of the RFLP marker (Figure 17-4). Other families with the disease will have different versions of the RFLP. If a researcher finds a RFLP that is invariably present in those members of a family with the disease, then it follows that this RFLP is close to the gene causing the disease. Once a RFLP that resided within a million or so nucleotides of the CF gene had been identified, investigators used other techniques to move from the RFLP into the unknown, adjoining regions of the DNA until they arrived at the CF gene itself.

➠ Once the gene responsible for cystic fibrosis had been located and its nucleotide sequence determined, samples of DNA from a variety of afflicted individuals, their parents, and other members of the population were analyzed. It turned out that 70 percent of the alleles responsible for cystic fibrosis in the United States contained the same genetic alteration—they were all missing three base pairs of DNA that coded for a phenylalanine at the 508th position within the polypeptide chain encoded by the CF gene. Individuals with alleles containing this small deletion were generally of Northern European descent, and it is likely that all of these individuals are descendants of a single Northern European who lived many generations ago.

The isolation of a disease-causing gene is a major step in understanding the underlying basis of the disease and may provide a foundation for new and innovative treatments. This is illustrated by some of the research described below which followed the announcement that the CF gene had been isolated.

Reverse Genetics: Working Backward from Genotype to Phenotype

Mendel began his studies with pea plants distinguishable by height and seed color; Morgan began his studies with fruit flies having different-colored eyes; geneticists working on sickle cell anemia learned decades ago that this disease was characterized by abnormal hemoglobin. In each of these cases, the investigator began with a relatively clear phenotypic difference. But what is the cause of traits such as cystic fibrosis, which affects many different organs in various ways? More specifically, what is the product of the CF gene, whose alteration has such devastating effects?

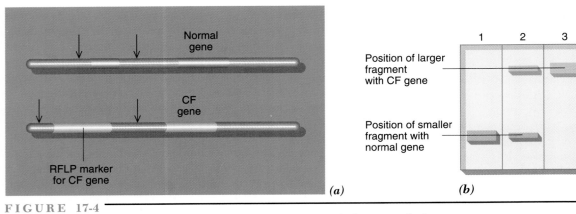

FIGURE 17-4

Linkage of the cystic fibrosis gene with a genetic marker. *(a)* The DNA of a heterozygous carrier for CF will have one normal allele and one mutant allele. The two alleles, located on homologous chromosomes, will inevitably be linked to different-sized restriction fragments. In this hypothetical example, the CF gene is closely linked to the green-colored RFLP marker. In this case, any member of the family who carries the CF allele will also have the longer restriction fragment. *(b)* Suppose two heterozygotes with the same homologues shown in *(a)* were to marry and have children. The children would be expected to have three different restriction fragment patterns. Lane 1 shows the DNA from a child who is homozygous normal and thus has two copies of the smaller fragment; lane 2 is from a heterozygous child who has one long and one short fragment and would be a carrier like the parents; and lane 3 is from a child who suffers from the disease and would have two copies of the longer fragment. Once a nearby marker is found, such as the green-colored RFLP shown in *(a)*, molecular biologists can work their way from the marker to the gene being sought.

⫸ Molecular biologists are now able to carry out a form of "reverse genetics." First, they isolate the gene and determine its nucleotide sequence. Then, they deduce the amino acid sequence of the protein using the universal genetic code and compare it to the amino acid sequence of other proteins whose function is known. Proteins of related (*homologous*) amino acid sequence have evolved from a common ancestral protein and often have a similar function. The amino acid sequence deduced for the CF gene suggested that the gene product was a membrane protein involved in the movement of chloride ions across the plasma membrane. Recall from Chapter 7 that water flows passively from regions of lower to regions of higher salt concentration. The blockage of chloride movement in people with CF is thought to have a secondary effect on the movement of water, causing a decreased water content—and thus an increased viscosity—in bodily secretions. The presence of DNA also contributes to the high viscosity of the mucus found in CF patients (see Figure 17-3).

CYSTIC FIBROSIS: A LESSON IN EVOLUTION

Why should one ethnic group have such a high incidence of a disease that is rare or nonexistent in other groups? One answer is simply by chance; the alleles for cystic fibrosis may simply have increased in frequency along with the rapid expansion of the Caucasian population over the past hundred years or so. It is also possible that some selective pressure has existed to maintain the CF alleles at a high frequency in the Caucasian population. Clearly, there is no selective advantage for homozygotes with cystic fibrosis, but there may be some advantage for the vastly larger number of heterozygous CF carriers. One recent hypothesis suggests that CF heterozygotes may be more resistant to cholera (a disease that causes the loss of body water and death by dehydration) than are homozygotes that lack a CF allele. Recall that the CF gene product is a chloride-transport protein that affects the movement of water in the body. A person having one CF allele would be expected to have a reduced capacity for chloride transport, which might result in a reduced loss of body water during a cholera infection.

The fact that a large percentage of the CF alleles in the American population are probably derived from a single Northern European ancestor illustrates how alleles can spread from one population to another as individuals move from place to place. As we will see in Chapter 33, evolution is characterized by a change in the percentages of alleles of a species. The movement of individuals from one population to another is one way to initiate such change.

Recent advances in DNA technology have led to techniques designed to locate specific disease-causing alleles on particular chromosomes, to determine the functions of the products encoded by these genes, and to offer hope that a cure for these devastating genetic disorders may be on the horizon. (See CTQ #3.)

PATTERNS OF TRANSMISSION OF GENETIC DISORDERS

⬤ The transmission of genetic characteristics in humans follows the same basic "Mendelian rules" described for pea plants and fruit flies. Disorders that result from mutations in specific genes are usually divided into three categories—autosomal recessive, autosomal dominant, and X-linked—depending on whether or not one defective allele is sufficient to cause the disease and on the type of chromosome on which the gene appears.

DISORDERS THAT RESULT FROM AUTOSOMAL RECESSIVE ALLELES

Recall that autosomes include all chromosomes except the sex chromosomes (X and Y). Since each cell has two copies of each autosome, a recessive mutation on one autosome will be masked by the presence of a normal allele on the homologous autosome. These heterozygotes are referred to as *carriers* because they harbor the mutant gene without being adversely affected, in most cases. You are probably a carrier for as many as five to ten lethal recessive alleles. Fortunately, recessive alleles only produce a disease phe-notype when they are present on both homologues; that is, in the homozygous recessive condition (Figure 17-5). Consequently, these disorders are usually rare. If two heterozygotes (carriers) have children, however, on the average, one in four of their offspring will suffer from the abnormal phenotype. Many homozygous recessive conditions lead to the death of the fetus and subsequent miscarriage, leaving the parents unaware that they are carriers of the defective alleles. Other homozygous recessive disorders, such as cystic fibrosis, phenylketonuria, and sickle cell anemia, lead to the birth of an infant with serious medical problems.

DISORDERS THAT RESULT FROM AUTOSOMAL DOMINANT ALLELES

Some genetic diseases result from *dominant* mutations (Figure 17-6) in which the defective gene codes for a product that directly causes the diseased condition. In other words, rather than the normal gene overriding the defective one, as in recessive conditions, the tables are turned, and the normal gene remains a "helpless bystander." Disease-causing dominant mutations can reside on either an autosome or a sex chromosome. Huntington's disease is an example of a condition that results from such an *autosomal dominant mutation*.

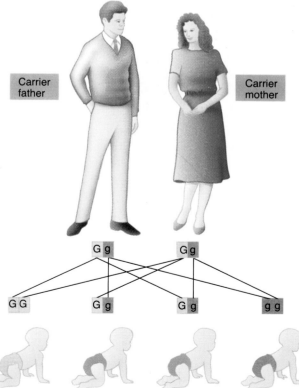

FIGURE 17-5

Inheritance of an autosomal recessive disorder. For these traits (including sickle cell anemia, phenylketonuria, and cystic fibrosis), both parents are carriers of the defective allele (indicated as g), but they do not show its ill effects due to the presence of the dominant normal allele (indicated as G). Each offspring has a one in four chance (25 percent) of inheriting two copies of the defective gene, resulting in the disease. Each offspring has a two in four chance of being a carrier and a one in four chance of inheriting two normal alleles.

Carrier father

Carrier mother

G g   G g

G G   G g   G g   g g

Normal   Carrier   Carrier   Affected

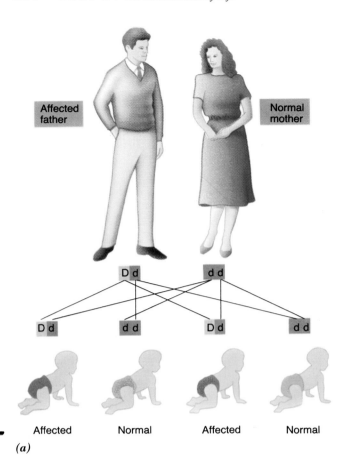

Affected father

Normal mother

D d   d d

D d   d d   D d   d d

Affected   Normal   Affected   Normal

(a)

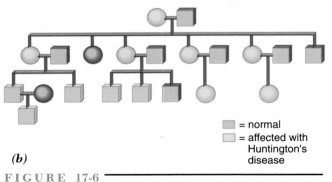

(b)

■ = normal
■ = affected with Huntington's disease

FIGURE 17-6

Inheritance of an autosomal dominant disorder. *(a)* For these traits, only one parent carries the defective allele (indicated as D), and that person exhibits the disease. On average, half the offspring will inherit the defective allele and develop the disorder. *(b)* Pedigree of a family with Huntington's disease.

Huntington's Disease

In 1872, Dr. George Huntington published his only scientific paper. It was titled "On Chorea," and it described a strange neurological disease that had plagued a large family living in New York. Victims of the disease exhibited no symptoms until they were 35 to 50 years old, at which time they began to show signs of unusual involuntary movements, loss of memory, depression, and irrational behavior. The disease progressed steadily and unerringly toward dementia, loss of motor control, and finally death. The underlying biochemical defect causing the neurological deterioration was then, and remains, a mystery. Huntington noted, however, that if one parent had the disease, "one or more offspring almost invariably suffer from it, if they live to adult age. But if the children go through life without it, the thread is broken and the grandchildren and great-grandchildren are free from the disease." Although unaware of it, Huntington was describing a trait that was inherited as an autosomal dominant mutation. Unlike autosomal recessive diseases, such as cystic fibrosis and PKU, only one parent has to have the defective allele for a child to inherit an autosomal dominant disease.

Huntington's disease received media attention when legendary songwriter Woody Guthrie died from the disease in 1967 (Figure 17-7*a*). Until very recently, children of Huntington's victims (such as Guthrie's son, Arlo) found themselves in a horrible position. Knowing that they faced a 50 percent risk of developing the disease, they had to wait until they were well along in their adult life before learning their destiny and possibly that of their children. Every time they forgot something or tripped over their own feet, they were reminded of their genetic inheritance and wondered whether it was the first sign of the onset of the disease.

Today, another name is associated with Huntington's disease—Columbia University's Nancy Wexler, psychologist and head of the Hereditary Diseases Foundation (Figure 17-7*b*). Nancy Wexler, herself at risk for Huntington's disease, has been the driving force behind an intensive research program aimed at isolating the gene responsible for the disease and determining the underlying malfunction. Wexler has managed to bring together researchers from several laboratories in an unprecedented collaborative hunt for the gene responsible for the disease.

The search for the Huntington's gene began in earnest in 1982, using RFLP technology. Armed with probes for 13 RFLPs and DNA samples from a small Iowa family with a history of Huntington's disease, James Gusella of Harvard University began his study of linkage analysis. The chance that any of the 13 RFLPs would be closely linked to the Huntington's gene was remote; Gusella's primary goal in this initial round of experiments was simply to work out the techniques needed for the analysis. Instead, Gusella hit the jackpot. The third RFLP he tested was not only on the same chromosome as the Huntington's gene (chromosome number 4) but was within about 5 million base pairs of the gene (the entire chromosome has about 200 million base pairs of

(a) *(b)*

FIGURE 17-7 ─────────────

(a) Woody Guthrie and *(b)* Nancy Wexler. Dr. Wexler is shown in front of a wall containing a pedigree of the inhabitants in the community on the shore of Lake Maracaibo, Venezuela.

DNA). For his good fortune, Gusella earned the nickname "Lucky Jim."

▐▶ Soon after his early finding, Gusella switched to the analysis of DNA samples that had been collected by Nancy Wexler during a series of trips to a number of remote fishing villages dotting the lagoons of Lake Maracaibo in Venezuela (Figure 17-7*b*). The residents of Lake Maracaibo were plagued with a strange malady they called *El Mal* ("the bad"), their term for Huntington's disease. According to their story, El Mal could be traced to a European sailor who had visited the area nearly 200 years earlier. By the 1980s, the Huntington's allele had spread rapidly through the local population; over 150 people were afflicted with the disease, and 1,100 more were at 50 percent risk. The Venezuelan community provided an ideal human pedigree for researchers to study, and the search began for a RFLP marker that was very close to the defective gene.

For ten years, over a dozen labs followed one lead after another, each leading to a dead end. Finally, in March, 1993, the gene was isolated and, appropriately, it was James Gusella's lab that found it. Sequencing studies revealed the basis for the defect. The Huntington's gene includes a region containing a repeating sequence of the trinucleotide, CAG, which specifies the amino acid glutamine. Whereas the normal gene contains about 10 to 20 copies of the trinucleotide, the altered gene that causes Huntington's disease contains over 40 copies of the triplet. The effect of these extra repeats is under intense investigation.

DISORDERS THAT RESULT FROM X-LINKED ALLELES

Some genetic disorders for which the defective allele is recessive are expressed in men even when only one copy of the gene is inherited. As we discussed on page 254, these conditions are due to X-linked alleles—faulty genes that lie on the X chromosome. X-linked disorders include hemo-philia, red–green colorblindness, congenital night blindness, ichthyosis (skin hardens to scalelike consistency), some forms of anemia (low hemoglobin content in the blood), and muscular dystrophy.

Muscular Dystrophy

Muscular dystrophy is an X-linked gene that strikes about 1 in every 4,000 males and is characterized by progressive muscle deterioration. In its severest form, known as *Duchenne muscular dystrophy*, the patient is confined to a wheelchair by age 12 and often dies of cardiac failure by age 20.

The gene responsible for muscular dystrophy was discovered piece by piece during the late 1980s. Unlike the other genes that had been sought, the gene causing muscular dystrophy is unusually large—spanning approximately 2 million base pairs of DNA, or nearly 0.1 percent of a person's total supply of genetic information! Most of this DNA consists of senseless introns, but the encoded protein, *dystrophin*, is also exceptionally large. Studies indicate that dystrophin contributes to the structure of the muscle cell's surface; in its absence, the membrane weakens and gradually deteriorates, accounting for the progression from seemingly normal muscles in an infant to the progressive loss of muscular tissue over the years. Curiously, a mouse strain that fails to produce dystrophin has been isolated, yet these animals do not exhibit the same type of muscle-wasting disease seen in humans.

The discovery of the dystrophin gene as the cause of muscular dystrophy has suggested a possible cure for the disease. What if the muscles of these sick children could somehow receive a supply of dystrophin? Could the deterioration be prevented or even reversed? In 1991, Peter Law, formerly of the University of Tennessee, conducted experiments with this goal in mind. In his study, 32 young boys with muscular dystrophy were injected with healthy, undifferentiated human muscle cells to determine whether these cells would participate in the formation of dystrophin-containing muscle tissue. The cells were injected into several muscles of the leg. Initial results, which have been met with controversy, indicate that the affected muscles were strengthened, but whether or not such a procedure can serve as an effective treatment remains uncertain.

COMPLEX DISEASES

Before you conclude that studies on the genetics of human disease only apply to those persons born with rare congenital disorders, consider the following: Many common diseases, including cancer, atherosclerosis, diabetes, Alzheimer's, manic depression, and even alcoholism, have a strong genetic component. For example, we have discussed in earlier chapters how mutations in oncogenes (page 202) or tumor suppressor genes (page 260) can lead to the development of cancer. As a result, those individuals born with mutations in certain genes are predisposed to develop particular types of cancer. This is the reason why women whose mothers have had breast cancer are at much higher risk of developing the disease than are members of the general population. Similarly, a man whose father develops prostate cancer is at increased risk of doing the same. It is becoming increasingly evident that the more we learn about human genetics, the better we will be able to understand the complex diseases that threaten our health.

Most genetic diseases are the result of single-gene defects and follow simple Mendelian patterns of inheritance, allowing risk assessments to be calculated based on simple probability laws. The genetic basis of more common diseases, including cancer, is much more complex, making risk assessments more difficult. (See CTQ #5.)

SCREENING HUMANS FOR GENETIC DEFECTS

The discovery of RFLPs (and their use in locating defective genes) has opened new vistas in the field of medical diagnostics. Couples can be tested so that they no longer have to wonder if they might be carriers for cystic fibrosis or sickle cell anemia or if their unborn fetus will be afflicted with the disease. A person whose mother or father is dying of Huntington's disease can now learn with 95 percent certainty whether or not he or she will be stricken with the disease. It may soon be possible to determine whether an infant has a predisposition for developing a particular form of cancer, heart disease, or diabetes. Let's examine how geneticists screen DNA in order to make these types of determinations.

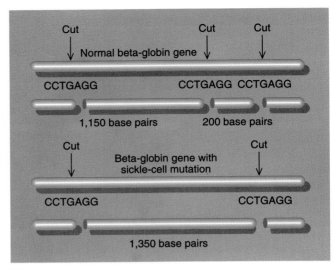

(a)

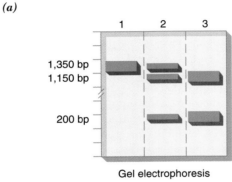

Gel electrophoresis

(b)

FIGURE 17-8

Screening for the sickle cell gene. *(a)* Restriction enzyme cuts the normal beta-globin gene at the enzyme's recognition sites (arrows), creating two fragments. The mutation responsible for sickle cell anemia changes the nucleotide sequence of the DNA in a way that eliminates a recognition site for the restriction enzyme. The sickle cell allele is therefore cut into only one fragment by this enzyme. This difference in restriction fragment length is readily revealed through electrophoresis, forming a basis for screening the population for the mutant allele. Part *(b)* shows the pattern produced by DNA from a person with sickle cell anemia (lane 1), a heterozygous carrier of the disease (lane 2), and a person who is homozygous normal (lane 3).

- *Sickle cell anemia*   Sickle cell anemia is an autosomal recessive disease that primarily afflicts individuals of central African ancestry. Sickle cell anemia is caused by a specific alteration in the beta-globin gene, which happens to lie directly within one of the sites on the DNA normally recognized by a restriction enzyme. DNA containing the normal globin allele is cut at this site, whereas DNA containing the sickle cell allele is not. As a result, RFLP analysis can determine if a person lacks the defective gene entirely; is a heterozygous carrier of the disease; or has two copies of the defective gene and, thus, the disease (Figure 17-8).

- *Cystic fibrosis*   Approximately 1 in 25 Caucasian Americans are carriers of the cystic fibrosis gene; thus, approximately 1 in 625 (25 × 25) Caucasian couples are at risk of having a child with the disease. Earlier, we noted that in 70 percent of the cases of CF in the United States, the alleles responsible are missing three particular nucleotides. Consequently, any screening test based on this single type of gene alteration would miss 30 percent of the carriers and more than half the couples at risk. Fortunately, screening tests have been developed that identify several other frequently occur-

ring types of gene alterations, so we can now screen for—and successfully detect—at least 90 percent of all abnormal CF genes.

- *Huntington's disease*   Even though the gene for Huntington's disease has just been isolated, a person at risk for the disease can usually determine whether or not he or she carries the defective allele. The current test for Huntington's disease depends on obtaining DNA from a close relative who has the disease. Analysis of this DNA can specifically identify the RFLP that is closely linked to the defective gene in the particular family being studied. If a person at risk in the family carries this same RFLP in his or her DNA, then the probability is very high that he or she has the same allele for Huntington's disease and will develop the disease.

SCREENING FETAL CELLS FOR GENETIC DISORDERS

Two commonly performed procedures allow physicians to obtain a sample of cells of an unborn fetus. These cells can then be screened for genetic deficiencies.

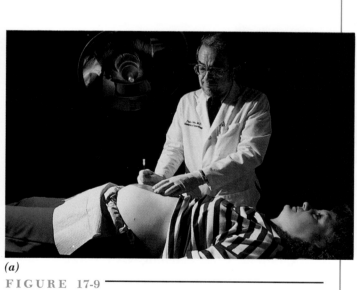

(a)

FIGURE 17-9

Sampling cells from the fetus. *(a)* Amniocentesis. *(b)* By employing the techniques of amniocentesis, or chorionic villus sampling, geneticists can remove a sample of fetal cells to determine whether the child will be born with a detectable chromosome abnormality (which can be determined by microscopic examination of the chromosomes of a fetal cell) or a defect in a gene (which can be determined by the ability of the fetal cells to grow in various media or by RFLP analysis).

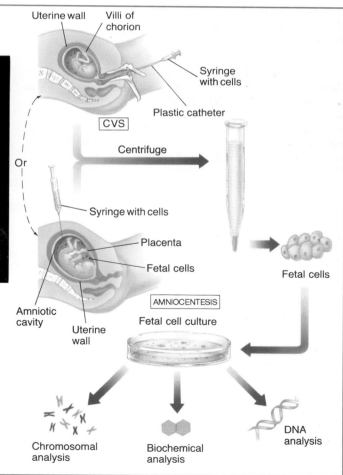

(b)

◁ THE HUMAN PERSPECTIVE ▷
Correcting Genetic Disorders by Gene Therapy

During the 1970s, a young boy named David captured the attention of the American public as "the boy in the plastic bubble" (Figure 1). The "bubble" was a sterile, enclosed environment in which David lived nearly his entire life. David required this extraordinary level of protection because he was born with a very rare inherited disease called *severe combined immunodeficiency disease* (SCID), which left him virtually lacking an immune system— the system that protects us from invading pathogens. The bubble protected David from viruses or bacteria that might infect and kill him, but it also kept him from any direct physical contact with the outside world, including his parents. In approximately 25 percent of cases, SCID results from the hereditary absence of a single enzyme, adenosine deaminase (ADA).

Gene therapy is the prospect by which a patient is cured by altering his or her genotype. The goal is to provide the patient with a normal gene, capable of pro-

FIGURE 1

- **Amniocentesis** is a procedure in which a hypodermic needle is inserted through the body wall of the pregnant woman into the fluid-filled amniotic sac that surrounds and cushions the growing fetus, and a sample of fluid is withdrawn (Figure 17-9*a*). Fetal cells that have been shed into the amniotic fluid are isolated and cultured in the laboratory for 1 or 2 weeks and are then subjected to genetic analysis. Amniocentesis is not usually performed before the fifteenth week of pregnancy, which leaves a relatively small amount of time in which the mother can choose to have a safe abortion should the tests reveal serious genetic problems in the fetus.

- **Chorionic villus sampling (CVS)** is a newer procedure, whereby a small sample of tissue is removed from the developing placenta. The fetal cells contained in this tissue sample can then be subjected to genetic analysis. Since CVS can be performed as early as the ninth week of pregnancy, it provides the mother with much more time to decide whether or not to terminate the pregnancy.

Both types of prenatal screening procedures carry some risk (0.5 to 1 percent) to the life of the fetus. Newer, safer techniques are being developed that will enable physicians to obtain fetal cells from the blood of the mother.

ducing the missing gene product. SCID is an excellent candidate for the development of such therapy for a number of reasons. First, there is no cure for the disease, which inevitably proves fatal at an early age. Second, SCID results from the absence of a single gene product (ADA). The ADA gene has been isolated and cloned and thus is available for treatment. Finally, the cells that normally express the ADA gene are white blood cells that can easily be removed from a patient, genetically modified, and then reintroduced into the patient by transfusion.

In 1990, a 4-year-old girl suffering from SCID became the first person authorized by the National Institutes of Health and the Food and Drug Administration (FDA) to receive gene therapy. In September 1990, the girl received a transfusion of her own white blood cells which had been genetically modified to carry normal copies of the ADA gene. It was hoped that the modified white blood cells would provide the girl with the necessary armaments to ward off future infections. Since white blood cells have a limited lifetime, the procedure must be repeated periodically to maintain the patient's immune capacity. At the time of this writing, the child is attending school and doing well; her immune system is apparently functioning normally.

Intensive effort is currently focused on isolating the *stem cells* that give rise to both red and white blood cells. If these stem cells can be isolated, and their genotype modified, patients with genetic blood cell diseases, such as SCID, will need only a single treatment since the genetically engineered stem cells will continue to provide healthy blood cells throughout the person's lifetime.

Cystic fibrosis is also a good candidate for gene therapy because the gene has been isolated and because the worst symptoms of the disease are caused by cells that line the airways and, therefore, are accessible to substances that can be delivered by inhalation of an aerosol (as in Figure 17-3). It is hoped that a similar aerosol approach can be used to deliver a normal CF gene directly to the defective cells of the respiratory tract of CF patients. Scientists are optimistic, based on the results of recent studies from Ronald Crystal's laboratory at the National Institutes of Health, in which the normal CF gene, attached to the DNA of a cold virus, was delivered to the lung cells of rats. The researchers found that the added CF gene was integrated into the chromosomes of the rat cells, and the human protein was produced for at least 6 weeks. Tests with humans will probably begin once the safety of the delivery procedure can be confirmed.

Another approach is being considered as a therapy for diseases that result when an abnormal product is carried in the blood. Hemophilia, for example, results from a deficiency of a particular clotting factor in the blood. The clotting factor is normally produced by the liver, but the source of the protein is not important to its function; its presence in the blood alone is required. Cells from the deep layer of the skin have been isolated and genetically modified so that they carry extra genes for this clotting factor. These cells might be reintroduced into the skin of a hemophiliac, where they will secrete the clotting factor into the blood, thus compensating for the abnormal clotting factor. Trial studies for this procedure are currently being carried out on dogs.

It is important to note that all of the procedures discussed above, as well as all of those being contemplated, involve the modification of *somatic cells*—those cells of the body that are not on the path to gamete formation. Modification of somatic cells will affect only the person being treated, and the modified chromosomes cannot be passed on to future generations. This would not be the case if the germ cells of the gonads were modified since these cells are part of the line that produces gametes. Thus far, the consensus is that no studies involving the modification of human germ line cells will be performed. Such studies would present risks for the genetic constitution of future generations and raise serious ethical questions about scientists tampering with human evolution.

Even though fetal cells are present in very small amounts in the mother's blood, techniques are available to separate them from the mother's cells and to amplify the fetal DNA by PCR (page 326). It will probably be several years before this procedure will be available, however.

The ability to obtain fetal cells allows medical geneticists to make several types of determinations (Figure 17-9*b*):

1. *Detection of chromosomal abnormalities* Amniocentesis and CVS are most commonly performed on older pregnant women who, because of their age, have an increased likelihood of giving birth to a child with Down syndrome. The determination is made by counting the number of chromosomes in the fetal cells in search of an extra chromosome number 21, which is readily detected by karyotype analysis (see Figure 10-3). The presence of some chromosomal aberrations can also be detected by analysis of fetal cells.

2. *Detection of metabolic deficiencies* Tests for over 200 genetic disorders that affect metabolic processes can also be carried out on cultured fetal cells. If fetal cells are missing a particular enzyme—such as HGPRT, whose deficiency causes Lesch-Nyhan syndrome—the condition may be revealed by the cells' inability to grow in certain culture media.

3. *Detection of defective alleles* With the development of RFLP analysis, more and more tests will become available for screening the DNA of fetal cells for genetic disorders that are not readily detected by standard microscopic or biochemical tests, such as cystic fibrosis and Huntington's disease.

ETHICAL CONSIDERATIONS

One of the byproducts of the search for the genes responsible for human diseases has been the development of techniques for diagnosing genetic disorders. These techniques have been hailed as modern medical "miracles." Consider, for example, how many people carrying a pair of PKU alleles have been saved from severe mental retardation. At the same time, serious unforseen ethical and psychological considerations have emerged from our recent breakthroughs. Consider the following situations.

• You have a 50 percent chance of developing Huntington's disease. Would you want to know as a young adult that your body will eventually (and inevitably) deteriorate and you will become mentally deranged? Or would you rather live with the uncertainty, holding onto the hope that you will not contract the disease? If you choose the latter route, would you have children, knowing that you might be passing on the defective gene? In fact, the test for Huntington's disease has been available for a number of years, and genetic counselors have been surprised at the relatively small number of people who have come forward to learn of their genetic fate.

• You find out that both you and the person whom you plan to marry carry the gene for cystic fibrosis. You were planning to get married and have a large family. Do you break off the relationship? Do you marry and have children, accepting the possibility that one out of four will have a disease that will be emotionally draining and expensive to treat and, in all likelihood, end in a premature death? If your wife becomes pregnant, do you submit to fetal testing and abort a fetus that carries a pair of defective alleles? Knowing that you face the possibility of having dependents with costly medical problems, should you be forced to reveal this information on applications for health or life insurance? Do insurance companies have the right to be informed of a person's genetic status?

• You learn that you are a heterozygous carrier of the sickle cell gene, which means that a portion of your hemoglobin molecules are abnormal. You have always wanted to be a pilot, and your doctor tells you this will not affect your ability to fulfill that goal, but you find there are barriers against sickle cell carriers in the airline industry because of the belief that these individuals will be adversely affected if a drop in cabin air pressure should occur. Do you inform the airline of your genetic status before accepting a job?

It is evident that DNA-based diagnostic tests have created new questions for which answers are not readily available. The greatest problems stem from the fact that these tests predict the presence or likelihood of diseases for which there are no cures. Given this circumstance, some argue it would be better not to employ such tests at all. Since current tests screen for relatively rare conditions, they have not had a major impact on most of our lives. But as we develop diagnostic tests to detect an individual's predisposition to develop far more common complex diseases, such as cancer, Alzheimer's, and heart disease, the problems will be compounded, and the entire matter of genetic screening is likely to come under careful scrutiny.

Just as a typographical error in a sentence can be detected by a proofreader, a change in the nucleotide sequence of a gene can be detected by screening procedures that use recently developed techniques in molecular biology. While these techniques offer great promise in identifying genetic abnormalities at an early age, they raise serious ethical considerations. (See CTQ #7.)

REEXAMINING THE THEMES

Biological Order, Regulation, and Homeostasis

↻ Most of the disorders discussed in this chapter result from a defect in a single gene, producing a severe breakdown of the body's normal functions. The fact that a deficiency in one gene product can have such devastating consequences illustrates an organism's need for a high degree of order. A deficiency in one process often leads to secondary failures, triggering a chain reaction that produces the visible consequences, such as life-threatening infections, muscle deterioration, digestive failure, or neurological degeneration.

Unity within Diversity

None of the diseases described in this chapter has any known counterpart in other mammals. This fact may be explained by our ignorance of animal afflictions, but it is more likely attributed to differences that exist even among closely related organisms. For example, a defect in the dystrophin gene has also been discovered in mice. But even though these mutant mice fail to produce dystrophin, the animals don't exhibit the deteriorative muscle wasting so characteristic of humans with muscular dystrophy.

Unity and diversity are also revealed by examination of nucleotide sequences in different people. While most sections of the genome are very similar from one person to the next, reflecting the genetic homogeneity within a species, large variations exist in the number of times certain sequences are repeated. These variations allow researchers to map genes by RFLP analysis and to develop various DNA screening procedures.

Evolution and Adaptation

Differences between species that arise during evolution result from underlying changes in the species' genetic composition. Such changes occur as new alleles arise by mutation and then spread throughout a species' population. In this chapter, we have seen two examples of how the apparent immigration of a single individual has promoted the spread of a particular allele into a new population—the cystic fibrosis allele in the United States, and the Huntington's allele in Venezuela. Since both alleles are deleterious, one would expect the selective disadvantage of possessing these alleles to restrict their spread. However, the spread of deleterious alleles has probably occurred more readily in recent years in human populations than in those of other organisms because of the explosive increase in the population in the past few hundred years and the medical advances in the treatment of inherited disorders. In the case of cystic fibrosis, it has been suggested that heterozygous carriers of the CF allele have a selective advantage over others in that they are more resistant to the effects of cholera.

SYNOPSIS

Human life is very sensitive to the proper number of chromosomes. Gametes containing an abnormal chromosome number are produced following nondisjunction during meiosis. In humans, zygotes having an abnormal chromosome number usually die in the womb or grow into individuals who are born with characteristic abnormalities, such as Down syndrome.

Recent advances in gene mapping have allowed geneticists to locate the chromosomal position of genes, even when the gene product has not been characterized. Gene mapping takes advantage of differences in DNA nucleotide sequences from one person to another—differences that result in variations (polymorphisms) in the length of restriction fragments (RFLPs) produced when human DNA is treated with restriction enzymes. The location of the gene in question is discovered by identifying a specific RFLP that is always present in the same members of a family that have a particular allele of the gene. Since the particular RFLP and the particular allele occur together in the relatives, the two must be closely linked.

Once a gene has been isolated, the amino acid sequence of the polypeptide can be deduced and the function of the polypeptide usually determined by comparing its amino acid sequence to that of proteins of known function. This is the technique used to determine that the CF gene codes for a chloride channel, providing insight into the molecular basis of the disease.

Genetic disorders are transmitted from generation to generation in several well-defined patterns. Disorders resulting from autosomal recessive alleles, such as cystic fibrosis and sickle cell anemia, occur only when both members of a pair of homologous autosomes carry the recessive alleles. In a heterozygous carrier, the normal allele produces a normal gene product that compensates for the defective product produced by the recessive allele. Disorders that result from autosomal dominant alleles, such as Huntington's disease, occur when only one member of a pair of homologous autosomes carries the defective allele. The abnormal gene product encoded by the allele causes the disorder, despite the presence of the normal product from the other homologue. Disorders that result from X-linked alleles, such as hemophilia and muscular dystrophy, occur almost exclusively in males. The allele producing the defective gene product is located on the X chromosome, and there is no compensating locus for the allele on the Y.

Recent advances in molecular genetics have initiated changes in the potential treatment of genetic-based disorders. The use of restriction enzymes and RFLP mapping has led to the development of techniques to screen fetuses and adults for the presence of certain genetic disorders and the *potential* to screen individuals for a genetic predisposition to developing cancer, heart disease, or mental illnesses. Gene therapy has moved from the drawing boards to clinical trials, as researchers are attempting to change the genotypes of somatic cells of individuals with certain genetic disorders, including SCID and cystic fibro-

sis. In addition, a 15-year project has begun to sequence the entire human genome. Information gained from this endeavor is expected to increase vastly our understanding of the genes that control all of our functions, to help in the treatment of genetic-based diseases, to provide an opportunity to manufacture many new protein products, and to help us better understand the course of human evolution.

Key Terms

nondisjunction (p. 336)
genetic marker (p. 338)
polymorphic gene (p. 338)

restriction fragment length
  polymorphism (RFLP) (p. 338)
amniocentesis (p. 348)

chorionic villus sampling (CVS) (p. 348)
gene therapy (p. 348)

Review Questions

1. Draw pedigrees that show the patterns by which autosomal recessive, autosomal dominant, and X-linked disorders are inherited.

2. Why does a person with an XXY chromosome abnormality develop into a male rather than a female? Which sex would you predict a person with XXXXY would exhibit?

3. Why do RFLPs make such useful genetic markers?

4. Give two different molecular bases for the occurrence of RFLPs. (Hint: check Figures 17-2a and 17-8a.)

5. Why do researchers believe that the majority of people carrying a CF allele have one particular Northern European ancestor? Why is it so difficult to identify all CF carriers by genetic screening?

6. Compare the extent of our knowledge about the molecular and physiological basis of PKU, CF, Huntington's disease, and muscular dystrophy.

Critical Thinking Questions

1. Diet soft drinks and other foods containing the artificial sweetener aspartame (trademark Nutrasweet) carry a warning for people with PKU. Aspartame is a dipeptide. Why do you suppose this warning might be necessary?

2. If the chromosome carrying a gene for Lesch–Nyhan syndrome were involved in nondisjunction, how might this affect the occurrence of the condition in female offspring? Explain your answer using what you have learned about gene dosage.

3. If you were a member of Congress and were asked to vote to appropriate money for the Human Genome Project, would you vote for it or against it? Why?

4. Suppose you have just isolated the gene responsible for causing Huntington's disease. What steps might you take to determine why the gene causes such devastating neurological damage?

5. How would you explain to someone why it may be easier to find a cure for muscular dystrophy than for atherosclerosis?

6. In this chapter, we noted that a person having a parent with Huntington's disease can learn with about 95 percent certainty whether he or she will develop the condition. Why do you suppose this person cannot learn this with 100 percent certainty?

7. In the 1960s, genetic sceening for sickle cell anemia, followed by genetic counseling, was tried in a small town in Greece in which the disease was prevalent. The program failed because people were unwilling to inform potential marriage partners if they were carriers, in case this would reduce their chances of marrying. When a similar program was proposed for screening African Americans for the disease, it failed because of charges that the program was racially motivated. What implications do the ethical and sociological questions raised by these cases have for the future of genetic diagnosis and screening?

Additional Readings

Baskin, Y. 1984. *The gene doctors*. New York: Morrow. (Introductory)

Bishop, J. E., and M. Waldholz. 1990. *Genome*. New York: Simon & Schuster. (Introductory)

Collins, F. S. 1992. Cystic fibrosis and therapeutic implications. *Science* 256:774-779. (Intermediate)

Culliton, B. J. 1991. Gene therapy on the move. *Nature* 354:429. (Intermediate)

Davies, K. 1991. Mapping the way forward. *Nature* 353:798–799. (Advanced)

Levitan, M. 1988. *Textbook of human genetics*. New York: Oxford University Press. (Advanced)

Maddox, J. 1991. The case for the human genome. *Nature* 352:11–14. (Intermediate)

Marx, J. 1992. Gene therapy for CF advances. *Science* 255:289. (Intermediate)

Nichols, E. K. 1988. *Human gene therapy*. Cambridge, MA: Harvard University Press. (Introductory)

Roberts, L. 1992. The Huntington's gene quest goes on. *Nature* 258:740-741. (Intermediate)

Thompson, L. 1992. Stem-cell gene therapy moves toward approval. *Science* 255:1072. (Intermediate)

Thompson, L. 1993. The first kids with new genes. *Time* June 7:50–53. (Introductory)

Verma, I. M. 1990. Gene therapy. *Sci. Amer.* Nov:68-84. (Intermediate)

Vogel, F., and A. G. Motulsky. 1987. *Human genetics*. New York: Springer-Verlag. (Intermediate)

Watson, J. D. 1990. The human genome project: past, present, and future. *Science* 248:44–49. (Intermediate)

Wingerson, L. 1990. *Mapping our genes*. New York: Dutton. (Introductory)

White, R., and J.-M. Lalouel. 1988. Chromosome mapping with DNA markers. *Sci. Amer.* 258:40–48. (Intermediate)

A P P E N D I X
◀ A ▶

Metric and Temperature Conversion Charts

Metric Unit (symbol)		*Metric to English*	*English to Metric*
Length			
kilometer (km)	= 1,000 (10^3) meters	1 km = 0.62 mile	1 mile = 1.609 km
meter (m)	= 100 centimeters	1 m = 1.09 yards	1 yard = 0.914 m
		= 3.28 feet	1 foot = 0.305 m
centimeter (cm)	= 0.01 (10^{-2}) meter	1 cm = 0.394 inch	1 inch = 2.54 cm
millimeter (mm)	= 0.001 (10^{-3}) meter	1 mm = 0.039 inch	1 inch = 25.4 mm
micrometer (μm)	= 0.000001 (10^{-6}) meter		
nanometer (nm)	= 0.000000001 (10^{-9}) meter		
angstrom (Å)	= 0.0000000001 (10^{-10}) meter		
Area			
square kilometer (km²)	= 100 hectares	1 km² = 0.386 square mile	1 square mile = 2.590 km²
hectare (ha)	= 10,000 square meters	1 ha = 2.471 acres	1 acre = 0.405 ha
square meter (m²)	= 10,000 square centimeters	1 m² = 1.196 square yards	1 square yard = 0.836 m²
		= 10.764 square feet	1 square foot = 0.093 m²
square centimeter (cm²)	= 100 square millimeters	1 cm² = 0.155 square inch	1 square inch = 6.452 cm²
Mass			
metric ton (t)	= 1,000 kilograms	1 t = 1.103 tons	1 ton = 0.907 t
	= 1,000,000 grams		
kilogram (kg)	= 1,000 grams	1 kg = 2.205 pounds	1 pound = 0.454 kg
gram (g)	= 1,000 milligrams	1 g = 0.035 ounce	1 ounce = 28.35 g
milligram (mg)	= 0.001 gram		
microgram (μg)	= 0.000001 gram		
Volume Solids			
1 cubic meter (m³)	= 1,000,000 cubic centimeters	1 m³ = 1.308 cubic yards	1 cubic yard = 0.765 m³
		= 35.315 cubic feet	1 cubic foot = 0.028 m³
1 cubic centimeter (cm³)	= 1,000 cubic millimeters	1 cm³ = 0.061 cubic inch	1 cubic inch = 16.387 cm³
Volume Liquids			
kiloliter (kl)	= 1,000 liters	1 kl = 264.17 gallons	
liter (l)	= 1,000 milliliters	1 l = 1.06 quarts	1 gal = 3.785 l
			1 qt = 0.94 l
			1 pt = 0.47 l
milliliter (ml)	= 0.001 liter	1 ml = 0.034 fluid ounce	1 fluid ounce = 29.57 ml
microliter (μl)	= 0.000001 liter		

TEMPERATURE

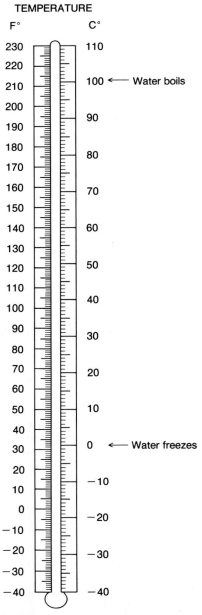

Fahrenheit to Centigrade: $°C = \frac{5}{9}(°F - 32)$
Centigrade to Fahrenheit: $°F = \frac{9}{5}(°C + 32)$

A P P E N D I X

◄ B ►

Microscopes: Exploring the Details of Life

Microscopes are the instruments that have allowed biologists to visualize objects that are vastly smaller than anything visible with the naked eye. There are broadly two types of specimens viewed in a microscope: whole mounts which consist of an intact subject, such as a hair, a living cell, or even a DNA molecule, and thin sections of a specimen, such as a cell or piece of tissue.

THE LIGHT MICROSCOPE

A light microscope consists of a series of glass lenses that bend (refract) the light coming from an illuminated specimen so as to form a visual image of the specimen that is larger than the specimen itself (a). The specimen is often stained with a colored dye to increase its visibility. A special phase contrast light microscope is best suited for observing unstained, living cells because it converts differences in the density of cell organelles, which are normally invisible to the eye, into differences in light intensity which can be seen.

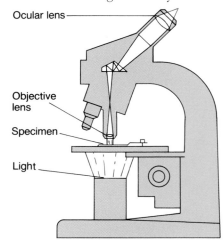

Ocular lens

Objective lens

Specimen

Light

(a)

All light microscopes have limited *resolving power*—the ability to distinguish two very close objects as being separate from each other. The resolving power of the light microscope is about 0.2 μm (about 1,000 times that of the naked eye), a property determined by the wave length of visible light. Consequently, objects closer to each other than 0.2 μm, which includes many of the smaller cell organ-

elles, will be seen as a single, blurred object through a light microscope.

THE TRANSMISSION ELECTRON MICROSCOPE

Appreciation of the wondrous complexity of cellular organization awaited the development of the transmission electron microscope (or TEM), which can deliver resolving powers 1000 times greater than the light microscope. Suddenly, biologists could see strange new structures, whose function was totally unknown—a breakthrough that has kept cell biologists busy for the past 50 years. The TEM (b) works by shooting a beam of electrons through very thinly sliced specimens that have been stained with heavy metals,

(b)

such as uranium, capable of deflecting electrons in the beam. The electrons that pass through the specimen undeflected are focused by powerful electromagnets (the lenses of a TEM) onto either a phosphorescent screen or high-contrast photographic film. The resolution of the TEM is so great—sufficient to allow us to see individual DNA molecules—because the wavelength of an electron beam is so small (about 0.0005 μm).

THE SCANNING ELECTRON MICROSCOPE

Specimens examined in the scanning electron microscope (SEM) are whole mounts whose surfaces have been coated with a thin layer of heavy metals. In the SEM, a fine beam of electrons scans back and forth across the specimen and the image is formed from electrons bouncing off the hills and valleys of its surface. The SEM produces a three-dimensional image of the surface of the specimen—which can

(c)   *(d)*

range in size from a virus to an insect head (c,d)—with remarkable depth and clarity. The SEM produces black and white images; the colors seen in many of the micrographs in the text have been added to enhance their visual quality. Note that the insect head (d) is that of an antennapedia mutant as described on p 687.

A P P E N D I X

◄ C ►

The Hardy-Weinberg Principle

If the allele for brown hair is dominant over that for blond hair, and curly hair is dominant over straight hair, then why don't all people by now have brown, curly hair? The **Hardy-Weinberg Principle** (developed independently by English mathematician G. H. Hardy and German physician W. Weinberg) demonstrates that the frequency of alleles remains the same from generation to generation unless influenced by outside factors. The outside factors that would cause allele frequencies to change are mutation, immigration and emigration (movement of individuals into and out of a breeding population, respectively), natural selection of particular traits, and breeding between members of a small population. In other words, unless one or more of these forces influence hair color and hair curl, the relative number of people with brown and curly hair will not increase over those with blond and straight hair.

To illustrate the Hardy-Weinberg Principle, consider a single gene locus with two alleles, *A* and *a*, in a breeding population. (If you wish, consider *A* to be the allele for brown hair and *a* to be the allele for blond hair.) Because there are only two alleles for the gene, the sum of the frequencies of *A* and *a* will equal 1.0. (By convention, allele frequencies are given in decimals instead of percentages.) Translating this into mathematical terms, if

p = the frequency of allele *A*, and
q = the frequency of allele *a*,

then $p + q = 1$.

If *A* represented 80 percent of the alleles in the breeding population ($p = 0.8$), then according to this formula the frequency of *a* must be 0.2 ($p + q = 0.8 + 0.2 = 1.0$).

After determining the allele frequency in a starting population, the predicted frequencies of alleles and genotypes in the next generation can be calculated. Setting up a

Punnett square with starting allele frequencies of $p = 0.8$ and $q = 0.2$:

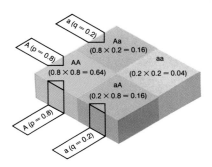

The chances of each offspring receiving any combination of the two alleles is the product of the probability of receiving one of the two alleles alone. In this example, the chances of an offspring receiving two *A* alleles is $p \times p = p^2$, or $0.8 \times 0.8 = 0.64$. A frequency of 0.64 means that 64 percent of the next generation will be homozygous dominant (*AA*). The chances of an offspring receiving two *a* alleles is $q^2 = 0.2 \times 0.2 = 0.04$, meaning 4 percent of the next generation is predicted to be *aa*. The predicted frequency of heterozygotes (*Aa* or *aA*) is 0.32 or $2pq$, the sum of the probability of an individual being *Aa*($p \times q = 0.8 \times 0.2 = 0.16$) plus the probability of an individual being *aA*($q \times p = 0.2 \times 0.8 = 0.16$). Just as all of the allele frequencies for a particular gene must add up to 1, so must all of the possible genotypes for a particular gene locus add up to 1. Thus, the Hardy-Weinberg Principle is

$$p^2 + 2pq + q^2 = 1$$
$$(0.64 + 0.32 + 0.04 = 1)$$

So after one generation, the frequency of possible genotypes is

$$AA = p^2 = 0.64$$
$$Aa = 2pq = 0.32$$
$$aa = q^2 = 0.04$$

Now let's determine the actual allele frequencies for A and a in the new generation. (Remember the original allele frequencies were 0.8 for allele A and 0.2 for allele a. If the Hardy-Weinberg Principle is right, there will be no change in the frequency of either allele.) To do this we sum the frequencies for each genotype containing the allele. Since heterozygotes carry both alleles, the genotype frequency must be divided in half to determine the frequency of each allele. (In our example, heterozygote Aa has a frequency of 0.32, 0.16 for allele A, plus 0.16 for allele a.) Summarizing then:

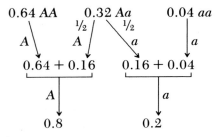

As predicted by the Hardy-Weinberg Principle, the frequency of allele A remained 0.8 and the frequency of allele a remained 0.2 in the new generation. Future generations can be calculated in exactly the same way, over and over again. As long as there are no mutations, no gene flow between populations, completely random mating, no natural selection, and no genetic drift, there will be no change in allele frequency, and therefore no evolution.

Population geneticists use the Hardy-Weinberg Principle to calculate a starting point allele frequency, a reference that can be compared to frequencies measured at some future time. The amount of deviation between observed allele frequencies and those predicted by the Hardy-Weinberg Principle indicates the degree of evolutionary change. Thus, this principle enables population geneticists to measure the rate of evolutionary change and identify the forces that cause changes in allele frequency.

APPENDIX
◄ D ►

Careers in Biology

Although many of you are enrolled in biology as a requirement for another major, some of you will become interested enough to investigate the career opportunities in life sciences. This interest in biology can grow into a satisfying livelihood. Here are some facts to consider:

- Biology is a field that offers a very wide range of possible science careers

- Biology offers high job security since many aspects of it deal with the most vital human needs: health and food

- Each year in the United States, nearly 40,000 people obtain bachelor's degrees in biology. But the number of newly created and vacated positions for biologists is increasing at a rate that exceeds the number of new graduates. Many of these jobs will be in the newer areas of biotechnology and bioservices.

Biologists not only enjoy job satisfaction, their work often changes the future for the better. Careers in medical biology help combat diseases and promote health. Biologists have been instrumental in preserving the earth's life-supporting capacity. Biotechnologists are engineering organisms that promise dramatic breakthroughs in medicine,

food production, pest management, and environmental protection. Even the economic vitality of modern society will be increasingly linked to biology.

Biology also combines well with other fields of expertise. There is an increasing demand for people with backgrounds or majors in biology complexed with such areas as business, art, law, or engineering. Such a distinct blend of expertise gives a person a special advantage.

The average starting salary for all biologists with a Bachelor's degree is $22,000. A recent survey of California State University graduates in biology revealed that most were earning salaries between $20,000 and $50,000. But as important as salary is, most biologists stress job satisfaction, job security, work with sophisticated tools and scientific equipment, travel opportunities (either to the field or to scientific conferences), and opportunities to be creative in their job as the reasons they are happy in their career.

Here is a list of just a few of the careers for people with degrees in biology. For more resources, such as lists of current openings, career guides, and job banks, write to Biology Career Information, John Wiley and Sons, 605 Third Avenue, New York, NY 10158.

A SAMPLER OF JOBS THAT GRADUATES HAVE SECURED IN THE FIELD OF BIOLOGY*

Agricultural Biologist	Bioanalytical Chemist	Brain Function	Environmental Center
Agricultural Economist	Biochemical/Endocrine	Researcher	Director
Agricultural Extension	Toxicologist	Cancer Biologist	Environmental Engineer
Officer	Biochemical Engineer	Cardiovascular Biologist	Environmental Geographer
Agronomist	Pharmacology Distributor	Cardiovascular/Computer	Environmental Law Specialist
Amino-acid Analyst	Pharmacology Technician	Specialist	Farmer
Analytical Biochemist	Biochemist	Chemical Ecologist	Fetal Physiologist
Anatomist	Biogeochemist	Chromatographer	Flavorist
Animal Behavior	Biogeographer	Clinical Pharmacologist	Food Processing Technologist
Specialist	Biological Engineer	Coagulation Biochemist	Food Production Manager
Anticancer Drug Research	Biologist	Cognitive Neuroscientist	Food Quality Control
Technician	Biomedical	Computer Scientist	Inspector
Antiviral Therapist	Communication Biologist	Dental Assistant	Flower Grower
Arid Soils Technician	Biometerologist	Ecological Biochemist	Forest Ecologist
Audio-neurobiologist	Biophysicist	Electrophysiology/	Forest Economist
Author, Magazines & Books	Biotechnologist	Cardiovascular Technician	Forest Engineer
Behavioral Biologist	Blood Analyst	Energy Regulation Officer	Forest Geneticist
Bioanalyst	Botanist	Environmental Biochemist	Forest Manager

Forest Pathologist
Forest Plantation Manager
Forest Products Technologist
Forest Protection Expert
Forest Soils Analyst
Forester
Forestry Information Specialist
Freeze-Dry Engineer
Fresh Water Biologist
Grant Proposal Writer
Health Administrator
Health Inspector
Health Scientist
Hospital Administrator
Hydrologist
Illustrator
Immunochemist
Immunodiagnostic
   Assay Developer
Inflammation Technologist
Landscape Architect
Landscape Designer
Legislative Aid
Lepidopterist
Liaison Scientist,
   Library of Medicine
   Computer Biologist
Life Science Computer
   Technologist
Lipid Biochemist
Livestock Inspector
Lumber Inspector

Medical Assistant
Medical Imaging Technician
Medical Officer
Medical Products Developer
Medical Writer
Microbial Physiologist
Microbiologist
Mine Reclamation Scientist
Molecular Endocrinologist
Molecular Neurobiologist
Molecular Parasitologist
Molecular Toxicologist
Molecular Virologist
Morphologist
Natural Products Chemist
Natural Resources Manager
Nature Writer
Nematode Control Biologist
Nematode Specialist
Nematologist
Neuroanatomist
Neurobiologist
Neurophysiologist
Neuroscientist
Nucleic Acids Chemist
Nursing Aid
Nutritionist
Occupational Health Officer
Ornamental Horticulturist
Paleontologist
Paper Chemist
Parasitologist

Pathologist
Peptide Biochemist
Pharmaceutical Writer
Pharmaceutical Sales
Pharmacologist
Physiologist
Planning Consultant
Plant Pathologist
Plant Physiologist
Production Agronomist
Protein Biochemist
Protein Structure & Design
   Technician
Purification Biochemist
Quantitative Geneticist
Radiation Biologist
Radiological Scientist
Regional Planner
Regulatory Biologist
Renal Physiologist
Renal Toxicologist
Reproductive Toxicologist
Research and Development
   Director
Research Technician
Research Liaison Scientist
Research Products Designer
Research Proposal Writer
Safety Assessment Sanitarian
Scientific Illustrator
Scientific Photographer
Scientific Reference Librarian

Scientific Writer
Soil Microbiologist
Space Station Life Support
   Technician
Spectroscopist
Sports Product Designer
Steroid Health Assessor
Taxonomic Biologist
Teacher
Technical Analyst
Technical Science Project
   Writer
Textbook Editor
Theoretical Ecologist
Timber Harvester
Toxicologist
Toxic Waste Treatment
   Specialist
Urban Planner
Water Chemist
Water Resources Biologist
Wood Chemist
Wood Fuel Technician
Zoning and Planning
   Manager
Zoologist
Zoo Animal Breeder
Zoo Animal Behaviorist
Zoo Designer
Zoo Inspector

*Results of one survey of California State University graduates. Some careers may require advanced degrees

Glossary

◄ A ►

Abiotic Environment Components of eco-systems that include all nonliving factors. (41)

Abscisic Acid (ABA) A plant hormone that inhibits growth and causes stomata to close. ABA may not be commonly involved in leaf drop. (21)

Abscission Separation of leaves, fruit, and flowers from the stem. (21)

Acclimation A physiological adjustment to environmental stress. (28)

Acetyl CoA Acetyl coenzyme A. A complex formed when acetic acid binds to a carrier coenzyme forming a bridge between the end products of glycolysis and the Krebs cycle in respiration. (9)

Acetylcholine Neurotransmitter released by motor neurons at neuromuscular junctions and by some interneurons. (23)

Acid Rain Occurring in polluted air, rain that has a lower pH than rain from areas with unpolluted air. (40)

Acids Substances that release hydrogen ions (H^+) when dissolved in water. (3)

Acid Snow Occurring in polluted air, snow that has a lower pH than snow from areas with unpolluted air. (40)

Acoelomates Animals that lack a body cavity between the digestive cavity and body wall. (39)

Acquired Immune Deficiency Syndrome (AIDS) Disease caused by infection with HIV (Human Immunodeficiency Virus) that destroys the body's ability to mount an immune response due to destruction of its helper T cells. (30, 36)

Actin A contractile protein that makes up the major component of the thin filaments of a muscle cell and the microfilaments of nonmuscle cells. (26)

Action Potential A sudden, dramatic reversal of the voltage (potential difference) across the plasma membrane of a nerve or muscle cell due to the opening of the sodium channels. The basis of a nerve impulse. (23)

Activation Energy Energy required to initiate chemical reaction. (6)

Active Site Region on an enzyme that binds its specific substrates, making them more reactive. (6)

Active Transport Movement of substances into or out of cells against a concentration gradient, i.e., from a region of lower concentration to a region of higher concentration. The process requires an expenditure of energy by the cell. (7)

Adaptation A hereditary trait that improves an organism's chances of survival and/or reproduction. (33)

Adaptive Radiation The divergence of many species from a single ancestral line. (33)

Adenosine Triphosphate (ATP) The molecule present in all living organisms that provides energy for cellular reactions in its phosphate bonds. ATP is the universal energy currency of cells. (6)

Adenylate Cyclase An enzyme activated by hormones that converts ATP to cyclic AMP, a molecule that activates resting enzymes. (25)

Adrenal Cortex Outer layer of the adrenal glands. It secretes steroid hormones in response to ACTH. (25)

Adrenal Medulla An endocrine gland that controls metabolism, cardiovascular function, and stress responses. (25)

Adrenocorticotropic Hormone (ACTH) An anterior pituitary hormone that stimulates the cortex of the adrenal glands to secrete cortisol and other steroid hormones. (25)

Adventitious Root System Secondary roots that develop from stem or leaf tissues. (18)

Aerobe An organism that requires oxygen to release energy from food molecules. (9)

Aerobic Respiration Pathway by which glucose is completely oxidized to CO_2 and H_2O, requiring oxygen and an electron transport system. (9)

Afferent (Sensory) Neurons Neurons that conduct impulses from the sense organs to the central nervous system. (23)

Age-Sex Structure The number of individuals of a certain age and sex within a population. (43)

Aggregate Fruits Fruits that develop from many pistils in a single flower. (20)

AIDS See Acquired Immune Deficiency Syndrome.

Albinism A genetic condition characterized by an absence of epidermal pigmentation that can result from a deficiency of any of a variety of enzymes involved in pigment formation. (12)

Alcoholic Fermentation The process in which electrons removed during glycolysis are transferred from NADH to form alcohol as an end product. Used by yeast during the commercial process of ethyl alcohol production. (9)

Aldosterone A hormone secreted by the adrenal cortex that stimulates reabsorption of sodium from the distal tubules and collecting ducts of the kidneys. (23)

Algae Any unicellular or simple colonial photosynthetic eukaryote. (37,38)

Algin A substance produced by brown algae harvested for human application because of its ability to regulate texture and consistency of

products. Found in ice cream, cosmetics, marsh-mellows, paints, and dozens of other products. (38)

Allantois Extraembryonic membrane that serves as a repository for nitrogenous wastes. In placental mammals, it helps form the vascular connections between mother and fetus. (32)

Allele Alternative form of a gene at a particular site, or locus, on the chromosome. (12)

Allele Frequency The relative occurrence of a certain allele in individuals of a population. (33)

Allelochemicals Chemicals released by some plants and animals that deter or kill a predator or competitor. (42)

Allelopathy A type of interaction in which one organism releases allelochemicals that harm another organism. (42)

Allergy An inappropriate response by the immune system to a harmless foreign substance leading to symptoms such as itchy eyes, runny nose, and congested airways. If the reaction occurs throughout the body (anaphylaxis) it can be life threatening. (30)

Allopatric Speciation Formation of new species when gene flow between parts of a population is stopped by geographic isolation. (33)

Alpha Helix Portion of a polypeptide chain organized into a defined spiral conformation. (4)

Alternation of Generations Sequential change during the life cycle of a plant in which a haploid (1N) multicellular stage (gametophyte) alternates with a diploid (2N) multicellular stage (sporophyte). (38)

Alternative Processing When a primary RNA transcript can be processed to form more than one mRNA depending on conditions. (15)

Altruism The performance of a behavior that benefits another member of the species at some cost to the one who does the deed. (44)

Alveolus A tiny pouch in the lung where gas is exchanged between the blood and the air; the functional unit of the lung where CO_2 and O_2 are exchanged. (29)

Alzheimer's Disease A degenerative disease of the human brain, particularly affecting acetylcholine-releasing neurons and the hippocampus, characterized by the presence of tangled fibrils within the cytoplasm of neurons and amyloid plaques outside the cells. (23)

Amino Acids Molecules containing an amino group ($-NH_2$) and a carboxyl group ($-COOH$) attached to a central carbon atom. Amino acids are the subunits from which proteins are constructed. (4)

Amniocentesis A procedure for obtaining fetal cells by withdrawing a sample of the fluid that surrounds a developing fetus (amniotic fluid) using a hypodermic needle and syringe. (17)

Amnion   Extraembryonic membrane that envelops the young embryo and encloses the amniotic fluid that suspends and cushions it. (32)

Amoeba   A protozoan that employs pseudopods for motility. (37)

Amphibia   A vertebrate class grouped into three orders: Caudata (tailed amphibians); Anura (tail-less amphibians); Apoda (rare worm-like, burrowing amphibians). (39)

Anabolic Steroids   Steroid hormones, such as testosterone, which promote biosynthesis (anabolism), especially protein synthesis. (25)

Anabolism   Biosynthesis of complex molecules from simpler compounds. Anabolic pathways are endergonic, i.e., require energy. (6)

Anaerobe   Organism that does not require oxygen to release energy from food molecules. (9)

Anaerobic Respiration   Pathway by which glucose is completely oxidized, using an electron transport system but requiring a terminal electron acceptor other than oxygen. (Compare with fermentation.) (9)

Analogous Structures (Homoplasies)   Structures that perform a similar function, such as wings in birds and insects, but did not originate from the same structure in a common ancestor. (33)

Anaphase   Stage of mitosis when the kinetochores split and the sister chromatids (now termed chromosomes) move to opposite poles of the spindle. (10)

Anatomy   Study of the structural characteristics of an organism. (18)

Angiosperm (Anthophyta)   Any plant having its seeds surrounded by fruit tissue formed from the mature ovary of the flowers. (38)

Animal   A mobile, heterotrophic, multicellular organism, classified in the Animal kingdom. (39)

Anion   A negatively charged ion. (3)

Annelida   The phylum which contains segmented worms (earthworms, leeches, and bristleworms). (39)

Annuals   Plants that live for one year or less. (18)

Annulus   A row of specialized cells encircling each sporangium on the fern frond; facilitates rupture of the sporangium and dispersal of spors. (38)

Antagonistic Muscles   Pairs of muscles whose contraction bring about opposite actions as illustrated by the biceps and triceps, which bends or straightens the arm at the elbow, respectively. (26)

Antenna Pigments   Components of photosystems that gather light energy of different wavelengths and then channel the absorbed energy to a reaction center. (8)

Anterior   In anatomy, at or near the front of an animal; the opposite of posterior. (39)

Anterior Pituitary   A true endocrine gland manufacturing and releasing six hormones

when stimulated by releasing factors from the hypothalamus. (25)

Anther   The swollen end of the stamen (male reproductive organ) of a flowering plant. Pollen grains are produced inside the anther lobes in pollen sacs. (20)

Antibiotic   A substance produced by a fungus or bacterium that is capable of preventing the growth of bacteria. (2)

Antibodies   Proteins produced by plasma cells. They react specifically with the antigen that stimulated their formation. (30)

Anticodon   Triplet of nucleotides in tRNA that recognizes and base pairs with a particular codon in mRNA. (14)

Antidiuretic Hormone (ADH)   One of the two hormones released by the posterior pituitary. ADH increases water reabsorption in the kidney, which then produces a more concentrated urine. (25)

Antigen   Specific foreign agent that triggers an immune response. (30)

Aorta   Largest blood vessel in the body through which blood leaves the heart and enters the systemic circulation. (28)

Apical Dominance   The growth pattern in plants in which axillary bud growth is inhibited by the hormone auxin, present in high concentrations in terminal buds. (21)

Apical Meristems   Centers of growth located at the tips of shoots, axillary buds, and roots. Their cells divide by mitosis to produce new cells for primary growth in plants. (18)

Aposematic Coloring   Warning coloration which makes an organism stand out from its surroundings. (42)

Appendicular Skeleton   The bones of the appendages and of the pectoral and pelvic girdles. (26)

Aquatic   Living in water. (40)

Archaebacteria   Members of the kingdom Monera that differ from typical bacteria in the structure of their membrane lipids, their cell walls, and some characteristics that resemble those of eukaryotes. Their lack of a true nucleus, however, accounts for their assignment to the Moneran kingdom. (36)

Archenteron   In gastrulation, the hollow core of the gastrula that becomes an animal's digestive tract. (32)

Arteries   Large, thick-walled vessels that carry blood away from the heart. (28)

Arterioles   The smallest arteries, which carry blood toward capillary beds. (28)

Arthropoda   The most diverse phylum on earth, so called from the presence of jointed limbs. Includes insects, crabs, spiders, centipedes. (39)

Ascospores   Sexual fungal spore borne in a sac. Produced by the sac fungi, Ascomycota. (37)

Asexual Reproduction   Reproduction without the union of male and female gametes. (31)

Association   In ecological communities, a major organization characterized by uniformity and two or more dominant species. (41)

Asymmetric   Referring to a body form that cannot be divided to produce mirror images. (39)

Atherosclerosis   Condition in which the inner walls of arteries contain a buildup of cholesterol-containing plaque that tends to occlude the channel and act as a site for the formation of a blood clot (thrombus). (7)

Atmosphere   The layer of air surrounding the Earth. (40)

Atom   The fundamental unit of matter that can enter into chemical reactions; the smallest unit of matter that possesses the qualities of an element. (3)

Atomic Mass   Combined number of protons and neutrons in the nucleus of an atom. (3)

Atomic Number   The number of protons in the nucleus of an atom. (3)

ATP   (see **Adenosine Triphosphate**)

ATPase   An enzyme that catalyzes a reaction in which ATP is hydrolyzed. These enzymes are typically involved in reactions where energy stored in ATP is used to drive an energy-requiring reaction, such as active transport or muscle contractility. (7, 26)

ATP Synthase   A large protein complex present in the plasma membrane of bacteria, the inner membrane of mitochondria, and the thylakoid membrane of chloroplasts. This complex consists of a baseplate in the membrane, a channel across the membrane through which protons can pass, and a spherical head (F_1 particle) which contains the site where ATP is synthesized from ADP and P_i (8, 9)

Atrioventricular (AV) Node   A neurological center of the heart, located at the top of the ventricles. (28)

Atrium   A contracting chamber of the heart which forces blood into the ventricle. There are two atria in the hearts of all vertebrates, except fish which have one atrium. (28)

Atrophy   The shrinkage in size of structure, such as a bone or muscle, usually as a result of disuse. (26)

Autoantibodies   Antibodies produced against the body's own tissue. (30)

Autoimmune Disease   Damage to a body tissue due to an attack by autoantibodies. Examples include thyroiditis, multiple sclerosis, and rheumatic fever. (30)

Autonomic Nervous System   The nerves that control the involuntary activities of the internal organs. It is composed of the parasympathetic system, which functions during normal activity, and the sympathetic system, which operates in times of emergency or prolonged exertion. (23)

Autosome   Any chromosome that is not a sex chromosome. (13)

Autotrophs Organisms that satisfy their own nutritional needs by building organic molecules photosynthetically or chemosynthetically from inorganic substances. (8)

Auxins Plant growth hormones that promote cell elongation by softening cell walls. (21)

Axial Skeleton The bones aligned along the long axis of the body, including the skull, vertebral column, and ribcage. (26)

Axillary Bud A bud that is directly above each leaf on the stem. It can develop into a new stem or a flower. (18)

Axon The long, sometimes branched extension of a neuron which conducts impulses from the cell body to the synaptic knobs. (23)

◄ **B** ►

Bacteriophage A virus attacking specific bacteria that multiplies in the bacterial host cell and usually destroys the bacterium as it reproduces. (36)

Balanced Polymorphism The maintenance of two or more alleles for a single trait at fairly high frequencies. (33)

Bark Common term for the periderm. A collective term for all plant tissues outside the secondary xylem. (18)

Base Substance that removes hydrogen ions (H^+) from solutions. (3)

Basidiospores Sexual spores produced by basidiomycete fungi. Often found by the millions on gills in mushrooms. (37)

Basophil A phagocytic leukocyte which also releases substances, such as histamine, that trigger an inflammatory response. (28)

Batesian Mimicry The resemblance of a good-tasting or harmless species to a species with unpleasant traits. (42)

Bathypelagic Zone The ocean zone beneath the mesopelagic zone, characterized by no light; inhabited by heterotrophic bacteria and benthic scavengers. (40)

B Cell A lymphocyte that becomes a plasma cell and produces antibodies when stimulated by an antigen. (30)

Benthic Zone The deepest ocean zone; the ocean floor, inhabitated by bottom dwelling organisms. (40)

Bicarbonate Ion HCO_3^-. (3, 29)

Biennials Plants that live for two years. (18)

Bilateral Symmetry The quality possessed by organisms whose body can be divided into mirror images by only one median plane. (39)

Bile Salts Detergentlike molecules produced by the liver and stored by the gallbladder that function in lipid emulsification in the small intestine. (27)

Binomial A term meaning "two names" or "two words". Applied to the system of nomenclature for categorizing living things with a genus and species name that is unique for each type of organism. (1)

Biochemicals Organic molecules produced by living cells. (4)

Bioconcentration The ability of an organism to accumulate substances within its' body or specific cells. (41)

Biodiversity Biological diversity of species, including species diversity, genetic diversity, and ecological diversity. (43)

Biogeochemical Cycles The exchanging of chemical elements between organisms and the abiotic environment. (41)

Biological Control Pest control through the use of naturally occurring organisms such as predators, parasites, bacteria, and viruses. (41)

Biological Magnification An increase in concentration of slowly degradable chemicals in organisms at successively higher trophic levels; for example, DDT or PCB's. (41)

Bioluminescence The capability of certain organisms to utilize chemical energy to produce light in a reaction catalyzed by the enzyme luciferase. (9)

Biomass The weight of organic material present in an ecosystem at any one time. (41)

Biome Broad geographic region with a characteristic array of organisms. (40)

Biosphere Zone of the earth's soil, water, and air in which living organisms are found. (40)

Biosynthesis Construction of molecular components in the growing cell and the replacement of these compounds as they deteriorate. (6)

Biotechnology A new field of genetic engineering; more generally, any practical application of biological knowledge. (16)

Biotic Environment Living components of the environment. (40)

Biotic Potential The innate capacity of a population to increase tremendously in size were it not for curbs on growth; maximum population growth rate. (43)

Blade Large, flattened area of a leaf; effective in collecting sunlight for photosynthesis. (18)

Blastocoel The hollow fluid-filled space in a blastula. (32)

Blastocyst Early stage of a mammalian embryo, consisting of a mass of cells enclosed in a hollow ball of cells called the trophoblast. (32)

Blastodisk In bird and reptile development, the stage equivalent to a blastula. Because of the large amount of yolk, cleavage produces two flattened layers of cells with a blastocoel between them. (32)

Blastomeres The cells produced during embryonic cleavage. (32)

Blastopore The opening of the archenteron that is the embryonic predecessor of the anus in vertebrates and some other animals. (32)

Blastula An early developmental stage in many animals. It is a ball of cells that encloses a cavity, the blastocoel. (32)

Blood A type of connective tissue consisting of red blood cells, white blood cells, platelets, and plasma. (28)

Blood Pressure Positive pressure within the cardiovascular system that propels blood through the vessels. (28)

Blooms are massive growths of algae that occur when conditions are optimal for algae proliferation. (37)

Body Plan The general layout of a plant's or animal's major body parts. (39)

Bohr Effect Increased release of O_2 from hemoglobin molecules at lower pH. (29)

Bone A tissue composed of collagen fibers, calcium, and phosphate that serves as a means of support, a reserve of calcium and phosphate, and an attachment site for muscles. (26)

Botany Branch of biology that studies the life cycles, structure, growth, and classification of plants. (18)

Bottleneck A situation in which the size of a species' population drops to a very small number of individuals, which has a major impact on the likelihood of the population recovering its earlier genetic diversity. As occurred in the cheetah population. (33)

Bowman's Capsule A double-layered container that is an invagination of the proximal end of the renal tubule that collects molecules and wastes from the blood. (28)

Brain Mass of nerve tissue composing the main part of the central nervous system. (23)

Brainstem The central core of the brain, which coordinates the automatic, involuntary body processes. (23)

Bronchi The two divisions of the trachea through which air enters each of the two lungs. (29)

Bronchioles The smallest tubules of the respiratory tract that lead into the alveoli of the lungs where gas exchange occurs. (29)

Bryophyta Division of non-vascular terrestrial plants that include liverworts, mosses, and hornworts. (38)

Budding Asexual process by which offspring develop as an outgrowth of a parent. (39)

Buffers Chemicals that couple with free hydrogen and hydroxide ions thereby resisting changes in pH. (3)

Bundle Sheath Parenchyma cells that surround a leaf vein which regulate the uptake and release of materials between the vascular tissue and the mesophyll cells. (18)

◄ **C** ►

C_3 Synthesis The most common pathway for fixing CO_2 in the synthesis reactions of photosynthesis. It is so named because the first detectable organic molecule into which CO_2 is incorporated is a 3-carbon molecule, phosphoglycerate (PGA). (8)

C_4 Synthesis   Pathway for fixing CO_2 during the light-independent reactions of photosynthesis. It is so named because the first detectable organic molecule into which CO_2 is incorporated is a 4-carbon molecule. (8)

Calcitonin   A thyroid hormone which regulates blood calcium levels by inhibiting its release from bone. (25)

Calorie   Energy (heat) necessary to elevate the temperature of one gram of water by one degree Centigrade ($1° C$). (6)

Calvin Cycle   The cyclical pathway in which CO_2 is incorporated into carbohydrate. See C_3 synthesis. (8)

Calyx   The outermost whorl of a flower, formed by the sepals. (20)

CAM   Crassulacean acid metabolism. A variation of the photosynthetic reactions in plants, biochemically identical to C_4 synthesis except that all reactions occur in the same cell and are separated by time. Because CAM plants open their stomates at night, they have a competitive advantage in hot, dry climates. (8)

Cambium   A ring or cluster of meristematic cells that increase the width of stems and roots when they divide to produce secondary tissues. (18)

Camouflage   Adaptations of color, shape and behavior that make an organism more difficult to detect. (42)

Cancer   A disease resulting from uncontrolled cell divisions. (10,13)

Capillaries   The tiniest blood vessels consisting of a single layer of flattened cells. (28)

Capillary Action   Tendency of water to be pulled into a small-diameter tube. (3)

Carbohydrates   A group of compounds that includes simple sugars and all larger molecules constructed of sugar subunits, e.g. polysaccharides. (4)

Carbon Cycle   The cycling of carbon in different chemical forms, from the environment to organisms and back to the environment. (41)

Carbon Dioxide Fixation   In photosynthesis, the combination of CO_2 with carbon-accepting molecules to form organic compounds. (8)

Carcinogen   A cancer-causing agent. (13)

Cardiac Muscle   One of the three types of muscle tissue; it forms the muscle of the heart. (26)

Cardiovascular System   The organ system consisting of the heart and the vessels through which blood flows. (28)

Carnivore   An animal that feeds exclusively on other animals. (42)

Carotenoid   A red, yellow, or orange plant pigment that absorbs light in 400-500 nm wavelengths. (8)

Carpels   Central whorl of a flower containing the female reproductive organs. Each separate carpel, or each unit of fused carpels, is called a pistil. (20)

Carrier Proteins   Proteins within the plasma membrane that bind specific substances and facilitate their movement across the membrane. (7)

Carrying Capacity   The size of a population that can be supported indefinitely in a given environment. (43)

Cartilage   A firm but flexible connective tissue. In the human, most cartilage originally present in the embryo is transformed into bones. (26)

Casparian Strip   The band of waxy suberin that surrounds each endodermal cell of a plant's root tissue. (18)

Catabolism   Metabolic pathways that degrade complex compounds into simpler molecules, usually with the release of the chemical energy that held the atoms of the larger molecule together. (6)

Catalyst   A chemical substance that accelerates a reaction or causes a reaction to occur but remains unchanged by the reaction. Enzymes are biological catalysts. (6)

Cation   A positively charged ion. (3)

Cecum   A closed-ended sac extending from the intestine in grazing animals lacking a rumen (e.g., horses) that enables them to digest cellulose. (27)

Cell   The basic structural unit of all organisms. (5)

Cell Body   Region of a neuron that contains most of the cytoplasm, the nucleus, and other organelles. It relays impulses from the dendrites to the axon. (23)

Cell Cycle   Complete sequence of stages from one cell division to the next. The stages are denoted G_1, S, G_2, and M phase. (10)

Cell Differentiation   The process by which the internal contents of a cell become assembled into a structure that allows the cell to carry out a specific set of activities, such as secretion of enzymes or contraction. (32)

Cell Division   The process by which one cell divides into two. (10)

Cell Fusion   Technique whereby cells are caused to fuse with one another producing a large cell with a common cytoplasm and plasma membrane. (5, 10)

Cell Plate   In plants, the cell wall material deposited midway between the daughter cells during cytokinesis. Plate material is deposited by small Golgi vesicles. (5, 10)

Cell Sap   Solution that fills a plant vacuole. In addition to water, it may contain pigments, salts, and even toxic chemicals. (5)

Cell Theory   The fundamental theory of biology that states: 1) all organisms are composed of one or more cells, 2) the cell is the basic organizational unit of life, 3) all cells arise from pre-existing cells. (5)

Cellular Respiration   (See **Aerobic respiration**)

Cellulose   The structural polysaccharide comprising the bulk of the plant cell wall. It is the most abundant polysaccharide in nature. (4, 5)

Cell Wall   Rigid outer-casing of cells in plants and other organisms which gives support, slows dehydration, and prevents a cell from bursting when internal pressure builds due to an influx of water. (5)

Central Nervous System   In vertebrates, the brain and spinal cord. (23)

Centriole   A pinwheel-shaped structure at each pole of a dividing animal cell. (10)

Centromere   Indented region of a mitotic chromosome containing the kinetochore. (10)

Cephalization   The clustering of neural tissues and sense organs at the anterior (leading) end of the animal. (39)

Cerebellum   A bulbous portion of the vertebrate brain involved in motor coordination. Its prominence varies greatly among different vertebrates . (23)

Cerebral Cortex   The outer, highly convoluted layer of the cerebrum. In the human, this is the center of higher brain functions, such as speech and reasoning. (23)

Cerebrospinal Fluid   Fluid present within the ventricles of the brain, central canal of the spinal cord, and which surrounds and cushions the central nervous system. (23)

Cerebrum   The most dominant part of the human forebrain, composed of two cerebral hemispheres, generally associated with higher brain functions. (23)

Cervix   The lower tip of the uterus. (31)

Chapparal   A type of shrubland in California, characterized by drought-tolerant and fire-adapted plants. (40)

Character Displacement   Divergence of a physical trait in closely related species in response to competition. (42)

Chemical Bonds   Linkage between atoms as a result of electrons being shared or donated. (3)

Chemical Evolution   Spontaneous synthesis of increasingly complex organic compounds from simpler molecules. (35)

Chemical Reaction   Interaction between chemical reactants. (6)

Chemiosmosis   The process by which a pH gradient drives the formation of ATP. (8, 9)

Chemoreceptors   Sensory receptors that respond to the presence of specific chemicals. (24)

Chemosynthesis   An energy conversion process in which inorganic substances (H, N, Fe, or S) provide energized electrons and hydrogen for carbohydrate formation (9, 36)

Chiasmata   Cross-shaped regions within a tetrad, occurring at points of crossing over or genetic exchange. (11)

Chitin Structural polysaccharide that forms the hard, strong external skeleton of many arthropods and the cell walls of fungi. (4)

Chlamydia Obligate intracellular parasitic bacteria that lack a functional ATP-generating system. (36)

Chlorophyll Pigments Major light-absorbing pigments of photosynthesis. (8)

Chlorophyta Green algae, the largest group of algae; members of this group were very likely the ancestors of the modern plant kingdom. (38)

Chloroplasts An organelle containing chlorophyll found in plant cells in which photosynthesis occurs. (5, 8)

Cholecystokinin (CCK) Hormone secreted by endocrine cells in the wall of the small intestine that stimulates the release of digestive products by the pancreas. (27)

Chondrocytes Living cartilage cells embedded within the protein-polysaccharide matrix they manufacture. (26)

Chordamesoderm In vertebrates, the block of mesoderm that underlies the dorsal ectoderm of the gastrula, induces the formation of the nervous system, and gives rise to the notochord. (32)

Chordate A member of the phylum Chordata possessing a skeletal rod of tissue called a notochord, a dorsal hollow nerve cord, gill slits, and a post-anal tail at some stage of its development. (39)

Chorion The outermost of the four extraembryonic membranes. In placental mammals, it forms the embryonic portion of the placenta. (32)

Chorionic Villus Sampling (CVS) A procedure for obtaining fetal cells by removing a small sample of tissue from the developing placenta of a pregnant woman. (17)

Chromatid Each of the two identical subunits of a replicated chromosome. (10)

Chromatin DNA-protein fibers which, during prophase, condense to form the visible chromosomes. (5, 10)

Chromatography A technique for separating different molecules on the basis of their solubility in a particular solvent. The mixture of substances is spotted on a piece of paper or other material, one end of which is then placed in the solvent. As the solvent moves up the paper by capillary action, each substance in the mixture is carried a particular distance depending on its solubility in the moving solvent. (8)

Chromosomes Dark-staining structures in which the organism's genetic material (DNA) is organized. Each species has a characteristic number of chromosomes. (5, 10)

Chromosome Aberrations Alteration in the structure of a chromosome from the normal state. Includes chromosome deletions, duplications, inversions, and translocations. (13)

Chromosome Puff A site on an insect polytene chromosome where the DNA has unraveled and is being transcribed. (15)

Cilia Short, hairlike structures projecting from the surfaces of some cells. They beat in coordinated ways, are usually found in large numbers, and are densely packed. (5)

Ciliated Mucosa Layer of ciliated epithelial cells lining the respiratory tract. The beating of cilia propels an associated mucous layer and trapped foreign particles. (29)

Circadian Rhythm Behavioral patterns that cycle during approximately 24 hour intervals.

Circulatory System The system that circulates internal fluids throughout an organism to deliver oxygen and nutrients to cells and to remove metabolic wastes. (28)

Class (Taxonomic) A level of the taxonomic hierarchy that groups together members of related orders. (1)

Classical Conditioning A form of learning in which an animal develops a response to a new stimulus by repeatedly associating the new stimulus with a stimulus that normally elicits the response. (44)

Cleavage Successive mitotic divisions in the early embryo. There is no cell growth between divisions. (32)

Cleavage Furow Constriction around the middle of a dividing cell caused by constriction of microfilaments. (10)

Climate The general pattern of average weather conditions over a long period of time in a specific region, including precipitation, temperature, solar radiation, and humidity. (40)

Climax Final or stable community of successional stages, that is more or less in equilibrium with existing environmental conditions for a long period of time. (41)

Climax Community Community that remains essentially the same over long periods of time; final stage of ecological succession. (41)

Clitoris A protrusion at the point where the labia minora merge; rich in sensory neurons and erectile tissue. (31)

Clonal Selection Mechanism The mechanism by which the body can synthesize antibodies specific for the foreign substance (antigen) that stimulated their production. (30)

Clones Offspring identical to the parent, produced by asexual processes. (15)

Closed Circulatory System Circulatory system in which blood travels throughout the body in a continuous network of closed tubes. (Compare with open circulatory system). (28)

Clumped Pattern Distribution of individuals of a population into groups, such as flocks or herds. (43)

Cnidaria A phylum that consists of radial symmetrical animals that have two cell layers. There are three classes: 1) Hydrozoa (hydra), 2) Scy-phozoa (jellyfish), 3) Anthozoa (sea anemones, corals). Most are marine forms that live in warm, shallow water. (39)

Cnidocytes Specialized stinging cells found in the members of the phylum Cnidaria. (39)

Coastal Waters Relatively warm, nutrient-rich shallow water extending from the high-tide mark on land to the sloping continental shelf. The greatest concentration of marine life are found in coastal waters. (40)

Coated Pits Indentations at the surfaces of cells that contain a layer of bristly protein (called clathrin) on the inner surface of the plasma membrane. Coated pits are sites where cell receptors become clustered. (7)

Cochlea Organ within the inner ear of mammals involved in sound reception. (24)

Codominance The simultaneous expression of both alleles at a genetic locus in a heterozygous individual. (12)

Codon Linear array of three nucleotides in mRNA. Each triplet specifies a particular amino acid during the process of translation. (14)

Coelomates Animals in which the body cavity is completely lined by mesodermally-derived tissues. (39)

Coenzyme An organic cofactor, typically a vitamin or a substance derived from a vitamin. (6)

Coevolution Evolutionary changes that result from reciprocal interactions between two species, e.g., flowering plants and their insect pollinators. (33)

Cofactor A non-protein component that is linked covalently or noncovalently to an enzyme and is required by the enzyme to catalyze the reaction. Cofactors may be organic molecules (coenzymes) or metals. (6)

Cohesion The tendency of different parts of a substance to hold together because of forces acting between its molecules. (3)

Coitus Sexual union in mammals. (31)

Coleoptile Sheath surrounding the tip of the monocot seedling, protecting the young stem and leaves as they emerge from the soil. (21)

Collagen The most abundant protein in the human body. It is present primarily in the extracellular space of connective tissues such as bone, cartilage, and tendons. (26)

Collenchyma Living plant cells with irregularly thickened primary cell walls. A supportive cell type often found inside the epidermis of stems with primary growth. Angular, lacunar and laminar are different types of collenchyma cells. (18)

Commensalism A form of symbiosis in which one organism benefits from the union while the other member neither gains nor loses. (42)

Community The populations of all species living in a given area. (41)

Compact Bone The solid, hard outer regions of a bone surrounding the honey-combed mass of spongy bone. (26)

Companion Cell Specialized parenchyma cell associated with a sieve-tube member in phloem. (18)

Competition Interaction among organisms that require the same resource. It is of two types: 1) intraspecific (between members of the same species); 2) interspecific (between members of different species). (42)

Competitive Exclusion Principle (Gause's Principle) Competition in which a winner species captures a greater share of resources, increasing its survival and reproductive capacity. The other species is gradually displaced. (42)

Competitive Inhibition Prevention of normal binding of a substrate to its enzyme by the presence of an inhibitory compound that competes with the substrate for the active site on the enzyme. (6)

Complement Blood proteins with which some antibodies combine following attachment to antigen (the surface of microorganisms). The bound complement punches the tiny holes in the plasma membrane of the foreign cell, causing it to burst. (28)

Complementarity The relationship between the two strands of a DNA molecule determined by the base pairing of nucleotides on the two strands of the helix. A nucleotide with guanine on one strand always pairs with a nucleotide having cytosine on the other strand; similarly with adenine and thymine. (14)

Complete Digestive Systems Systems that have a digestive tract with openings at both ends—a mouth for entry and an anus for exit. (27)

Complete Flower A flower containing all four whorls of modified leaves—sepels, petals, stamen, and carpels. (20)

Compound Chemical substances composed of atoms of more than one element. (3)

Compound Leaf A leaf that is divided into leaflets, with two or more leaflets attached to the petiole. (18)

Concentration Gradient Regions in a system of differing concentration representing potential energy, such as exist in a cell and its environment, that cause molecules to move from areas of higher concentration to lower concentration. (7)

Conditioned Reflex A reflex ("automatic") response to a stimulus that would not normally have elicited the response. Conditioned reflexes develop by repeated association of a new stimulus with an old stimulus that normally elicits the response. (44)

Conformation The three-dimensional shape of a molecule as determined by the spatial arrangement of its atoms. (4)

Conformational Change Change in molecular shape (as occurs, for example, in an en-zyme as it catalyzes a reaction, or a myosin molecule during contraction). (6)

Conjugation A method of reproduction in single-celled organisms in which two cells link and exchange nuclear material. (11)

Connective Tissues Tissues that protect, support, and hold together the internal organs and other structures of animals. Includes bone, cartilage, tendons, and other tissues, all of which have large amounts of extracellular material. (22)

Consumers Heterotrophs in a biotic environment that feed on other organisms or organic waste. (41)

Continental Drift The continuous shifting of the earth's land masses explained by the theory of plate tectonics. (35)

Continuous Variation An inheritance pattern in which there is graded change between the two extremes in a phenotype (compare with discontinuous variation). (12)

Contraception The prevention of pregnancy. (31)

Contractile Proteins Actin and myosin, the protein filaments that comprise the bulk of the muscle mass. During contraction of skeletal muscle, these filaments form a temporary association and slide past each other, generating the contractile force. (26)

Control (Experimental) A duplicate of the experiment identical in every way except for the one variable being tested. Use of a control is necessary to demonstrate cause and effect. (2)

Convergent Evolution The evolution of similar structures in distantly related organisms in response to similar environments. (33)

Cork Cambium In stems and roots of perennials, a secondary meristem that produces the outer protective layer of the bark. (18)

Coronary Arteries Large arteries that branch immediately from the aorta, providing oxygen-rich blood to the cardiac muscle. (28)

Corpus Callosum A thick cable composed of hundreds of millions of neurons that connect the right and left cerebral hemispheres of the mammalian brain. (23)

Corpus Luteum In the mammalian ovary, the structure that develops from the follicle after release of the egg. It secretes hormones that prepare the uterine endometrium to receive the developing embryo. (31)

Cortex In the stem or root of plants, the region between the epidermis and the vascular tissues. Composed of ground tissue. In animals, the outermost portion of some organs. (18)

Cotyledon The seed leaf of a dicot embryo containing stored nutrients required for the germinated seed to grow and develop, or a food digesting seed leaf in a monocot embryo. (20)

Countercurrent Flow Mechanism for increasing the exchange of substances or heat from one stream of fluid to another by having the two fluids flow in opposite directions. (29)

Covalent Bonds Linkage between two atoms which share the same electrons in their outermost shells. (3)

Cranial Nerves Paired nerves which emerge from the central stalk of the vertebrate brain and innervate the body. Humans have 12 pairs of cranial nerves. (23)

Cranium The bony casing which surrounds and protects the vertebrate brain. (23)

Cristae The convolutions of the inner membrane of the mitochondrion. Embedded within them are the components of the electron transport system and proton channels for chemiosmosis. (9)

Crossing Over During synapsis, the process by which homologues exchange segments with each other. (11)

Cryptic Coloration A form of camouflage wherein an organism's color or patterning helps it resemble its background. (42)

Cutaneous Respiration The uptake of oxygen across virtually the entire outer body surface. (29)

Cuticle 1) Waxy layer covering the outer cell walls of plant epidermal cells. It retards water vapor loss and helps prevent dehydration. (18) 2) Outer protective, nonliving covering of some animals, such as the exoskeleton of anthropods. (26, 39)

Cyanobacteria A type of prokaryote capable of photosynthesis using water as a source of electrons. Cyanobacteria were responsible for initially creating an O_2-containing atmosphere on earth. (35, 36)

Cyclic AMP (Cyclic adenosine monophosphate) A ring-shaped molecular version of an ATP minus two phosphates. A regulatory molecule formed by the enzyme adenylate cyclase which converts ATP to cAMP. A second messenger. (25)

Cyclic Pathways Metabolic pathways in which the intermediates of the reaction are regenerated while assisting the conversion of the substrate to product. (9)

Cyclic Photophosphorylation A pathway that produces ATP, but not NADPH, in the light reactions of photosynthesis. Energized electrons are shuttled from a reaction center, along a molecular pathway, back to the original reaction center, generating ATP en route. (8)

Cysts Protective, dormant structure formed by some protozoa. (37)

Cytochrome Oxidase A complex of proteins that serves as the final electron carrier in the mitochondrial electron transport system, transferring its electrons to O_2 to form water. (9)

Cytokinesis Final event in eukaryotic cell division in which the cell's cytoplasm and the new nuclei are partitioned into separate daughter cells. (10)

Cytokinins Growth-producing plant hormones which stimulate rapid cell division. (21)

Cytoplasm General term that includes all parts of the cell, except the plasma membrane and the nucleus. (5)

Cytoskeleton Interconnecting network of microfilaments, microtubules, and intermediate filaments that serves as a cell scaffold and provides the machinery for intracellular movements and cell motility. (5)

Cytotoxic (Killer) T Cells A class of T cells capable of recognizing and destroying foreign or infected cells. (30)

◄ D ►

Day Neutral Plants Plants that flower at any time of the year, independent of the relative lengths of daylight and darkness. (21)

Deciduous Trees or shrubs that shed their leaves in a particular season, usually autumn, before entering a period of dormancy. (40)

Deciduous Forest Forests characterized by trees that drop their leaves during unfavorable conditions, and leaf out during warm, wet seasons. Less dense than tropical rain forests. (40)

Decomposers (Saprophytes) Organisms that obtain nutrients by breaking down organic compounds in wastes and dead organisms. Includes fungi, bacteria, and some insects. (41)

Deletion Loss of a portion of a chromosome, following breakage of DNA. (13)

Denaturation Change in the normal folding of a protein as a result of heat, acidity, or alkalinity. Such changes result in a loss of enzyme functioning. (4)

Dendrites Cytoplasmic extensions of the cell body of a neuron. They carry impulses from the area of stimulation to the cell body. (23)

Denitrification The conversion by denitrifying bacteria of nitrites and nitrates into nitrogen gas. (41)

Denitrifying Bacteria Bacteria which take soil nitrogen, usable to plants, and convert it to unusable nitrogen gas. (41)

Density-Dependent Factors Factors that control population growth which are influenced by population size. (43)

Density-Independent Factors Factors that control population growth which are not affected by population size. (43)

Deoxyribonucleic Acid (DNA) Double-stranded polynucleotide comprised of deoxyribose (a sugar), phosphate, and four bases (adenine, guanine, cytosine, and thymine). Encoded in the sequence of nucleotides are the instructions for making proteins. DNA is the genetic material in all organisms except certain viruses. (14)

Depolarization A decrease in the potential difference (voltage) across the plasma membrane of a cell typically due to an increase in the movement of sodium ions into the cell. Acts to excite a target cell. (23)

Dermal Bone Bones of vertebrates that form within the dermal layer of the skin, such as the scales of fishes and certain bones of the skull. (26)

Dermal Tissue System In plants, the epidermis in primary growth, or the periderm in secondary growth. (18)

Dermis In animals, layer of cells below the epidermis in which connective tissue predominates. Embedded within it are vessels, various glands, smooth muscle, nerves, and follicles. (26)

Desert Biome characterized by intense solar radiation, very little rainfall, and high winds. (40)

Detrivore Organism that feeds on detritus, dead organisms or their parts, and living organisms' waste. (41)

Deuterostome One path of development exhibited by coelomate animals (e.g., echinoderms and chordates). (39)

Diabetes Mellitus A disease caused by a deficiency of insulin or its receptor, preventing glucose from being absorbed by the cells. (25)

Diaphragm A sheet of muscle that separates the thoracic cavity from the abdominal wall. (29)

Diastolic Pressure The second number of a blood pressure reading; the lowest pressure in the arteries just prior to the next heart contraction. (28)

Diatoms are golden-brown algae that are distinguished most dramatically by their intricate silica shells. (37)

Dicotyledonae (Dicots) One of the two classes of flowering plants, characterized by having seeds with two cotyledons, flower parts in 4s or 5s, net-veined leaves, one main root, and vascular bundles in a circular array within the stem. (Compare with Monocotylenodonae). (18)

Diffusion Tendency of molecules to move from a region of higher concentration to a region of lower concentration, until they are uniformly dispersed. (7)

Digestion The process by which food particles are disassembled into molecules small enough to be absorbed into the organism's cells and tissues. (27)

Digestive System System of specialized organs that ingests food, converts nutrients to a form that can be distributed throughout the animal's body, and eliminates undigested residues. (27)

Digestive Tract Tubelike channel through which food matter passes from its point of ingestion at the mouth to the elimination of indigestible residues from the anus. (27)

Dihybrid Cross A mating between two individuals that differ in two genetically-determined traits. (12)

Dimorphism Presence of two forms of a trait within a population, resulting from diversifying selection. (33)

Dinoflagellates Single-celled photosynthesizers that have two flagella. They are members of the pyrophyta, phosphorescent algae that sometimes cause red tide, often synthesizing a neurotoxin that accumulates in plankton eaters, causing paralytic shellfish poisoning in people who eat the shellfish. (37)

Dioecious Plants that produce either male or female reproductive structures but never both. (38)

Diploid Having two sets of homologous chromosomes. Often written 2N. (10, 13)

Directional Selection The steady shift of phenotypes toward one extreme. (33)

Discontinuous Variation An inheritance pattern in which the phenomenon of all possible phenotypes fall into distinct categories. (Compare with continuous variation). (12)

Displays The signals that form the language by which animals communicate. These signals are species specific and stereotyped and may be visual, auditory, chemical, or tactile. (44)

Disruptive Coloration Coloration that disguises the shape of an organism by breaking up its outline. (42)

Disruptive Selection The steady shift toward more than one extreme phenotype due to the elimination of intermediate phenotypes as has occurred among African swallowtail butterflies whose members resemble more than one species of distasteful butterfly. (33)

Divergent Evolution The emergence of new species as branches from a single ancestral lineage. (33)

Diversifying Selection The increasing frequency of extreme phenotypes because individuals with average phenotypes die off. (33)

Diving Reflex Physiological response that alters the flow of blood in the body of diving mammals that allows the animal to maintain high levels of activity without having to breathe. (29)

Division (or Phylum) A level of the taxonomic hierarchy that groups together members or related classes. (1)

DNA (see **Deoxyribonucleic Acid**)

DNA Cloning The amplification of a particular DNA by use of a growing population of bacteria. The DNA is initially taken up by a bacterial cell—usually as a plasmid—and then replicated along with the bacteria's own DNA. (16)

DNA Fingerprint The pattern of DNA fragments produced after treating a sample of DNA with a particular restriction enzyme and separating the fragments by gel electrophoresis. Since different members of a population have DNA with a different nucleotide sequence, the pattern of DNA fragments produced by this method can be used to identify a particular individual. (16)

DNA Ligase The enzyme that covalently joins DNA fragments into a continuous DNA strand. The enzyme is used in a cell during replication to seal newly-synthesized fragments and by biotechnologists to form recombinant DNA molecules from separate fragments. (14, 16)

DNA Polymerase Enzyme responsible for replication of DNA. It assembles free nucleotides, aligning them with the complementary ones in the unpaired region of a single strand of DNA template. (14)

Dominant The form of an allele that masks the presence of other alleles for the same trait. (12)

Dormancy A resting period, such as seed dormancy in plants or hibernation in animals, in which organisms maintain reduced metabolic rates. (21)

Dorsal In anatomy, the back of an animal. (39)

Double Blind Test A clinical trial of a drug in which neither the human subjects or the researchers know who is receiving the drug or placebo. (2)

Down Syndrome Genetic disorder in humans characterized by distinct facial appearance and mental retardation, resulting from an extra copy of chromosome number 21 (trisomy 21) in each cell. (11, 17)

Duodenum First part of the human small intestine in which most digestion of food occurs. (27)

Duplication The repetition of a segment of a chromosome. (13)

◄ **E** ►

Ecdysis Molting process by which an arthropod periodically discards its exoskeleton and replaces it with a larger version. The process is controlled by the hormone ecydysone. (39)

Ecdysone An insect steroid hormone that triggers molting and metamorphosis. (15)

Echinodermata A phylum composed of animals having an internal skeleton made of many small calcium carbonate plates which have jutting spines. Includes sea stars, sea urchins, etc. (39)

Echolocation The use of reflected sound waves to help guide an animal through its environment and/or locate objects. (24)

Ecological Equivalent Organisms that occupy similar ecological niches in different regions or ecosystems of the world. (41)

Ecological Niche The habitat, functional role(s), requirements for environmental resources and tolerance ranges for each abiotic condition in relation to an organism. (41)

Ecological Pyramid Illustration showing the energy content, numbers of organisms, or biomass at each trophic level. (41)

Ecology The branch of biology that studies interactions among organisms as well as the interactions of organisms , and their physical environment. (40)

Ecosystem Unit comprised of organisms interacting among themselves and with their physical environment. (41)

Ecotypes Populations of a single species with different, genetically fixed tolerance ranges. (41)

Ectoderm In animals, the outer germ cell layer of the gastrula. It gives rise to the nervous system and integument. (32)

Ectotherms Animals that lack an internal mechanism for regulating body temperature. "Cold-blooded" animals. (28)

Edema Swelling of a tissue as the result of an accumulation of fluid that has moved out of the blood vessels. (28)

Effectors Muscle fibers and glands that are activated by neural stimulation. (23)

Efferent (Motor) Nerves The nerves that carry messages from the central nervous system to the effectors, the muscles, and glands. They are divided into two systems: somatic and autonomic. (23)

Egg Female gamete, also called an ovum. A fertilized egg is the product of the union of female and male gametes (egg and sperm cells). (32)

Electrocardiogram (EKG) Recording of the electrical activity of the heart, which is used to diagnose various types of heart problems. (28)

Electron Acceptor Substances that are capable of accepting electrons transferred from an electron donor. For example, molecular oxygen (O_2) is the terminal electron acceptor during respiration. Electron acceptors also receive electrons from chlorophyll during photosynthesis. Electron acceptors may act as part of an electron transport system by transferring the electrons they receive to another substance. (8, 9)

Electron Carrier Substances (such as NAD^+ and FAD) that transport electrons from one step of a metabolic pathway to the next or from metabolic reactions to biosynthetic reactions. (8, 9)

Electrons Negatively charged particles that orbit the atomic nucleus. (3)

Electron Transport System Highly organized assembly of cytochromes and other proteins which transfer electrons. During transport, which occurs within the inner membranes of mitochondria and chloroplasts, the energy extracted from the electrons is used to make ATP. (8, 9)

Electrophoresis A technique for separating different molecules on the basis of their size and/or electric charge. There are various ways the technique is used. In gel electrophoresis, proteins or DNA fragments are driven through a porous gel by their charge, but become separated according to size; the larger the molecule, the slower it can work its way through the pores in the gel, and the less distance it travels along the gel. (16)

Element Substance composed of only one type of atom. (3)

Embryo An organism in the early stages of development, beginning with the first division of the zygote. (32)

Embryo Sac The fully developed female gametophyte within the ovule of the flower. (20)

Emigration Individuals permanently leaving an area or population. (43)

Endergonic Reactions Chemical reactions that require energy input from another source in order to occur. (6)

Endocrine Glands Ductless glands, which secrete hormones directly into surrounding tissue fluids and blood vessels for distribution to the rest of the body by the circulatory system. (25)

Endocytosis A type of active transport that imports particles or small cells into a cell. There are two types of endocytic processes: phagocytosis, where large particles are ingested by the cell, and pinocytosis, where small droplets are taken in. (7)

Endoderm In animals, the inner germ cell layer of the gastrula. It gives rise to the digestive tract and associated organs and to the lungs. (32)

Endodermis The innermost cylindrical layer of cortex surrounding the vascular tissues of the root. The closely pressed cells of the endodermis have a waxy band, forming a waterproof layer, the Casparian strip. (18)

Endogenous Plant responses that are controlled internally, such as biological clocks controlling flower opening. (21)

Endometrium The inner epithelial layer of the uterus that changes markedly with the uterine (menstrual) cycle in preparation for implantation of an embryo. (31)

Endoplasmic Reticulum (ER) An elaborate system of folded, stacked and tubular membranes contained in the cytoplasm of eukaryotic cells. (5)

Endorphins (Endogenous Morphinelike Substances) A class of peptides released from nerve cells of the limbic system of the brain that can block perceptions of pain and produce a feeling of euphoria. (23)

Endoskeleton The internal support structure found in all vertebrates and a few invertebrates (sponges and sea stars). (26)

Endosperm Nutritive tissue in plant embryos and seeds. (20)

Endosperm Mother Cell A binucleate cell in the embryo sac of the female gametophyte, occurring in the ovule of the ovary in angiosperms. Each nucleus is haploid; after fertilization, nutritive endosperm develops. (20)

Endosymbiosis Theory A theory to explain the development of complex eukaryotic cells by proposing that some organelles once were free-living prokaryotic cells that then moved into another larger such cell, forming a beneficial union with it. (5)

Endotherms Animals that utilize metabolically produced heat to maintain a constant, elevated body temperature. "Warm-blooded" animals. (28)

End Product The last product in a metabolic pathway. Typically a substance, such as an amino acid or a nucleotide, that will be used as a monomer in the formation of macromolecules. (6)

Energy The ability to do work. (6)

Entropy Energy that is not available for doing work; measure of disorganization or randomness. (6)

Environmental Resistance The factors that eventually limit the size of a population. (43)

Enzyme Biological catalyst; a protein molecule that accelerates the rate of a chemical reaction. (6)

Eosiniphil A type of phagocytic white blood cell. (28)

Epicotyl The portion of the embryo of a dicot plant above the cotyledons. The epicotyl gives rise to the shoot. (20)

Epidermis In vertebrates, the outer layer of the skin, containing superficial layers of dead cells produced by the underlying living epithelial cells. In plants, the outer layer of cells covering leaves, primary stem, and primary root. (26, 18)

Epididymis Mass of convoluted tubules attached to each testis in mammals. After leaving the testis, sperm enter the tubules where they finish maturing and acquire motility. (31)

Epiglottis A flap of tissue that covers the glottis during swallowing to prevent food and liquids from entering the lower respiratory tract. (29)

Epinephrine (Adrenalin) Substance that serves both as an excitatory neurotransmitter released by certain neurons of the CNS and as a hormone released by the adrenal medulla that increases the body's ability to combat a stressful situation. (25)

Epipelagic Zone The lighted upper ocean zone, where photosynthesis occurs; large populations of phytoplankton occur in this zone. (40)

Epiphyseal Plates The action centers for ossification (bone formation). (26)

Epistasis A type of gene interaction in which a particular gene blocks the expression of another gene at another locus. (12)

Epithelial Tissue Continuous sheets of tightly packed cells that cover the body and line its tracts and chambers. Epithelium is a fundamental tissue type in animals. (22)

Erythrocytes Red blood cells. (28)

Erythropoietin A hormone secreted by the kidney which stimulates the formation of erythrocytes by the bone marrow. (28)

Essential Amino Acids Eight amino acids that must be acquired from dietary protein. If even one is missing from the human diet, the synthesis of proteins is prevented. (27)

Essential Fatty Acids Linolenic and linoleic acids, which are required for phospholipid construction and must be acquired from a dietary source. (27)

Essential Nutrients The 16 minerals essential for plant growth, divided into two groups: macronutrients, which are required in large quantities, and micronutrients, which are needed in small amounts. (19)

Estrogen A female sex hormone secreted by the ovaries when stimulated by pituitary gonadotrophins. (31)

Estuaries Areas found where rivers and streams empty into oceans, mixing fresh water with salt water. (40)

Ethology The study of animal behavior. (44)

Ethylene Gas A plant hormone that stimulates fruit ripening. (21)

Etiolation The condition of rapid shoot elongation, small underdeveloped leaves, bent shoot-hook, and lack of chlorophyll, all due to lack of light. (21)

Eubacteria Typical procaryotic bacteria with peptidoglycan in their cell walls. The majority of monerans are eubacteria. (36)

Eukaryotic Referring to organisms whose cellular anatomy includes a true nucleus with a nuclear envelope, as well as other membrane-bound organelles. (5)

Eusocial Species Social species that have sterile workers, cooperative care of the young, and an overlap of generations so that the colony labor is a family affair. (44)

Eutrophication The natural aging process of lakes and ponds, whereby they become marshes and, eventually, terrestrial environments.

Evolution A process whereby the characteristics of a species change over time, eventually leading to the formation of new species that go about life in new ways. (33)

Evolutionarily Stable Strategy (ESS) A behavioral strategy or course of action that depends on what other members of the population are doing. By definition, an ESS cannot be replaced by any other strategy when most of the members of the population have adopted it. (44)

Excitatory Neurons Neurons that stimulate their target cells into activity. (23)

Excretion Removal of metabolic wastes from an organism. (28)

Excretory System The organ system that eliminates metabolic wastes from the body. (28)

Exergonic Reactions Chemical reactions that occur spontaneously with the release of energy. (6)

Exocrine Glands Glands which secrete their products through ducts directly to their sites of action, e.g., tear glands. (26)

Exocytosis A form of active transport used by cells to move molecules, particles, or other cells contained in vesicles across the plasma membrane to the cell's environment. (5)

Exogenous Plant responses that are controlled externally, or by environmental conditions. (21)

Exons Structural gene segments that are transcribed and whose genetic information is subsequently translated into protein. (15)

Exoskeletons Hard external coverings found in some animals (e.g., lobsters, insects) for protection, support, or both. Such organisms grow by the process of molting. (26)

Exploitative Competition A competition in which one species manages to get more of a resource, thereby reducing supplies for a competitor. (42)

Exponential Growth An increase by a fixed percentage in a given time period; such as population growth per year. (43)

Extensor Muscle A muscle which, when contracted, causes a part of the body to straighten at a joint. (26)

External Fertilization Fertilization of an egg outside the body of the female parent. (31)

Extinction The loss of a species. (33)

Extracellular Digestion Digestion occurring outside the cell; occurs in bacteria, fungi, and multicellular animals. (27)

Extracellular Matrix Layer of extracellular material residing just outside a cell. (5)

◄ **F** ►

F_1 First filial generation. The first generation of offspring in a genetic cross. (12)

F_2 Second filial generation. The offspring of an F_1 cross. (12)

Facilitated Diffusion The transport of molecules into cells with the aid of "carrier" proteins embedded in the plasma membrane. This carrier-assisted transport does not require the expenditure of energy by the cell. (7)

FAD Flavin adenine dinucleotide. A coenzyme that functions as an electron carrier in metabolic reactions. When it is reduced to $FADH_2$, this molecule becomes a cellular energy source. (9)

Family A level of the taxonomic hierarchy that groups together members of related genera. (1)

Fast-Twitch Fibers Skeletal muscle fibers that depend on anaerobic metabolism to produce ATP rapidly, but only for short periods of time before the onset of fatigue. Fast-twitch fibers generate greater forces for shorter periods than slow-twitch fibers. (9)

Fat A triglyceride consisting of three fatty acids joined to a glycerol. (4)

Fatty Acid A long unbranched hydrocarbon chain with a carboxyl group at one end. Fatty acids lacking a double bond are said to be saturated. (4)

Fauna The animals in a particular region.

Feedback Inhibition (Negative Feedback) A mechanism for regulating enzyme activity by temporarily inactivating a key enzyme in a biosynthetic pathway when the concentration of the end product is elevated. (6)

Fermentation The direct donation of the electrons of NADH to an organic compound without their passing through an electron transport system. (9)

Fertility Rate In humans, the average number of children born to each woman between 15 and 44 years of age. (43)

Fertilization The process in which two haploid nuclei fuse to form a zygote. (32)

Fetus The term used for the human embryo during the last seven months in the uterus. During the fetal stage, organ refinement accompanies overall growth. (32)

Fibrinogen A rod-shaped plasma protein that, converted to fibrin, generates a tangled net of fibers that binds a wound and stops blood loss until new cells replace the damaged tissue. (28)

Fibroblasts Cells found in connective tissues that secrete the extracellular materials of the connective tissue matrix. These cells are easily isolated from connective tissues and are widely used in cell culture. (22)

Fibrous Root System Many approximately equal-sized roots; monocots are characterized by a fibrous root system. Also called diffuse root system. (18)

Filament The stalk of a stamen of angiosperms, with the anther at its tip. Also, the threadlike chain of cells in some algae and fungi. (20)

Filamentous Fungus Multicellular members of the fungus kingdom comprised mostly of living threads (hyphae) that grow by division of cells at their tips (see molds). (37)

Filter Feeders Aquatic animals that feed by straining small food particles from the surrounding water. (27, 39)

Fitness The relative degree to which an individual in a population is likely to survive to reproductive age and to reproduce. (33)

Fixed Action Patterns Motor responses that may be triggered by some environmental stimulus, but once started can continue to completion without external stimuli. (44)

Flagella Cellular extensions that are longer than cilia but fewer in number. Their undulations propel cells like sperm and many protozoans, through their aqueous environment. (5)

Flexor Muscle A muscle which, when contracted, causes a part of the body to bend at a joint. (26)

Flora The plants in a particular region. (21)

Florigen Proposed A chemical hormone that is produced in the leaves and stimulates flowering. (21)

Fluid Mosaic Model The model proposes that the phospholipid bilayer has a viscosity similar to that of light household oil and that globular proteins float like icebergs within this bilayer. The now favored explanation for the architecture of the plasma membrane. (5)

Follicle (Ovarian) A chamber of cells housing the developing oocytes. (31)

Food Chain Transfers of food energy from organism to organism, in a linear fashion. (41)

Food Web The map of all interconnections between food chains for an ecosystem. (41)

Forest Biomes Broad geographic regions, each with characteristic tree vegetation: 1) tropical rain forests (lush forests in a broad band around the equator), 2) deciduous forests (trees and shrubs drop their leaves during unfavorable seasons), 3) coniferous forest (evergreen conifers). (40)

Fossil Record An entire collection of remains from which paleontologists attempt to reconstruct the phylogeny, anatomy, and ecology of the preserved organisms. (34)

Fossils The preserved remains of organisms from a former geologic age. (34)

Fossorial Living underground.

Founder Effect The potentially dramatic difference in allele frequency of a small founding population as compared to the original population. (33)

Founder Population The individuals, usually few, that colonize a new habitat. (33)

Frameshift Mutation The insertion or deletion of nucleotides in a gene that throws off the reading frame. (14)

Free Radical Atom or molecule containing an unpaired electron, which makes it highly reactive. (3)

Freeze-Fracture Technique in which cells are frozen into a block which is then struck with a knife blade that fractures the block in two. Fracture planes tend to expose the center of membranes for EM examination. (5)

Fronds The large leaf-like structures of ferns. Unlike true leaves, fronds have an apical meristem and clusters of sporangia called sori. (38)

Fruit A mature plant ovary (flower) containing seeds with plant embryos. Fruits protect seeds and aid in their dispersal. (20)

Fruiting Body A spore-producing structure that extends upward in an elevated position from the main mass of a mold or slime mold. (37)

FSH Follicle stimulating hormone. A hormone secreted by the anterior pituitary that prepares a female for ovulation by stimulating the primary follicle to ripen or stimulates spermatogenesis in males. (31)

Functional Groups Accessory chemical entities (e.g., —OH, —NH$_2$, —CH$_3$), which help determine the identity and chemical properties of a compound. (4)

Fundamental Niche The potential ecological niche of a species, including all factors affecting that species. The fundamental niche is usually never fully utilized. (41)

Fungus Yeast, mold, or large filamentous mass forming macroscopic fruiting bodies, such as mushrooms. All fungi are eukaryotic nonphotosynthetic heterotrophics with cell walls. (37)

◀ **G** ▶

G$_1$ Stage The first of three consecutive stages of interphase. During G$_1$, cell growth and normal functions occur. The duration of this stage is most variable. (10)

G$_2$ Stage The final stage of interphase in which the final preparations for mitosis occur. (10)

Gallbladder A small saclike structure that stores bile salts produced by the liver. (27)

Gamete A haploid reproductive cell—either a sperm or an egg. (10)

Gas Exchange Surface Surface through which gases must pass in order to enter or leave the body of an animal. It may be the plasma membrane of a protistan or the complex tissues of the gills or the lungs in multicellular animals. (29)

Gastrovascular Cavity In cnidarians and flatworms, the branched cavity with only one opening. It functions in both digestion and transport of nutrients. (39)

Gastrula The embryonic stage formed by the inward migration of cells in the blastula. (32)

Gastrulation The process by which the blastula is converted into a gastrula having three germ layers (ectoderm, mesoderm, and endoderm). (32)

Gated Ion Channels Most passageways through a plasma membrane that allow ions to pass contain "gates" that can occur in either an open or a closed conformation. (7, 23)

Gel Electrophoresis (See **Electrophoresis**)

Gene Pool All the genes in all the individuals of a population. (33)

Gene Regulatory Proteins Proteins that bind to specific sites in the DNA and control the transcription of nearby genes. (15)

Genes Discrete units of inheritance which determine hereditary traits. (12, 14)

Gene Therapy Treatment of a disease by alteration of the person's genotype, or the genotype of particular affected cells. (17)

Genetic Carrier A heterozygous individual who shows no evidence of a genetic disorder but, because they possess a recessive allele for a disorder, can pass the mutant gene on to their offspring. (17)

Genetic Code The correspondence between the various mRNA triplets (codons, e.g., UGC) and the amino acid that the triplet specifies (e.g., cysteine). The genetic code includes 64 possible three-letter words that constitute the genetic language for protein synthesis. (14)

Genetic Drift Random changes in allele frequency that occur by chance alone. Occurs primarily in small populations. (33)

Genetic Engineering The modification of a cell or organism's genetic composition according to human design. (16)

Genetic Equilibrium A state in which allele frequencies in a population remain constant from generation to generation. (33)

Genetic Mapping Determining the locations of specific genes or genetic markers along particular chromosomes. This is typically accomplished using crossover frequencies; the more often alleles of two genes are separated during crossing over, the greater the distance separating the genes. (13)

Genetic Recombination The reshuffling of genes on a chromosome caused by breakage of DNA and its reunion with the DNA of a homologoue. (11)

Genome The information stored in all the DNA of a single set of chromosomes. (17)

Genotype An individual's genetic makeup. (12)

Genus Taxonomic group containing related species. (1)

Geologic Time Scale The division of the earth's 4.5 billion-year history into eras, periods, and epochs based on memorable geologic and biological events. (35)

Germ Cells Cells that are in the process of or have the potential to undergo meiosis and form gametes. (11, 31)

Germination The sprouting of a seed, beginning with the radicle of the embryo breaking through the seed coat. (21)

Germ Layers Collective name for the endoderm, ectoderm, and mesoderm, from which all the structures of the mature animal develop. (32)

Gibberellins More than 50 compounds that promote growth by stimulating both cell elongation and cell division. (21)

Gills Respiratory organs of aquatic animals. (29)

Globin The type of polypeptide chains that make up a hemoglobin molecule.

Glomerular Filtration The process by which fluid is filtered out of the capillaries of the glomerulus into the proximal end of the nephron. Proteins and blood cells remain behind in the bloodstream. (28)

Glomerulus A capillary bundle embedded in the double-membraned Bowman's capsule, through which blood for the kidney first passes. (28)

Glottis Opening leading to the larynx and lower respiratory tract. (29)

Glucagon A hormone secreted by the Islets of Langerhans that promotes glycogen breakdown to glucose. (25)

Glucocorticoids Steroid hormones which regulate sugar and protein metabolism. They are secreted by the adrenal cortex. (25)

Glycogen A highly branched polysaccharide consisting of glucose monomers that serves as a storage of chemical energy in animals. (4)

Glycolysis Cleavage, releasing energy, of the six-carbon glucose molecule into two molecules of pyruvic acid, each containing three carbons. (9)

Glycoproteins Proteins with covalently-attached chains of sugars. (5)

Glycosidic Bond The covalent bond between individual molecules in carbohydrates. (4)

Golgi Complex A system of flattened membranous sacs, which package substances for secretion from the cell. (5)

Gonadotropin-Releasing Hormone (GnRH) Hypothalmic hormone that controls the secretion of the gonadotropins FSH and LH. (31)

Gonadotropins Two anterior pituitary hormones which act on the gonads. Both FSH (follicle-stimulating hormone) and LH (luteinizing hormone) promote gamete development and stimulate the gonads to produce sex hormones. (25)

Gonads Gamete-producing structures in animals: ovaries in females, testes in males. (31)

Grasslands Areas of densely packed grasses and herbaceous plants. (40)

Gravitropisms (Geotropisms) Changes in plant growth caused by gravity. Growth away from gravitational force is called negative gravitropism; growth toward it is positive. (21)

Gray Matter Gray-colored neural tissue in the cerebral cortex of the brain and in the butterfly-shaped interior of the spinal cord. Composed of nonmyelinated cell bodies and dendrites of neurons. (23)

Greenhouse Effect The trapping of heat in the Earth's troposphere, caused by increased levels of carbon dioxide near the Earth's surface; the carbon dioxide is believed to act like glass in a greenhouse, allowing light to reach the Earth, but not allowing heat to escape. (41)

Ground Tissue System All plant tissues except those in the dermal and vascular tissues. (18)

Growth An increase in size, resulting from cell division and/or an increase in the volume of individual cells. (10)

Growth Hormone (GH) Hormone produced by the anterior pituitary; stimulates protein synthesis and bone elongation. (25)

Growth Ring In plants with secondary growth, a ring formed by tracheids and/or vessels with small lumens (late wood) during periods of unfavorable conditions; apparent in cross section. (18)

Guard Cells Specialized epidermal plant cells that flank each stomated pore of a leaf. They regulate the rate of gas diffusion and transpiration. (18)

Guild Group of species with similar ecological niches. (41)

Guttation The forcing of water and mineral completely out to the tips of leaves as a result of positive root pressure. (19)

Gymnosperms The earliest seed plants, bearing naked seeds. Includes the pines, hemlocks, and firs. (38)

◄ **H** ►

Habitat The place or region where an organism lives. (41)

Habituation The phenomenon in which an animal ceases to respond to a repetitive stimulus. (23, 44)

Hair Cells Sensory receptors of the inner ear that respond to sound vibration and bodily movement. (24)

Half-Life The time required for half the mass of a radioactive element to decay into its stable, non-radioactive form. (3)

Haplodiploidy A genetic pattern of sex determination in which fertilized eggs develop into females and non-fertilized eggs develop into males (as occurs among bees and wasps). (44)

Haploid Having one set of chromosomes per cell. Often written as 1N. (10)

Hardy-Weinberg Law The maintenance of constant allele frequencies in a population from one generation to the next when certain conditions are met. These conditions are the absence of mutation and migration, random mating, a large population, and an equal chance of survival for all individuals. (33)

Haversian Canals A system of microscopic canals in compact bone that transport nutrients to and remove wastes from osteocytes. (26)

Heart An organ that pumps blood (or hemolymph in arthropods) through the vessels of the circulatory system. (28)

Helper T Cells A class of T cells that regulate immune responses by recognizing and activating B cells and other T cells. (30)

Hemocoel In arthropods, the unlined spaces into which fluid (hemolymph) flows when it leaves the blood vessels and bathes the internal organs. (28)

Hemoglobin The iron-containing blood protein that temporarily binds O_2 and releases it into the tissues. (4, 29)

Hemophilia A genetic disorder determined by a gene on the X chromosome (an X-linked trait) that results from the failure of the blood to form clots. (13)

Herbaceous Plants having only primary growth and thus composed entirely of primary tissue. (18)

Herbivore An organism, usually an animal, that eats primary producers (plants). (42)

Herbivory The term for the relationship of a secondary consumer, usually an animal, eating primary producers (plants). (42)

Heredity The passage of genetic traits to offspring which consequently are similar or identical to the parent(s). (12)

Hermaphrodites Animals that possess gonads of both the male and the female. (31)

Heterosporous Higher vascular plants producing two types of spores, a megaspore which grows into a female gametophyte and a microspore which grows into a male gametophyte. (38)

Heterozygous A term applied to organisms that possess two different alleles for a trait. Often, one allele (A) is dominant, masking the presence of the other (a), the recessive. (12)

High Intertidal Zone In the intertidal zone, the region from mean high tide to around just below sea level. Organisms are submerged about 10% of the time. (40)

Histones Small basic proteins that are complexed with DNA to form nucleosomes, the basic structural components of the chromatin fiber. (14)

Homeobox That part of the DNA sequence of homeotic genes that is similar (homologous) among diverse animal species. (32)

Homeostasis Maintenance of fairly constant internal conditions (e.g., blood glucose level, pH, body temperature, etc.) (22)

Homeotic Genes Genes whose products act during embryonic development to affect the spatial arrangement of the body parts. (32)

Hominids Humans and the various groups of extinct, erect-walking primates that were either our direct ancestors or their relatives. Includes the various species of *Homo* and *Australopithecus*. (34)

Homo the genus that contains modern and extinct species of humans. (34)

Homologous Structures Anatomical structures that may have different functions but develop from the same embryonic tissues, suggesting a common evolutionary origin. (34)

Homologues Members of a chromosome pair, which have a similar shape and the same sequence of genes along their length. (10)

Homoplasy (see **Analogous Structures**)

Homosporous Plants that manufacture only one type of spore, which develops into a gametophyte containing both male and female reproductive structures. (38)

Homozygous A term applied to an organism that has two identical alles for a particular trait. (12)

Hormones Chemical messengers secreted by ductless glands into the blood that direct tissues to change their activities and correct imbalances in body chemistry. (25)

Host The organism that a parasite lives on and uses for food. (42)

Human Chorionic Gonadotropin (HCG) A hormone that prevents the corpus luteum from degenerating, thereby maintaining an adequate level of progesterone during pregnancy. It is produced by cells of the early embryo. (25)

Human Immunodeficiency Virus (HIV) The infectious agent that causes AIDS, a disease in which the immune system is seriously disabled. (30, 36)

Hybrid An individual whose parents possess different genetic traits in a breeding experiment or are members of different species. (12)

Hybridization Occurs when two distinct species mate and produce hybrid offspring. (33)

Hybridoma A cell formed by the fusion of a malignant cell (a myeloma) and an antibody-producing lymphocyte. These cells proliferate indefinitely and produce monoclonal antibodies. (30)

Hydrogen Bonds Relatively weak chemical bonds formed when two molecules share an atom of hydrogen. (3)

Hydrologic Cycle The cycling of water, in various forms, through the environment, from Earth to atmosphere and back to Earth again. (41)

Hydrolysis Splitting of a covalent bond by donating the H+ or OH- of a water molecule to the two components. (4)

Hydrophilic Molecules Polar molecules that are attracted to water molecules and readily dissolve in water. (3)

Hydrophobic Interaction When nonpolar molecules are "forced" together in the presence of a polar solvent, such as water. (3)

Hydrophobic Molecules Nonpolar substances, insoluble in water, which form aggregates to minimize exposure to their polar surroundings. (3)

Hydroponics The science of growing plants in liquid nutrient solutions, without a solid medium such as soil. (19)

Hydrosphere That portion of the Earth composed of water. (40)

Hydrostatic Skeletons Body support systems found usually in underwater animals (e.g., marine worms). Body shape is protected against gravity and other physical forces by internal hydrostatic pressure produced by contracting muscles encircling their closed, fluid-filled chambers. (26)

Hydrothermal Vents Fissures in the ocean floor where sea water becomes superheated. Chemosynthetic bacteria that live in these vents serve as the autotrophs that support a diverse community of ocean-dwelling organisms. (8)

Hyperpolarization An increase in the potential difference (voltage) across the plasma membrane of a cell typically due to an increase in the movement of potassium ions out of the cell. Acts to inhibit a target cell. (23)

Hypertension High blood pressure (above about 130/90). (28)

Hypertonic Solutions Solutions with higher solute concentrations than found inside the cell. These cause a cell to lose water and shrink. (7)

Hypervolume In ecology, a multidimensional area which includes all factors in an organism's ecological niche, or its' potential niche. (41)

Hypocotyl Portion of the plant embryo below the cotyledons. The hypocotyl gives rise to the root and, very often, to the lower part of the stem. (20)

Hypothalamus The area of the brain below the thalamus that regulates body temperature, blood pressure, etc. (25)

Hypothesis A tentative explanation for an observation or a phenomenon, phrased so that it can be tested by experimentation. (2)

Hypotonic Solutions Solutions with lower solute concentrations than found inside the cell. These cause a cell to accumulate water and swell. (7)

◄ I ►

Immigration Individuals permanently moving into a new area or population. (43)

Immune System A system in vertebrates for the surveillance and destruction of disease-causing microorganisms and cancer cells. Composed of lymphocytes, particularly B cells and T cells, and triggered by the introduction of antigens into the body which makes the body, upon their destruction, resistant to a recurrence of the same disease. (30)

Immunoglobulins (IGs) Antibody molecules. (30)

Imperfect Flowers Flowers that contain either stamens or carpels, making them male or female flowers, respectively. (20)

Imprinting A type of learning in which an animal develops an association with an object after exposure to the object during a critical period early in its life. (44)

Inbreeding When individuals mate with close relatives, such as brothers and sisters. May occur when population sizes drastically shrink and results in a decrease in genetic diversity. (33)

Incomplete Digestive Tract A digestive tract with only one opening through which food is taken in and residues are expelled. (27)

Incomplete (Partial) Dominance A phenomenon in which heterozygous individuals are phenotypically distinguishable from either homozygous type. (12)

Incomplete Flower Flowers lacking one or more whorls of sepals, petals, stamen, or pistils. (20)

Independent Assortment The shuffling of members of homologous chromosome pairs in meiosis I. As a result, there are new chromosome combinations in the daughter cells, which later produce offspring with random mixtures of traits from both parents. (11, 12)

Indoleatic Acid (IAA) An auxin responsible for many plant growth responses including apical dominance, a growth pattern in which shoot tips prevent axillary buds from sprouting. (21)

Induction The process in which one embryonic tissue induces another tissue to differentiate along a pathway that it would not otherwise have taken. (32) Stimulation of transcription of a gene in an operon. Occurs when the repressor protein is unable to bind to the operator. (15)

Inflammation A body strategy initiated by the release of chemicals following injury or infection which brings additional blood with its protective cells to the injured area. (30)

Inhibitory Neurons Neurons that oppose a response in the target cells. (23)

Inhibitory Neurotransmitters Substances released from inhibitory neurons where they synapse with the target cell. (23)

Innate Behavior Actions that are under fairly precise genetic control, typically species-specific, highly stereotyped, and that occur in a complete form the first time the stimulus is encountered. (44)

Insight Learning The sudden solution to a problem without obvious trial-and-error procedures. (44)

Insulin One of the two hormones secreted by endocrine centers called Islets of Langerhans; promotes glucose absorption, utilization, and storage. Insulin is secreted by them when the concentration of glucose in the blood begins to exceed the normal level. (25)

Integumentary System The body's protective external covering, consisting of skin and subcutaneous tissue. (26)

Integuments Protective covering of the ovule. (20)

Intercellular Junctions Specialized regions of cell-cell contact between animal cells. (5)

Intercostal Muscles Muscles that lie between the ribs in humans whose contraction expands the thoracic cavity during breathing. (29)

Interference Competition One species' direct interference by another species for the same limited resource; such as aggressive animal behavior. (42)

Internal Fertilization Fertilization of an egg within the body of the female. (31)

Interneurons Neurons situated entirely within the central nervous system. (23)

Internodes The portion of a stem between two nodes. (18)

Interphase Usually the longest stage of the cell cycle during which the cell grows, carries out normal metabolic functions, and replicates its DNA in preparation for cell division. (10)

Interstitial Cells Cells in the testes that produce testosterone, the major male sex hormone. (31)

Interstitial Fluid The fluid between and surrounding the cells of an animal; the extracellular fluid. (28)

Intertidal Zone The region of beach exposed to air between low and high tides. (40)

Intracellular Digestion Digestion occurring inside cells within food vacuoles. The mode of

digestion found in protists and some filter-feeding animals (such as sponges and clams). (27)

Intraspecific Competition Individual organisms of one species competing for the same limited resources in the same habitat, or with overlapping niches. (42)

Intrinsic Rate of Increase (r_m) the maximum growth rate of a population under conditions of maximum birth rate and minimum death rate. (43)

Introns Intervening sequences of DNA in the middle of structural genes, separating exons. (15)

Invertebrates Animals that lack a vertebral column, or backbone. (39)

Ion An electrically charged atom created by the gain or loss of electrons. (3)

Ionic Bond The noncovalent linkage formed by the attraction of oppositely charged groups. (3)

Islets of Langerhans Clusters of endocrine cells in the pancreas that produce insulin and glucagon. (25)

Isolating Mechanisms Barriers that prevent gene flow between populations or among segments of a single population. (33)

Isotopes Atoms of the same element having a different number of neutrons in their nucleus. (3)

Isotonic Solutions Solutions in which the solute concentration outside the cell is the same as that inside the cell. (7)

◄ **J** ►

Joints Structures where two pieces of a skeleton are joined. Joints may be flexible, such as the knee joint of the human leg or the joints between segments of the exoskeleton of the leg of an insect, or inflexible, such as the joints (sutures) between the bones of the skull. (26)

J-Shaped Curve A curve resulting from exponential growth of a population. (43)

◄ **K** ►

Karyotype A visual display of an individual's chromosomes. (10)

Kidneys Paired excretory organs which, in humans, are fist-sized and attached to the lower spine. In vertebrates, the kidneys remove nitrogenous wastes from the blood and regulate ion and water levels in the body. (28)

Killer T Cells A type of lymphocyte that functions in the destruction of virus-infected cells and cancer cells. (30)

Kinases Enzymes that catalyze reactions in which phosphate groups are transferred from ATP to another molecule. (6)

Kinetic Energy Energy in motion. (6)

Kinetochore Part of a mitotic (or meiotic) chromosome that is situated within the centromere and to which the spindle fibers attach. (10)

Kingdom A level of the taxonomic hierarchy that groups together members of related phyla or divisions. Modern taxonomy divides all organisms into five Kingdoms: Monera, Protista, Fungi, Plantae, and Animalia. (1)

Klinefelter Syndrome A male whose cells have an extra X chromosome (XXY). The syndrome is characterized by underdeveloped male genitalia and feminine secondary sex characteristics. (17)

Krebs Cycle A circular pathway in aerobic respiration that completely oxidizes the two pyruvic acids from glycolysis. (9)

K-Selected Species Species that produce one or a few well-cared for individuals at a time. (43)

◄ **L** ►

Lacteal Blind lymphatic vessel in the intestinal villi that receives the absorbed products of lipid digestion. (27)

Lactic Acid Fermentation The process in which electrons removed during glycolysis are transferred from NADH to pyruvic acid to form lactic acid. Used by various prokaryotic cells under oxygen-deficient conditions and by muscle cells during strenuous activity. (9)

Lake Large body of standing fresh water, formed in natural depressions in the Earth. Lakes are larger than ponds. (40)

Lamella In bone, concentric cylinders of calcified collagen deposited by the osteocytes. The laminated layers produce a greatly strengthened structure. (26)

Large Intestine Portion of the intestine in which water and salts are reabsorbed. It is so named because of its large diameter. The large instestine, except for the rectum, is called the colon. (27)

Larva A self-feeding, sexually, and developmentally immature form of an animal. (32)

Larynx The short passageway connecting the pharynx with the lower airways. (29)

Latent (hidden) Infection Infection by a microorganism that causes no symptoms but the microbe is well-established in the body. (36)

Lateral Roots Roots that arise from the pericycle of older roots; also called branch roots or secondary roots. (18)

Law of Independent Assortment Alleles on nonhomologous chromosomes segregate independently of one another. (12)

Law of Segregation During gamete formation, pairs of alleles separate so that each sperm or egg cell has only one gene for a trait. (12)

Law of the Minimum The ecological principle that a species' distribution will be limited by whichever abiotic factor is most deficient in the environment. (41)

Laws of Thermodynamics Physical laws that describe the relationship of heat and mechanical energy. The first law states that energy cannot be created or destroyed, but one form can change into another. The second law states that the total energy the universe decreases as energy conversions occur and some energy is lost as heat. (6)

Leak Channels Passageways through a plasma membrane that do not contain gates and, therefore, are always open for the limited diffusion of a specific substance (ion) through the membrane. (7, 23)

Learning A process in which an animal benefits from experience so that its behavior is better suited to environmental conditions. (44)

Lenticels Loosely packed cells in the periderm of the stem that create air channels for transferring CO_2, H_2O, and O_2. (18)

Leukocytes White blood cells. (28)

LH Luteinizing hormone. A hormone secreted by the anterior pituitary that stimulates testosterone production in males and triggers ovulation and the transformation of the follicle into the corpus luteum in females. (31)

Lichen Symbiotic associations between certain fungi and algae. (37)

Life Cycle The sequence of events during the lifetime of an organism from zygote to reproduction. (39)

Ligaments Strong straps of connective tissue that hold together the bones in articulating joints or support an organ in place. (26)

Light-Dependent Reactions First stage of photosynthesis in which light energy is converted to chemical energy in the form of energy-rich ATP and NADPH. (8)

Light-Independent Reactions Second stage of photosynthesis in which the energy stored in ATP and NADPH formed in the light reactions is used to drive the reactions in which carbon dioxide is converted to carbohydrate. (8)

Limb Bud A portion of an embryo that will develop into either a forelimb or hindlimb. (32)

Limbic System A series of an interconnected group of brain structures, including the thalamus and hypothalamus, controlling memory and emotions. (23)

Limiting Factors The critical factors which impose restraints of the distribution, health, or activities of an organism. (41)

Limnetic Zone Open water of lakes, through which sunlight penetrates and photosynthesis occurs. (40)

Linkage The tendency of genes of the same chromosome to stay together rather than to assort independently. (13)

Linkage Groups Groups of genes located on the same chromosome. The genes of each linkage group assort independently of the genes of other linkage groups. In all eukaryotic organisms, the number of linkage groups is equal to the haploid number of chromosomes. (13)

Lipids A diverse group of biomolecules that are insoluble in water. (4)

Lithosphere The solid outer zone of the Earth; composed of the crust and outermost portion of the mantle. (40)

Littoral Zone Shallow, nutrient-rich waters of a lake, where sunlight reaches the bottom; also the lakeshore. Rooted vegetation occurs in this zone. (40)

Locomotion The movement of an organism from one place to another. (26)

Locus The chromosomal location of a gene. (13)

Logistic Growth Population growth producing a sigmoid, or S-shaped, growth curve. (43)

Long-Day Plants Plants that flower when the length of daylight exceeds some critical period. (21)

Longitudinal Fission The division pattern in flagellated protozoans, where division is along the length of the cell.

Loop of Henle An elongated section of the renal tubule that dips down into the kidney's medulla and then ascends back out to the cortex. It separates the proximal and distal convoluted tubules and is responsible for forming the salt gradient on which water reabsorption in the kidney depends. (28)

Low Density Lipoprotein (LDL) Particles that transport cholesterol in the blood. Each particle consists of about 1,500 cholesterol molecules surrounded by a film of phospholipids and protein. LDLs are taken into cells following their binding to cell surface LDL receptors. (7)

Low Intertidal Zone In the intertidal zone, the region which is uncovered by "minus" tides only. Organisms are submerged about 90% of the time. (40)

Lumen A space within an hollow organ or tube. (28)

Luminescence (see **Bioluminescence**)

Lungs The organs of terrestrial animals where gas exchange occurs. (29)

Lymph The colorless fluid in lymphatic vessels. (28)

Lymphatic System Network of fluid-carrying vessels and associated organs that participate in immunity and in the return of tissue fluid to the main circulation. (28)

Lymphocytes A group of non-phagocytic white blood cells which combat microbial invasion, fight cancer, and neutralize toxic chemicals. The two classes of lymphocytes, B cells and T cells, are the heart of the immune system. (28, 30)

Lymphoid Organs Organs associated with production of blood cells and the lymphatic system, including the thymus, spleen, appendix, bone marrow, and lymph nodes. (30)

Lysis (1) To split or dissolve. (2) Cell bursting.

Lysomes A type of storage vesicle produced by the Golgi complex, containing hydrolytic (digestive) enzymes capable of digesting many kinds of macromolecules in the cell. The membrane around them keeps them sequestered. (5)

◀ **M** ▶

M Phase That portion of the cell cycle during which mitosis (nuclear division) and cytokinesis (cytoplasmic division) takes place. (10)

Macroevolution Evolutionary changes that lead to the appearance of new species. (33)

Macrofungus Filamentous fungus so named for the large size of its fleshy sexual structures; a mushroom, for example. (37)

Macromolecules Large polymers, such as proteins, nucleic acids, and polysaccharides. (4)

Macronutrients Nutrients required by plants in large amounts: carbon, oxygen, hydrogen, nitrogen, potassium, calcium, phosphorus, magnesium, and sulfur. (19)

Macrophages Phagocytic cells that develop from monocytes and present antigen to lymphocytes. (30)

Macroscopic Referring to biological observations made with the naked eye or a hand lens.

Mammals A class of vertebrates that possesses skin covered with hair and that nourishes their young with milk from mammary glands. (39)

Mammary Glands Glands contained in the breasts of mammalian mothers that produce breast milk. (39)

Marsupials Mammals with a cloaca whose young are born immature and complete their development in an external pouch in the mother's skin. (39)

Mass Extinction The simultaneous extinction of a multitude of species as the result of a drastic change in the environment. (33, 35)

Maternal Chromosomes The set of chromosomes in an individual that were inherited from the mother. (11)

Mechanoreceptors Sensory receptors that respond to mechanical pressure and detect motion, touch, pressure, and sound. (24)

Medulla The center-most portion of some organs. (23)

Medusa The motile, umbrella-shaped body form of some members of the phylum Cnidaria, with mouth and tentacles on the lower, concave service. (Compare with polyp.) (39)

Megaspores Spores that divide by mitosis to produce female gametophytes that produce the egg gamete. (20)

Meiosis The division process that produces cells with one-half the number of chromosomes in each somatic cell. Each resulting daughter cell is haploid (1N) (11)

Meiosis I A process of reductional division in which homologous chromosomes pair and then segregate. Homologues are partitioned into separate daughter cells. (11)

Meiosis II Second meiotic division. A division process resembling mitosis, except that the haploid number of chromosomes is present. After the chromosomes line up at the meta-phase plate, the two sister chromatids separate. (11)

Melanin A brown pigment that gives skin and hair its color (12)

Melanoma A deadly form of skin cancer that develops from pigment cells in the skin and is promoted by exposure to the sun. (14)

Memory Cells Lymphocytes responsible for active immunity. They recall a previous exposure to an antigen and, on subsequent exposure to the same antigen, proliferate rapidly into plasma cells and produce large quantities of antibodies in a short time. This protection typically lasts for many years. (30)

Mendelian Inheritance Transmission of genetic traits in a manner consistent with the principles discovered by Gregor Mendel. Includes traits controlled by simple dominant or recessive alleles; more complex patterns of transmission are referred to as Nonmendelian inheritance. (12)

Meninges The thick connective tissue sheath which surrounds and protects the vertebrate brain and spinal cord. (23)

Menstrual Cycle The repetitive monthly changes in the uterus that prepare the endometrium for receiving and supporting an embryo. (31)

Meristematic Region New cells arise from this undifferentiated plant tissue; found at root or shoot apical meristems, or lateral meristems. (18)

Meristems In plants, clusters of cells that retain their ability to divide, thereby producing new cells. One of the four basic tissues in plants. (18)

Mesoderm In animals, the middle germ cell layer of the gastrula. It gives rise to muscle, bone, connective tissue, gonads, and kidney. (32)

Mesopelagic Zone The dimly lit ocean zone beneath the epipelagic zone; large fishes, whales and squid occupy this zone; no phytoplankton occur in this zone. (40)

Mesophyll Layers of cells in a leaf between the upper and lower epidermis; produced by the ground meristem. (18)

Messenger RNA (mRNA) The RNA that carries genetic information from the DNA in the nucleus to the ribosomes in the cytoplasm, where the sequence of bases in the mRNA is translated into a sequence of amino acids. (14)

Metabolic Intermediates Compounds produced as a substrate are converted to end product in a series of enzymatic reactions. (6)

Metabolic Pathways Set of enzymatic reactions involved in either building or dismantling complex molecules. (6)

Metabolic Rate A measure of the level of activity of an organism usually determined by measuring the amount of oxygen consumed by an individual per gram body weight per hour. (22)

Metabolic Water Water produced as a product of metabolic reactions. (28)

Metabolism The sum of all the chemical reactions in an organism; includes all anabolic and catabolic reactions. (6)

Metamorphosis Transformation from one form into another form during development. (32)

Metaphase The stage of mitosis when the chromosomes line-up along the metaphase plate, a plate that usually lies midway between the spindle poles. (10)

Metaphase Plate Imaginary plane within a dividing cell in which the duplicated chromosomes become aligned during metaphase. (10)

Microbes Microscopic organisms. (36)

Microbiology The branch of biology that studies microorganisms. (36)

Microevolution Changes in allele frequency of a species' gene pool which has not generated new species. Exemplified by changes in the pigmentation of the peppered moth and by the acquisition of pesticide resistance in insects. (33)

Microfibrils Bundles formed from the intertwining of cellulose molecules, i.e., long chains of glucose molecules in the cell walls of plants. (5)

Microfilaments Thin actin-containing protein fibers that are responsible for maintenance of cell shape, muscle contraction and cyclosis. (5)

Micrometer One millionth (1/1,000,000) of a meter.

Micronutrients Nutrients required by plants in small amounts: iron, chlorine, copper, manganese, zinc, molybdenum, and boron. (19)

Micropyle A small opening in the integuments of the ovule through which the pollen tube grows to deliver sperm. (21)

Microspores Spores within anthers of flowers. They divide by mitosis to form pollen grains, the male gametophytes that produce the plant's sperm. (20)

Microtubules Thin, hollow tubes in cells; built from repeating protein units of tubulin. Microtubules are components of cilia, flagella, and the cytoskeleton. (5)

Microvilli The small projections on the cells that comprise each villus of the intestinal wall, further increasing the absorption surface area of the small intestine. (27)

Middle Intertidal Zone In the intertidal zone, the region which is covered and uncovered twice a day, the zero of tide tables. Organisms are submerged about 50% of the time. (40)

Migration Movements of a population into or out of an area. (44)

Mimicry A defense mechanism where one species resembles another in color, shape, behavior, or sound. (42)

Mineralocorticoids Steroid hormones which regulate the level of sodium and potassium in the blood. (25)

Mitochondria Organelles that contain the biochemical machinery for the Krebs cycle and the electron transport system of aerobic respiration. They are composed of two membranes, the inner one forming folds, or cristae. (9)

Mitosis The process of nuclear division producing daughter cells with exactly the same number of chromosomes as in the mother cell. (10)

Mitosis Promoting Factor (MPF) A protein that appears to be a universal trigger of cell division in eukaryotic cells. (10)

Mitotic Chromosomes Chromosomes whose DNA-protein threads have become coiled into microscopically visible chromosomes, each containing duplicated chromatids ready to be separated during mitosis. (10)

Molds Filamentous fungi that exist as colonies of threadlike cells but produce no macroscopic fruiting bodies. (37)

Molecule Chemical substance formed when two or more atoms bond together; the smallest unit of matter that possesses the qualities of a compound. (3)

Mollusca A phylum, second only to Arthropoda in diversity. Composed of three main classes: 1) Gastropoda (spiral-shelled), 2) Bivalvia (hinged shells), 3) Cephalopoda (with tentacles or arms and no, or very reduced shells). (39)

Molting (Ecdysis) Shedding process by which certain arthropods lose their exoskeletons as their bodies grow larger. (39)

Monera The taxonomic kingdom comprised of single-celled prokaryotes such as bacteria, cyanobacteria, and archebacteria. (36)

Monoclonal Antibodies Antibodies produced by a clone of hybridoma cells, all of which descended from one cell. (30)

Monocotyledae (Monocots) One of the two divisions of flowering plants, characterized by seeds with a single cotyledon, flower parts in 3s, parallel veins in leaves, many roots of approximately equal size, scattered vascular bundles in its stem anatomy, pith in its root anatomy, and no secondary growth capacity. (18)

Monocytes A type of leukocyte that gives rise to macrophages. (28)

Monoecious Both male and female reproductive structures are produced on the same sporophyte individual. (20, 38)

Monohybrid Cross A mating between two individuals that differ only in one genetically-determined trait. (12)

Monomers Small molecular subunits which are the building blocks of macromolecules. The macromolecules in living systems are constructed of some 40 different monomers. (4)

Monotremes A group of mammals that lay eggs from which the young are hatched. (39)

Morphogenesis The formation of form and internal architecture within the embryo brought about by such processes as programmed cell death, cell adhesion, and cell movement. (32)

Morphology The branch of biology that studies form and structure of organisms.

Mortality Death rate in a population or area. (43)

Motile Capable of independent movement.

Motor Neurons Nerve cells which carry outgoing impulses to their effectors, either glands or muscles. (23)

Mucosa The cell layer that lines the digestive tract and secretes a lubricating layer of mucus. (27)

Mullerian Mimicry Resemblance of different species, each of which is equally obnoxious to predators. (42)

Multicellular Consisting of many cells. (35)

Multichannel Food Chain Where the same primary producer supplies the energy for more than one food chain. (41)

Multiple Allele System Three or more possible alleles for a given trait, such as ABO blood groups in humans. (12)

Multiple Fission Division of the cell's nucleus without a corresponding division of cytoplasm.

Multiple Fruits Fruits that develop from pistils of separate flowers. (20)

Muscle Fiber A muiltinucleated skeletal muscle cell that results from the fusion of several pre-muscle cells during embryonic development. (26)

Muscle Tissue Bundles and sheets of contractile cells that shorten when stimulated, providing force for controlled movement. (26)

Mutagens Chemical or physical agents that induce genetic change. (14)

Mutation Random heritable changes in DNA that introduce new alleles into the gene pool. (14)

Mutualism The symbiotic interaction in which both participants benefit. (42)

Mycology The branch of biology that studies fungi. (37)

Mycorrhizae An association between soil fungi and the roots of vascular plants, increasing the plant's ability to extract water and minerals from the soil. (19)

Myelin Sheath In vertebrates, a jacket which covers the axons of high-velocity neurons, thereby increasing the speed of a neurological impulse. (23)

Myofibrils In striated muscle, the banded fibrils that lie parallel to each other, constituting the bulk of the muscle fiber's interior and powering contraction. (26)

Myosin A contractile protein that makes up the major component of the thick filaments of a muscle cell and is also present in nonmuscle cells. (26)

◄ **N** ►

NADPH Nicotinamide adenine dinucleotide phosphate. NADPH is formed by reduction of $NADP^+$, and serves as a store of electrons for use in metabolism (see Reducing Power). (9)

NAD⁺ Nicotinamide adenine dinucleotide. A coenzyme that functions as an electron carrier in metabolic reactions. When reduced to NADH, the molecule becomes a cellular energy source. (9)

Natality Birthrate in a population or area. (43)

Natural Killer (NK) Cells Nonspecific, lymphocytelike cells which destroy foreign cells and cancer cells. (30)

Natural Selection Differential survival and reproduction of organisms with a resultant increase in the frequency of those best adapted to the environment. (33)

Neanderthals A subspecies of Homo sapiens different from that of modern humans that were characterized by heavy bony skeletons and thick bony ridges over the eyes. They disappeared about 35,000 years ago. (34)

Nectary Secretory gland in flowering plants containing sugary fluid that attracts pollinators as a food source. Usually located at the base of the flower. (20)

Negative Feedback Any regulatory mechanism in which the increased level of a substance inhibits further production of that substance, thereby preventing harmful accumulation. A type of homeostatic mechanism. (22, 25)

Negative Gravitropism In plants, growth against gravitational forces, or shoot growth upward. (21)

Nematocyst Within the stinging cell (cnidocyte) of cnidarians, a capsule that contains a coiled thread which, when triggered, harpoons prey and injects powerful toxins. (39)

Nematoda The widespread and abundant animal phylum containing the roundworms. (39)

Nephridium A tube surrounded by capillaries found in an organism's excretory organs that removes nitrogenous wastes and regulates the water and chemical balance of body fluids. (28)

Nephron The functional unit of the vertebrate kidney, consisting of the glomerulus, Bowman's capsule, proximal and distal convoluted tubules, and loop of Henle. (28)

Nerve Parallel bundles of neurons and their supporting cells. (23)

Nerve Impulse A propagated action potential. (23)

Nervous Tissue Excitable cells that receive stimuli and, in response, transmit an impulse to another part of the animal. (23)

Neural Plate In vertebrates, the flattened plate of dorsal ectoderm of the late gastrula that gives rise to the nervous system. (32)

Neuroglial Cells Those cells of a vertebrate nervous system that are not neurons. Includes a variety of cell types including Schwann cells. (23)

Neuron A nerve cell. (23)

Neurosecretory Cells Nervelike cells that secrete hormones rather than neurotransmitter substances when a nerve impulse reaches the distal end of the cell. In vertebrates, these cells arise from the hypothalamus. (25)

Neurotoxins Substances, such as curare and tetanus toxin, that interfere with the transmission of neural impulses. (23)

Neurotransmitters Chemicals released by neurons into the synaptic cleft, stimulating or inhibiting the post-synaptic target cell. (23)

Neurulation Formation by embryonic induction of the neural tube in a developing vertebrate embryo. (32)

Neutrons Electrically neutral (uncharged) particles contained within the nucleus of the atom. (3)

Neutrophil Phagocytic leukocyte, most numerous in the human body. (28)

Niche An organism's habitat, role, resource requirements, and tolerance ranges for each abiotic condition. (42)

Niche Breadth Relative size and dimension of ecological niches; for example, broad or narrow niches. (41)

Niche Overlap Organisms that have the same habitat, role, environmental requirements, or needs. (41)

Nitrogen Fixation The conversion of atmospheric nitrogen gas N_2 into ammonia (NH_3) by certain bacteria and cyanobacteria. (19)

Nitrogenous Wastes Nitrogen-containing metabolic waste products, such as ammonia or urea, that are produced by the breakdown of proteins and nucleic acids. (28)

Nodes The attachment points of leaves to a stem. (18)

Nodes of Ranvier Uninsulated (nonmyelinated) gaps along the axon of a neuron. (23)

Noncovalent Bonds Linkages between two atoms that depend on an attraction between positive and negative charges between molecules or ions. Includes ionic and hydrogen bonds. (3)

Non-Cyclic Photophosphorylation The pathway in the light reactions of photosynthesis in which electrons pass from water, through two photosystems, and then ultimately to NADP⁺. During the process, both ATP and NADPH are produced. It is so named because the electrons do not return to their reaction center. (8)

Nondisjunction Failure of chromosomes to separate properly at meiosis I or II. The result is that one daughter will receive an extra chromosome and the other gets one less. (11, 13)

Nonpolar Molecules Molecules which have an equal charge distribution throughout their structure and thus lack regions with a localized positive or negative charge. (3)

Notochord A flexible rod that is below the dorsal surface of the chordate embryo, beneath the nerve cord. In most chordates, it is replaced by the vertebral column. (32)

Nuclear Envelope A double membrane pierced by pores that separates the contents of the nucleus from the rest of the eukaryotic cell. (5)

Nucleic Acids DNA and RNA; linear polymers of nucleotides, responsible for the storage and expression of genetic information. (4, 14)

Nucleoid A region in the prokaryotic cell that contains the genetic material (DNA). It is unbounded by a nuclear membrane. (36)

Nucleoplasm The semifluid substance of the nucleus in which the particulate structures are suspended. (5)

Nucleosomes Nuclear protein complex consisting of a length of DNA wrapped around a central cluster of 8 histones. (14)

Nucleotides Monomers of which DNA and RNA are built. Each consists of a 5-carbon sugar, phosphate, and a nitrogenous base. (4)

Nucleous (pl. nucleoli) One or more darker regions of a nucleus where each ribosomal subunit is assembled from RNA and protein. (5)

Nucleus The large membrane-enclosed organelle that contains the DNA of eukaryotic cells. (5)

Nucleus, Atomic The center of an atom containing protons and neutrons. (3)

◄ O ►

Obligate Symbiosis A symbiotic relationship between two organisms that is necessary for the survival or both organisms. (42)

Olfaction The sense of smell. (24)

Oligotrophic Little nourished, as a young lake that has few nutrients and supports little life. (40)

Omnivore An animal that obtains its nutritional needs by consuming plants and other animals. (42)

Oncogene A gene that causes cancer, perhaps activated by mutation or a change in its chromosomal location. (10)

Oocyte A female germ cell during any of the stages of meiosis. (31)

Oogenesis The process of egg production. (31)

Oogonia Female germ cells that have not yet begun meiosis. (31)

Open Circulatory System Circulatory system in which blood travels from vessels to tissue spaces, through which it percolates prior to returning to the main vessel (compare with closed circulatory system). (28)

Operator A regulatory gene in the operon of bacteria. It is the short DNA segment to which the repressor binds, thus preventing RNA polymerase from attaching to the promoter. (15)

Operon A regulatory unit in prokaryotic cells that controls the expression of structural genes. The operon consists of structural genes that produce enzymes for a particular metabolic pathway, a regulator region composed of a promoter and an operator, and R (regulator) gene that produces a repressor. (15)

Order A level of the taxonomic hierarchy that groups together members of related families. (1)

Organ Body part composed of several tissues that performs specialized functions. (22)

Organelle A specialized part of a cell having some particular function. (5)

Organic Compounds Chemical compounds that contain carbon. (4)

Organism A living entity able to maintain its organization, obtain and use energy, reproduce, grow, respond to stimuli, and display homeostatis. (1)

Organogenesis Organ formation in which two or more specialized tissue types develop in a precise temporal and spatial relationship to each other. (32)

Organ System Group of functionally related organs. (22)

Osmoregulation The maintenance of the proper salt and water balance in the body's fluids. (28)

Osmosis The diffusion of water through a differentially permeable membrane into a hypertonic compartment. (7)

Ossification Synthesis of a new bone. (26)

Osteoclast A type of bone cell which breaks down the bone, thereby releasing calcium into the bloodstream for use by the body. Osteoclasts are activated by hormones released by the parathyroid glands. (26)

Osteocytes Living bone cells embedded within the calcified matrix they manufacture. (26)

Osteoporosis A condition present predominantly in postmenopausal women where the bones are weakened due to an increased rate of bone resorption compared to bone formation. (26)

Ovarian Cycle The cycle of egg production within the mammalian ovary. (31)

Ovarian Follicle In a mammalian ovary, a chamber of cells in which the oocyte develops. (31)

Ovary In animals, the egg-producing gonad of the female. In flowering plants, the enlarged base of the pistil, in which seeds develop. (20)

Oviduct (Fallopian Tube) The tube in the female reproductive organ that connects the ovaries and uterus and where fertilization takes place. (31)

Ovulation The release of an egg (ovum) from the ovarian follicle. (31)

Ovule In seed plants, the structure containing the female gametophyte, nucellus, and integuments. After fertilization, the ovule develops into a seed. (20, 38)

Ovum An unfertilized egg cell; a female gamete. (31)

Oxidation The removal of electrons from a compound during a chemical reaction. For a carbon atom, the fewer hydrogens bonded to a carbon, the greater the oxidation state of the atom. (6)

Oxidative Phosphorylation The formation of ATP from ADP and inorganic phosphate that occurs in the electron-transport chain of cellular respiration. (8, 9)

Oxyhemoglobin A complex of oxygen and hemoglobin, formed when blood passes through the lungs and is dissociated in body tissues, where oxygen is released. (29)

Oxytocin A female hormone released by the posterior pituitary which triggers uterine contractions during childbirth and the release of milk during nursing. (25)

◄ P ►

P680 Reaction Center (P = Pigment) Special chlorophyll molecule in Photosystem II that traps the energy absorbed by the other pigment molecules. It absorbs light energy maximally at 680 nm. (8)

Palisade Parenchyma In dicot leaves, densely packed, columnar shaped cells functioning in photosynthesis. Found just beneath the upper epidermis. (18)

Pancreas In vertebrates, a large gland that produces digestive enzymes and hormones. (27)

Parallel Evolution When two species that have descended from the same ancestor independently acquire the same evolutionary adaptations. (33)

Parapatric Speciation The splitting of a population into two species' populations under conditions where the members of each population reside in adjacent areas. (33)

Parasite An organism that lives on or inside another called a host, on which it feeds. (39, 42)

Parasitism A relationship between two organisms where one organism benefits, and the other is harmed. (42)

Parasitoid Parasitic organisms, such as some insect larvae, which kill their host. (42)

Parasympathetic Nervous System Part of the autonomic nervous system active during relaxed activity. (23)

Parathyroid Glands Four glands attached to the thyroid gland which secrete parathyroid hormone (PTH). When blood calcium levels are low, PTH is secreted, causing calcium to be released from bone. (25)

Parenchyma The most prevalent cell type in herbaceous plants. These thin-walled, polygonal-shaped cells function in photosynthesis and storage. (18)

Parthenogenesis Process by which offspring are produced without egg fertilization. (31)

Passive Immunity Immunity achieved by receiving antibodies from another source, as occurs with a newborn infant during nursing. (30)

Paternal Chromosomes The set of chromosomes in an individual that were inherited from the father. (11)

Pathogen A disease-causing microorganism. (36)

Pectoral Girdle In humans, the two scapulae (shoulder blades) and two clavicles (collarbones) which support and articulate with the bones of the upper arm. (26)

Pedicel A shortened stem carrying a flower. (20)

Pedigree A diagram showing the inheritance of a particular trait among the members of the family. (13)

Pelagic Zone The open oceans, divided into three layers: 1) photo- or epipelagic (sunlit), 2) mesopelagic (dim light), 3) aphotic or bathypelagic (always dark). (40)

Pelvic Girdle The complex of bones that connect a vertebrate's legs with its backbone. (26)

Penis An intrusive structure in the male animal which releases male gametes into the female's sex receptacle. (31)

Peptide Bond The covalent bond between the amino group of one amino acid and the carboxyl group of another. (4)

Peptidoglycan A chemical component of the prokaryotic cell wall. (36)

Percent Annual Increase A measure of population increase; the number of individuals (people) added to the population per 100 individuals. (43)

Perennials Plants that live longer than two years. (18)

Perfect Flower Flowers that contain both stamens and pistils. (20)

Perforation Plate In plants, that portion of the wall of vessel members that is perforated, and contains an area with neither primary nor secondary cell wall; a "hole" in the cell wall. (18)

Pericycle One or more layers of cells found in roots, with phloem or xylem to its' inside, and the endodermis to its' outside. Functions in producing lateral roots and formation of the vascular cambium in roots with secondary growth. (18)

Periderm Secondary tissue that replaces the epidermis of stems and roots. Consists of cork, cork cambium, and an internal layer of parenchyma cells. (18)

Peripheral Nervous System Neurons, excluding those of the brain and spinal cord, that permeate the rest of the body. (23)

Peristalsis Sequential waves of muscle contractions that propel a substance through a tube. (27)

Peritoneum The connective tissue that lines the coelomic cavities. (39)

Permeability The ability to be penetrable, such as a membrane allowing molecules to pass freely across it. (7)

Petal The second whorl of a flower, often brightly colored to attract pollinators; collectively called the corolla. (20)

Petiole The stalk leading to the blade of a leaf. (18)

pH A scale that measures the concentration of hydrogen ions in a solution. The pH scale extends from 0 to 14. Acidic solutions have a pH of less than 7; alkaline solutions have a pH above 7; neutral solutions have a pH equal to 7. (3)

Phagocytosis Engulfing of food particles and foreign cells by amoebae and white blood cells. A type of endocytosis. (5)

Pharyngeal Pouches In the vertebrate embryo, outgrowths from the walls of the pharynx that break through the body surface to form gill slits. (32)

Pharynx The throat; a portion of both the digestive and respiratory system just behind the oral cavity. (29)

Phenotype An individual's observable characteristics that are the expression of its genotype. (12)

Pheromones Chemicals that, when released by an animal, elicit a particular behavior in other animals of the same species. (44)

Phloem The vascular tissue that transports sugars and other organic molecules from sites of photosynthesis and storage to the rest of the plant. (18)

Phloem Loading The transfer of assimilates to phloem conducting cells, from photosynthesizing source cells. (19)

Phloem Unloading The transfer of assimilates to storage (sink) cells, from phloem conducting cells. (19)

Phospholipids Lipids that contain a phosphate and a variable organic group that form polar, hydrophilic regions on an otherwise nonpolar, hydrophobic molecule. They are the major structural components of membranes. (4)

Phosphorylation A chemical reaction in which a phosphate group is added to a molecule or atom. (6)

Photoexcitation Absorption of light energy by pigments, causing their electrons to be raised to a higher energy level. (8)

Photolysis The splitting of water during photosynthesis. The electrons from water pass to Photosystem II, the protons enter the lumen of the thylakoid and contribute to the proton gradient across the thylakoid membrane, and the oxygen is released into the atmosphere. (8)

Photon A particle of light energy. (8)

Photoperiod Specific lengths of day and night which control certain plant growth responses to light, such as flowering or germination. (21)

Photoperiodism Changes in the behavior and physiology of an organism in response to the relative lengths of daylight and darkness, i.e., the photoperiod. (21)

Photoreceptors Sensory receptors that respond to light. (24)

Photorespiration The phenomenon in which oxygen binds to the active site of a CO_2-fixing enzyme, thereby competing with CO_2 fixation, and lowering the rate of photosynthesis. (8)

Photosynthesis The conversion by plants of light energy into chemical energy stored in carbohydrate. (8)

Photosystems Highly organized clusters of photosynthetic pigments and electron/hydrogen carriers embedded in the thylakoid membranes of chloroplasts. There are two photosystems, which together carry out the light reactions of photosynthesis. (8)

Photosystem I Photosystem with a P700 reaction center; participates in cyclic photophosphorylation as well as in noncyclic photophosphorylation. (8)

Photosystem II Photosystem activated by a P680 reaction center; participates only in noncyclic photophosphorylation and is associated with photolysis of water. (8)

Phototropism The growth responses of a plant to light. (21)

Phyletic Evolution The gradual evolution of one species into another. (33)

Phylogeny Evolutionary history of a species. (35)

Phylum The major taxonomic divisions in the Animal kingdom. Members of a phylum share common, basic features. The Animal kingdom is divided into approximately 35 phyla. (39)

Physiology The branch of biology that studies how living things function. (22)

Phytochrome A light-absorbing pigment in plants which controls many plant responses, including photoperiodism. (21)

Phytoplankton Microscopic photosynthesizers that live near the surface of seas and bodies of fresh water. (37)

Pineal Gland An endocrine gland embedded within the brain that secretes the hormone melatonin. Hormone secretion is dependent on levels of environmental light. In amphibians and reptiles, melatonin controls skin coloration. In humans, pineal secretions control sexual maturation and daily rhythms. (25)

Pinocytosis Uptake of small droplets and dissolved solutes by cells. A type of endocytosis. (5)

Pistil The female reproductive part and central portion of a flower, consisting of the ovary, style and stigma. May contain one carpel, or one or more fused carpels. (20)

Pith A plant tissue composed of parenchyma cells, found in the central portion of primary growth stems of dicots, and monocot roots. (18)

Pith Ray Region between vascular bundles in vascular plants. (18)

Pituitary Gland (see **Posterior and Anterior Pituitary**).

Placenta In mammals (exclusive of marsupials and monotremes), the structure through which nutrients and wastes are exchanged between the mother and embryo/fetus. Develops from both embryonic and uterine tissues. (32)

Plant Multicellular, autotrophic organism able to manufacture food through photosynthesis. (38)

Plasma In vertebrates, the liquid portion of the blood, containing water, proteins (including fibrinogen), salts, and nutrients. (28)

Plasma Cells Differentiated antibody-secreting cells derived from B lymphocytes. (30)

Plasma Membrane The selectively permeable, molecular boundary that separates the cytoplasm of a cell from the external environment. (5)

Plasmid A small circle of DNA in bacteria in addition to its own chromosome. (16)

Plasmodesmata Openings between plant cell walls, through which adjacent cells are connected via cytoplasmic threads. (19)

Plasmodium Genus of protozoa that causes malaria. (37)

Plasmodium A huge multinucleated "cell" stage of a plasmodial slime mold that feeds on dead organic matter. (37)

Plasmolysis The shrinking of a plant cell away from its cell wall when the cell is placed in a hypertonic solution. (7)

Platelets Small, cell-like fragments derived from special white blood cells. They function in clotting. (28)

Plate Tectonics The theory that the earth's crust consists of a number of rigid plates that rest on an underlying layer of semimolten rock. The movement of the earth's plates results from the upward movement of molten rock into the solidified crust along ridges within the ocean floor. (35)

Platyhelminthes The phylum containing simple, bilaterally symmetrical animals, the flatworms. (39)

Pleiotropy Where a single mutant gene produces two or more phenotypic effects. (12)

Pleura The double-layered sac which surrounds the lungs of a mammal. (29, 39)

Pneumocytis Pneumonia (PCP) A disease of the respiratory tract caused by a protozoan that strikes persons with immunodeficiency diseases, such as AIDS. (30)

Point Mutations Changes that occur at one point within a gene, often involving one nucleotide in the DNA. (14)

Polar Body A haploid product of meiosis of a female germ cell that has very little cytoplasm and disintegrates without further function. (31)

Polar Molecule A molecule with an unequal charge distribution that creates distinct positive and negative regions or poles. (3)

Pollen The male gametophyte of seed plants, comprised of a generative nucleus and a tube nucleus surrounded by a tough wall. (20)

Pollen Grain The male gametophyte of conifers and angiosperms, containing male gametes. In angiosperms, pollen grains are contained in the pollen sacs of the anther of a flower. (20)

Pollination The transfer of pollen grains from the anther of one flower to the stigma of another. The transfer is mediated by wind, water, insects, and other animals. (20)

Polygenic Inheritance An inheritance pattern in which a phenotype is determined by two or more genes at different loci. In humans, examples include height and pigmentation. (12)

Polymer A macromolecule formed of monomers joined by covalent bonds.. Includes proteins, polysaccharides, and nucleic acids. (4)

Polymerase Chain Reaction (PCR) Technique to amplify a specific DNA molecule using a temperature-sensitive DNA polymerase obtained from a heat-resistant bacterium. Large numbers of copies of the initial DNA molecule can be obtained in a short period of time, even when the starting material is present in vanishingly small amounts, as for example from a blood stain left at the scene of a crime. (16)

Polymorphic Property of some protozoa to produce more than one stage of organism as they complete their life cycles. (37)

Polymorphic Genes Genes for which several different alleles are known, such as those that code for human blood type. (17)

Polyp Stationary body form of some members of the phylum Cnidaria, with mouth and tentacles facing upward. (Compare with medusa.) (39)

Polypeptide An unbranched chain of amino acids covalently linked together and assembled on a ribosome during translation. (4)

Polyploidy An organism or cell containing three or more complete sets of chromosomes. Polyploidy is rare in animals but common in plants. (33)

Polysaccharide A carbohydrate molecule consisting of monosaccharide units. (4)

Polysome A complex of ribosomes found in chains, linked by mRNA. Polysomes contain the ribosomes that are actively assembling proteins. (14)

Polytene Chromosomes Giant banded chromosomes found in certain insects that form by the repeated duplication of DNA. Because of the multiple copies of each gene in a cell, polytene chromosomes can generate large amounts of a gene product in a short time. Transcription occurs at sites of chromosome puffs. (13)

Pond Body of standing fresh water, formed in natural depressions in the Earth. Ponds are smaller than lakes. (40)

Population Individuals of the same species inhabiting the same area. (43)

Population Density The number of individual species living in a given area. (43)

Positive Gravitropism In plants, growth with gravitational forces, or root growth downward. (21)

Posterior Pituitary A gland which manufactures no hormones but receives and later releases hormones produced by the cell bodies of neurons in the hyopthalamus. (25)

Potential Energy Stored energy, such as occurs in chemical bonds. (6)

Preadaptation A characteristic (adaptation) that evolved to meet the needs of an organism in one type of habitat, but fortuitously allows the organism to exploit a new habitat. For example, lobed fins and lungs evolved in ancient fishes to help them live in shallow, stagnant ponds, but also facilitated the evolution of terrestrial amphibians. (33, 39)

Precells Simple forerunners of cells that, presumably, were able to concentrate organic molecules, allowing for more frequent molecular reactions. (35)

Predation Ingestion of prey by a predator for energy and nutrients. (42)

Predator An organism that captures and feeds on another organism (prey). (42)

Pressure Flow In the process of phloem loading and unloading, pressure differences resulting from solute increases in phloem conducting cells and neighboring xylem cells cause the flow of water to phloem. A concentration gradient is created between xylem and phloem cells. (19)

Prey An organism that is captured and eaten by another organism (predator). (42)

Primary Consumer Organism that feeds exclusively on producers (plants). Herbivores are primary consumers. (41)

Primary Follicle In the mammalian ovary, a structure composed of an oocyte and its surrounding layer of follicle cells. (31)

Primary Growth Growth from apical meristems, resulting in an increase in the lengths of shoots and roots in plants. (18)

Primary Immune Response Process of antibody production following the first exposure to an antigen. There is a lag time from exposure until the appearance in the blood of protective levels of antibodies. (30)

Primary Oocyte Female germ cell that is either in the process of or has completed the first meiotic division. In humans, germ cells may remain in this stage in the ovary for decades. (31)

Primary Producers All autotrophs in a biotic environment that use sunlight or chemical energy to manufacture food from inorganic substances. (41)

Primary Sexual Characteristics Gonads, reproductive tracts, and external genitals. (31)

Primary Spermatocyte Male germ cell that is either in the process of or has completed the first meiotic division. (31)

Primary Succession The development of a community in an area previously unoccupied by any community; for example, a "bare" area such as rock, volcanic material, or dunes. (41)

Primary Tissues Tissues produced by primary meristems of a plant, which arise from the shoot and root apical meristems. In general, primary tissues are a result of an increase in plant length. (18)

Primary Transcript An RNA molecule that has been transcribed but not yet subjected to any type of processing. The primary transcript corresponds to the entire stretch of DNA that was transcribed. (15)

Primates Order of mammals that includes humans, apes, monkeys, and lemurs. (39)

Primitive An evolutionary early condition. Primitive features are those that were also present in an early ancestor, such as five digits on the feet of terrestrial vertebrates. (34)

Prions An infectious particle that contains protein but no nucleic acid. It causes slow diseases of animals, including neurological disease of humans. (36)

Processing-Level Control Control of gene expression by regulating the pathway by which a primary RNA transcript is processed into an mRNA. (15)

Products In a chemical reaction, the compounds into which the reactants are transformed. (6)

Profundal Zone Deep, open water of lakes, where it is too dark for photosynthesis to occur. (40)

Progesterone A hormone produced by the corpus luteum within the ovary. It prepares and maintains the uterus for pregnancy, participates in milk production, and prevents the ovary from releasing additional eggs late in the cycle or during pregnancy. (25)

Prokaryotic Referring to single-celled organisms that have no membrane separating the DNA from the cytoplasm and lack membrane-enclosed organelles. Prokaryotes are confined to the kingdom Monera; they are all bacteria. (36)

Prokaryotic Fission The most common type of cell division in bacteria (prokaryotes). Duplicated DNA strands are attached to the plasma membrane and become separated into two cells following membrane growth and cell wall formation. (10, 36)

Prolactin A hormone produced by the anterior pituitary, stimulating milk production by mammary glands. (25)

Promoter A short segment of DNA to which RNA polymerase attaches at the start of transcription. (15)

Prophase Longest phase of mitosis, involving the formation of a spindle, coiling of chromatin fibers into condensed chromosomes, and movement of the chromosomes to the center of the cell. (10)

Prostaglandins Hormones secreted by endocrine cells scattered throughout the body responsible for such diverse functions as contraction of uterine muscles, triggering the inflammatory response, and blood clotting. (25)

Prostate Gland A muscular gland which produces and releases fluids that make up a substantial portion of the semen. (31)

Proteins Long chains of amino acids, linked together by peptide bonds. They are folded into specific shapes essential to their functions. (4)

Prothallus The small, heart-shaped gametophyte of a fern. (38)

Protists A member of the kingdom Protista; simple eukaryotic organisms that share broad taxonomic similarities. (36, 37)

Protocooperation Non-compulsory interactions that benefit two organisms, e.g., lichens. (42)

Proton Gradient A difference in hydrogen ion (proton) concentration on opposite sides of a membrane. Proton gradients are formed during photosynthesis and respiration and serve as a store of energy used to drive ATP formation. (8, 9)

Protons Positively charged particles within the nucleus of an atom. (3)

Protostomes One path of development exhibited by coelomate animals (e.g., mollusks, annelids, and arthropods). (39)

Protozoa Member of protist kingdom that is unicellular and eukaryotic; vary greatly in size, motility, nutrition and life cycle. (37)

Provirus DNA copy of a virus' nucleic acid that becomes integrated into the host cell's chromosome. (36)

Pseudocoelamates Animals in which the body cavity is not lined by cells derived from mesoderm. (39)

Pseudopodia (psuedo = false, pod = foot). Pseudopodia are fingerlike extensions of cytoplasm that flow forward from the "body" of an amoeba; the rest of the cell then follows. (37)

Puberty Development of reproductive capacity, often accompanied by the appearance of secondary sexual characteristics. (31)

Pulmonary Circulation The loop of the circulatory system that channels blood to the lungs for oxygenation. (28)

Punctuated Equilibrium Theory A theory to explain the phenomenon of the relatively sudden appearance of new species, followed by long periods of little or no change. (33)

Punnett Square Method A visual method for predicting the possible genotypes and their expected ratios from a cross. (12)

Pupa In insects, the stage in metamorphosis between the larva and the adult. Within the pupal case, there is dramatic transformation in body form as some larval tissues die and others differentiate into those of the adult. (32)

Purine A nitrogenous base found in DNA and RNA having a double ring structure. Adenine and guanine are purines. (14)

Pyloric Sphincter Muscular valve between the human stomach and small intestine. (27)

Pyrimidine A nitrogenous base found in DNA and RNA having a single ring structure. Cytosine, thymine, and uracil are pyrimidines. (14)

Pyramid of Biomass Diagrammatic representation of the total dry weight of organisms at each trophic level in a food chain or food web. (41)

Pyramid of Energy Diagrammatic representation of the flow of energy through trophic levels in a food chain or food web. (41)

Pyramid of Numbers Similar to a pyramid of energy, but with numbers of producers and consumers given at each trophic level in a food chain or food web. (41)

◄ Q ►

Quiescent Center The region in the apical meristem of a root containing relatively inactive cells. (18)

◀ R ▶

R-Group The variable portion of a molecule. (4)

r-Selected Species Species that possess adaptive strategies to produce numerous offspring at once. (43)

Radial Symmetry The quality possessed by animals whose bodies can be divided into mirror images by more than one median plane. (39)

Radicle In the plant embryo, the tip of the hypocotyl that eventually develops into the root system. (20)

Radioactivity A property of atoms whose nucleus contains an unstable combination of particles. Breakdown of the nucleus causes the emission of particles and a resulting change in structure of the atom. Biologists use this property to track labeled molecules and to determine the age of fossils. (3)

Radiodating The use of known rates of radioactive decay to date a fossil or other ancient object. (3, 34)

Radioisotope An isotope of an element that is radioactive. (3)

Radiolarian A prozoan member of the protistan group Sarcodina that secretes silicon shells through which it captures food.

Rainshadow The arid, leeward (downwind) side of a mountain range. (40)

Random Distribution Distribution of individuals of a population in a random manner; environmental conditions must be similar and individuals do not affect each other's location in the population. (43)

Reactants Molecules or atoms that are changed to products during a chemical reaction. (6)

Reaction A chemical change in which starting molecules (reactants) are transformed into new molecules (products). (6)

Reaction Center A special chlorophyll molecule in a photosystem (P_{700} in Photosystem I, P_{680} in Photosystem II). (8)

Realized Niche Part of the fundamental niche of an organism that is actually utilized. (41)

Receptacle The base of a flower where the flower parts are attached; usually a widened area of the pedicel. (20)

Receptor-Mediated Endocytosis The uptake of materials within a cytoplasmic vesicle (endocytosis) following their binding to a cell surface receptor. (7)

Receptor Site A site on a cell's plasma membrane to which a chemical such as a hormone binds. Each surface site permits the attachment of only one kind of hormone. (5)

Recessive An allele whose expression is masked by the dominant allele for the same trait. (12)

Recombinant DNA A DNA molecule that contains DNA sequences derived from different biological sources that have been joined together in the laboratory. (16)

Recombination The rejoining of DNA pieces with those of a different strand or with the same strand at a point different from where the break occurred. (11, 13)

Red Marrow The soft tissue in the interior of bones that produces red blood cells. (26)

Red Tide Growth of one of several species of reddish brown dinoflagellate algae so extensive that it tints the coastal waters and inland lakes a distinctive red color. Often associated with paralytic shellfish poisoning (see dinoflagellates). (37)

Reducing Power A measure of the cell's ability to transfer electrons to substrates to create molecules of higher energy content. Usually determined by the available store of NADPH, the molecule from which electrons are transferred in anabolic (synthetic) pathways. (6)

Reduction The addition of electrons to a compound during a chemical reaction. For a carbon atom, the more hydrogens that are bonded to the carbon, the more reduced the atom. (6)

Reduction Division The first meiotic division during which a cell's chromosome number is reduced in half. (11)

Reflex An involuntary response to a stimulus. (23)

Reflex Arc The simplest example of central nervous system control, involving a sensory neuron, motor neuron, and usually an interneuron. (23)

Regeneration Ability of certain animals to replace injured or lost limbs parts by growth and differentiation of undifferentiated stem cells. (15)

Region of Elongation In root tips, the region just above the region of cell division, where cells elongate and the root length increases. (18)

Region of Maturation In root tips, the region above the region of elongation; cells differentiate and root hairs occur in this region. (18)

Regulatory Genes Genes whose sole function is to control the expression of structural genes. (15)

Releaser A sign stimulus that is given by an individual to another member of the same species, eliciting a specific innate behavior. (44)

Releasing Factors Hormones secreted by the tips of hypothalmic neurosecretory cells that stimulate the anterior pituitary to release its hormones. GnRH, for example, stimulates the release of gonadotropins. (25)

Renal Referring to the kidney. (28)

Replication Duplication of DNA, usually prior to cell division. (14)

Replication Fork The site where the two strands of a DNA helix are unwinding during replication. (14)

Repression Inhibition of transcription of a gene which, in an operon, occurs when repressor protein binds to the operator. (15)

Repressor Protein encoded by a bacterial regulatory gene that binds to an operator site of an operon and inhibits transcription. (15)

Reproduction The process by which an organism produces offspring. (31)

Reproductive Isolation Phenomenon in which members of a single population become split into two populations that no longer interbreed. (33)

Reproductive System System of specialized organs that are utilized for the production of gametes and, in some cases, the fertilization and/or development of an egg. (31)

Reptiles Members of class Reptilia, scaly, air-breathing, egg-laying vertebrates such as lizards, snakes, turtles, and crocodiles. (39)

Resolving Power The ability of an optical instrument (eye, microscopes) to discern whether two very close objects are separate from each other. (APP.)

Resource Partitioning Temporal or spatial sharing of a resource by different species. (42)

Respiration Process used by organisms to exchange gases with the environment; the source of oxygen required for metabolism. The process organisms use to oxidize glucose to CO_2 and H_2O using an electron transport system to extract energy from electrons and store it in the high-energy bonds of ATP. (29)

Respiratory System The specialized set of organs that function in the uptake of oxygen from the environment. (29)

Resting Potential The electrical potential (voltage) across the plasma membrane of a neuron when the cell is not carrying an impulse. Results from a difference in charge across the membrane. (23)

Restriction Enzyme A DNA-cutting enzyme found in bacteria. (16)

Restriction Fragment Length Polymorphism (RFLP) Certain sites in the DNA tend to have a highly variable sequence from one individual to another. Because of these differences, restriction enzymes cut the DNA from different individuals into fragments of different length. Variations in the length of particular fragments (RFLPs) can be used as genetic signposts for the identification of nearby genes of interest. (17)

Restriction Fragments The DNA fragments generated when purified DNA is treated with a particular restriction enzyme. (16)

Reticular Formation A series of interconnected sites in the core of the brain (brainstem) that selectively arouse conscious activity. (23)

Retroviruses RNA viruses that reverse the typical flow of genetic information; within the infected cell, the viral DNA serves as a template for synthesis of a DNA copy. Examples include HIV, which causes AIDS, and certain cancer viruses. (36)

Reverse Genetics Determining the amino acid sequence and function of a polypeptide from the nucleotide sequence of the gene that codes for that polypeptide. (17)

Reverse Transcriptase An enzyme present in retroviruses that transcriibes a strand of DNA, using viral RNA as the template. (36)

Rhizoids Slender cells that resemble roots but do not absorb water or minerals. (36)

Rhodophyta Red algae; seaweeds that can absorb deeper penetrating light rays than most aquatic photosynthesizers. (36)

Rhyniophytes Ancient plants having vascular tissue which thrived in marshy areas during the Silurian period.

Ribonucleic Acid (RNA) Single-stranded chain of nucleotides each comprised of ribose (a sugar), phosphate, and one of four bases (adenine, guanine, cytosine, and uracil). The sequence of nucleotides in RNA is dictated by DNA, from which it is transcribed. There are three classes of RNA: mRNA, tRNA, and rRNA, all required for protein synthesis. (4, 14)

Ribosomal RNA (rRNA) RNA molecules that form part of the ribosome. Included among the rRNAs is one that is thought to catalyze peptide bond formation. (14)

Ribosomes Organelles involved in protein synthesis in the cytoplasm of the cell. (14)

Ribozymes RNAs capable of catalyzing a chemical reaction, such as peptide bond formation or RNA cutting and splicing. (15)

Rickettsias A group of obligate intracellular parasites, smaller than the typical prokaryote. They cause serious diseases such as typhus. (36)

River Flowing body of surface fresh water; rivers are formed from the convergence of streams. (40)

RNA Polymerase The enzyme that directs transcription and assembling RNA nucleotides in the growing chain. (14)

RNA Processing The process by which the intervening (noncoding) portions of a primary RNA transcript are removed and the remaining (coding) portions are spliced together to form an mRNA. (15)

Root Cap A protective cellular helmet at the tip of a root that surrounds delicate meristematic cells and shields them from abrasion and directs the growth downward. (18)

Root Hairs Elongated surface cells near the tip of each root for the absorption of water and minerals. (18)

Root Nodules Knobby structures on the roots of certain plants. They house nitrogen-fixing bacteria which supply nitrogen in a form that can be used by the plant. (19)

Root Pressure A positive pressure as a result of continuous water supply to plant roots that assists (along with transpirational pull) the pushing of water and nutrients up through the xylem. (19)

Root System The below-ground portion of a plant, consisting of main roots, lateral roots, root hairs, and associated structures and systems such as root nodules or mycorrhizae. (18)

Rough ER (RER) Endoplasmatic reticulum with many ribosomes attached. As a result, they appear rough in electron micrographs. (5)

Ruminant Grazing mammals that possess an additional stomach chamber called rumen which is heavily fortified with cellulose-digesting microorganisms. (27)

◀ S ▶

S Phase The second stage of interphase in which the materials needed for cell division are synthesized and an exact copy of cell's DNA is made by DNA replication. (10)

Sac Body The body plan of simple animals, like cnidarians, where there is a single opening leading to and from a digestion chamber.

Saltatory Conduction The "hopping" movement of an impulse along a myelinated neuron from one Node of Ranvier to the next one. (23)

Sap Fluid found in xylem or sieve of phloem. (20)

Saprophyte Organisms, mainly fungi and bacteria, that get their nutrition by breaking down organic wastes and dead organisms, also called decomposers. (42)

Saprobe Organism that obtains its nutrients by decomposing dead organisms. (37)

Sarcolemma The plasma membrane of a muscle fiber. (26)

Sarcomere The contractile unit of a myofibril in skeletal muscle. (26)

Sarcoplasmic Reticulum (SR) In skeletal muscle, modified version of the endoplasmic reticulum that stores calcium ions. (26)

Savanna A grassland biome with alternating dry and rainy seasons. The grasses and scattered trees support large numbers of grazing animals. (40)

Scaling Effect A property that changes disproportionately as the size of organisms increase. (22)

Scanning Electron Microscope (SEM) A microscope which operates by showering electrons back and forth across the surface of a specimen prepared with a thin metal coating. The resultant image shows three-dimensional features of the specimen's surface. (APP.)

Schwann Cells Cells which wrap themselves around the axons of neurons forming an insulating myelin sheath composed of many layers of plasma membrane. (23)

Sclereids Irregularly-shaped sclerenchyma cells, all having thick cell walls; a component of seed coats and nuts. (18)

Sclerenchyma Component of the ground tissue system of plants. They are thick walled cells of various shapes and sizes, providing support or protection. They continue to function after the cell dies. (18)

Sclerenchyma Fibers Non-living elongated plant cells with tapering ends and thick secondary walls. A supportive cell type found in various plant tissues. (18)

Sebaceous Glands Exocrine glands of the skin that produce a mixture of lipids (sebum) that oil the hair and skin. (26)

Secondary Cell Wall An additional cell wall that improves the strength and resiliency of specialized plant cells, particularly those cells found in stems that support leaves, flowers, and fruit. (5)

Secondary Consumer Organism that feeds exclusively on primary consumers; mostly animals, but some plants. (41)

Secondary Growth Growth from cambia in perennials; results in an increase in the diameter of stems and roots. (18)

Secondary Meristems (vascular cambium, cork cambrium) Rings or clusters of meristematic cells that increase the width of stems and roots when the divide. (18)

Secondary Sex Characteristics Those characteristics other than the gonads and reproductive tract that develop in response to sex hormones. For example, breasts and pubic hair in women and a deep voice and pubic hair in men. (31)

Secondary Succession The development of a community in an area previously occupied by a community, but which was disturbed in some manner; for example, fire, development, or clear-cutting forests. (41)

Secondary Tissues Tissues produced to accommodate new cell production in plants with woody growth. Secondary tissues are produced from cambia, which produce vascular and cork tissues, leading to an increase in plant girth. (18)

Second Messenger Many hormones, such as glucagon and thyroid hormone, evoke a response by binding to the outer surface of a target cell and causing the release of another substance (which is the second messenger). The best-studied second messenger is cyclic AMP which is formed by an enzyme on the inner surface of the plasma membrane following the binding of a hormone to the outer surface of the membrane. The cyclic AMP diffuses into the cell and activates a protein kinase. (25)

Secretion The process of exporting materials produced by the cell. (5)

Seed A mature ovule consisting of the embryo, endosperm, and seed coat. (20)

Seed Dormancy Metabolic inactivity of seeds until favorable conditions promote seed germination. (20)

Secretin Hormone secreted by endocrine cells in the wall of the intestine that stimulates the release of digestive products from the pancreas. (25)

Segmentation A condition in which the body is constructed, at least in part, from a series of repeating parts. Segmentation occurs in annelids, arthropods, and vertebrates (as revealed during embryonic development). (39)

Selectively Permeable A term applied to the plasma membrane because membrane proteins control which molecules are transported. Enables a cell to import and accumulate the molecules essential for normal metabolism. (7)

Semen The fluid discharged during a male orgasm. (31)

Semiconsevative Replication The manner in which DNA replicates; half of the original DNA strand is conserved in each new double helix. (14)

Seminal Vesicles The organs which produce most of the ejaculatory fluid. (31)

Seminiferous Tubules Within the testes, highly coiled and compacted tubules, lined with a self-perpetuating layer of spermatogonia, which develop into sperm. (31)

Senescence Aging and eventual death of an organism, organ or tissue. (3, 18)

Sense Strand The one strand of a DNA double helix that contains the information that encodes the amino sequence of a polypeptide. This is the strand that is selectively transcribed by RNA polymerase forming an mRNA that can be properly translated. (14)

Sensory Neurons Neurons which relay impulses to the central nervous system. (23)

Sensory Receptors Structures that detect changes in the external and internal environment and transmit the information to the nervous system. (24)

Sepal The outermost whorl of a flower, enclosing the other flower parts as a flower bud; collectively called the calyx. (20)

Sessile Sedentary, incapable of independent movement. (39)

Sex Chromosomes The one chromosomal pair that is not identical in the karyotypes of males and females of the same animal species. (10, 13)

Sex Hormones Steroid hormones which influence the production of gametes and the development of male or female sex characteristics. (25)

Sexual Dimorphism Differences in the appearance of males and females in the same species. (33)

Sexual Reproduction The process by which haploid gametes are formed and fuse during fertilization to form a zygote. (31)

Sexual Selection The natural selection of adaptations that improve the chances for mating and reproducing. (33)

Shivering Involuntary muscular contraction for generating metabolic heat that raises body temperature. (28)

Shoot In angiosperms, the system consisting of stems, leaves, flowers and fruits. (18)

Shoot System The above-ground portion of an angiosperm plant consisting of stems with nodes, including branches, leaves, flowers and fruits. (18)

Short-Day Plants Plants that flower in late summer or fall when the length of daylight becomes shorter than some critical period. (21)

Shrubland A biome characterized by densely growing woody shrubs in mediterranean type climate; growth is so dense that understory plants are not usually present. (40)

Sickle Cell Anemia A genetic (recessive autosomal) disorder in which the beta globin genes of adult hemoglobin molecules contain an amino acid substitution which alters the ability of hemoglobin to transport oxygen. During times of oxygen stress, the red blood cells of these individuals may become sickle shaped, which interferes with the flow of the cells through small blood vessels. (4, 17)

Sieve Plate Found in phloem tissue in plants, the wall between sieve-tube members, containing perforated areas for passage of materials. (18)

Sieve-Tube Member A living, food-conducting cell found in phloem tissue of plants; associated with a companion cell. (18)

Sigmoid Growth Curve An S-shaped curve illustrating the lag phase, exponential growth, and eventual approach of a population to its carrying capacity. (43)

Sign Stimulus An object or action in the environment that triggers an innate behavior. (44)

Simple Fruits Fruits that develop from the ovary of one pistil. (20)

Simple Leaf A leaf that is undivided; only one blade attached to the petiole. (18)

Sinoatrial (SA) Node A collection of cells that generates an action potential regulating heart beat; the heart's pacemaker. (28)

Skeletal Muscles Separate bundles of parallel, striated muscle fibers anchored to the bone, which they can move in a coordinated fashion. They are under voluntary control. (26)

Skeleton A rigid form of support found in most animals either surrounding the body with a protective encasement or providing a living girder system within the animal. (26)

Skull The bones of the head, including the cranium. (26)

Slow-Twitch Fibers Skeletal muscle fibers that depend on aerobic metabolism for ATP production. These fibers are capable of undergoing contraction for extended periods of time without fatigue, but generate lesser forces than fast-twitch fibers. (9)

Small Intestine Portion of the intestine in which most of the digestion and absorption of nutrients takes place. It is so named because of its narrow diameter. There are three sections: duodenum, jejunum, and ilium. (27)

Smell Sense of the chemical composition of the environment. (24)

Smooth ER (SER) Membranes of the endoplasmic reticulum that have no ribosomes on their surface. SER is generally more tubular than the RER. Often acts to store calcium or synthesize steroids. (5)

Smooth Muscle The muscles of the internal organs (digestive tract, glands, etc.). Composed of spindle-shaped cells that interlace to form sheets of visceral muscle. (26)

Social Behavior Behavior among animals that live in groups composed of individuals that are dependent on one another and with whom they have evolved mechanisms of communication. (44)

Social Learning Learning of a behavior from other members of the species. (44)

Social Parasitism Parasites that use behavioral mechanisms of the host organism to the parasite's advantage, thereby harming the host. (42)

Solute A substance dissolved in a solvent. (3)

Solution The resulting mixture of a solvent and a solute. (3)

Solvent A substance in which another material dissolves by separating into individual molecules or ions. (3)

Somatic Cells Cells that do not have the potential to form reproductive cells (gametes). Includes all cells of the body except germ cells. (11)

Somatic Nervous System The nerves that carry messages to the muscles that move the skeleton either voluntarily or by reflex. (23)

Somatic Sensory Receptors Receptors that respond to chemicals, pressure, and temperature that are present in the skin, muscles, tendons, and joints. Provides a sense of the physiological state of the body. (24)

Somites In the vertebrate embryo, blocks of mesoderm on either side of the notochord that give rise to muscles, bones, and dermis. (32)

Speciation The formation of new species. Occurs when one population splits into separate populations that diverge genetically to the point where they become separate species. (33)

Species Taxonomic subdivisions of a genus. Each species has recognizable features that distinguish it from every other species. Members of one species generally will not interbreed with members of other species. (33)

Specific Epithet In taxonomy, the second term in an organism's scientific name identifying its species within a particular genus. (1)

Spermatid Male germ cell that has completed meiosis but has not yet differentiated into a sperm. (31)

Spermatogenesis The production of sperm. (31)

Spermatogonia Male germ cells that have not yet begun meiosis. (31)

Spermatozoa (Sperm) Male gametes. (31)

Sphinctors Circularly arranged muscles that close off the various tubes in the body.

Spinal Cord A centralized mass of neurons for processing neurological messages and linking the brain with that part of peripheral nervous system not reached by the cranial nerves. (23)

Spinal Nerves Paired nerves which emerge from the spinal cord and innervate the body. Humans have 31 pairs of spinal nerves. (23)

Spindle Apparatus In dividing eukaryotic cells, the complex rigging, made of microtubules, that aligns and separates duplicated chromosomes. (10)

Splash Zone In the intertidal zone, the uppermost region receiving splashes and sprays of water to the mean of high tides. (40)

Spleen One of the organs of the lymphatic system that produces lymphocytes and filters blood; also produces red blood cells in the human fetus. (28)

Splicing The step during RNA processing in which the coding segments of the primary transcript are covalently linked together to form the mRNA. (15)

Spongy Parenchyma In monocot and dicot leaves, loosely arranged cells functioning in photosynthesis. Found above the lower epidermis and beneath the palisade parenchyma in dicots, and between the upper and lower epidermis in monocots. (18)

Spontaneous Generation Disproven concept that living organisms can arise directly from inanimate materials. (2)

Sporangiospores Black, asexual spores of the zygomycete fungi. (37)

Sporangium A hollow structure in which spores are formed. (37)

Spores In plants, haploid cells that develop into the gametophyte generation. In fungi, an asexual or sexual reproductive cell that gives rise to a new mycelium. Spores are often lightweight for their dispersal and adapted for survival in adverse conditions. (37)

Sporophyte The diploid spore producing generation in plants. (38)

Stabilizing Selection Natural selection favoring an intermediate phenotype over the extremes. (33)

Starch Polysaccharides used by plants to store energy. (4)

Stamen The flower's male reproductive organ, consisting of the pollen-producing anther supported by a slender stalk, the filament. (20)

Stem In plants, the organ that supports the leaves, flowers, and fruits. (18)

Stem Cells Cells which are undifferentiated and capable of giving rise to a variety of different types of differentiated cells. For example, hematopoietic stem cells are capable of giving rise to both red and white blood cells. (17)

Steroids Compounds classified as lipids which have the basic four-ringed molecular skeleton as represented by cholesterol. Two examples of steroid hormones are the sex hormones; testosterone in males and estrogen in females. (4, 25)

Stigma The sticky area at the top of each pistil to which pollen adheres. (20)

Stimulus Any change in the internal or external environment to which an organism can respond. (24)

Stomach A muscular sac that is part of the digestive system where food received from the esophagus is stored and mixed, some breakdown of food occurs, and the chemical degradation of nutrients begins. (27)

Stomates (Pl. Stomata) Microscopic pores in the epidermis of the leaves and stems which allow gases to be exchanged between the plant and the external environment. (18)

Stratified Epithelia Multicellular layered epithelium. (22)

Stream Flowing body of surface fresh water; streams merge together into larger streams and rivers. (40)

Stretch Receptors Sensory receptors embedded in muscle tissue enabling muscles to respond reflexively when stretched. (23, 24)

Striated Referring to the striped appearance of skeletal and cardiac muscle fibers. (26)

Strobilus In lycopids, terminal, cone-like clusters of specialized leaves that produce sporangia.

Stroma The fluid interior of chloroplasts. (8)

Stromatolites Rocks formed from masses of dense bacteria and mineral deposits. Some of these rocky masses contain cells that date back over three billion years revealing the nature of early prokaryotic life forms. (35)

Structural Genes DNA segments in bacteria that direct the formation of enzymes or structural proteins. (15)

Style The portion of a pistil which joins the stigma to the ovary. (20)

Substrate-Level Phosphorylation The formation of ATP by direct transfer of a phosphate group from a substrate, such as a sugar phosphate, to ADP. ATP is formed without the involvement of an electron transport system. (9)

Substrates The reactants which bind to enzymes and are subsequently converted to products. (6)

Succession The orderly progression of communities leading to a climax community. It is one of two types: primary, which occurs in areas where no community existed before; and secondary, which occurs in disturbed habitats where some soil and perhaps some organisms remain after the disturbance. (41)

Succulents Plants having fleshy, water-storing stems or leaves. (40)

Suppressor T Cells A class of T cells that regulate immune responses by inhibiting the activation of other lymphocytes. (30)

Surface Area-to-Volume Ratio The ratio of the surface area of an organism to its volume, which determines the rate of exchange of materials between the organism and its environment. (22)

Surface Tension The resistance of a liquid's surface to being disrupted. In aqueous solutions, it is caused by the attraction between water molecules. (3)

Survivorship Curve Graph of life expectancy, plotted as the number of survivors versus age. (43)

Sweat Glands Exocrine glands of the skin that produce a dilute salt solution, whose evaporation cools in the body. (26)

Symbiosis A close, long-term relationship between two individuals of different species. (42)

Symmetry Referring to a body form that can be divided into mirror image halves by at least one plane through its body. (39)

Sympathetic Nervous System Part of the autonomic nervous system that tends to stimulate bodily activities, particularly those involved with coping with stressful situations. (23)

Sympatric Speciation Speciation that occurs in populations with overlapping distributions. It is common in plants when polyploidy arises within a population. (33)

Synapse Juncture of a neuron and its target cell (another neuron, muscle fiber, gland cell). (23)

Synapsis The pairing of homologous chromosomes during prophase of meiosis I. (11)

Synaptic Cleft Small space between the synaptic knobs of a neuron and its target cell. (23)

Synaptic Knobs The swellings that branch from the end of the axon. They deliver the neurological impulse to the target cell. (23)

Synaptonemal Complex Ladderlike structure that holds homologous chromosomes together as a tetrad during crossing over in prophase I of meiosis. (11)

Synovial Cavities Fluid-filled sacs around joints, the function of which is to lubricate and separate articulating bone surfaces. (26)

Systemic Circulation Part of the circulatory system that delivers oxygenated blood to the tissues and routes deoxygenated blood back to the heart. (28)

Systolic Pressure The first number of a blood pressure reading; the highest pressure attained in the arteries as blood is propelled out of the heart. (28)

◀ T ▶

Taiga A biome found south of tundra biomes; characterized by coniferous forests, abundant precipitation, and soils that thaw only in the summer. (40)

Tap Root System Root system of plants having one main root and many smaller lateral roots. Typical of conifers and dicots. (18)

Taste Sense of the chemical composition of food. (24)

Taxonomy The science of classifying and grouping organisms based on their morphology and evolution. (1)

T Cell Lymphocytes that carry out cell-mediated immunity. They respond to antigen stimulation by becoming helper cells, killer cells, and memory cells. (30)

Telophase The final stage of mitosis which begins when the chromosomes reach their spindle poles and ends when cytokinesis is completed and two daughter cells are produced. (10)

Tendon A dense connective tissue cord that connects a skeletal muscle to a bone. (26)

Teratogenic Embryo deforming. Chemicals such as thalidomide or alcohol are teratogenic because they disturb embryonic development and lead to the formation of an abnormal embryo and fetus. (32)

Terminal Electron Acceptor In aerobic respiration, the molecule of O_2 which removes the electron pair from the final cytochrome of the respiratory chain. (9)

Terrestrial Living on land. (40)

Territory (Territoriality) An area that an animal defends against intruders, generally in the protection of some resource. (42, 44)

Tertiary Consumer Animals that feed on secondary consumers (plant or animal) or animals only. (41)

Test Cross An experimental procedure in which an individual exhibiting a dominant trait is crossed to a homozygous recessive to determine whether the first individual is homozygous or heterozygous. (12)

Testis In animals, the sperm-producing gonad of the male. (23)

Testosterone The male sex hormone secreted by the testes when stimulated by pituitary gonadotropins. (31)

Tetrad A unit of four chromatids formed by a synapsed pair of homologous chromosomes, each of which has two chromatids. (11)

Thallus In liverworts, the flat, ground-hugging plant body that lacks roots, stems, leaves, and vascular tissues. (38)

Theory of Tolerance Distribution, abundance and existence of species in an ecosystem are determined by the species' range of tolerance of chemical and physical factors. (41)

Thermoreceptors Sensory receptors that respond to changes in temperature. (24)

Thermoregulation The process of maintaining a constant internal body temperature in spite of fluctuations in external temperatures. (28)

Thigmotropism Changes in plant growth stimulated by contact with another object, e.g., vines climbing on cement walls. (21)

Thoracic Cavity The anterior portion of the body cavity in which the lungs are suspended. (39)

Thylakoids Flattened membrane sacs within the chloroplast. Embedded in these membranes are the light-capturing pigments and other components that carry out the light-dependent reactions of photosynthesis. (8)

Thymus Endocrine gland in the chest where T cells mature. (30)

Thyroid Gland A butterfly-shaped gland that lies just in front of the human windpipe, producing two metabolism-regulating hormones, thyroxin and triodothyronine. (25)

Thyroid Hormone A mixture of two iodinated amino acid hormones (thyroxin and triiodothyronine) secreted by the thyroid gland. (25)

Thyroid Stimulating Hormone (TSH) An anterior pituitary hormone which stimulates secretion by the thyroid gland. (25)

Tissue An organized group of cells with a similar structure and a common function. (22)

Tissue System Continuous tissues organized to perform a specific function in plants. The three plant tissue systems are: dermal, vascular, and ground (fundamental). (18)

Tolerance Range The range between the maximum and minimum limits for an environmental factor that is necessary for an organism's survival. (41)

Totipotent The genetic potential for one type of cell from a multicellular organism to give rise to any of the organism's cell types, even to generate a whole new organism. (15)

Trachea The windpipe; a portion of the respiratory tract between the larynx and bronchii. (29)

Tracheal Respiratory System A network of tubes (tracheae) and tubules (tracheoles) that carry air from the outside environment directly to the cells of the body without involving the circulatory system. (29, 39)

Tracheid A type of conducting cell found in xylem functioning when a cell is dead to transport water and dissolved minerals through its hollow interior. (18)

Tracheophytes Vascular plants that contain fluid-conducting vessels. (38)

Transcription The process by which a strand of RNA assembles along one of the DNA strands. (14)

Transcriptional-Level Control Control of gene expression by regulating whether or not a specific gene is transcribed and how often. (15)

Transduction A type of genetic recombination resulting from transfer of genes from one organism to another by a virus.

Transfer RNA (tRNA) A type of RNA that decodes mRNA's codon message and translates it into amino acids. (14)

Transgenic Organism An organism that possesses genes derived from a different species. For example, a sheep that carries a human gene and secretes the human protein in its milk is a transgenic animal. (16)

Translation The cell process that converts a sequence of nucleotides in mRNA into a sequence of amino acids in a polypeptide. (14)

Translational-Level Control Control of gene expression by regulating whether or not a specific mRNA is translated into a polypeptide. (15)

Translocation The joining of segments of two nonhomologous chromosomes (13)

Transmission Electron Microscope (TEM) A microscope that works by shooting electrons through very thinly sliced specimens. The result is an enormously magnified image, two-dimensional, of remarkable detail. (App.)

Transpiration Water vapor loss from plant surfaces. (19)

Transpiration Pull The principle means of water and mineral transport in plants, initiated by transpiration. (19)

Transposition The phenomenon in which certain DNA segments (mobile genetic elements, or jumping genes) tend to move from one part of the genome to another part. (15)

Transverse Fission The division pattern in ciliated protozoans where the plane of division is perpendicular to the cell's length.

Trimester Each of the three stages comprising the 266-day period between conception and birth in humans. (32)

Triploid Having three sets of chromosomes, abbreviated 3N. (11)

Trisomy Three copies of a particular chromosome per cell. (17)

Trophic Level Each step along a feeding pathway. (41)

Trophozoite The actively growing stage of polymorphic protozoa. (37)

Tropical Rain Forest Lush forests that occur near the equator; characterized by high annual rainfall and high average temperature. (40)

Tropical Thornwood A type of shrubland occurring in tropical regions with a short rainy season. Plants lose their small leaves during dry seasons, leaving sharp thorns. (40)

Tropic Hormones Hormones that act on endocrine glands to stimulate the production and release of other hormones. (25)

Tropisms Changes in the direction of plant growth in response to environmental stimuli, e.g., light, gravity, touch. (21)

True-Breeder Organisms that, when bred with themselves, always produce offspring identical to the parent for a given trait. (12)

Tubular Reabsorption The process by which substances are selectively returned from the fluid in the nephron to the bloodstream. (28)

Tubular Secretion The process by which substances are actively and selectively transported from the blood into the fluid of the nephron. (28)

Tumor-Infiltrating Lymphocytes (TILs) Cytotoxic T cells found within a tumor mass that have the capability to specifically destroy the tumor cells. (30)

Tumor-Suppressor Genes Genes whose products act to block the formation of cancers. Cancers form only when both copies of these genes (one on each homologue) are mutated. (13)

Tundra The marshy, unforested biome in the arctic and at high elevations. Frigid temperatures for most of the year prevent the subsoil from thawing, which produces marshes and ponds. Dominant vegetation includes low growing plants, lichens, and mosses. (40)

Turgor Pressure The internal pressure in a plant cell caused by the diffusion of water into the cell. Because of the rigid cell wall, pressure can increase to where it eventually stops the influx of more water. (7)

Turner Syndrome A person whose cells have only one X chromosome and no second sex chromosome (XO). These individuals develop as immature females. (17)

◀ U ▶

Ultimate (Top) Consumer The final carnivore trophic level organism, or organisms that escaped predation; these consumers die and are eventually consumed by decomposers. (41)

Ultracentrifuge An instrument capable of spinning tubes at very high speeds, delivering centrifugal forces over 100,000 times the force of gravity. (9)

Unicellular The description of an organism where the cell is the organism. (35)

Uniform Pattern Distribution of individuals of a population in a uniform arrangement, such as individual plants of one species uniformly spaced across a region. (43)

Urethra In mammals, a tube that extends from the urinary bladder to the outside. (28)

Urinary Tract The structures that form and export urine: kidneys, ureters, urinary bladder, and urethra. (28)

Urine The excretory fluid consisting of urea, other nitrogenous substances, and salts dissolved in water. It is formed by the kidneys. (28)

Uterine (Menstrual) Cycle The repetitive monthly changes in the uterus that prepare the endometrium for receiving and sustaining an embryo. (31)

Uterus An organ in the female reproductive system in which an embryo implants and is maintained during development. (31)

◀ V ▶

Vaccines Modified forms of disease-causing microbes which cannot cause disease but retain the same antigens of it. They permit the immune system to build memory cells without diseases developing during the primary immune response. (30)

Vacoconstriction Reduction in the diameter of blood vessels, particularly arterioles. (28)

Vacuole A large organelle found in mature plant cells, occupying most of the cell's volume, sometimes more than 90% of it. (5)

Vagina The female mammal's copulatory organ and birth canal. (31)

Variable (Experimental) A factor in an experiment that is subject to change, i.e., can occur in more than one state. (2)

Vascular Bundles Groups of vascular tissues (xylem and phloem) in the shoot of a plant. (19)

Vascular Cambrium In perennials, a secondary meristem that produces new vascular tissues. (18)

Vascular Cylinder Groups of vascular tissues in the central region of the root. (18)

Vascular Plants Plants having a specialized conducting system of vessels and tubes for transporting water, minerals, food, etc., from one region to another. (18)

Vascular Tissue System All the vascular tissues in a plant, including xylem, phloem, and the vascular cambium or procambium. (18)

Vasodilation Increase in the diameter of blood vessels, particularly arterioles. (28)

Veins In plants, vascular bundles in leaves. In animals, blood vessels that return blood to the heart. (28)

Venation The pattern of vein arrangement in leaf blades. (18)

Ventricle Lower chamber of the heart which pumps blood through arteries. There is one ventricle in the heart of lower vertebrates and two ventricles in the four-chambered heart of birds and mammals. (28)

Venules Small veins that collect blood from the capillaries. They empty into larger veins for return to the heart. (28)

Vertebrae The bones that form the backbone. In the human there are 33 bones arranged in a gracefully curved line along the bone, cushioned from one another by disks of cartilage. (26)

Vertebral Column The backbone, which encases and protects the spinal cord. (26)

Vertebrates Animals with a backbone. (39)

Vesicles Small membrane-enclosed sacs which form from the ER and Golgi complex. Some store chemicals in the cells; others move to the surface and fuse with the plasma membrane to secrete their contents to the outside. (5)

Vessel A tube or connecting duct containing or circulating body fluids. (18)

Vessel Member A type of conducting cell in xylem functioning when the cell is dead to transport water and dissolved minerals through its hollow interior; also called a vessel element. (18)

Vestibular Apparatus A portion of the inner ear of vertebrates that gathers information about the position and movement of the head for use in maintaining balance and equilibrium. (24)

Vestigial Structure Remains of ancestral structures or organs which were, at one time, useful. (34)

Villi Finger-like projections of the intestinal wall that increase the absorption surface of the intestine. (27)

Viroids are associated with certain diseases of plants. Each viroid consists solely of a small single-stranded circle of RNA unprotected by a protein coat. (36)

Virus Minute structures composed of only heredity information (DNA or RNA), surrounded by a protein or protein/lipid coat. After infection, the viral nucleic acid subverts the metabolism of the host cell, which then manufactures new virus particles. (36)

Visible Light The portion of the electromagnetic spectrum producing radiation from 380 nm to 750 nm detectable by the human eye.

Vitamins Any of a group of organic compounds essential in small quantities for normal metabolism. (27)

Vocal Cords Muscular folds located in the larynx that are responsible for sound production in mammals. (29)

Vulva The collective name for the external features of the human female's genitals. (31)

◄ W ►

Water Vascular System A system for locomotion, respiration, etc., unique to echinoderms. (39)

Wavelength The distance separating successive crests of a wave. (8)

Waxes A waterproof material composed of a number of fatty acids linked to a long chain alcohol. (4)

White Matter Regions of the brain and spinal cord containing myelinated axons, which confer the white color. (23)

Wild Type The phenotype of the typical member of a species in the wild. The standard to which mutant phenotypes are compared. (13)

Wilting Drooping of stems or leaves of a plant caused by water loss. (7)

Wood Secondary xylem. (18)

◄ X ►

X Chromosome The sex chromosome present in two doses in cells of a female, and in one dose in the cells of a male. (13)

X-Linked Traits Traits controlled by genes located on the X chromosome. These traits are much more common in males than females. (13)

Xylem The vascular tissue that transports water and minerals from the roots to the rest of the plant. Composed of tracheids and vessel members. (18)

◄ Y ►

Y Chromosome The sex chromosome found in the cells of a male. When Y-carrying sperm unite with an egg, all of which carry a single X chromosome, a male is produced. (13)

Y-Linked Inheritance Genes carried only on Y chromosomes. There are relatively few Y-linked traits; maleness being the most important such trait in mammals. (13)

Yeast Unicellular fungus that forms colonies similar to those of bacteria. (37)

Yolk A deposit of lipids, proteins, and other nutrients that nourishes a developing embryo. (32)

Yolk Sac A sac formed by an extraembryonic membrane. In humans, it manufactures blood cells for the early embryo and later helps to form the umbilical cord. (32)

◄ Z ►

Zero Population Growth In a population, the result when the combined positive growth factors (births and immigration). (43)

Zooplankton Protozoa, small crustaceans and other tiny animals that drift with ocean currents and feed on phytoplankton. (37, 40)

Zygospore The diploid spores of the zygomycete fungi, which include Rhizopus, a common bread mold. After a period of dormancy, the zygospore undergoes meiosis and germinates. (36)

Zygote A fertilized egg. The diploid cell that results from the union of a sperm and egg. (32)

Photo Credits

Part 1 Opener Norbert Wu. **Chapter 1** Fig. 1.1: Jeff Gnass. Fig. 1.2a: Frans Lanting/Minden Pictures, Inc. Fig. 1.2b: David Muench. Fig. 1.3a: Courtesy Dr. Alan Cheetham, National Museum of National History, Smithsonian Institution. Fig. 1.3b: Manfred Kage/Peter Arnold. Fig. 1.3c: Stephen Dalton/NHPA. Fig. 1.3d: Charles Summers, Jr. Fig. 1.3e: Larry West/Photo Researchers. Fig. 1.3f: Bianca Lavies. Fig. 1.3g: Steve Allen/Peter Arnold. Fig. 1.3h: George Grall. Fig. 1.3i: Michio Hoshino/Minden Pictures, Inc. Fig. 1.6a: CNRI/Science Photo Library/Photo Researchers. Fig. 1.6b: M. Abbey/Visuals Unlimited. Fig. 1.6c: Steve Kaufman/Peter Arnold. Fig. 1.6d: Willard Clay. Fig. 1.6e: Jim Bradenburg/Minden Pictures, Inc. Fig. 1.7: (pages 16-17) Charles A. Mauzy; (page 16, top) Anthony Mercieca/Natural Selection; (page 16, center) Wolfgang Bayer/Bruce Coleman, Inc.; (page 16, bottom) Dr. Jeremy Burgess/Science Photo Library/Photo Researchers; (page 17, top) Dr. Eckart Pott/Bruce Coleman, Inc.; (page 17, bottom) Art Wolfe. Fig. 1..8: Paul Chesley/Photographers Aspen. Fig. 1.9a: Rainbird/Robert Harding Picture Library. Fig. 1.11: Dick Luria/FPG International. **Chapter 2** Fig. 2.2a: Couresy Institut Pasteur. Fig. 2.3a: Topham/The Image Works. Fig. 2.3b: Leonard Lessin/Peter Arnold. Fig. 2.3c: Bettmann Archive. Fig. 2.5a: Laurence Gould/Earth Scenes/Animals Animals. Fig. 2.5b: Ted Horowitz/The Stock Market. **Part 2 Opener:** Nancy Kedersha. **Chapter 3** Fig. 3.1: Franklin Viola. Fig. 3.2: Courtesy Pachyderm Scientific Industries. Bioline: Jacan-Yves Kerban/The Image Bank. Fig. 3.7: Courtesy Stephen Harrison, Harvard Biochemistry Department. Fig. 3.11: Gary Milburn/Tom Stack & Associates. Fig. 3.12: Courtesy R. S. Wilcox, Biology Department, SUNY Binghampton. **Chapter 4** Fig. 4.1: Alastair Black/Tony Stone World Wide. Fig. 4.5a: Don Fawcett/Visuals Unlimited. Fig. 4.5b: Jeremy Burgess/Photo Researchers. Fig. 4.5c: Cabisco/Visuals Unlimited. Fig. 4.6: Tony Stone World Wide. Fig. 4.7: Zig Leszcynski/Animals Animals. Fig. 4.8: Robert & Linda Mitchell. Human Perspective: (a) Photofest (b) Bill Davila/Retna. Fig. 4.12a: Frans Lanting/AllStock, Inc. Fig. 4.12b: Mantis Wildlife Films/Oxford Scientific Films/Animals Animals. Fig. 4.17: Stanley Flegler/Visuals Unlimited. Fig. 4.18a: Courtesy Nelson Max, Lawrence Livermore Laboratory. Fig. 4.18b: Tsuned Hayashida/The Stock Market. **Chapter 5** Fig. 5.1a: The Granger Collection. Fig. 5.1b: Bettmann Archive. Fig. 5.2a: Dr. Jeremey Burgess/Photo Researchers. Fig. 5.2b: CNRI/Photo Researchers. Fig. 5.4: Omikron/Photo Researchers. Fig. 5.6a: Courtesy Richard Chao, California State University at Northridge. Fig. 5.6c: Courtesy Daniel Branton, University of Berkeley. Fig. 5.7: Courtesy G. F. Bohr. Fig. 5.8: Courtesy

Michael Mercer, Zoology Department, Arizona State University. Fig. 5.9: Courtesy D. W. Fawcett, Harvard Medical School. Fig. 5.10a: Courtesy U.S. Department of Agriculture, Fig. 5.11: Courtesy Dr. Birgit H. Satir, Albert Einstein College of Medicine. Fig. 5.13: Courtesy Lennart Nilsson, from *A Child Is Born*. Fig. 5.14a: K. R. Porter/Photo Researchers. Fig. 5.14c: Courtesy Lennart Nilsson, BonnierAlba. Fig. 5.14d: Courtesy Lennart Nilsson, From *A Child Is Born*. Fig. 5.15a: Courtesy J. Elliot Weier. Fig. 5.16a: Courtesy J.V. Small. Fig. 5.17a: Peter Parks/Animals Animals. Fig. 5.17b: Courtesy Dr. Manfred Hauser, RUHR-Universitat Rochum. Fig. 5.18: Courtesy Jean Paul Revel, Division of Biology, California Institute of Technology. Fig. 5.19a: Courtesy E. Vivier, from *Paramecium* by W.J. Wagtendonk, Elsevier North Holland Biomedical Press, 1974. Fig. 5.19b: David Phillips/Photo Researchers. Fig. 5.20: Courtesy C.J. Brokaw and T.F. Simonick, *Journal of Cell Biology*, 75:650 (1977). Reproduced with permission. Fig. 5.21: Courtesy L.G. Tilney and K. Fujiwara. Fig. 5.22a: Courtesy W. Gordon Whaley, University of Texas, Austin. Fig. 5.22c: Courtesy R.D. Preston, University of Leeds, London. **Chapter 6** Fig. 6.1a: Alex Kerstitch. Fig. 6.1b: Peter Parks/Earth Scenes. Fig. 6.2: Marty Stouffer/Animals Animals. Fig. 6.3: Kay Chernush/The Image Bank. Fig. 6.4b: Courtesy Computer Graphics Laboratory, University of California, San Francisco. Fig. 6.8: Courtesy Stan Koszelek, Ph.D., University of California, Riverside. Fig. 6.10: Swarthout/The Stock Market. **Chapter 7** Fig. 7.5: Ed Reschke. Fig. 7.10a: Lennart Nilsson, ©Boehringer Ingelheim, International Gmbh; from *The Incredible Machine*. Human Perspective: (Fig. 1a) Martin Rotker/Phototake; (Fig. 1b) Cabisco/Visuals Unlimited. Fig. 7.11a: Courtesy Dr. Ravi Pathak, Southwestern Medical Center, University of Texas. **Chapter 8** Fig. 8.1a: Carr Clifton. Fig. 8.1b: Shizuo Lijima/Tony Stone World Wide. Fig. 8.5: Joe Englander/Viesti Associates, Inc. Fig. 8.6: Courtesy T. Elliot Weier. Fig. 8.10: Courtesy Lawrence Berkeley Laboratory, University of California. Fig. 8.13: Pete Winkel/Atlanta/Stock South. Bioline: Courtesy Woods Hole Oceanographic Institution. **Chapter 9** Fig. 9.1: Stephen Frink/AllStock, Inc. Bioline: (Fig. 1) Frank Oberle/Bruce Coleman, Inc.; (Fig. 2) Movie Stills Archive. Fig. 9.6: Topham/The Image Works. Fig. 9.9: H. Fernandez-Moran. Human Perspective: (top and bottom insets) Courtesy MacDougal; (top) Jerry Cooke; (bottom) Richard Kane/Sportchrome East/West. Fig. 9.11: Peter Parks/NHPA. Fig. 9.12a: Grafton M. Smith/The Image Bank. **Chapter 10** Fig. 10.1: Sipa Press. Fig. 10.2a: Institut Pasteur/CNRI/Phototake. Fig. 10.3a: Dr. R. Vernon/Phototake.

Fig. 10.3b: CNRI/Science Photo Library/Photo Researchers. Fig. 10.6: Courtesy Dr. Andrew Bajer, University of Oregon, Fig. 10.7: CNRI/Science Photo Library/Photo Researchers. Fig. 10.8: Dr. G. Shatten/Science Photo Library/Photo Researchers. Fig. 10.9: Courtesy Professor R.G. Kessel. Fig. 10.10: David Phillips/Photo Researchers. **Chapter 11** Fig. 11.3: Courtesy Science Software Systems. Fig. 11.5a: Courtesy Dr. A.J. Solari, from *Chromosoma*, vol. 81, p. 330 (1980), Human Perspective: Donna Zweig/Retna. Fig. 11.7: Cabisco/Visuals Unlimited. **Part 3 Opener:** David Scharf. **Chapter 12** Fig. 12.1: Courtesy Dr. Ing Jaroslav Krizenecky. Fig. 12.6a: Doc Pele/Retna. Fig. 12.6b: FPG International. Fig. 12.9: Sydney Freelance/Gamma Liaison. Fig. 12.11 Hans Reinhard/Bruce Coleman, Inc. **Chapter 13** Fig. 13.2a: Robert Noonan. Fig. 13.6: Biological Photo Service. Bioline: Norbert Wu. Fig. 13.9: Historical Pictures Service. Fig. 13.11a: Courtesy M.L. Barr. Fig. 13.11b: Jean Pragen/Tony Stone World Wide. Fig. 13.12: Courtesy Lawrence Livermore National Laboratory. **Chapter 14** Fig. 14.1b: Lee D. Simon/Science Photo Library/Photo Researchers. Fig. 14.4: David Leah/Science Photo Library/Photo Researchers. Fig. 14.5: Dr. Gopal Murti/Science Photo Library/Photo Researchers. Fig. 14.6b: Fawcett/Olins/Photo Researchers. Fig. 14.7: Courtesy U.K. Laemmli. Fig. 14.10a: From M. Schnos and R.B. Inman, *Journal of Molecular Biology*, 51:61-73 (1970), ©Academic Press. Fig. 14.10b: Courtesy Professor Joel Huberman, Roswell Park Memorial Institute. Human Perspective: (Fig. 2) Courtesy Skin Cancer Foundation; (Fig. 3) Mark Lewis/Gamma Liaison. Fig. 14.17: Courtesy Dr. O.L. Miller, Oak Ridge National Laboratory. **Chapter 15** Fig. 15.1: Courtesy Richard Goss, Brown University. Fig. 15.2a: (top left) Oxford Scientific Films/Animals Animals; (top right) F. Stuart Westmorland/Tom Stack & Associates. Fig. 15.2b: Courtesy Dr. Cecilio Barrera, Department of Biology, New Mexico State University. Fig. 15.3b: Courtesy Michael Pique, Research Institute of Scripps Clinic. Fig. 15.7b: Courtesy Wen Su and Harrison Echols, University of California, Berkeley. Fig. 15.8a: Courtesy Stephen Case, University of Mississippi Medical Center. Fig. 15.9: Roy Morsch/The Stock Market. **Chapter 16** Fig. 16.1a: David M. Dennis/Tom Stack & Associates. Fig. 16.1b: Courtesy Lakshmi Bhatnagor, Ph.D., Michigan Biotechnology Institute. Fig. 16.2: Art Wolfe/All Stock, Inc. Fig. 16.3: Ken Graham. Fig. 16.4a: Courtesy R.L. Brinster, Laboratory for Reproductive Physiology, University of Pennsylvania. Fig. 16.4b: John Marmaras/Woodfin Camp & Associates. Fig. 16.5: Courtesy Robert Hammer, School of Veterinary Medicine, University of

Photo Researchers; (Fig. 2) Courtesy Lennart Nilsson, Boehringer Ingelheim International GmbH. **Chapter 31** Fig. 31.1a: R. La Salle/Valan Photos. Fig. 31.1b: Richard Campbell/Biological Photo Service. Fig. 31.1c: Robert Harding Picture Library. Fig. 31.2a: Doug Perrine/DRK Photo. Fig. 31.2b: George Grall. Fig. 31.2c: Biological Photo Service. Fig. 31.2d: Densey Clyne/Ocford Scientific Films/Animals Animals. Fig. 31.3a: Chuck Nicklin. Fig. 31.3b: Robert & Linda Mitchell. Bioline: Hans Pfletschinger/Peter Arnold. Fig. 31.5: Courtesy Richard Kessel and Randy Kandon, from *Tissues and Organs*, W.H. Freeman. Fig. 31.9: C. Edelmann/Photo Researchers. Fig. 31.13: From A.P. McCauley and J.S. Geller, "Guide to Norplant Counseling", *Population Reports*, Sept. '92, Series K, No. 4, p. 4; photo courtesy Johns Hopkins University Population Information Program. **Chapter 32** Fig. 32.1a: Courtesy Jonathan Van Blerkom, University of Colorado, Boulder. Fig. 32.1b: Doug Perrine/DRK Photo. Fig. 32.2a: G. Shih and R. Kessel/Visuals Unlimited. Fig. 32.2b: Courtesy A.L. Colwin and L.H. Colwin. Fig. 32.3: Courtesy E.M. Eddy and B.M. Shapiro. Fig. 32.4: Courtesy Richard Kessel and Gene Shih. Fig. 32.10b: Courtesy L. Saxen and S. Toivonen. Fig. 32.13b: Courtesy K. Tosney, from *Tissue Interactions and Development* by Norman Wessels. Fig. 32.14: Dwight Kuhn. Fig. 32.17: Courtesy Lennart Nilsson, from *A Child Is Born*. **Part 6 Opener:** David Doubilet/National Geographic Society. **Chapter 33** Fig. 33.1: Ivan Polunin/NHPA. Fig. 33.3: Courtesy of Victor A. McKusick, Medical Genetics Department, Johns Hopkins University. Fig. 33.4a: Sharon Cummings/Dembinsky Photo Associates. Fig. 33.4b: © Ron Kimball Studios. Fig. 33.5b-c: Courtesy Professor Lawrence Cook, University of Manchester. Fig. 33.8a: COMSTOCK, Inc. Fig. 33.8b: Tom McHugh/AllStock, Inc. Fig. 33.8c: Kevin Schafer & Martha Hill/Tom Stack & Associates; (inset) Courtesy K.W. Barthel, Museum beim Solenhofer Aktienverein, Germany. Bioline: (Fig. a) Robert Shallenberger; (Fig. b) Scott Camazine/Photo Researchers; (Fig. c) Jane Burton/Bruce Coleman, Inc.; (Fig. d) R. Konig/Jacana/The Image Bank; (Fig. e) Nancy Sefton/Photo Researchers; (Fig. f) Seaphoto Limited/Planet Earth Pictures. Fig. 33.10a: John Garrett/Tony Stone World Wide. Fig. 33.10b: Heather Angel. Fig. 33.11a: Frans Lanting/Minden Pictures, Inc. Fig. 33.11b: S. Nielsen/DRK Photo. Fig. 33.12: Zeisler/AllStock, Inc. Fig. 33.13a: Zig Leszczynski/Animals Animals. Fig. 33.13b: Art Wolfe/AllStock, Inc. Fig. 33.16a: Gary Milburn/Tom Stack & Associates. Fig. 33.16b: Tom McHugh/Photo Researchers. Fig. 33.16c: Nick Bergkessel/Photo Researchers. Fig. 33.16d: C.S. Pollitt/Australasian Nature Transparencies. **Chapter 34** Fig. 34.1a: Leonard Lee Rue III/Earth Scenes/Animals Animals. Fig. 34.1b: Bradley Smith/Earth Scenes/Animals Animals. Fig. 34.4a: Courtesy Professor George Poinar, University of California, Berkeley. Fig. 34.4b: David Muench Photography. Fig. 34.5a: William E. Ferguson. Fig.

34.7a: Courtesy Merlin Tuttle. Fig. 34.7b: Stephen Dalton/Photo Researchers. Fig. 34.9: David Brill. Fig. 34.10: Courtesy Institute of Human Origins. Fig. 34.12: John Reader/Science Photo Library/Photo Researchers. **Chapter 35** Fig. 35.2a: Courtesy Dr. S.M. Awramik, University of California, Santa Barbara. Fig. 35.2b: William E. Ferguson. Fig. 35.4: Kim Taylor/Bruce Coleman, Inc. Fig. 35.5: Courtesy Professor Seilacher. Biuoline: Louie Psihoyos/Matrix. Fig. 35.11 and 35.13 Carl Buell. **Part 7 Opener:** Y. Arthus/Peter Arnold. **Chapter 36** Fig. 36.1: Courtesy Zoological Society of San Diego. Fig. 36.2a-b: David M. Phillips/Visuals Unlimited. Fig. 36.2c: Omikron/Science Source/Photo Researchers. Fig. 36.3a: Courtesy Wellcom Institute for the History of Medicine. Fig. 36.3b: Courtesy Searle Corporation. Fig. 36.5a: A.M. Siegelman/Visuals Unlimited. Fig. 36.5b: Science Vu/Visuals Unlimited. Fig. 36.5c: John D. Cunningham/Visuals Unlimited. Fig. 36.6: (top) Courtesy Dr. Edward J. Bottone, Mount Sinai Hospital, New York; (bottom) Courtesy R.S. Wolfe and J.C. Ensign. Fig. 36.7: CNRI/Science Source/Photo Researchers. Fig. 36.8: Courtesy C.C. Remsen, S.W. Watson, J.N. Waterbury and H.S. Tuper, from *J. Bacteriology*, vol 95, p. 2374, 1968. Human Perspective: Rick Rickman/Duomo Photography, Inc. Fig. 36.9: Sinclair Stammers/Science Source/Photo Researchers. Fig. 36.10: Courtesy R.P. Blakemore and Nancy Blakemore, University of New Hampshire. Bioline: (Fig. 1) Courtesy B. Ben Bohlool, NiftAL; (Fig. 2) Courtesy Communication Arts. Fig. 36.11: (left) Courtesy Dr. Russell, Steere, Advanced Biotechnologies, Inc.; (inset) CNRI/Science Source/Photo Researchers. 36.12b: Richard Feldman/Phototake. **Chapter 37** Fig. 37.1a: Jerome Paulin/Visuals Unlimited. Fig. 37.1b: Stanley Flegler/Visuals Unlimited. Fig. 37.1c: Victor Duran/Sharnoff Photos. Fig. 37.1d: Michael Fogden/DRK Photo. Figure 37.2: R. Kessel and G. Shih/Visuals Unlimited. Fig. 37.3: Courtesy Romano Dallai, Department of Biology, Universitá de Siena. Fig. 37.5a: M. Abbey/Visuals Unlimited. Fig. 37.5b: James Dennis/CNRI/Phototake. Fig. 37.8a: Manfred Kage/Peter Arnold. Fig. 37.8b: Eric Grave/Science Source/Photo Researchers. Fig. 37.9: Mark Conlin. Fig. 37.10: John D. Cunningham/Visuals Unlimited. Fig. 37.11a: David M. Phillips/Visuals Unlimited. Fig. 37.11b: Kevin Schafer/Tom Stack & Associates. Fig. 37.12a: E.R. Degginger. Fig. 37.12b: Waaland/Biological Photo Service. Fig. 37.13: Dr. Jeremy Burgess/Science Source/Photo Researchers. Fig. 37.14: Sylvia Duran Sharnoff. Fig. 37.15: Herb Charles Ohlmeyer/Fran Heyl Associates. Fig. 37.16 and 37.17a: Michael Fogden/Earth Scenes/Animals Animals. Fig. 37.17b: Stephen Dalton/NHPA. **Chapter 38** Fig. 38.1a: (left) Courtesy Dr. C.H. Muller, University of California, Santa Barbara; (inset) William E. Ferguson. Fig. 38.1b: Doug Wechsler/Earth Scenes. Fig. 38.2: T. Kitchin/Tom Stack & Associates. Fig. 38.3a: Michael P. Gadomski/Earth Scenes/Animals Animals. Fig. 38.3b: Courtesy Edward S. Ross, California Academy of Sciences. Fig.

38.3c: Rod Planck/ Photo Researchers. Fig. 38.3d: Kurt Coste/Tony Stone World Wide. Bioline: (Fig. a) Michael P. Gadomski/Photo Researchers; (Fig. b-c) E.R. Degginger; (Fig. d) COMSTOCK, Inc. Fig. 38.6a: Kim Taylor/Bruce Coleman, Inc. Fig. 38.6b: Breck Kent. Fig. 38.6c: Flip Nicklin/Minden Pictures, Inc. Fig. 38.6d: D.P. Wilson/Eric & David Hosking/Photo Researchers. Fig. 38.7: J.R. Page/Valan Photos; (inset) Stan E. Elems/Visuals Unlimited. Fig. 38.8a: Runk/Schoenberger/Grant Heilman Photography. Fig. 38.8b: John Gerlach/Tom Stack & Associates. Fig. 38.8c: Robert A. Ross/R.A.R.E. Photography. Fig. 38.11: John H. Trager/Visuals Unlimited. Fig. 38.13: John D. Cunningham/Visuals Unlimited. Fig. 38.14: Milton Rand/Tom Stack & Associates. Fig. 38.17: Cliff B. Frith/Bruce Coleman, Inc. Fig. 38.18: Leonard Lee Rue/Photo Researchers. Fig. 38.19: Anthony Bannister/Earth Scenes. Fig. 38.21a: J. Carmichael/The Image Bank. Fig. 38.21b: Sebastio Barbosa/The Image Bank. Fig. 38.21c: Holt Studios/Earth Scenes. Fig. 38.21d: Gwen Fidler/COMSTOCK, Inc. **Chapter 39** Fig. 39.1a: Courtesy Karl G. Grell, Universität Tübingen Institut für Biologie. Fig. 39.1b: François Gohier/Photo Researchers. Fig. 39.2a: Larry Ulrich/DRK Photo. Fig. 39.2b: Christopher Newbert/Four by Five/SUPERSTOCK. Fig. 39.6a: Chuck Davis. Fig. 39.7a: Robert & Linda Mitchell. Fig. 39.7b: Runk/Schoenberger/Grant Heilman Photography. Fig. 39.7c: Fred Bavendam/Peter Arnold. Fig. 39.10a: Carl Roessler/Tom Stack & Associates. Fig. 39.10b: Goivaux Communication/Phototake. Fig. 39.10c: Biomedia Associates. Fig. 39.14a: Gary Milburn/Tom Stack & Associates. Fig. 39.14b: E.R. Degginger. Fig. 39.14c: Christopher Newbert/Four by Five/SUPERSTOCK. Fig. 39.15: Courtesy D. Phillips. Fig. 39.16a: Marty Snyderman/Visuals Unlimited. Fig. 39.16b: Robert & Linda Mitchell. Fig. 39.18a: John Cancalosi/Tom Stack & Associates. Fig. 39.18b: John Shaw/Tom Stack & Associates. Fig. 39.18c: Patrick Landman/Gamma Liaison. Fig. 39.20a: Robert & Linda Mitchell. Fig. 39.20b: Edward S. Ross, National Geographic, *The Praying of Predators*, February 1984, p. 277. Fig. 39.20c: Maria Zorn/Animals Animals. Fig. 39.21: Courtesy Philip Callahan. Fig. 39.22a: Biological Photo Service. Fig. 39.22b: Kim Taylor/Bruce Coleman, Inc. Fig. 39.22c: Tom McHugh/Photo Researchers. Fig. 39.23a: E.R. Degginger. Fig. 39.23b: Christopher Newbert/Four by Five/SUPERSTOCK. Fig. 39.26b: Dave Woodward/Tom Stack & Associates. Fig. 39.26c: Terry Ashley/Tom Stack & Associates. Fig. 39.28a: Russ Kinne/COMSTOCK, Inc. Fig. 39.28b: Ken Lucas/Biological Photo Service. Fig. 39.30: Marc Chamberlain/Tony Stone World Wide. Fig. 39.31a: W. Gregory Brown/Animals Animals. Fig. 39.31b: Chris Newbert/Four by Five/SUPERSTOCK. Fig. 39.32a: Zig Leszcynski/Animals Animals. Fig. 39.32b: Chris Mattison/Natural Science Photos. Fig. 39.33a: Michael Fogden/Animals Animals. Fig. 39.33b: Jonathan Blair/Woodfin Camp & Associates. Fig. 39.34a: Bob and Clara Calhoun/Bruce Coleman, Inc. Fig.

Index

Note: A t following a page number denotes a table,
f denotes a figure, and n denotes a footnote.

◄ A ►

Abiotic environment, 934–936, 936–937
ABO blood group antigens, 240
"Abortion" pill, 679
Abscisic acid, 427t, 433
Abscission, 433
Absorption of nutrients, 568, 574
Accidental scientific discovery, 35–37
Accommodation, 503
Acetylcholine, 475t
Acetyl coenzyme A (acetyl CoA), 181
Acid rain, 392, 918
Acids, 61–62
Acoelomate, 870
Acquired immune deficiency syndrome (AIDS), 657,
    794, 796–797
Acromegaly, 526, 527f
Acrosome, 670
ACTH (adrenocorticotropic hormone), 522t, 526
Actin, 555
Action potential, 471–473
Activation energy, 123, 124f
Active immunity, 653
Active site of enzyme, 124f, 125
Active transport, 142–144
"Adam's apple," 625
Adaptation, 4–5, 728–729
    and evolution, 4–5, 18
    to food type, 577, 581
    and nitrogenous wastes, 613
    and osmoregulation, 611–613
    and photosynthesis, 164–166
    of respiratory system, 632–635
Adaptive radiation, 734
Addison's disease, 528–529
Adenine, 269, 270f
    base pairing
        in DNA, 271f, 272
        in RNA, 278
Adenosine triphosphate, *see* ATP
Adenylate cyclase, 534, 535f
ADH, *see* Antidiuretic hormone
Adipocytes, 76
Adrenal gland, 522t, 526–529
    cortex, 522t, 526, 528–529
    medulla, 522t, 529
Adrenaline (epinephrine), 522t, 529
Adrenocorticotropic hormone (ACTH), 522t, 526
Adventitious root system, 373, 374f
Aerobes, appearance on earth, 174
Aerobic respiration, 180–184
    versus fermentation, 175, 178
Aestivation, 927
Agar, 841
Age of Mammals, 774
Age of Reptiles, 771
Agent Orange, 431
Age–sex structure of population, 988
    for humans, 999, 1000f
Aggregate fruits, 412, 413f
Agnatha, 886–887
AIDS (acquired immune deficiency syndrome), 642–
    643, 657, 794, 796–797
AIDS virus, *see* Human immunodeficiency virus

"Air breathers," 633, 635
Alarm calls, 1023
Albinism, 240
Alcohol, in pregnancy, 704
Alcoholic fermentation, 178, 179
Aldosterone, 522t, 528, 610–611
Algae, 816
    as bacteria, 792
    as lichens, 821
    as plants, 834, 835–842
    as protists, 816–819
Algin, 840
Alkalinity, 61
Allantois, 703
Allele frequency, 717–718
Alleles, 234–235, 240, 241
    linkage groups and, 249, 250
Allelochemicals, 972
Allelopathy, 974–975
Allergies, 656
Alligators, 889
Allopatric speciation, 732
Allosteric site, 130f
Alpha-helix, 67, 80
Alpine tundra, 922
Alternation of generations, 833
Altruism, 1022–1026
Alveolar-capillary gas exchange, 625–626, 627
Alveoli, 625–626
Alzheimer's disease, 101, 483, 484
Ames test, 286–287
Amino acids, 79–80
    assembly into proteins, 78f, 277, 278, 284–285,
        286f
    and diet, 578
    and genetic code, 278–279, 282f
    sequencing, 66–67, 84, 750
Amino group, 78f, 79
Ammonia, excretion, 607, 613
Amniocentesis, 347f, 348
Amnion, 703
Amniotic fluid, 703
Amoebas, 813, 816
Amphibians, 888–889
    heart in, 605f
Anabolic pathways, 128
Anabolic reactions, electron transfer in, 128–129
Anaerobes, 174, 791
Anaerobic respiration, 179n
Analogous, versus homologous, 458, 744–746
Analogous features, 744
Anaphase, of mitosis, 203f, 206
Anaphase I, of meiosis, 217f, 218–219
Anaphase II, of meiosis, 217f, 221
Anaphylaxis, 656
Anatomy, and evolution, 749–750
Androgen-insensitivity syndrome, 305
Anecdotal evidence, 31
Angiosperms, 849, 852–853
Angiotensin, 591
Animal behavior, 1006–1030
Animal kingdom, 858–897
    phyla, 863, 864t
    phylogenetic relations, 865f
    phylum Annelida, 864t, 875–876
    phylum Arthropoda, 864t, 876–882
    phylum Chordata, 864t, 884–886
    phylum Cnidaria, 864t, 867–869

    phylum Echinodermata, 864t, 882–884
    phylum Mollusca, 864t, 874–875
    phylum Nematoda, 864t, 871–873
    phylum Platyhelminthes, 864t, 869–871
    phylum Porifera, 864t, 864–867
Animal models, 320
Animals:
    basic characteristics, 860–861
    body plan, 861–863
    cognition in, 1024
    desert adaptations, 926–927
    evolution, 861, 865f, 873
    form and function, 448–464
    growth and development, 686–709
Annelida, 864t, 875–876, 877f
Annual plants, 361
Annulus of fern, 847, 848f
Antarctic ozone hole, 902–903
Anteater, 893
Antennae of insects, 879
Antenna pigments, 158f, 159
Anther, 403
Antheridium, 842
Anthophyta, 839t, 852–853
Anthropoids, 894f
Antibiotics, discovery of, 37
Antibodies, 646, 647, 650–654
    formation of, 650–652
        DNA rearrangement, 651, 652f
    monoclonal, 654
    to oneself (autoantibodies), 654
    production of, 652–653
    specificity, 650
    structure, 650, 651f
Anticodon, 279–280, 283f
Antidiuretic hormone (ADH; vasopressin), 522t, 525,
    610, 611
Antigen-binding site, 650, 651f
Antigens, 650
    and antibody production, 652
    defined, 646
    presentation of, 652, 654f
Anti-oncogenes, 260
Antioxidants, and free radicals, 56
Aorta, 595
Apes, 893
    genetic relation to human, 259, 329
Aphids:
    and ants, 962f
    and phloem research, 394
Apical dominance, 430, 431f
Apical meristem, 362, 372f, 373, 375f
Aposematic coloring, 970
Appendicitis, 575
Appendicular skeleton, 548–550
Appendix, 574–575, 576f
Aquatic animals, respiration in, 620–621, 632–633
Aquatic ecosystems, 908–916
Archaebacteria, 791, 796–798
Archegonium, 842
Archenteron, 693f
Arctic tundra, 922
Arteries, 589–590, 595
Arterioles, 590
Arthropoda, 864t, 876–882
Artificial selection, 721
Asbestosis, 101

Ascaris, 873
Ascocarp, 823
Ascomycetes, 822t, 823–824
Ascorbic acid (vitamin C), 540–541, 579t
    and cancer, 31, 39–40
    and colds, 30–31
Ascospores, 823
Ascus, 823
Asexual reproduction, 199, 664–665
    in flowering plants, 417–418
Association, defined, 933
Asthma, 656
Atherosclerosis, 75, 146, 590
Athletes and steroids, 528
Atmosphere, 904
Atolls, 912
Atomic basis of life, 48–64
Atomic mass, 52
Atomic number, 52
Atoms, structure of, 51–55
ATP (adenosine triphosphate), 121–123
    in active transport, 142–143, 144f
    chemical structure, 121f
    formation of
        by chemiosmosis, 161, 184
        by electron transport, 182–184
        by glycolysis, 175–178
        by Krebs cycle, 180f, 181, 182f
        energy yield, 178, 179, 180, 182, 185, 188
        proton gradient and, 161, 184
        site of, 172–173, 184f
    and glucose formation, 163–164
    hydrolysis of, 121f, 122–123
    and muscle, 186–187, 552, 556
    photosynthesis and, 160f, 161–163
ATPase, 142
ATP synthase, 160f, 161, 173, 184
Atrioventricular (AV) node, 599
Atrioventricular (AV) valves, 597
Atrium of heart, 593f, 594
Australopithecines, 752, 754, 755f
Australopithecus species, 743, 752, 754, 755f
Autoantibodies, 654
Autoimmune diseases, 656–657
Autonomic nervous system, 490–491
Autosomal dominant disorders, 343–345
Autosomal recessive disorders, 343
Autosomes, 253
Autotoxicity, 974
Autotrophs, 154–155, 764
Auxins, 425, 427, 430–431
Aves, 889–892
AV (atrioventricular) node, 599
AV (atrioventricular) valves, 597
Axial skeleton, 548, 549f
Axillary bud, 377
Axon, 468, 469f
AZT (azidothymidine), 657

◄ B ►

Backbone, 886
    origin of, 694
Bacteria, 786–800. *See also* Prokaryotes
    beneficial, 799
    chemosynthetic, 167–168, 798
    in digestive tract, 575, 581
    disease-causing, 792
    enzyme inhibition in, 126
    in fermentation, 179
    in genetic engineering, 314, 316, 324
    in genetics research, 266–267
    gene transfer between, 804
    growth rate, 788
    metabolic diversity, 791

motility, 792
    and nitrogen fixation, 390, 799, 949–951
    structure of, 793f
    taxonomic criteria, 788–791
    ubiquity of, 799
Bacterial chlorophyll, 791–792
Bacteriophage, 268, 269f, 801
Balance, 506, 508–509
Ball-and-socket joint, 550
Bark, 369
Barnacles, 876
Barr body, 256f
Base pairing:
    in DNA, 271f, 272
    in RNA, 278
Bases, acids and, 61–62
Basidia, 824
Basidiocarp, 824–825
Basidiomycetes, 822t, 824–825
Basidiospores, 824
Basophils, 601–602
Bats, 498–499, 750, 751f
B cells, 647, 652–653
Beagle (HMS), voyage of, 20f, 21
Bean seedling development, 416, 417f
Beards, and gene activation, 305
Bees:
    dances, 1019, 1020f
    haplodiploidy in, 1025–1026
Behavioral adaptation, 729
Behavior of animals, 1006–1030
    development of, 1014–1015
    genes and, 1008–1009, 1010–1011
    mechanisms, 1008–1011
Beri-beri, 566–567
Beta-pleated sheet, 80
Bicarbonate ion, in blood, 629
Biennial plants, 361
Bile salts, 574
Binomial system of nomenclature, 13
Biochemicals, defined, 68
Biochemistry, 66–86, 750
Bioconcentration, 946
Biodiversity Treaty, 997
Bioethics, 25, 40–41. *See also* Ethical issues
Biogeochemical cycles, 947–952
Biogeography, 750–752
Biological classification, 11–15
Biological clock, of plants, 434, 438
Biological magnification, 946
Biological organization, 7, 11, 12f
Biological species concept, 732, 830
Biology, defined, 6
Biomass, 943
    pyramid of, 944
Biomes, 907f, 916–927, 934
Biosphere, 11, 902–930
    boundaries, 904
Biotechnology:
    applications, 314–321
    DNA "fingerprints," 328, 329
    genetic engineering techniques, 321–329
    opposition to, 321
    patenting genetic information, 315
    recombinant DNA techniques, 321–325. *See also*
        Recombinant DNA
    without recombinant DNA, 325–329
Biotic environment, 934–936, 936–937
Biotic (reproductive) potential, 986, 989–990
Biotin, 579t
Birds, 889–892
    fossil, 746–747, 748f
Birth:
    canal, 673
    control, 677–681

control pills, 678–679
    in human, 705
Bivalves, 874
Blade of leaf, 377
Blastocoel, 690
Blastocyst, 700
Blastodisk, 691f
Blastomere, 690, 692
Blastopore, 693f, 873
Blastula, 690, 691f
Blastulation, 690
"Blind" test, 39
Blood:
    clotting, 450, 602–603
    composition, 600–603
    flow
        through circulatory system, 594f
        through heart, 593f
    gas exchange in, 627–629
Blood-brain barrier, 110
Blood:
    pressure, 589–590. *See also* Hypertension
    type, 94, 240
    vessels, 588–593
Blooms of algae, 818
Blue-green algae, *see* Cyanobacteria
Body:
    cavities, 861
    defense mechanisms
        nonspecific, 644–645
        specific, 646–650
    fluids:
        compartments, 588
        regulation of, 606–613
    mechanics, 558–559
    organization, 696–698
        genetic control of, 686–687
    plan, 861–863
    size, 460
    surface area, 460–462
        and respiration, 622–623, 626
    symmetry, 861, 862f
    temperature, 449, 450, 451f, 588, 614
    weight, 76, 460
Body contact and social bonds, 1021
Body planes, 862f
Body segmentation, 863, 875–876
Bohr effect, 629
Bonding, 1021–1022
Bones, 546–548
    human skeleton, 548, 549f
    repair and remodeling, 550
Bony fishes, 887, 888f
Bottleneck, genetic, 720
Botulism, 478
Bowman's capsule, 608–609
Brain, 481–487
    evolution, 492
    left versus right, 504
    and sensory stimuli, 510–511
Brainstem, 481f, 486
Breast-feeding, and immunity, 655
Breathing:
    mechanics of, 626–627
    regulation of, 629, 632
Bristleworms, 875–876
Bronchi, 625
Bronchioles, 625
Brown algae, 838–840
Bryophyta, 839t
Bryophytes, 835, 842–843
Budding, in animals, 664
Buffers, 62
Bundle-sheath cells, 165f, 166
Bursitis, 551

◄ C ►

C₃ synthesis, 163–164
C₄ synthesis, 163t, 164–166
Caecilians, 888–889
Calciferol (vitamin D), 579t
Calcitonin, 522t, 531
Calcium:
    blood levels, 520, 530f, 531
    in diet, 580, 580t
    in muscle contraction, 557
Calorie, 60
Calvin–Benson cycle, 164
Calyx of flower, 403
Cambia, 362, 372–373
Camel's hump, 927
Camouflage, 967–968
cAMP (cyclic AMP), 533–534, 535f, 819
CAM synthesis, 163t, 166
Cancer, *see also* Carcinogens
    and cell cycle, 202
    and chromosomal aberrations, 259, 260
    genetic predisposition to, 346
    immunotherapy for, 648
    and lymphatic system, 606
    of skin, 280–281
    and smoking, 630
    viruses causing, 803
Capillaries, 587, 590–593
    lymphatic, 606
Capillary action, 60
Capsid, viral, 801
Capsule, bacterial, 792
Carbohydrases, 573
Carbohydrates, 70–74
    as energy source, 188
    in human diet, 578
    of plasma membrane, 94
    synthesis in plants, 163–166
Carbon:
    biological importance, 68–69
    in oxidation and reduction, 128
    properties, 68–69
Carbon cycle, 948–949
Carbon dating of fossils, 54
Carbon dioxide:
    and bicarbonate ions, 629
    in respiration, 628f, 629, 632
Carbon dioxide fixation, 163–164
    and leaf structure, 165–166
Carbonic acid, 629
Carbonic anhydrase, 629
Carbon monoxide poisoning, 627
Carboxyl group, 78f, 79
Carcinogens, 286–287
    testing for, 37–39, 286–287
Cardiac muscle, 558, 559f
Cardiac sphincter, 569
Cardiovascular system, *see* Circulatory system
Careers in biology, Appendix D
Carnivore, 966
Carotenoids, 157
Carpels, 403
Carrier of genetic trait, 254–255, 343
Carrier proteins of plasma membrane, 141, 143f
Carrying capacity (*K*) of environment, 992, 999–1001
Carson, Rachel, 714–715
Cartilage, 548
Cartilaginous fishes, 887
Casparian strips, 374, 376f, 387, 389f
Catabolic pathways, 128
Catalysts, enzymes as, 123
Cavemen, 742
Cecum, 574–575, 576f, 581
Cell body of neuron, 468, 469f
Cell cycle, 199–202

Cell death, and development, 697
Cell differentiation, 302–303, 698–699
Cell division, 194–210, 212–225. *See also* Meiosis;
    Mitosis
    and cancer, 202
    defined, 196
    in eukaryotes, 197–199
    in prokaryotes, 197
    regulation of, 201–202
    triggering factor, 194–195
    types, 196–197
Cell fate, 690–691
    induction and, 695
Cell fusion, 89
Cell-mediated immunity, 647
Cell plate, in plant cytokinesis, 206
Cells, 88–114. *See also* Eukaryotic cells; Prokaryotic
    cells
    discovery of, 90–91
    exterior structures, 108–110
    as fundamental unit of life, 91
    genetically engineered, 314
    how large particles enter, 136–137, 144–147
    how molecules enter, 138–144
    junctions between, 109–110
    locomotion, 106–107
    membrane, *see* Plasma membrane
    as property of living organism, 7, 11
    size limits, 92–93
    specialization, 302–303
    totipotent, 297
Cell surface receptor, 94, 147–148, 533–534, 535f
Cell theory, 90–91
Cellular defenses
    nonspecific, 645
    specific (T cells), 647, 649–650
Cellular slime molds, 819
Cellulose, 73, 119f
    in diet, 578, 581
Cell wall, 108
    of bacteria, 787
    of plants, 108–109
Cenozoic Era, 774
Centipedes, 876, 881–882
Central nervous system (CNS), 479, 481–489
    brain, *see* Brain
    spinal cord, 488–489
Centromere, 198, 204f, 218f
    versus kinetochore, 205n
Cephalization, 492, 861
Cephalopods, 874–875
Cerebellum, 481f, 486
Cerebral cortex, 481f, 482
Cerebrospinal fluid, 482
Cerebrum, 482–485
Certainty and science, 40–41
Cervix, 673
Cestodes, 869, 870f, 871
Chain (polypeptide) elongation and termination,
    285
Chain of reactions, 1009
Chaparral, 924, 925f
Character displacement, 965
Chargaff's rules, 272
Charging enzymes, 280, 286f
Cheetah, 730
Chelicerates, 876, 878f, 881
Chemical bonds, 55–58
Chemical defenses, 972, 974–975
Chemical energy, from light energy, 156–163
Chemical evolution, 760–761
Chemical reactions:
    electron transfer in, 128–129
    endergonic, 122–123
    energy in, 121–123

exergonic, 122
favored versus unfavorable, 122
Chemiosmosis, 161
Chemoreception, 509–510, 512
Chemoreceptors, 500, 501t
    and breathing control, 632
Chemosynthesis, 167–168
Chemosynthetic bacteria, 167–168, 798
Chest cavity, 626
Chiasmata (of chromosomes), 218f
Chimpanzee, 894f
Chitin, 73
Chlamydia infections, 795
Chlorophylls, 157
    bacterial, 791–792
Chlorophyta, 835, 837
Chloroplasts, 104
    forerunners of, 792, 793f
    and origin of eukaryote cell, 110, 765, 962
    in photosynthesis, 155, 156, 158f
Cholecystokinin, 522t, 573
Cholera, 573
Cholesterol, 75, 77f, 136–137
    and atherosclerosis, 75, 146
Chondrichthyes, 888f
Chondrocytes, 548
Chordamesoderm, 694–695
Chordata, 864t, 884–886
Chorion, 700, 703
Chorionic villus sampling, 347f, 348
Chromatids, 198, 204f, 216, 218
Chromatin, 97
    and DNA packing, 274f
Chromosomal exchange, 250, 292
Chromosomal puffs, 305
Chromosomes, 97, 246–264, 334–353. *See also* Sex
    chromosomes; Homologous chromosomes;
    Meiosis; Mitosis
    abnormalities, 221, 257–259
        deletions, 258
        duplications, 258
        inversions, 259
        translocations, 259
        wrong number, 221, 336–337
            detection in fetus, 349
            nondisjunction, 236
            polyploidy, 260–261
    banding patterns, 198f, 253f
    condensation, 203–204
    discovery of functions, 246–247
    DNA in, 273, 274f, 275f
    in eukaryotes, 273
    giant (polytene), 252–253, 305
    mapping, 252, 338
        Human Genome Project, 340
    mitotic, 198–199, 204
    number, 199
        in humans, 212–213
    in prokaryotes, 273
    structure, 273, 274f, 275f
Chrysophytes, 817–818
Chyme, 570
Chymotrypsin, 573
Cigarette smoking, 630–631, 704
Cilia:
    of cell, 106–107
    of protozoa, 814–816
Ciliata, 815t
Circadian plant activities, 438
Circulatory (cardiovascular) systems, 455, 588–605
    discovery of, 586–587
    evolution of, 603–605
    in humans, 456f, 588–600
    open versus closed, 605
Circumcision, 669

Citric acid cycle, 181
Clams, 864t, 874
Class, in taxonomic hierarchy, 15
Classical conditioning, 1011–1012
Cleavage, 690
Cleavage furrow, in cell division, 206
Climate, 905–907, 916
Climax community, 952
Clitoris, 673
Clock, biological, 434, 438
Clonal deletion, 654
Clonal selection theory, 651–653
Clones, 296–297
Clotting factors, 602
Clotting of blood, 450, 602–603
Club fungi, 824
Club mosses, 845–846
Clumped distribution patterns, 987
Cnidaria, 864t, 867–869
CNS, *see* Central nervous system; Brain
Coagulation (clotting) of blood, 450, 602–603
Coastal water habitats, 908, 910–912
Cobalamin (vitamin B$_{12}$), 579t
Cocaine, 487
Cochlea, 507f, 508
Codominance, 239–240
Codons, 278
    role in mutations, 286
    role in translation, 279–281, 284–285
Coelacanth fish, 859
Coelom, 861, 863f, 871–872, 875–876
Coenzyme, 127
Coenzyme A (CoA), 181
Coevolution, 734f, 735
Cofactors, 127
Cognition in animals, 1024
Coherence theory of truth, 41
Cohesion of water, 60
Coitus interruptus, 681
Cold (disease), 803
Coleoptile, 417
Coleoptile tip, 424–425
Collagen, 453–454, 541
Collecting duct, 608–609
Collenchyma cells, 364
Colon, 576f
Color adaptations, 968, 970
Colorblindness, 255f, 256f
Combat, 1017, 1018–1019
Commensalism, 976
Communication, 1018–1022
    functions, 1018–1019
    types of signals, 1019–1021
Communities, 11, 932–933, 960–982
    changes in, 952
    climax, 952
    pioneer, 952
Compact bone, 546, 547f
Companion cell of plant, 372
"Compass" bacterium, 798
Competition, 960–961, 963–965
Competitive exclusion, 941, 961, 964
Complement, 645
Complementarity in DNA structure, 272
Complete flower, 407
Composite flower, 406f
Compound (chemical), 55
Compound leaf, 377
Concentration gradient, 138, 142, 144f
    and energy storage, 143–144, 145f
    of protons, 160f
Condensation of molecules, 69, 70f
Conditioned response, 1011–1012
Conditioning, 1011–1012
Condoms, 681
Cones of gymnosperms, 849

Cones of retina, 503, 505f, 506
Confirmation in scientific method, 33
Conformational changes in proteins, 81
Coniferophyta, 839t, 851–852
Coniferous forests, 917, 920, 921f
Conjugation in protists, 222
Connective tissue, 453–454
    and vitamin C, 541
Consumers, 937, 942
Continental drift, 767–770
    and natal homing, 312–313
Continuous variation, 240
Contraception, 677–681
Control group, 33, 37
Controlled experiments, 37–39
Convergent evolution, 734f, 735, 736f
Conversion charts, Appendix A
Cooperation between animals, 1023–1024
Copepods, 881
Coral reefs, 911–912
Corals, 864t, 867–869
Corepressor of operon, 300f
Cork cambium, 362
Corn seedling development, 417, 418f
Corolla of flower, 403
Coronary arteries, 595, 596f
Coronary bypass surgery, 597
Corpolites, 746
Corpus callosum, 481f, 482, 504
Corpus luteum, 676f, 677
Cortex of plants, 372
Cortisol (hydrocortisone), 522t, 526, 528
Cotyledons, 411
Countercurrent flow, 633
Coupling, in chemical reactions, 122
Courtship behavior, 668, 1018
Covalent bonds, 55
    in oxidation and reduction, 128
Covalent modification of enzyme activity, 129
Cowper's gland, 672
Crabs, 864t, 876, 881
Cranial nerves, 489–490
Cranium, skull, 482
Crassulacean acid metabolism, 163t, 166
Cretinism, 531
Crick, Francis, 269–272
Cri du chat syndrome, 258
Crinoids, 864t
Crocodiles, 889
Crops, biotechnology and, 316–319, 418–419
Crosses, genetic
    dihybrid, 238
    monohybrid, 233, 234f
    test cross, 237
Crossing over, 216, 218f, 250, 251f, 292
    and genetic mapping, 252
Crustaceans, 876, 878f, 881
Cryptic coloration, 968
Curare, 478
Cutaneous respiration, 622
Cuticle of arthropods, 545
Cuticle of plants, 366
    and transpiration, 392
Cyanide, 184
Cyanobacteria, 791, 792, 796
    role in evolution, 174, 764
Cycadophyta (cycads), 839t, 849
Cyclic AMP (cAMP), 533–534, 535f, 819
Cyclic photophosphorylation, 161–163
Cystic fibrosis, 240, 337t, 338, 341f
    and evolution, 342
    gene for, 338, 341, 342f
    gene therapy for, 349
    screening test for, 347
Cysts, protozoal, 814
Cytochrome oxidase, 184

Cytokinesis:
    in meiosis, 217f, 219
    in mitosis, 206
Cytokinins, 427t, 432
Cytoplasm:
    components of, 97–105
    receptors in, 534, 535f
Cytosine, 269, 270f
    base pairing, 271f, 272
Cytoskeleton, 104–105
Cytotoxic (killer) T cells, 648

◀ D ▶

Dancing bees, 1019, 1020f
Dark (light-independent) photosynthetic reactions, 155f, 156, 158f, 163–166
Darwin, Charles, 19–24, 721–722, 750–752
    and age of earth, 746
    and childhood sexuality, 662–663
    and plant growth, 424–425
Darwin's finches, 21–23
Daughter cells, 196
Davson-Danielli model of membrane structure, 89
Day-neutral plants, 437
DDT, 714–715, 946
Deciduous forests, 917, 919–920, 921f
Decomposers, 937
Deep-sea-diving mammals, 620–621
Defecation reflex, 574
Defense adaptations, 966t, 967–972
Defense mechanisms of body
    nonspecific, 644–645
    specific, 646–650
Denaturing of protein, 81, 127
Dendrites, 468, 469f
Denitrifying bacteria, 950
Density-dependent and -independent population-control factors, 996
Deoxyribonucleic acid, *see* DNA
Deoxyribose, 278
Derived homologies, 746
Dermal bone, 543
Dermal tissue system of plants, 365, 366–369, 367f
Dermis, 542
Deserts, 926
    and mycorrhizae, 388
    and osmoregulation, 613
    plant adaptations to, 164–166, 926
Detritovores, 937, 942–943
Deuteromycetes, 822t, 825–826
Deuterostomes, 873
Development, 10, 686–709
    embryonic, *see* Embryo, development
    postembryonic, 699–700
Developmental defects, 686
Developmental "fate," 690–691, 695
Diabetes, 531–532
Dialysis, 611
Diaphragm
    anatomic, 626
    contraceptive, 681
Diatoms, 817–818
Dichloro-diphenyl-trichloro-ethane (DDT), 714–715, 946
Dicotyledons (dicots), 361, 411, 853
    seedling development, 416, 417f
    versus monocots, 361t, 372
Dieback of population, 992
Diet, *see* Food; Nutrients
Dietary deficiencies, 566–567
Diffusion, 138–139
    facilitated, 141, 143f
DiGeorge's syndrome, 687
Digestion, 568
    intracellular versus extracellular, 576

Digestive systems, 455, 566–584
  evolution of, 576–577, 581
  incomplete versus complete, 576
  in human, 456f, 568–575
    hormones of, 522t
Digestive tract, 568, 861
Dihybrid cross, 238
Dimer of thymine bases, 280
Dimorphism, 727
  sexual, 727–728
Dinoflagellates, 818
Dinosaurs, 771, 772–773
Dioecious plants, 407, 849
Diphosphoglycerate (DPG), 129
Diploid (2N) cells, 199, 214, 833
Directional selection, 725–726
Disaccharides, 70
Discontinuous variation, 240
Disruptive coloration, 968
Disruptive selection, 726–727
Distal convoluted tubule, 608–609, 610
Divergent evolution, 734
Division, in taxonomic hierarchy, 15
DNA, 82, 266–290, 312–332. *See also* Nucleic acids;
    Recombinant DNA
  amplification, 326–327, 329
  base pairs in, 271f, 272
  cloning, 324
  damage to, 280–281, 286–287
  functions, 272–285
    discovery of, 266–267
    gene expression, 277–285
    information storage, 272–275
    inheritance, 275–277
    protein synthesis, 277–285, 286f. *See also*
      Transcription
    as template, 275
  in mitosis, 203–204
  noncovalent bonds in, 57f, 58
  packaging in chromosomes, 273, 274f
  rearrangement, 651, 652f
  regulatory sites, 304
  repeating sequences, 338, 339f
  replication, 275–277
  sequencing, 325–326
    and evolution, 329, 750
    legal aspects, 315
    and turtle migration, 313
  splicing, 321
  structure, 268–272
    double helix, 271f
    Watson-Crick model, 271–272
  unexpressed, 303
DNA "fingerprints," 328
DNA ligase, in recombinant DNA, 323
DNA polymerase, 275–277
DNA probe, 326
DNA repair enzymes, 287
DNA restriction fragments, 325
Domain, in taxonomic hierarchy, 15n
Dominance, 234
  incomplete (partial), 239–240
Dominant allele, characteristic, trait, 234–235,
    236–237
  and genetic disorders, 343–345
Dopamine, 475t, 487
Dormancy in plants, 434
Dorsal hollow nerve cord, 884
Double-blind test, 39
Double bonds, 56, 75
"Double fertilization," 410
Double helix of DNA, 271f
Doves and hawks, 1017
Down syndrome (trisomy 21), 213, 221, 336, 337t
DPG (diphosphoglycerate), 129
Drosophila melanogaster, use in genetics, 249–253

Drugs, *see also* Medicines
  mood-altering, 487
  and pregnancy, 704
Duchenne muscular dystrophy, 337t, 345
Duodenum, 573
Dystrophin, 345

◁ E ▷

Ear, 506–509
  bones of, evolution, 459–460
Earth's formation, 762–763
Earthworms, 875–876, 877f
Eating, 568
Ecdysone, 305–306
Echinodermata, 864t, 882–884
Echolocation (sonar), 498–499
Eclipse phase in viruses, 800
Ecological equivalents, 942
Ecological niches, 939–941
Ecological pyramids, 943–946
Ecology, 902–930, 932–958, 960–982, 984–1004
  defined, 905
  ethical issues, 909
  and evolution, 905
Ecosystems, 11, 932–958
  aquatic, 908–916
  energy flow in, 942–946
  linkage between, 934
  nutrient recycling in, 947–952
  structure, 934–939
  succession in, 952–955
Ecotype, 831, 938
Ectoderm, 693
Ectotherms, 614
Edema, 593
Egg (ovum), 669
  human, 673–675
Egg-laying mammals, 892
Ehlers-Danlos syndrome, 337t, 541
Ejaculation, 672
Ejaculatory fluid, 672
Elaters, 846
Electrocardiogram (EKG), 599f
Electron acceptor, 158f, 159
Electron carriers:
  and NAD + conversion to NADH, 177f
  and NADP + conversion to NADPH, 160–161
Electron donor, 129
Electrons, 51, 52–55
  in oxidation, 128–129
  "photoexcited," 156, 159
  in reduction, 128–129
Electron transfer in chemical reactions, 128–129
Electron transport system, 160–161
  and ATP formation, 182–184
  in glycolysis, 177
  in photosynthesis, 160–161, 162
  in respiration, 182–185
Elements, 50–51
Elephantiasis, 873
Ellis–van Creveld syndrome, 720f, 721
Embryo, 686–709
  defined, 688
  development, 688–699
    blastulation, 690
    cleavage, 690
    energy source for, 688
    and evolution, 705
    fertilization, 689
    gastrulation, 692–693
    genetic control of, 686–687
    in humans, 700–705
    implantation, 677
    induction of, 695
    metamorphosis, 699–700

  morphogenesis, 696–697
  neurulation, 694–695
  organogenesis, 696
  in flowering plants, 411
  frozen, 679, 680
  risks to, 704
Embryology, and evolution, 750
Embryonic disk, 703
Embryo sacs of flowering plants, 408
Emigration, 988
Endangered species, 995–996
Endergonic reactions, 122–123
Endocrine glands, 521f, 522t
Endocrine systems, 457, 518–538
  digestion and, 572f
  evolution of, 534, 536
  functions, 520–521
  in humans, 456f, 523–534
  nervous system and, 523–524
  properties, 521–523
Endocytosis, 144–148
Endoderm, 693
Endodermis, 373–374, 376f
Endogenous, defined, 438
Endoplasmic reticulum (ER), 98
Endorphins, 487
Endoskeletons, 546–548
Endosperm, 408, 411
Endosperm mother cell, 408
Endospores, 792
Endosymbiosis hypothesis (endosymbiont theory), 110,
    111f, 962
Endothelin, 591
Endotherms, 614
Energy:
  acquisition and use, 10, 18, 118–123
  biological transfers of, 119f
  in chemical reactions, 121–123
  conversions of, 119, 156–163
    genetic engineering and, 316
  defined, 118
  in electrons, 53–55
  expenditure by organ systems, 457
  flow through ecosystems, 942–946
  forms of, 118
  in human diet, 578
  and membrane transport, 142–144, 145f
  pyramid of, 943–944, 946
  storage of, 118
    and ATP, 121–122, 143–144, 161
    and ionic gradients, 143–144, 145f
  yield:
    from aerobic respiration, 185, 188
    from fermentation, 179
    from glycolysis, 178, 185
    from substrate-level phosphorylation, 185
Energy conservation, law of, 119
Enkephalins, 487
Entropy, 120–121
Environment:
  adaptation to, 4–5
    photosynthesis and, 164–166
  influence on genetics, 241
  and plant growth, 427, 434–438
Environmental resistance to population growth,
    991–992
Environmental science, 909
Enzymes, 10, 123–127
  discovery of, 116–117
  effect of heat on, 127
  effect of pH on, 127
  and evolution, 131
  how they work, 124f, 125
  inhibition of, 126
  interaction with substrate, 124f, 125
  in lysosomes, 100–102

Enzymes (Cont'd)
  of pancreas, 573
  regulation of, 129–130
  of stomach, 570
  synthetic modifications of, 131
Eosinophils, 601–602
Epicotyl, 411
Epidermal cells of plants, 369
Epidermal hairs of plants, 393f
Epidermis of skin, 542
Epididymis, 670
Epiglottis, 569f, 624
Epinephrine (adrenaline), 522t, 529
Epiphyseal plates, 547f
Epiphytes, 845
Epistasis, 240
Epithelial tissue, 452–453
Epithelium, 452
ER (endoplasmic reticulum), 98
Erectile tissue, 669, 670f
Erection, 669
Erythrocytes (red blood cells), 600–601, 627
Erythropoietin, 522t
Escape adaptations, 966t, 969–970
Esophagus, 569
Essential amino acids, 578
Essential nutrients, in plants, 388, 390t
Estivation, 927
Estrogen, 75, 77f, 522t, 532, 675, 676f
Estrus, 668
Estuary habitats, 912–914, 915f
Ethical issues:
  anencephalic organ donors, 489
  facts versus values, 909
  fertility practices, 680
  gene therapy, 349
  genetic screening tests, 350
  patenting genetic sequences, 315
Ethics and biology, 25, 40–41
Ethylene gas, 427t, 432
Etiolation, 438
Eubacteria, 791–796
Euglenophyta, 818–819
Eukaryota (proposed kingdom), 809
Eukaryotes:
  chromosomes in, 273
  evolution of, 764–765
  gene regulation in, 302–308
  versus prokaryotes, 91–92, 787
Eukaryotic cells, 91–93, 95f
  cell cycle in, 201f
  cell division in, 197–199
  components, 91, 94t, 95–107
  evolutionary origin, 110, 111f, 764–765, 787–788,
    789f, 962
  glucose oxidation in, 183f
  respiration in, 180f
Eusociality, 1025–1026
Eutrophic lakes, 916, 917f
Evergreens, 851
Evolution, 4–5, 18–25, 716–740, 742–758, 760–779
  amino acid sequencing and, 84, 750
  of Animal kingdom, 861, 865f, 873
  basis of, 717–731
  chemical, 760–761
  Darwin's principles, 24–25
  DNA sequencing and, 329, 750
  of eukaryotic cell, 110, 111f, 764–766, 787–788,
    789f, 962
  evidence for, 742–758
  of fungi, 826
  genetic changes and, 258, 259, 287, 293, 303
  of humans, 742–743, 752–756, 755f
  of mammals, 892
  of organ systems, 457–460
    circulatory, 603–605

digestive, 576–577, 581
endocrine, 534, 536
excretory, 611–612
immune, 655
integument, 543–544
nervous, 491–493
respiratory, 636
skeletomuscular, 560–561
  of osmoregulation, 611–613
  patterns of, 734–735
  of plants, 833–834, 835f
  of primates, 894f
  prokaryotes and, 787, 789f
  protists and, 811–812
  viruses and, 803–804
Evolutionary relationships, 744–746
Evolutionary stable strategy, 1017
Excitatory neurons, 468
Excretion, 607
Excretory systems, 455, 606–613
  evolution of, 611–612
  in humans, 456f, 607–611
Exercise, muscle metabolism in, 186–187
Exergonic reactions, 122
Exocytosis, 99
Exogenous, defined, 438
Exons, 302f, 303, 307
Exoskeletons, 544–546, 877
Experimental group, 33, 37
Experimentation, 32
  controlled, 37–39
Exploitative competition, 963–964
Exponential growth of population, 990–991
Extensor muscles, 552–553
External fertilization, 665
Extinction, 736, 995–996
  acceleration of, 997
  Permian, 770–771
Extracellular digestion, 576
Extracellular matrix, 108, 546
Eye, 503, 505f, 506
Eyeball, 503

◀ F ▶

Facilitated diffusion, 141, 143f
Facultative anaerobes, 791
FAD, FADH₂ (flavin adenine dinucleotide), 181,
    182f, 185
Fallopian tube (oviduct), 673
  ligation (tieing) of, 681
Familial hypercholesterolemia, 146, 337t
Family, in taxonomic hierarchy, 14
Fat cells, 76
Fats, 75, 188
Fatty acids, 75, 578
Favored chemical reaction, 122
Feathers, 889–890
Feces, 574
Feedback mechanisms, 450
  and enzyme activity, 129–130
  and hormone regulation, 523
Female reproductive system, 673–677
Female sex hormones, 675, 676f, 677
Fermentation, 178–179
  and exercise, 187
  versus aerobic respiration, 175, 178
Ferns, 847, 848f
Fertility rate, 999
Fertilization, 689
  in flowering plants, 408–410
  in humans, 677
Fertilization membrane, 689
Fetal alcohol syndrome, 704
Fetus, 700
  genetic screening tests, 347–350

risks to, 704
Fever blisters, 796, 803
Fibrin, 602
Fibrinogen, 602
Fibrous root system, 373, 374f
Fighting, 1017, 1018–1019
Filament of flower, 403
Filamentous fungi, 820–821, 822
Filial imprinting, 1013–1014
Filter feeders, 577, 581
  and evolution of gills, 636
Fire algae, 818
First Family of humans, 754, 755f
Fishes, 886–887
  air-breathing, 636
  heart of, 605f
  sense organs of, 512
Fission, 664
Fitness, 1022
Fixed action patterns, 1009
Flagella:
  of bacteria, 792, 793f
  of cell, 106–107
  of protozoa, 814–816
Flatworms, 864t, 869–871, 872
Flavin adenine dinucleotide (FAD, FADH₂), 181,
    182f, 185
Fleming, Sir Alexander, 35–37
Flexor muscles, 552–553
Flight, adaptations for, 890, 891f
Florigen, 434
Flowering, 400–401, 437
Flowering plants, 852–853
  asexual (vegetative) reproduction, 417–418
  pollination, 402–407
  sexual reproduction, 402–417
Flowers, 378
  color, 239
  structure, 403
  types, 407
  why they wilt, 391
Fluid compartments of body, 588
Fluid-mosaic model of membrane structure, 89
Flukes, 869, 870f, 871, 872
Folic acid, 579t
Follicle of ovary, 673–674
Follicle-stimulating hormone (FSH), 522t, 526
  in females, 675, 676f
  in males, 672
Food, see also Nutrients
  adaptations to, 577, 581
  conversion to nutrients, 568
  sharing of, 1024–1025
  transport in plants, 393–396
Food chains, 942
Food crops, biotechnology and, 316–319, 418–419
Food webs, 942–943, 944f
"Foreskin," 669
Forests, 917–920
Form, relation to function, 15
Fossil dating, 54, 746
Fossil fuels, 949, 950
Fossil record, 746, 747, 749
Founder effect, 720–721
Frameshift mutations, 286
Free radicals, 56, 174
Freeze-fracturing, 89
Freshwater habitats, 914–916
  and osmoregulation, 612–613
Freud, Sigmund, 663
Frogs, 888–889
Fronds, 847
Fructose, 70, 71f
Fruit:
  development of, 412
  dispersal of, 412, 415f

and human civilization, 414
protective properties, 833
ripening, 432
types, 412, 413f
Fruit fly, in genetics, 249–253
Fruiting body of slime mold, 819
FSH, *see* Follicle-stimulating hormone
Function, relation to form, 15
Functional groups, in molecules, 69
Fungi, 820–826
diseases from, 822, 823–824, 825–826
evolution of, 826
imperfect, 825
nutrition of, 822
reproduction in, 822
uses of, 820
Fungus kingdom, 820–826
Fur, 542

◄ G ►

GABA (gamma-aminobutyric acid), 475t
Galactosemia, 337t
Galen, 586–587
Gallbladder secretions, 574
Gametangia, 833
Gametes, 199
in flowering plants, 407–408, 409f
in human female, 674–675
in life cycle, 214, 215f
in meiosis, 217f
Gametophytes, 408, 409f, 833
Gamma-aminobutyric acid (GABA), 475t
Gaseous nutrient cycles, 947–951
Gas exchange, 622–623
alveolar-capillary, 625–626
in lung, 627
in tissues, 627–629
Gastric juice, 570, 572
Gastrin, 522t
Gastrointestinal tract, 568. *See also* Digestive systems;
Digestive tract
hormones of, 522t
Gastropods, 874
Gastrula, 692
Gastrulation, 692–693
in humans, 703
Gated ion channels, 471
Gause's principle of competitive exclusion, 961, 964
Gel electrophoresis, 325, 326f
Gender, *see* Sex
Gene dosage, 256–257, 337
Gene flow, 719
Gene frequency, 717–718
Genentech, 314
Gene pool, 717
Gene products, 239, 338
Generative cell, 408, 409f
Gene regulatory proteins, 295, 298
in eukaryotes, 304
in prokaryotes, 301
Genes, 246–264, 266–290, 292–310. *See also* Alleles;
Mutations
and behavior, 1008–1009, 1010–1011
duplication, 258
exchange of, *see* Crossing over
expression of, 277–285
visualization of, 305–306
homeotic, 686–687
and hormone action, 534, 535f
mapping, 252, 338, 341, 342f
Human Genome Project, 340
rearrangements, 651, 652f
regulation of expression, 292–310
in eukaryotes, 302–308
need for, 294

in prokaryotes, 298–301
selective expression, 294, 295f
split, 302f, 303
structural, 299, 300f-301f
transfer between bacteria, 804
transposition, 293
Gene therapy, 348–349
Genetic code, 272, 278–279, 282f
Genetic disorders, 334–350, 337t. *See also*
Chromosomes, abnormalities
gene therapy for, 348–349
tests for, 346–350
in fetus, 347–350
X-linked, 254
Genetic drift, 719, 720–721
Genetic engineering, *see* Biotechnology;
Recombinant DNA
Genetic equilibrium, 718
Genetic information
and living organism, 10
storage in DNA, 272–275
universality of, 18
Genetic markers, 249
in gene mapping, 338, 342f
Genetic mosaic, 256f, 257
Genetic recombination, 216, 218f, 219f
Genetics, 230–244, 246–264, 266–290, 292–310,
312–332, 334–353. *See also* Inheritance
Mendelian, 232–239
Genetic variability, 5
and meiosis, 214–216, 219
and sexual reproduction, 665
Genital herpes, 796
Genitalia, in human:
female, 673
male, 669
Genome, defined, 340
Genotype, 235
Genus:
in binomial system, 13
in taxonomic hierarchy, 14
Geography, and evolution, 750–752
Geologic time scale, 764, 765f
German measles, and embryo, 704
Germ cells, 214
Germination, 416
Germ layers, 693
Germ theory of disease, 785
Giant (polytene) chromosomes, 252–253, 305
Gibberellins, 427t, 431–432
Gibbon, 894f
Gill filaments, 633
Gill(s), 632–633, 634f
evolution of, 636
slits, 706, 884
Ginkgophyta, 839t, 849
Girdling of plants, 384–385
Gizzard, 570
Glans penis, 669
Global warming, 950
Glomerular filtration, 609–610
Glomerulus, 608
Glottis, 569f, 624
Glucagon, 522t, 531, 532f, 533–534
Glucocorticoids, 526, 528
Glucoreceptors, 572
Glucose, 70, 71f
blood levels, 531, 532f
energy yield from, 175, 185, 188
oxidation, 174–188
in photosynthesis, 163–164
Glycerol, 75
Glycine, 475t
Glycogen, 72–73
Glycolysis, 175–178, 176f
energy yield from, 185

Glycoproteins of plasma membrane, 94
Gnetophyta, 839t, 849
Goiter, 529, 531
Golgi complex, 98–99, 100
Gonadotropic hormones (gonadotropins), 522t, 526
in females, 675, 676f
in males, 672
Gonadotropin-releasing hormone (GnRH):
in females, 675, 676f
in males, 672
Gonads, 532
Gondwanaland, 769
Gonorrhea, 794, 795
Gorilla, 893f, 894f
Gradualism, 737
Graft rejection, 650
Grain, and human civilization, 414
Granulocytes, 601–602
Grasshopper, 878, 879f
Grasslands, 922–923
Graves' disease, 531
Gravitropism, 438
Gray crescent, 692
Gray matter:
of brain, 481–482, 481f
of spinal cord, 488f, 489
Green algae, 835, 837–838
Greenhouse effect, 798, 950
Grooming of others, 1022
Ground tissue system of plants, 365, 367f, 372
Group courtship, 1018
Group defense responses, 970
Group living, 1018
Growth, 10
in animals, 686–709
in plants, 362, 424–440
of populations, 986, 988–1001
Growth curves, 991, 993–994
Growth factor, 202
Growth hormone (GH; somatotropin), 522t, 525,
527f, 534f
athletes and, 528
Growth regulators, 432
Growth ring, 370
Guanine, 269, 270f
base pairing, 271f, 272
Guard cells, 368
Guilds, 941
Gut, 568, 861
Guthrie test for phenylketonuria (PKU), 335
Guttation, 391
Gymnosperms, 847, 849

◄ H ►

Habituation, 483, 1011
Hagfish, 886–887
Hair, 542, 892
Hair cells of ear, 506, 507f, 508
Half-life of radioisotopes, 54
Halophilic bacteria, 796
Haplodiploidy, 1025–1026
Haploid (1N) cells, 199, 214, 216, 833
"Hardening" (narrowing) of the arteries, 75, 146
Hardy-Weinberg law, principle, 718, Appendix C
Harvey, William, 586–587
Haversian canals, 547f
Hawks and doves, 1017
Hearing, 506–508
in fishes, 512
Heart, 593–600
blood flow in, 593f
changes at birth, 705
evolution of, 605f
excitation and contraction, 598–599
pulse rate, 599–600

Heart (Cont'd)
  valves of, 597
Heart attack, 590, 597, 602–603
Heartbeat, 597–600
"Heartburn," 569
Heat, plant adaptations to, 164–166
Helper T cells, 649
Helping behavior, 1023–1024
Hemocoel, 861, 878
Hemoglobin:
  and gene duplication, 258, 302f
  as oxygen carrier, 627
  shape, 80f
  in sickle cell anemia, 81, 82f
Hemophilia, 337t, 602
  gene therapy for, 349
  inheritance of, 254–255
Herbaceous plants, 362
Herbivore, 966
Herbivory, 966
Heredity, defined, 232. *See also* Inheritance
Hermaphrodite, 667
Heroin, 487
Herpes virus, 796, 803
Heterocyst, 796
Heterosporous plants, 845
Heterotrophs, 155
Heterozygous individual, 235
Hierarchy, taxonomic, 14–15
Hierarchy of life, 11
Hinge joint, 550
Hippocampus, 483, 486
Hirudineans, 875–876
Histocompatibility antigens, 650
Histones, 273, 274f
HIV *see* Human immunodeficiency virus
Homeobox, 687
Homeostasis, 10, 15, 448–449
  and circulatory system, 588
  effect of environment, 455
  and energy expenditure, 457
  mechanisms for, 450
Homeotic genes, 686–687
Hominids, 742–743, 752–756, 755f
Homo erectus, 743, 752, 755f
Homo habilis, 753, 755f
Homologous, versus analogous, 458, 744–746
Homologous chromosomes, 199
  in genetic recombination, 216, 218f
  independent assortment, 219, 220f
Homology, 705, 744–746
Homoplasy, 744
Homo sapiens, 755f
Homo sapiens neanderthalensis, 755f
Homo sapiens sapiens, 752f, 755f
Homosexuality, hypothalamus and, 663
Homosporous plants, 843
Homozygous individual, 235
Hookworms, 873
Hormones, 518–538
  for birth control, 678–679
  discovery of, 521–522
  in labor, 705
  list of, 522t
  of plants, *see* Plants, hormones
  receptors for, 533–534, 535f
  regulation of, 523
  in reproduction, 669
    female, 675, 676f
    male, 672–673
  types, 522
  and urine formation, 610–611
Horsetails, 846–847
Hot dry environment, 164–166. *See also* Desert
Human chorionic gonadotropin (HCG), 677
Human Genome Project, 340

Human immunodeficiency virus (HIV; AIDS virus), 643, 794, 797, 803
  discovery of, 642–643
  and helper T cell, 649
  in pregnancy, 704
Humans:
  body cavities, 863f
  body systems, 456f
    circulatory, 456f, 588–600
    digestive, 456f, 568–575
    endocrine, 456f, 523–534
    excretory, 456f, 607–611
    immune, 456f, 642–660
    integumentary, 456f
    lymphoid (lymphatic), 456f, 646f
    nervous, 456f, 481–491
    reproductive, 456f, 667–677
    respiratory, 456f, 623–627
  chromosome number, 212–213
  embryonic development, 700–705
  evolution, 742–743, 752–756, 755f
  First Family, 754, 755f
  genetically engineered proteins, 314, 315t
  genetic disorders, 334–350
  nutrition, 578–580
  population growth, 997–1001
    doubling time, 998
    earth's carrying capacity and, 999–1001
    fertility rate and, 999
  as research subjects, 39–40
  skeleton, 548–551
  vestigial structures, 749
Hunger, 572
Huntington's disease, 337t, 343–345
  screening test for, 347
Hybridization, 733–734
Hybridomas, 654
Hybrids, 233
Hydras, 864t, 867–869
Hydrocortisone (cortisol), 522t, 526, 528
Hydrogen bonds, 58
  and properties of water, 59
Hydrogen ion, and acidity, 61–62
Hydrologic cycle, 947–948
Hydrolysis, 69, 70f
Hydrophilic molecules, 59
Hydrophobic molecules, 57f, 58
Hydroponics, 387–388
Hydrosphere, 904
Hydrostatic skeletons, 544
Hydrothermal vent communities, 167
Hypercholesterolemia, 146, 337t
  and membrane transport, 136–137
Hypertension, 590, 591, 610
Hypertonic solution, 140, 141f
Hyperventilation, 632
Hyphae, 820
Hypocotyl, 411
Hypoglycemia, 578
Hypothalamus, 481f, 486
  and body temperature, 449
  in human sexuality, 663
  and pituitary, 524, 525f
  and urine volume, 611
Hypothesis, scientific, 32, 33
Hypotonic solution, 140, 141f

◀ I ▶

Ice, properties of, 60–61
Ice-nucleating protein, 316
Icons for themes in this book, 15
Ileocecal valve, 576f
Immigration, 988
Immune response, 650
  secondary, 653

Immune system, 457, 642–660
  components, 647
  disorders of, 656–657
  evolution of, 655
  and fetus, 705
  mechanics of, 646–647, 649–650
  and viral diseases, 803
Immune tolerance, 654
Immunity:
  active, 653
  passive, 655
Immunization, 646, 654–655
Immunodeficiency disorders, 657, *See also* AIDS
Immunoglobulins, 650. *See also* Antibodies
Immunological memory, 652–653
Immunological recall, 653
Immunotherapy for cancer, 648
Imperfect flower, 407
Imperfect fungi, 825
Implantation of embryo, 677
Impotence, 669
Imprinting, 1013–1014
Inbreeding, 730–731
Inclusive fitness, 1022
Incomplete (partial) dominance, 239–240
Incomplete flower, 407
Incubation period, 643
Independent assortment:
  of homologous chromosomes, 219, 220f
  Mendel's law of, 238–239
Indoleacetic acid, 430
Inducer of operon, 300f
Inducible operon, 300f, 301
Inductive reasoning, 40
Indusium, 847
Industrial products, genetic engineering and, 316
Infertility, 677, 679, 680
Inflammation, 645
Inheritance:
  autosomal dominant, 343–345
  autosomal recessive, 343
  and genetic disorders, 343–346
  polygenic, 240–241
  predicting, 235–237
  sex and, 253–257
  X-linked, 345–346
Inhibiting factors of hypothalamus, 524
Inhibitory neurons, 468
Initiation (start) codon, 281, 284
Initiator tRNA, 284
Innate behavior, 1009–1010
Insecticides, *see also* Pesticides
  genetically engineered, 317
  neurotoxic, 478
  plant hormones as, 434, 975
Insects, 864t, 876, 878–881
  metamorphosis in, 305–306
  respiration in, 633, 635f
Insight learning, 1012–1013
Insulin, 522t, 531, 532f
  discovery of, 518–519
  genetically engineered, 314
Integument, 542–544
Integumentary systems, 456f, 457
Interactions between animals, types and outcomes, 963t
Intercellular junctions, 109–110
Intercostal muscles, 627
Interference competition, 963–964
Interferon, 645
Intergene spacers, 302f
Interkinesis, in meiosis, 219
Interleukin II, 649
Internal fertilization, 665
Interneurons, 470
Internode of plant stem, 377
Intertidal habitats, 912, 913f, 914f

Intestinal secretions, 572–573
Intestine:
 large 574–575
 small, 572–574
Intracellular digestion, 576
Intrauterine device (IUD), 679, 681
Intrinsic rate of increase (r_o), 990
Introns, 302f, 303, 306–307
Invagination hypothesis of eukaryote origin, 110, 111f, 765
Invertebrates, 863–886
In vitro fertilization, 679
Iodine, and thyroid, 529
Ion, defined, 58
Ion channels of plasma membrane, 139, 140f, 471
Ionic bonds, 57, 58
Ionic gradient, 143–144, 145f
Irish potato blight, 820
Iron in diet, 580, 580t
Islets of Langerhans, 518, 531, 532f
Isolating mechanism, 732, 733t
Isotonic solution, 140, 141f
Isotope, 52
IUD (intrauterine device), 679, 681

◀ J ▶

Jaundice, 574
Java Man, 743
Jaws, 887
Jellyfish, 864t, 867–869
Jenner, Edward, 646
"Jet lag," 533
Joints, 550–551
J-shaped curve, 991
"Jumping genes," 293
Junctions between cells, 109–110

◀ K ▶

K (carrying capacity) of environment, 992
 K-selected reproductive strategies, 995–996
Kangaroo, 893
Kelp, commercial uses, 840
Kidney, 607–611
 artificial, 611
 function, 609–610
  regulation of, 610–611
 hormones of, 522t
Kidney failure, 611
Killer (cytotoxic) T cells, 648
Kinetic energy, 118
Kinetochore, 204f, 205–206, 217f, 219f
 versus centromere, 205n
Kingdoms, 13f, 14f, 15
 Animal, 858–897
 Fungus, 820–826
 Plant, 830–856
 Monera (prokaryotes), 786–800
 Protist (Protoctista), 810–819
 taxonomy of, 808–809
Kin selection, 1022–1023
Kinsey, Alfred, 663
Klinefelter syndrome, 337, 337t
Knee joint, 551f
Koala, 892f, 893
Koch, Robert, 785
Krebs cycle, 180–182
Kwashiorkor, 579

◀ L ▶

Labia, 673
Labor, 705
Lac (lactose) operon, 300f, 301
Lactase, 574
Lacteals, 574
Lactic acid fermentation, 178, 187

Lactose, 70
Lake habitats, 915–916, 917f, 918
Lamellae:
 of bone, 547f
 of gills, 633
Lampreys, 886
Lancelets, 864t, 884–886
Language, 485
Lanugo, 703
Large intestine, 574–575
Larva, 688
Larynx, 624–625
Latent viral infection, 803
Lateral roots, 373, 374f
Laurasia, 769
Laws, principles, rules:
 Chargaff's, 272
 competitive exclusion, 941, 961
 energy conservation, 119
 Hardy-Weinberg, 718, Appendix C
 independent assortment, 238–239
 the minimum, 938–939
 natural selection, 24–25
 segregation, 235
 stratigraphy, 746
 thermodynamics, 119–120
LDLs (low-density lipoproteins), 136–137, 146
 receptors for, 137, 146, 147f
Leaf, 377–378
 and carbohydrate synthesis, 165–166
 simple versus compound, 377
Leaflets, 377
Leaf nodules, 396
Leakey: Louis, Mary, Richard, 753
Learning, 483, 485, 1011–1014
Leeches, 875–876
Lek display, 1018
Lemur, 894f
Lens of eye, 503, 505f
Lenticels, 369
Lesch-Nyhan syndrome, 336, 337t
Leukocytes (white blood cells), 601–602
LH, see Luteinizing hormone
Lichens, 821
Liebig's law of the minimum, 938–939
Life:
 origin of, 760–761, 763–764
 properties of, 6–7, 8f-9f
Life cycle, 214, 215f
Life expectancy, and population density, 989
Ligaments, 551
Light:
 energy from, 156–163
 and plant growth, 435–438
Light-absorbing pigments:
 of eye, 506
 of plants, 156, 157–159
Light-dependent photosynthetic reactions, 155f, 156–163
Light-independent (dark) photosynthetic reactions, 155f, 156, 158f, 163–166
Lignin, 109
Limbic system, 486–487
Limiting factors, 938–939
Linkage groups, 248–249
Lipases, 573
Lipid bilayer of plasma membrane, 89, 138
Lipids, 71t, 75
 in human diet, 578
Lithosphere, 904
Liver, 574
Living organisms
 classification of, 11–15
 properties of, 6–11
Lizards, 889
Lobes of cerebrum, 482
Lobsters, 881

Locomotion, 558–559
Logistic growth, 994
Long-day plants, 400, 437
Loop of Henle, 608–609, 610
Loris, 894f
Lotka-Volterra equation, 961
Low-density lipoproteins (LDLs), 136–137, 146
 receptors for, 137, 146, 147f
Lucy (hominid), 753–754, 755f
Lungs, 623, 625–626, 635
 changes at birth, 705
 evolution of, 636
 gas exchange in, 627
Lupus (systemic lupus erythematosus), 657
Luteinizing hormone (LH), 522t, 526
 in females, 675, 676f
 in males, 672
Lycophyta (lycopods), 839t, 845–846
Lyme disease, 792
Lymphatic systems, 456f, 606
Lymph nodes, 606
Lymphocytes, 647. See also B cells; T cells
Lymphoid tissues, 646f, 647
Lysosomes, 100–102

◀ M ▶

Macroevolution, 731
Macrofungi, 821
Macromolecules, 69, 70f, 71t
 evolution of, 83–84
Macronutrients, in plants, 388, 390t
Macrophages, 602, 647
Magnesium, in diet, 580, 580t
Malaria, 813, 815f
 and sickle cell anemia, 723f, 724, 812
Male reproduction, 669–673
 hormonal control, 672–673
Male sex hormones, 672–673
Malignancy, see Cancer; Tumor
Mammals, 892–894
 Age of, 774
 egg-laying, 892
 evolution of, 892
 heart in, 605f
Mapping of genes, 252, 338, 341, 342f
 Human Genome Project, 340
Marine habitats, 908–912
 and osmoregulation, 612
Marrow of bones, 546, 547f
Marsupials, 892f, 893
Mastigophora, 815t
Mating, nonrandom, 719, 727–728, 730–731
Mating types of algae, 837
Matter, defined, 51
McClintock, Barbara, 292–293
Mechanoreceptor, 500, 501t
Medicines, genetically engineered, 314, 321
Medulla, 481f
Medusas, 868–869
Megaspores, 407, 408
Meiosis, 199, 212–225
 chromosomal abnormalities and, 221
 defined, 214
 and genetic variability, 214–216, 219
 in human female, 674–675
 importance, 214–216
 reduction division in, 216
 stages, 216–221
 versus mitosis, 200f, 216, 222
Meiosis I, 216, 217f
Meiosis II, 217f, 219, 221
Meissner's corpuscles, 502f
Melanoma, 280–281
Melatonin, 522t, 533
Membrane invagination hypothesis, 110, 111f, 765

Membrane potential, 470–471
Membranes, 136–150
 of cell, *see* Plasma membrane
 in cytoplasm, 97–98
 of ear (eardrum), 507f, 508
 of mitochondria, 102–104
 of nucleus (nuclear envelope), 97
 organization in, 160
 transport across, 136–144
  active, 142–144
  passive diffusion, 138–139
Memory, 483
 cells, 653
Mendel, Gregor, 232–235
Mendelian genetics, 232–239
 exceptions to, 239–241
Mendel's law of independent assortment, 238–239
Mendel's law of segregation, 235
Meninges, 482
Menopause, 675
Menstrual (uterine) cycle, 675, 677
Menstruation, 676f, 677
Meristem, 362–363
 of root, 373, 375f
 of stem, 372
Mesoderm, 693
Mesophyll, 377
Mesophyll cells, and photosynthesis, 165f, 166
Mesotrophic lake, 916, 917f
Mesozoic Era, 771, 774
Messenger RNA (mRNA)
 transcription, 278, 279f
  primary transcript processing, 306–307
 translation, 279–280, 284–285, 286f
  and control of gene expression, 307–308
Metabolic intermediates, 127
Metabolic ledger, 185, 188
Metabolic pathways, 127–130
 and glucose oxidation, 188–189
Metabolism:
 as property of living organism, 10
 regulation of, 129–130
Metamorphosis, 699–700
 in insects, 881
  ecdysone and, 305–306
Metaphase, of mitosis, 203f, 205–206
Metaphase I, of meiosis, 217f, 218–219
Metaphase II, of meiosis, 217f, 221
Metaphase plate, 217f, 219
Methane, 55f, 56
Methane-producing bacteria, 796, 798
Metric conversion chart, Appendix A
Microevolution, 731
Micronutrients, in plants, 388, 390t
Micropyle, 408
Microscopes, Appendix B
Microspores, 407, 408, 409f
Microvilli, 574, 575f
Midbrain, 481f
Milk, 892
 intolerance of, 574
Millipedes, 876, 882
Mimicry, 971
Mineralocorticoids, 528
Minerals:
 absorption in plants, 387–391
 in human diet, 580
Minimum, Liebig's law of the, 938–939
"Missing link," 743
Mites, 876
Mitochondria, 102–104
 and ATP synthesis, 172–173, 184
 and origin of eukaryote cell, 110, 765, 962
Mitosis:
 defined, 199
 phases, 202–206

versus meiosis, 200f, 216, 222
Mitosis promoting factor (MPF), 195, 202
Mitotic chromosome, 273, 274f, 275f
Molds, 821
Molecular biology, 268, 750
Molecular defenses
 nonspecific, 645
 specific (antibodies), 647
Molecules, 55, 57
 chemical structure, 69
Mollusca, 864t, 874–875
Molting, 878
Monera kingdom, 786–800. *See also* Bacteria; Prokaryotes
Monkeys, 893, 894f
Monoclonal antibodies, 654
Monocotyledons (monocots), 361, 411, 853
 seedling development, 417, 418f
 versus dicots, 361t, 372
Monocytes, 602
Monoecious plants, 407, 851
Monohybrid cross, 233, 234f
Monomer, 69, 70f
Monosaccharides, 70
Monotremes, 892–893
Mood-altering drugs, 487
Morgan, Thomas Hunt, 249–250
"Morning after" pill, 679
Morphine, 487
Morphogenesis, 696–698
Morphogenic induction, 695
Morphological adaptation, 728
Mortality, and population density, 988, 989
Mosaic, genetic, 256f, 257
Mother cell, 196
Motor cortex, 485
Motor neurons, 470
Mouth, in digestion, 568–569
MPF (mitosis promoting factor), 195, 202
mRNA, *see* Messenger RNA
Mucosa of digestive tract, 568
Mud flats, 912, 913f
Multicellular organisms, origin of, 765–766
Multichannel food chains, 942, 944f
Multiple fruits, 412, 413f
Multiple sclerosis, 473
Muscle fiber, 186–187, 553, 554f
Muscles, 454, 551–558
 cardiac, 558, 559f
 contraction of, 555–557
 in nerve impulse transmission, 474f, 475t, 476f,
  479–480, 556
 skeletal, *see* Skeletal muscle
 smooth, 558
Muscular dystrophy, 345–346
Muscular systems, 456f, 457
Mushrooms, 824–825
Mussels, 874
Mutagens, 252, 286–287
Mutant alleles, 241
Mutations, 241–242, 719
 and genetic diversity, 242
 molecular basis, 286–287
 point mutations, 241
Mutualism, 978
Mycorrhizae, 386–387, 388
Myelin sheath of axon, 469f, 470, 473
Myofibrils, 553, 554f
Myosin, 555
Myriapods, 876, 881–882

◄ N ►

NAD⁺, NADH (nicotinamide adenine dinucleotide):
 energy yield from, 185
 in fermentation versus respiration, 178
 in glycolysis, 176f, 177

in Krebs cycle, 181, 182f
NADP⁺, NADPH (nicotinamide adenine dinucleotide phosphate):
 as electron donor, 129
 in photosynthesis, 159f, 161
 as reducing power, 189
Names, scientific, 13
Nasal cavity, 624
Natal homing, 312–313
Natality, 988
Natural killer (NK) cells, 645
Natural selection, 23–25, 719, 721–727
 patterns of, 724–727
 pesticides and, 715
 principles of, 24–25
Neanderthals, 742, 755f
Nectaries, 405
Nematoda, 864t, 871–873
Neodarwinism, 717
Nephron, 607–609
Nerve cells, *see* Neurons
Nerve gas, 478
Nerve growth factor (NGF), 466–467
Nerve impulse transmission, 470–473, 556
Nerve net, 492
Nerves, 479
Nerve tissue, 454f, 455
Nervous systems, 457, 466–496, 498–516
 central, 479, 481–489
 evolution, 491–493
 of human, 456f, 481–491
 origin of, 694–695
 peripheral, 479, 489–491
 role in digestion, 572
 of vertebrates, 479–481
Neural circuits, 479–481
Neural plate, 694
Neural tube, 694, 884
Neuroendocrine system, 523–524
Neuroglial cells, 470
Neurons, 468–470
 form and function, 468, 469f
 target cells, 468
 types, 470
Neurosecretory cells, 523, 525f
 in evolution, 534, 536
Neurotoxins, 478, 818
Neurotransmission (synaptic transmission), 474–479
Neurotransmitters, 474–477, 475t
Neurulation, 694–695
Neutrons, 51
Neutrophils, 601–602
Niche overlap, 941, 963
Niches, 939–941
Nicotinamide adenine dinucleotide, *see* NAD⁺,
 NADH
Nicotinamide adenine dinucleotide phosphate, *see*
 NADP⁺, NADPH
Nicotine, 630
Nicotinic acid (niacin), 579t
Nitrifying bacteria, 950
Nitrogen, excretion and, 607, 613
Nitrogen cycle, 949–951
Nitrogen fixation, 390–391
 by bacteria, 390, 799, 949–951
 by cyanobacteria, 796
 genetic engineering and, 316
Nitrogenous bases:
 in DNA, 269–271
 in RNA, 82, 83f, 278
Nodes of plant stem, 377
Nodes of Ranvier, 469f, 470, 473
Nomenclature, binomial system, 13
Noncovalent bonds, 57–58
Noncyclic photophosphorylation, 160f, 161, 162t
Nondisjunction, 336

Nongonococcal urethritis, 795
Nonpolar molecules, 56–57
Nonrandom mating, 719, 727–728, 730–731
Nonsense (stop) codons, 281, 285
Nonvascular plants, 839t
Norepinephrine (noradrenaline), 475t, 522t, 529
Norplant, 678
Nostrils, 624
Notochord, 694, 884
Nucleases, 573
Nucleic acids, 71t, 81–83. *See also* DNA; RNA
    sequencing of, 325
    structure, 82–83, 269–271
    in viruses, 800
Nucleoid of prokaryotic cell, 92
Nucleoli, 97
Nucleoplasm, 97
Nucleosome, 273, 274f
Nucleotides, 82–83
    in DNA, 269–271
    and genetic code, 278
    and genetic information, 272
    and protein assembly, 277
    in RNA, 278
    structure, 269–271
Nucleus, 92, 95–97
    envelope (nuclear membrane), 97
    matrix, 97
    transplantation of, 297
Nutrient cycles, 947–952
Nutrients, *see also* Food
    in human diet, 578–580
    in plants, 388, 390t
Nutritional deficiencies:
    in humans, 566–567
    in plants, 390t

◄ O ►

Obesity, 76
Obligate anaerobes, 791
Ocean:
    ecosystems, 908
    floor, hydrothermal vent communities in, 167
Octopuses, 874–875
Oils, 75
Olfaction, 509, 510
Oligochaetes, 875–876
Oligotrophic lake, 916, 917f
Omnivore, 966
Oncogenes, 202
Oocyte, 673, 674
Oogenesis, in humans, 673–675
Oogonia, 674, 675f
Open growth of plants, 362
Operant conditioning, 1012
Operator region of operon, 299
Operon, 299–301
Opiates, 487
Opossum, 893
Optimality theory, 1016–1017
Oral contraceptives, 678–679
Orangutan, 894f
Order, in taxonomic hierarchy, 15
Organ donations, ethical issues, 489
Organelles, 91, 95–107
Organic chemicals, defined, 68
Organogenesis, 696, 697–698
Organs, defined, 455
Organ systems, 455–460
    energy expenditure, 457
    evolution of, 457–460
    in humans, *see* Humans, body systems
    types, 455–457
Organ transplants, 650
Orgasm, 672

Origin of Species, 24
Osmoregulation, 606–613
    adaptations of, 611–613
    defined, 607
Osmoregulatory (excretory, urinary) systems, 455,
    456f
Osmosis, 140–141
Osteichthyes, 888f
Osteoblasts, 550
Osteoclasts, 550
Osteocytes, 546, 547f
Osteoporosis, 550
Ovarian cycle, 675
Ovaries of flower, 403
Ovaries of human, 522t, 532, 673–674
Overstory of forest, 917
Oviduct (fallopian tube), 673
Ovulation, 674, 676f
Ovules, 408, 852
Ovum, 669
    of human, 673–675
Oxidation, 128–129
Oxygen:
    appearance on earth, 174, 764
    as electron acceptor, 182, 184
    release into tissues, 594f, 627–629
    toxic effects, 174
    transport, 600–601
    uptake in lungs, 594f, 627
Oxyhemoglobin, 627
Oxytocin, 522t, 525
Oysters, 874
Ozone layer, 902–903

◄ P ►

PABA (para-aminobenzoic acid), 126
Pacemaker of heart, 110, 598, 599
Pacinian corpuscle, 502
Pain receptor, 500, 501t
Paleozoic Era, 767–771
Palisade parenchyma, 378
Pancreas, 522t, 531–532, 573
Pancreatitis, 573
Pangaea, 769
Panting, 614
Pantothenic acid, 579t
Paper chromatography, 153
Parallel evolution, 734f, 735
Parapatric speciation, 732
Parasitism, 872, 972–974
Parasitoids, 972
Parasympathetic nervous system, 490–491
Parathyroid glands, 522t, 530f, 531
Parathyroid hormone (PTH), 522t, 531
Parenchyma cells, 363, 364f
Parthenocarpy, 418
Parthenogenesis, 664
Partial (incomplete) dominance, 239–240
Parturition (birth), 705
Passive immunity, 655
Pathogenic bacteria, 792
Pauling, Linus, 30–31, 39–40, 67
Pedicel, 403
Pedigree, 255, 256f
Peking Man, 743
Pelagic zone, 908
Penicillin, 35–37, 126
Penis, 665, 669, 670f
PEP (phosphoenolpyruvate) carboxylase, 166
Pepsin, 570
Pepsinogen, 570
Peptic ulcer, 570, 572
Peptide bonds, 78f, 80
Peptidoglycan, 787

Perennial plants, 361, 362–363, 366
Perfect flower, 407
Perforation plate in xylem, 370
Perforins, 647
Pericycle, 373, 376f, 377
Periderm, 369
Periosteum, 547f
Peripheral nervous system, 479, 489–491
Peristalsis, 569, 571f
Permafrost, 922
Pesticides, *see also* Insecticides
    bioconcentration of, 946
    fungi as, 823
    genetically engineered, 317
    natural selection and, 715
Petals, 403
Petiole of leaf, 377
PGA (phosphoglyceric acid), 163–164
PGAL (phosphoglyceraldehyde), 129, 164
pH, 61–62
    and blood gas exchange, 629
Phaeophyta, 838
Phage (bacteriophage), 268, 269f, 801
Phagocytic cells, 601–602, 644f, 645
Phagocytosis, 144–147
Pharyngeal gill slits, 706, 884
Pharynx, 569, 624, 636
Phenotype, 235
    environment and, 241
Phenylketonuria (PKU), 334–335, 337t
Pheromones, 1021
Philadelphia chromosome, 259, 260
Phimosis, 669
Phloem, 369–370, 372
    and food transport, 393–396
Phosphoenolpyruvate (PEP) carboxylase, 166
Phosphoglyceraldehyde (PGAL), 129, 164
Phosphoglyceric acid (PGA), 163–164
Phospholipids, 71t, 75, 77f
    in plasma membrane, 77f, 89, 138
Phosphorus, in diet, 580, 580t
Phosphorus cycle, 951–952
Phosphorylation, 175, 177–178. *See also*
    Photophosphorylation
"Photoexcited" electrons, 156, 159
Photolysis, 160
Photons, 156
Photoperiod, 434
    plant responses to, 400–401, 435–438
    and reproduction, 668
Photoperiodism, 435
Photophosphorylation, 161
    cyclic, 161–163, 162t
    noncyclic, 160f, 161, 162t
Photoreceptor, 500, 501t
Photorespiration, 166
Photosynthesis, 152–170
    in algae, 816
    ATP in, 160f, 161–162
    in bacteria, 791–792
    central role, 154
    early studies, 153–154
    and energy of electrons, 54f, 55
    light-dependent reactions, 155f, 156–163
    light-independent (dark) reactions, 155f, 156, 158f,
        163–166
    overview, 155–156, 158f
    as two-stage process, 155f, 156
Photosynthetic autotrophs, 154
Photosynthetic pigments, 156, 157–159
Photosystems, 159
Phototropism, 438
Phyletic speciation, 732
Phylogenetic relationships, 14
Phylum, in taxonomic hierarchy, 15
Physical defenses, 970–971

Phytochrome, 435–436
 and flowering, 437
 and photoperiodism, 436
 and seed germination, 438
 and shoot development, 438
Phytoplankton, 816, 908
Pigmentation in humans, genetics of, 240, 241f
Pigments:
 light-absorbing:
  in eye, 506
  in plants, 156, 157–159
 photosynthetic, 156, 157–159
Piltdown Man, 743, 752–753
Pineal gland, 522t, 533
Pinocytosis, 144, 147–148
Pioneer community, 952
Pistil, 403
Pith of plants, 372
Pith rays, 372
Pituitary gland, 481f, 522t, 524–526
Placebo, 39
Placenta, 700
Placentals, 893
Planaria, 869–871
Planes of body symmetry, 862f
Plankton, 881, 908
Plant fracture properties, 375
Plant kingdom, 830–856
Plants, 358–382, 384–398, 400–422, 424–443,
  830–856. *See also* Flowering plants
 annual, biennial, perennial, 361
 basic design, 360–361, 362f
 carbohydrate synthesis in, 163–166
 cell vacuoles, 102
 cell wall, 108–109
 circulatory system, 384–398
 classification, 839t
 communities, 932–933
 competition between, 961
 cytokinesis in, 206
 desert adaptations, 164–166, 926
 evolution, 833–834, 835f
 fossil, 843, 844f
 genetic engineering of, 316–319, 375
 growth and development, 424–443
  control of, 426–427
  timing of, 434–438
 growth signal in, 424–425
 hormones, 427–434
  discovery of, 424–425
  as insecticides, 434, 975
 longevity, 362–363, 366
 nonphotosynthesizing, 834, 836
 nutrients, 388, 390t
  transport of, 393–396
 organs, 372–378
  flower, 378
  leaf, 377–378
  root, 373–377
  stem, 372–373
 osmosis in, 140–141, 142f
 polyploidy in, 260–261
 sex chromosomes in, 254
 sexual reproduction in, 402–417
 species differences, 830
 transport in, 384–398
Plant tissues, 362–365
 cambium, 362, 372–373
 collenchyma, 364
 parenchyma, 363
 primary, 362
  of root, 373–374, 376f, 377
  of stem, 372
 sclerenchyma, 364–365
 secondary, 362
  of root, 377

 of stem, 372
Plant tissue systems, 365–372, 367f
 dermal, 365, 366–369
 ground, 365, 372
 vascular, 365, 369–372
Plaques, in arteries, 75, 603
Plasma, 600
Plasma cells, 647, 653
Plasma membrane (cell membrane):
 carrier proteins of, 141, 143f
 components, 93–94
 formation, 99
 functions, 93–94
 ion channels, 139, 140f, 471
 permeability, 138, 139
 receptors on, 94, 146, 147–148, 533–534, 535f
 structure, 88–89, 93–94, 138
 transport across, 136–144
  active transport, 142–144
  passive diffusion, 138–139
Plasmid, 321, 322f
Plasmodesmata, 387
Plasmodial slime molds, 819
Plasmodium, 813, 815f
Plasmolysis, 141, 142f
Platelets, 602
Plate tectonics, 767–769
Platyhelminthes, 864t, 869–871
Platypus, 892f, 893
Play, 1013
Pleiotropy, 240
Pleura, 626
Pleurisy, 626
Pneumococcus, role in molecular genetics, 266–267
Pneumocystis infection, 813
Point mutation, 286
Poison ivy, 656
Polar body, 675
Polarity in neurons, 471–473
Polar molecules, 56–57
Pollen, 403
 protective properties, 833
Pollen:
 sacs, 403f
 tube, 409f, 410
Pollination, 402–407, 410f
Pollution
 consequences, 918, 992, 993
 genetically engineered degradation, 316–317
Polychaetes, 875–876
Polygenic inheritance, 240–241
Polymer, 69, 70f
Polymerase chain reaction, 326–327, 329
Polymorphic genes, 338
Polymorphism, 727
Polypeptide chain, 78f, 80
 assembly, *see* Proteins, synthesis
Polyploidy, 260–261
 and speciation, 733
Polyps, 868–869
Polysaccharides, 70–74
Polysome, 285
Polytene (giant) chromosomes, 252–253, 305
Polyunsaturated fats, 75
Pond habitats, 915–916
Pons, 481f
Population density, 986, 987, 996
Population ecology, 984–1004
Population growth, 988–1001
 and earth's carrying capacity, 999–1001
 factors affecting, 988–997
 fertility and, 999
 in humans, 997–1001
  doubling time, 998
  percent annual increase, 998
 rate, 986

Populations, 11, 717
 age–sex structure, 988, 999, 1000f
 defined, 986
 distribution patterns, 987–988
 structure, 986–988
Porifera, 864–867, 864t
Porpoises, and echolocation (sonar), 499
Positional information, 696
Postembryonic development, 699–700
Potassium:
 in diet, 580, 580t
 transport across cell membrane, 142, 144f
Potential (stored) energy, 118
Pragmatism, 41
Prairies, 923
Predation, 966–972
Predator, 966
 adaptations, 967–972
 group behavior, 1018
Predator–prey dynamics, 966–967
Pregnancy, *see also* Embryo; Fetus
 prevention of, 677–681
 stages of, 700–705
Prepuce, 669
Pressure-flow transport of plant nutrients, 395–396
Pressure gradient, and respiration, 626–627
Prey, 966
 defenses, 966t, 967–972
 group behavior, 1018
Primary endosperm cell, 410, 411f
Primary growth of plants, 362
Primary plant tissues, 362
Primary organizer, 695
Primary producers, 937
Primary root tissues, 373–374, 376f, 377
Primary stem tissues, 372
Primary succession, 952, 953–955
Primary transcript, 306–307
Primates, 893–894
 evolution, 894f
Primitive homologies, 746
Prion, 803
Probability, in genetics, 236
Progesterone, 522t, 676f, 677
Prokaryotes, 787–800. *See also* Bacteria
 activities of, 798–800
 fossilized, 763
 gene regulation in, 298–301
 as universal ancestor, 789f
 versus eukaryotes, 91–92, 787
Prokaryotic cell, 91–92, 793f
 cell division in, 197
 chromosomes in, 273
 and origin of eukaryote cell, 110, 765, 787–788
Prokaryotic fission, 197
Prolactin, 308, 522t, 525
Promoter, 299
Prophase, of mitosis, 203–205
Prophase I, of meiosis, 216–218, 217f
Prophase II, of meiosis, 217f, 219
Prosimians, 894f
Prostaglandins, 532, 705
Prostate gland, 672
Prostate specific antigen (PSA), 654
Protein kinases, 129, 195, 202
Proteins, 71t, 79–81
 conformational changes in, 81
 denaturing of, 127
 as energy source, 188
 in human diet, 578
 of plasma membrane, 94, 138
 structure, 66–67, 78f, 79–81
  primary, 80
  quaternary, 80f, 81
  secondary, 80
  tertiary, 80f, 81

synthesis, 277–285, 286f
  chain elongation, 285
  chain termination, 285
  initiation, 284–285
  site of, 98
  transcription in, 277, 278, 279f, 286f
  translation in, 277, 279, 286f
Proteolytic enzymes, 573
Proterozoic Era, 764–766
Prothallus, 847, 848f
Protist kingdom, 810–819
  taxonomic problems, 810
Protists:
  evolution and, 811–812
  reproduction in, 222, 811
Protochordates, 884–886
Protocooperation, 977–978
Protoctista, 809n
Proton gradient, and ATP formation, 160f, 161, 184
Protons, 51, 61
Protoplast manipulation, 418–419
Protostomes, 873
Protozoa, 812–816
  classification, 815t
  motility, 814–816
  pathogenic, 813
  polymorphic, 813–814
Proximal convoluted tubule, 608–609, 610
Pseudocoelom, 861, 871–872
Pseudopodia, 814–816
Psilophyta, 839t, 844f, 845
Psychoactive drugs, 487
Pterophyta, 839t, 847
Pterosaurs, 771
Puberty, in females, 675
Puffballs, 824
Puffing in chromosomes, 305
Pulmonary circulation, 594–595
Punctuated equilibrium, 737
Punnett square, 235–236
Pupa, 881
Purines, 269, 270f
Pus, 645
Pyloric sphincter, 572
Pyramid of biomass, 944
Pyramid of energy, 943–944, 946
Pyramid of numbers, 944
Pyridoxine (vitamin B$_6$), 579t
Pyrimidines, 269, 270f
Pyrrophyta, 818
Pyruvic acid:
  energy yield from, 185
  in glycolysis, 176f, 178
  in Krebs cycle, 181, 182f

◀ R ▶

r (rate of increase), 990
  r-selected reproductive strategies, 995–996
R group, of molecule, 71t, 79–80
  and enzyme-substrate interaction, 125
Radiation, effect on DNA, 280–281, 287
Radicle, 411
Radioactivity, 52
  discovery of, 48–49
Radioisotopes, 54, 746
Rainfall, 906
Rain forests, 428–429, 917, 919, 920f
Rainshadow, 926
Random distribution of population, 987–988
Rate of increase, intrinsic (r_o), 990
Rays, 887, 888f
Reactant, 123
Reaction center, 159
Reaction-center pigments, 158f, 159
Reactions, see Chemical reactions

Reading frame, 284
  frameshift mutations, 286
Receptacle of flower, 403
Receptor-mediated endocytosis, 147
Receptors:
  in cytoplasm, 534, 535f
  for hormones, 533–534, 535f
  for LDLs, 137, 146, 147f
  on plasma membrane, 94, 147–148, 533–534, 535f
    and medical treatment, 146
  sensory, 479, 500–502
Recessive allele, characteristic, trait, 234–235, 237
  and genetic disorders, 343, 345–346
Recessiveness, 234
Reciprocal altruism, 1023
Recombinant DNA, see also Biotechnology
  amplification by cloning, 324
  defined, 314
  formation and use, 321–325
  medicines from, 314, 315t, 321
Recruitment, 1019
Rectum, 574, 576f
Recycling nutrients in ecosystems, 947–952
Red algae, 838f–839f, 841–842
Red blood cells (erythrocytes), 600–601, 627
Red tide, 818
Reducing power of cell, 129
Reduction, 128–129
Reduction division, in meiosis, 216
Reefs, 911–912
Reflex, 479
  conditioned, 1011
Reflex arc, 479–480
Region of elongation, 372, 375f
Region of maturation, 372, 375f
Regulation of biological activity, 15
Regulatory gene, 299
Releaser of fixed action pattern, 1009
Releasing factors of hypothalamus, 524
Renal tubule, 607–609
Renin, 522t, 591
Replication fork, 276, 277f
Replication of DNA, 275–277
Repressible operon, 300f, 301
Repressor protein, 299, 300f, 301
Reproduction:
  asexual, 199
  asexual versus sexual, 664–667
  communication and, 1018
  prevention of, 677–681
  as property of living organism, 10
Reproductive adaptation, 729
Reproductive isolation, 732
Reproductive (biotic) potential, 986, 989–990
Reproductive strategies, 995
Reproductive systems, 457, 662–684
  human, 456f, 667–677
    female, 673–677
    male, 669–673
Reproductive timing, 668
Reptiles, 889, 890f
  Age of, 771
Research, scientific, 32–35
Resource partitioning, 964–965
Respiration, 174–192
  aerobic, 180–184
    versus fermentation, 175, 178
  and surface area, 622–623, 626
  versus breathing, 175n
Respiratory center, 632
Respiratory surfaces, 622–623
Respiratory systems, 455, 620–640
  adaptation of, 632–635
  evolution of, 636
  in human, 456f, 623–627
Respiratory tract, 623–624

Response to stimuli, as property of living organism, 10
Resting potential, 471
Restriction enzymes, and recombinant DNA, 321–323
Restriction fragment length polymorphisms (RFLPs), 338, 339f
  use in gene mapping, 341, 342f
Restriction fragments, 325
Reticular formation, 486
Retina, 503, 505f, 506
Retinoblastoma, 260
Retinol (vitamin A), 579t
Retrovirus, 643, 803
Rheumatic fever, 657
Rhizoids, 823, 842
Rhodophyta, 841
Rhyniophytes, 843
Rhythm method of contraception, 681
Riboflavin (vitamin B$_2$), 579t
Ribonucleic acid, see RNA
Ribose, 71f, 278
Ribosomal RNA (rRNA), 281–282
Ribosomes, 97, 98
  assembly, 283f
  in protein synthesis, 281–284, 286f
  structure, 282–284
Ribozymes, 306, 307
Ribulose biphoshpate (RuBP), 164
Rigor mortis, 556
River habitats, 914–915
RNA, see also Nucleic acids; Transcription; Translation
  enzymatic action, 306, 307
  processing, 306–307
    and control of gene expression, 304, 307, 308f
    self-processing, 307
  splicing, 307
  structure, 82, 278
RNA polymerase, 278, 286f
  and operon, 299
Rods of retina, 503, 505f, 506
Root cap, 373, 375f
Root hairs, 373, 374f
Root nodules, 389–390
Root pressure, 391
Root system of plants, 361, 362f, 373–377
Rough endoplasmic reticulum (RER), 98, 100
Roundworms, 864t, 871–873
rRNA (ribosomal RNA), 281–282
RU486, 678–679
Rubella virus, risk to embryo, 704
RuBP (ribulose biphoshpate), 164
Rumen, 581
Ruminants, 581, 813
"Runner's high," 487

◀ S ▶

Saguaro cactus decline, 984–985
Salamanders, 888–889
Saliva, 568–569
Salivary glands, 568
Saltatory conduction, 473
Sand dollars, 882
Sandy beaches, 912, 913f
Sanitation, origins of, 784–785
SA (sinoatrial) node, 598
Saprobes, 822
Saprophytic decomposers, 798
Sarcodina, 815t
Sarcolemma, 554f
Sarcomeres, 553–556
Sarcoplasmic reticulum, 556–557
Saturated fats, 75
Savannas, 923–924
Scales of fish, 543

Scaling effects, 460
Scallops, 874
Schistosomes, 870f, 871
Schizophrenia, 230–231
Schwann cells, 469f, 470
SCID (severe combined immunodeficiency), 337t,
  348–349
Science and certainty, 40–41
Scientific method, 32–35, 39f
  caveats, 40–41
  facts versus values, 909
  practical applications, 35–40
Scientific names, 13
Scientific theory, 35
  versus truth, 40–41
Sclereids, 365
Sclerenchyma cells, fibers, 364–365
Scorpions, 876
Scrotum, 669
Scurvy, 540–541
Sea anemones, 867–869
Sea cucumbers, 882
Sea stars, 864t, 882, 883f, 884
Sea turtles, natal homing, 312–313
Sea urchins, 864t, 882
Seasonal affective disorder syndrome (SADS), 533
Seaweeds, 838, 841–842
Secondary growth of plants, 362
Secondary plant tissues, 362
Secondary root tissues, 377
Secondary sex characteristics
  female, 675
  male, 672–673
Secondary stem tissues, 372
Secondary succession, 952–953, 955
"Second messenger," 533, 535f
Secretin, 522t, 573
Secretory pathway (chain), 99–100, 101f
Sedimentary nutrient cycles, 947, 951–952
Seedless vascular plants, 845–847
Seedling development, 416–417
Seed plants, 847–853
Seeds, 411, 412f
  development, 411
  dispersal, 412, 415f, 416
  dormancy, 416
  germination, 438
  protective properties, 833
  viability, 416
Segmentation of body, 863, 875–876
Segmented worms, 864t, 875–876, 877f
Segregation, Mendel's law of, 235
Selective gene expression, 294, 295f
Self-recognition, 654
Semen, 670, 672
Semicircular canals, 507f, 508
Semiconservative replication, 275
Seminal vesicle, 672
Seminiferous tubule, 669, 671f
Semmelweis, Ignaz, 784–785
Senescence in plants, 432
Sense organs, 498–516
  diversity of, 511–513
  in fishes, 512
  how they work, 500–515
Sense strand of DNA, 278
Sensory cortex, 502, 503f, 511
Sensory neurons, 470
Sensory receptors, 479, 500–502
  types, 500, 501t
Sepals, 403
Sequencing:
  of amino acids, 66–67, 84, 750
  of DNA, 325–326
    and evolution, 329, 750
    and frameshift mutations, 286

legal aspects, 315
  and turtle migration, 313
  of nucleic acid, 325
Sere, 952
Serosa of digestive tract, 568
Serotonin, 475t
Severe combined immunodeficiency (SCID), 337t,
  348–349
Sex:
  age–sex structure, 988, 999, 1000f
  determination of, 253–254, 255
Sex changes in fish, 254
Sex chromosomes, 253
  abnormal number, 336–337
  and gender determination, 255
  and gene dosage, 256–257
Sex linkage, 254–255, 345–346
Sexual dimorphism, 727–728
Sexual imprinting, 1014
Sexually transmitted diseases, 794–797
Sexual orientation, 663
Sexual reproduction
  advantages, 664–665
  in flowering plants, 402–417
  and meiosis, 214, 215f
  strategies, 665–667
Sexual rituals, 668
Sexual selection, 727–728
Sharing food, 1024–1025
Sharks, 887
Shell, 545
Shellfish, 874
  poisoning by, 818
Shivering, 614
Shoot system of plants, 361, 362f
Short-day plants, 400, 437
Shrew, 893
Shrimp, 876
Shrublands, 924, 925f
Sickle cell anemia, 81, 82f, 337t, 346f, 347
  and allele frequency, 717
  codon change in, 286
  and malaria, 723f, 724, 812
  and natural selection, 723–724
Sieve plates, 370
Sieve-tube members, 370
Sight, 502–506, 512
Sigmoid (S-shaped) growth curve, 993–994
Sign stimulus, 1009
Silent Spring, 715
Silicosis, 101
Silk, 881
Simple fruits, 412, 413f
Simple leaf, 377
Singer-Nicolson model of membrane, 89
Sinks, and plant nutrient transport, 394
Sinoatrial (SA) node, 598
Skates, 887
Skeletal muscle, 552–557
  cells in, 553–555
  evolution, 561
  fiber types in, 186–187
  metabolism in, 186–187
Skeletal systems, 457, 544–551
  evolution of, 560–561
  in human, 456f, 548–551
Skin, 542–544
  color in humans, 240, 241f
  sensory receptors in, 502
Skull (cranium), 482
Slime molds, 819
Slugs, 819, 874
Small intestine, 572–574
Smell, 509, 510
Smoking, 630–631, 704
Smooth endoplasmic reticulum (SER), 98

Smooth muscle, 558
Snails, 864t, 874, 875f
Snakes, 889
Social behavior, 1018–1022
Social bonds, 1021–1022
Social learning, 1013
Social parasitism, 974
Sodium:
  in diet, 580, 580t
  transport across cell membrane, 142–144
Sodium—potassium pump, 142, 144f
Soft palate, 569f
Soil, formation of, 953
Solar energy, 154, 156f
  dangers of, 280–281
Solutions, molecular properties, 59–60
Solvent, 59
Somatic cells versus germ cells, 214
Somatic division of peripheral nervous system, 490
Somatic sensation, 500–502
Somatic sensory cortex, 502, 503f
Somatotropin, see Growth hormone
Sori, 847
Speciation, 731–734
Species:
  attributes, 731–732, 830
  competition, 960–961
  diversity and biomes, 917
  origin of, 731–734
  in taxonomic hierarchy, 14
Specific epithet, 13
Sperm, 669, 689, 670
Spermatid, 669
Spermatocytes, 669
Spermatogenesis, 669–670, 671f
Spermatogonia, 669
Spermatozoa, 669
Spermicides, 679
Sphenophyta, 839t, 846–847
Spiders, 864t, 876, 881
  territorial behavior of, 1006–1007
Spina bifida, 694
Spinal cord, 488–489
Spinal nerve roots, 488f
Spinal nerves, 489–490
Spindle apparatus, 204–205
Spindle fibers, 204–205
Spine, see Vertebral column
Spiracles, 633, 635f
Split brain, 504
Split genes, 302f, 303
Sponges, 864t, 866–867, 864t
Spongy bone, 546, 547f
Spongy parenchyma, 378
Spontaneous generation, 32, 33
Sporangiospores, 823
Sporangium, 823, 847, 848f
Spores of bacteria, 792
Spores of fungi, 822
Spores of plants, 214, 407, 833, 843, 845
Sporophyll, 846
Sporophyte, 833
Sporozoa, 815t
Squids, 864t, 874
Stabilizing selection, 724–725
Stamens, 403
Starch, 73
Starfish (sea star), 882, 883f, 884
Start (initiation) codon, 281, 284
Stem of plants, 372–373
Sterilization, sexual, 681
Steroids, 75, 77f
  and athletes, 528
  and gene activation, 304–305
  receptors for, 534, 535f
"Sticky ends" in recombinant DNA, 322f, 323

Stigma of flower, 403
Stimulus:
    perception of, 511
    receptors for, 500
    sign stimulus, 1009
Stomach, 570–572
    secretory cells of, 571f, 572
Stomates (stomatal pores), 165f, 166, 368, 392–393, 833
Stop (nonsense) codons, 281, 285
Stratigraphy, law of, 746
Stream habitats, 914–915
Streptococcus, in pregnancy, 704
Stretch reflex, 479, 480f
Strobilus, 846
Stroke, 590
Stromatolites, 763
Structural genes, 299, 300f
Style, of flower, 403
Subatomic particles, 51
Substrate, 124f, 125
Substrate-level phosphorylation, 176f, 177–178, 185
Succession in ecosystems, 952–955
Succulents, 926
Sucrose, 70
Sugars, 70–74. See also Glucose
Sulfa drugs, 126
Sun, formation of, 762
Sunlight:
    and climate, 905–906
    dangers of, 280–281
    energy from, 154, 156f
    undersea, 908
Superoxide radical, 56
Suppressor T cells, 649
Surface area, and respiration, 622–623, 626
Surface area–volume ratio, 460–462
Surface tension, 60, 61f
Surfactant, 60
Surrogate motherhood, 679
Survivorship curve, 989
Suspensor, of plant embryo, 411
Sutures of skull, 550
Swallowing, 569
Sweating, 614, 927
"Swollen glands," 606
Symbiosis, 962–963
    and origin of eukaryote cell, 110, 765
Symmetry of body, 861, 862f
Sympathetic nervous system, 490–491
Sympatric speciation, 732
Synapses, 474–478
    nerve poisons and, 478
Synapsis, 216, 217f, 218
Synaptic cleft, 474
Synaptic knob, 468, 469f, 474f
Synaptic transmission (neurotransmission), 474–479
Synaptic vesicles, 474
Synaptonemal complex, 218, 219f
Synovial cavities, 551
Syphilis, 795–796
Systemic circulation, 594f, 595
Systemic lupus erythematosus, 657

◄ T ►

Tail, in chordates, 884
Tapeworms, 869, 870f, 871, 872
Tap root system, 373, 374f
Tarsier, 894f
Taste, 509–510
Taung Child (hominid), 743, 752
Taxonomy, 11–15, 808–809
Tay-Sachs syndrome, 337t
T cell receptors, 647
T cells, 647, 648, 649–650
Tectonic plates, 767–769

Teeth, 570
Telophase, of mitosis, 203f, 206
Telophase I, of meiosis, 217f, 219
Telophase II, of meiosis, 217f, 221
Temperature
    of body, 449, 450, 451f, 588, 614
    and chemical reactions, 127
    and plant growth, 434–435
Temperature conversion chart, Appendix A
Template in DNA replication, 275
Tendon, 552
Teratogens, 704
Territorial behavior, 1006–1007, 1016–1017
Test cross, 237
Testes (testicles), 255, 522t, 532, 669
Testosterone, 75, 77f, 522t, 532, 672–673
    dangers of, 528
Testosterone receptor, 304
Testosterone-receptor protein, 305
Test population, in experiments, 37
"Test-tube babies," 679
Tetanus, 478
Tetrad (chromosomal), 216, 218f, 219
Thalamus, 481f, 486
Thalassemia, 286
Thallus, 835
Themes recurring in this book, 15
Theory, scientific, 35
    versus truth, 40–41
Theory of tolerance, 938
Therapsids, 774
Thermodynamically unfavorable reactions, 122
Thermodynamics, laws of, 119–120
Thermophilic bacteria, 167, 796
Thermoreceptor, 500, 501t
Thermoregulation, 449, 450, 451f, 588, 614
Thiamine (vitamin B₁), 567, 579t
Thigmotropism, 439
Thoracic cavity, 626
Thornwood, 924
Threatened species, 995–996
Throat (pharynx), 569, 624, 636
Thrombin, 602
Thrombus, 603
Thylakoids, 104, 156
    and electron transport system, 160
    photosynthesis in, 158f, 159, 160–162
Thymine, 269, 270f
    base pairing, 271f, 272
Thymine dimer, 280
Thymus gland, 646f, 649
Thyroid gland, 522t, 529–531
Thyroid hormone, 529, 531
Thyroiditis, 657
Thyroid-stimulating hormone (TSH), 522t, 526
Thyroxine (T-4), 522t, 529
Ticks, 876
Tide pools, 912, 913f
Timberline, 920
Tissue plasminogen activator (TPA), 131, 603
Tissues:
    in animals, 451, 452
        gas exchange in, 627–629
    in plants, 362–365
Tocopherols (vitamin E), 579t
Tolerance range, 938
Tomato, genetically engineered, 318
Top carnivore, 942
Totipotency, 297
Touching, 1021
Toxic chemicals, 946, 992, 993
Toxicity, testing for, 37–39
Toxic shock syndrome, 681
TPA (issue plasminogen activator), 131, 603
Trace elements in diet, 580, 580t
Trachea, in arthropods, 633, 635f, 878

Trachea, in humans, 625
Tracheids, 370
Tracheoles, 633, 635f
Tracheophytes, 835, 839t, 843–853
Transcription, 277, 278, 279f, 286f
    and control of gene expression, 304–306
    primary transcript processing, 306–307
Transfer RNA (tRNA), 279–280, 282f, 283f
    role in translation, 284–285, 286f
Transgenic animals, 318f, 319, 321
Translation of RNA, 277, 279–285, 286f
    chain elongation, 285
    chain termination, 285
    initiation, 284–285
    and regulation of gene expression, 304, 307–308
Transpiration, 386, 391–393
Transpiration pull, 391–392
Transplanted organs, 489, 650
Transport:
    across plasma membrane, 136–144
    of electrons, 160–161, 182–185
    of oxygen, 600–601, 627
    in plants, 386–396
Transposition of genes, 293
Transverse tubules of muscle fiber, 556
Trematodes, 869, 870f, 871
Trichinosis, 873
Triglycerides, 71t
Triiodothyronine (T-3), 522t, 529
Trilobites, 878f
Trimesters of pregnancy, 700
Trisomy 21 (Down syndrome), 213, 221, 336, 337t
tRNA, see Transfer RNA
Trophic levels, 942
Trophoblast, 700, 701f, 703
Trophozoites, 813–814
Tropical rain forests, 428–429, 917, 919, 920f
Tropical thornwood, 924
Tropic hormones, 525, 526f
Tropisms in plants, 438–439
Trp (tryptophan) operon, 300f, 301
True bacteria, 791–796
Truth, and scientific theory, 40–41
Trypsin, 573
Tuatara, 889
Tubal ligation, 681
Tube cell, 408, 409f
Tubeworms, 875–876
Tubular reabsorption, 610
Tubular secretion, 610
Tubules of flatworm, 612f
Tubules of kidney, 607–609
Tumor-infiltrating lymphocytes (TILs), 648
Tumor-specific antigens, 648
Tumor-suppressor genes, 260
Tundra, 920, 922
Tunicates, 864t, 884, 885f
Turbellaria, 869–871
Turgor pressure, 141, 142f
Turner syndrome, 337
Turtles, 889
Tympanic membrane (eardrum), 507f, 508

◄ U ►

Ultimate carnivore, 942
Ultraviolet radiation, see also Sunlight
    effect on DNA, 280–281
Umbilical cord, 700
Uncertainty and scientific theories, 40–41
Understory of forest, 917
Uniform distribution patterns, 987
Uniramiates, 878f
Unity of life, 18
Unsaturated (double) bonds, 75
Uracil, 278, 279f

Ureters, 607
Urethra, 607, 669, 673
Urinary bladder, 607
Urinary (excretory, osmoregulatory) systems, 455, 456f
Urine, 607
    formation of, 609–611
Urkaryote, 798
Uterine (menstrual) cycle, 675, 677
Uterus (womb), 673

◄ V ►

Vaccines, 654–655. *See also* Immunization
    for AIDS, 657
    for contraception, 677
Vacuoles, 102
Vagina, 673
Vaginal sponge, 681
Variable, in scientific experiment, 32
Vascular cambium, 362
Vascular plants, 361, 839t, 843–853
    higher (seed plants), 839t, 847–853
    lower (seedless), 839t, 845–847
Vascular tissue system of plants, 365, 367f, 369–372
Vas deferens, 670
Vasectomy, 681
Vasopressin, *see* Antidiuretic hormone
Vector, in recombinant DNA, 321
Vegetables, 412
Vegetative (asexual) reproduction in flowering plants,
    417–418
Veins, 593, 595
Venation of leaf, 378
Venereal diseases, 794–797
Ventricles:
    of brain, 481f, 482
    of heart, 593f, 594
Venules, 593
Vernalization, 435
Vernix caseosa, 703
Vertebral column (backbone), 886
    origin of, 694

Vertebrates, 863, 864t, 884, 886–894
Vesicles, 97
Vessel members of xylem, 370
Vestibular apparatus, 507f, 508
Vestigial structures, 706, 749–750
Villi, chorionic, 700
Villi, intestinal, 574, 575f
Viroid, 803
Viruses, 787, 800–804
    diseases from, 794, 796–797, 803, 804
    evolutionary significance, 803–804
    gene transfer by, 804
    how they work, 801–802
    life cycle, 802f
    properties, 800–801
    as renegade genes, 803
    structure, 801
Vision, 502–506
    in fishes, 512
Visual cortex, 506
Vitamins, 567, 579–580
    as coenzymes, 127
    list of, 579t
Vocal cords, 624–625
Vulva, 673

◄ W ►

Water:
    absorption in plants, 386–387, 389f
    biogeologic cycle, 947–948
    osmosis and, 140–141
    properties, 58f, 59–61
    photosynthesis and, 155, 160
"Water breathers," 632–633
Water molds, 822
Watson, James, 269–272
Waxes, 71t, 75
Weeds, 940–941
    control of, 941
    germination, 438
Whisk ferns, 845

White blood cells (leukocytes), 601–602
White matter:
    of brain, 481f, 482
    of spinal cord, 488f, 489
Wild type alleles, 241
Winds, prevailing, 906–907
Womb (uterus), 673
Wood, 370
Woody plants, 362
Worms:
    flat, 864t, 869–871, 872
    round, 864t, 871–873
    segmented, 864t, 875–876, 877f

◄ X ►

Xanthophyta, 817
X chromosome, 253
    and gene dosage, 256–257
    X-linked characteristics, 254
    X-linked disorders, 345–346
Xeroderma pigmentosum, 280, 337t
X-ray crystallography, 67
Xylem, 369, 370
    water and mineral transport, 386–393

◄ Y ►

Y chromosome, 253
    Y-linked characteristics, 255
Yeasts, 820
    in fermentation, 179
    infection from, 826
    reproduction in, 822
Yolk, 688
Yolk sac, 703

◄ Z ►

Zidovudine (AZT), 657
Zooplankton, 813, 908
Zygomycetes, 822–823, 822t
Zygospore, 823
Zygote, 214, 215f, 690